Dimitris Tsoukalas
Christos Tsamis (eds.)

Simulation of Semiconductor Processes and Devices 2001

SISPAD 01

SpringerWienNewYork

Dr. Dimitris Tsoukalas
Dr. Christos Tsamis
Institute of Microelectronics
"Demokritos" National Center for Scientific Research
Athens, Greece

Printed in Austria

Typesetting: Camera ready by authors
Printing: Druckerei Theiss GmbH, A-9400 Wolfsberg
Printed on acid-free and chlorine-free bleached paper
SPIN 10845347

With 474 Figures

ISBN 3-211-83708-6 Springer-Verlag Wien New York

Editorial

This volume contains the proceedings of the International Conference on Simulation of Semiconductor Devices and Processes, SISPAD '01, held on September 5-7, 2001, in Athens. The conference is organized annually and takes place in turn in Europe, Japan and the USA. SISPAD '99 was organized at Kyoto Research Park, Japan, and SISPAD '00 took place in Seattle, Washington, USA.

The conference provided an open forum for the presentation of the latest results and trends in process and device simulation. The trend towards shrinking device dimensions and increasing complexity in process technology demands the continuous development of advanced models describing the basic physical phenomena involved. New simulation tools are developed to complete the hierarchy in the Technology Computer-Aided Design simulation chain between microscopic and macroscopic approaches.

The conference program featured 8 invited papers, 60 papers for oral presentation and 34 papers for poster presentation, selected from a total of 165 abstracts from 30 countries around the world. These papers disclose new and interesting concepts for simulating processes and devices.

The proceedings were printed from the authors camera-ready manuscripts. We would like to express our sincere appreciation to the authors for their high-quality contributions, their cooperation and efforts.

Also we would like to thank the members of the conference committee for reviewing the manuscripts carefully.

SISPAD'02 will be held on September 4-6, 2002 in Kobe, Japan. Certain to continue the tradition of being the most important conference in its field, we once again look forward to a most successful and exciting event.

Dimitris Tsoukalas
Christos Tsamis

Steering Committee

K. De Meyer	IMEC	Belgium
J. Faricelli	Compaq	USA
M. Law	University of Florida	USA
P. Leon	PetaLogic Corp.	USA
N. Nakayama	STARC	Japan
S. Odanaka	Osaka University	Japan
H. Ryssel	FhG-IIS-B	Germany
D. Tsoukalas	NCSR 'Demokritos'	Greece
A. Yoshii	Japan Women's University	Japan

Honorary Committee

R. Dutton	Stanford University	USA
S. Selberherr	TU Vienna	Austria
K. Taniguchi	Osaka University	Japan

Technical Program Committee

Y. Awano	Fujitsu Lab	Japan
G. Baccarani	University of Bologna	Italy
R. Brunetti	University of Modena	Italy
N. Cowern	Philips	The Netherlands
M. Duane	Applied Materials	USA
S. Dunham	University of Washington	USA
N. Goldsman	University of Maryland	USA
M. Hane	NEC	Japan
H. Jaouen	ST Microelectronics	France
M. Jaraiz	University of Valladolid	Spain
S. Jones	Marconi Caswell Ltd	UK
J. B. Kuo	University of Waterloo	Canada
P. Leon	PetaLogic Corp.	USA
J. Lorenz	FhG-IIS-B	Germany
M. Miura-Mattausch	Hiroshima University	Japan
W. Molzer	Infineon Technologies	Germany
K. Nishi	SELETE	Japan
P. Oldiges	IBM	USA
M. Profirescu	TU Bucharest	Romania
K. Rahmat	IBM	USA
S. Selberherr	TU Vienna	Austria
W. Shoenmaker	IMEC	Belgium
T. Toyabe	Toyo University	Japan
G. Wachutka	TU Munich	Germany

Contents

Macroscopic Quantum Carrier Transport Modeling

Zhiping Yu, Robert W. Dutton, and Daniel W. Yergeau
Center for Integrated Systems
Stanford University, Stanford, CA 94305, USA
yu@ee.stanford.edu

Mario G. Ancona
Naval Research Laboratory, Washington, D.C. 20375, USA

Abstract

It has been established [1]-[4] that the density gradient (DG) model is the lowest order, in terms of $\hbar$, approximation of the Wigner function approach to including quantum mechanical (QM) effects in carrier transport. In this paper, we report a five-equation PDE system (reduced to three-equation at thermal equilibrium) which preserves the numerical stability of classical drift-diffusion (DD) model, yet faithfully manifests QM corrections. Tunneling through the gate oxide (or barrier region) is modeled by ballistic transport with each type of carrier (electrons or holes) further split into forward and backward moving species and solved for separately. The entire device, including semiconductor and barrier regions, is solved self-consistently. Terminal characteristics, either *dc* or small signal *ac*, for realistic, multi-dimensional device structures can be simulated using this model. An SOI device example is simulated and the comparison with microscopic (Schrödinger/Poisson) results is excellent. A DG prediction of a dipole in the poly gate near the poly/gate-oxide interface is also confirmed by microscopic simulation. Both *I*-*V* and *C*-*V* for MOS devices including SOI are shown.

1 Introduction

Quantum mechanical (QM) effects in semiconductor devices are manifestations of the wave nature of highly confined electrons. For scaled MOSFETs, two QM effects are most prominent:

1. Quantum confinement in the inversion or accumulation layer in the substrate adjacent to the Si-SiO_2 interface.
2. Tunneling through the thin gate oxide (for thicknesses below approximately 2nm).

An "exact" analysis of these phenomena requires quantum mechanics. In its simplest form, a microscopic treatment of the confinement effects involves an equilibrium problem in one-dimension and can be solved using the Hartree approximation. This is the conventional approach; one finds that the energy bands split into discrete sub-bands with the wave functions associated with each energy level spread over the breadth of the well. Because of the large chemical potential barrier at the Si-SiO_2 interface, the wave functions essentially vanish at that interface; the peak carrier concentration is thus located away from the substrate surface. Qualitatively, all of this remains true in a real device but the details change because the device is actually three-dimensional, and the degree of confinement varies with position along the channel.

The microscopic treatment of gate tunneling is even more problematic because it is in essence a non-equilibrium problem. One-dimensional, one-electron treatments have long been used but these lack self-consistency. Much more sophisticated one-dimensional treatments such as those based on non-equilibrium Green's functions [6] capture the physics quite well for one-dimension but are too computationally demanding to be extended to multiple dimensions and to be used for everyday engineering analysis.

For many important problems there is a way out of this dilemma that comes with the recognition that a full-scale treatment of the quantum mechanics is overkill. For MOS scaling problems, quantum interference phenomena are typically not important. Furthermore, in many device situations, a large number of the sub-bands are filled. Therefore, the details of sub-band structure have little consequence. This is the motivation for developing a macroscopic transport model with the lowest-order QM corrections. One such approach is the so-called density gradient (DG) model [1].

2 The evolution of DG model from Wigner function approach

The DG model can be developed using either macroscopic arguments [2] or from quantum mechanics [1]. The macroscopic approach introduces lowest-order effects of non-locality by having the internal energy of the electron gas depend not only on the density of electrons, but also on the gradient of the density. Alternatively, one can derive the DG model from the Wigner function, which is the probability density function of particles in real and momentum space. The Wigner function is constructed from the wave function as [7]

$$f(\boldsymbol{r},\boldsymbol{p}) = \frac{1}{(\hbar\pi)^l}\int_{-\infty}^{\infty} \Psi^*(\boldsymbol{r}+\boldsymbol{r}')\Psi(\boldsymbol{r}-\boldsymbol{r}')e^{2i\boldsymbol{p}\cdot\boldsymbol{r}'/\hbar}d\boldsymbol{r}' \tag{1}$$

where $\boldsymbol{p}$ is the momentum, Ψ is the wave function, and l is the dimensionality, e.g., for 3D, $l = 3$. One can view the above expression as a partial Fourier transform of the density matrix.

The Wigner function is real, but not necessarily positive, which poses difficulties if it is used directly to find carrier concentration, etc. To derive the DG model from the Wigner function, one notes that if Ψ is the solution to the Schrödinger equation then the Wigner function must satisfy

$$\frac{\partial f}{\partial t} + \boldsymbol{v}\cdot\nabla_{\boldsymbol{r}} f - \frac{1}{\hbar}\nabla_{\boldsymbol{r}} U\cdot\nabla_{\boldsymbol{k}} f + \sum_{\alpha=1}^{\infty}\frac{(-1)^{\alpha+1}}{\hbar 4^{\alpha}(2\alpha+1)!}(\nabla_{\boldsymbol{r}}\cdot\nabla_{\boldsymbol{k}})^{2\alpha+1}Uf = \left.\frac{\partial f}{\partial t}\right|_C \tag{2}$$

where U is the potential, and all other symbols have their conventional meaning. By utilizing only the lowest order term of the series expansion in the above equation, one obtains

$$\frac{\partial f}{\partial t} + \boldsymbol{v}\cdot\nabla_{\boldsymbol{r}} f - \frac{1}{\hbar}\left[\nabla_{\boldsymbol{r}} U\cdot\nabla_{\boldsymbol{k}} f - \frac{1}{24}(\nabla_{\boldsymbol{r}}\cdot\nabla_{\boldsymbol{k}})^3 Uf\right] = \left.\frac{\partial f}{\partial t}\right|_C \tag{3}$$

Or under further assumptions such as isotropic effective mass, this equation can be recast as

$$\frac{\partial f}{\partial t} + \frac{\hbar}{m^*}\boldsymbol{k} \cdot \nabla_{\boldsymbol{r}} f - \frac{1}{\hbar}\nabla_{\boldsymbol{r}} \left(U - \frac{\hbar^2}{8m^*}\nabla_{\boldsymbol{r}}^2 \ln n \right) \cdot \nabla_{\boldsymbol{k}} f = \left.\frac{\partial f}{\partial t}\right|_C \tag{4}$$

This equation is called the quantum corrected Boltzmann transport equation (QBTE) because of its close resemblance to the conventional Boltzmann transport equation (BTE). In fact, by comparing the above equation to the conventional BTE and taking the zeroth moment of the equation, one gets a carrier continuity equation which is identical to the classical drift-diffusion (DD) model, except that the carrier flux now has the form of

$$\boldsymbol{F}_n = D_n \nabla n - \mu_n n \nabla \psi - 2\mu_n n b_n \nabla \left(\frac{\nabla^2 \sqrt{n}}{\sqrt{n}} \right) \tag{5}$$

where

$$b_n = \hbar^2/(4lqm_n^*) \tag{6}$$

where l can again be viewed as the space dimensionality (although it needs not be an integer value). The additional term in the drift part of the model is due to the quantum correction and is sometimes lumped into the potential with the term

$$2b_n \frac{\nabla^2 \sqrt{n}}{\sqrt{n}} \tag{7}$$

being the so-called quantum potential. Note that the unusual $\sqrt{n}$ enters this expression by virtue of the transformation

$$-\frac{1}{2}\frac{1}{n^2}\nabla n \cdot \nabla n + \frac{1}{n}\nabla^2 n = 2\frac{\nabla^2 \sqrt{n}}{\sqrt{n}} \tag{8}$$

3 DG model formulation

3.1 Five-equation set in semiconductor regions

The quantum correction of DG theory obviously raises the order of the differential system, and in order to avoid higher order (no higher than 2) spatial mathematical operators, the carrier continuity equation (for either electrons or holes) is split into two equations with two independent variables, n and ϕ_n. Thus, the conventional DD set of semiconductor equations, which include Poisson's equation and the carrier continuity equations, is expanded from three to five equations, all being second order PDEs. The following five-equation PDE system is solved for in terms of the five independent variables, ψ, $\sqrt{n}$ (or n), $\sqrt{p}$ (or p), ϕ_n, and ϕ_p:

$$\nabla \cdot (\epsilon \nabla \psi) + q(p - n + N_D^+ - N_A^-) = 0 \tag{9}$$

$$\nabla \cdot (b_n \nabla \sqrt{n}) + \frac{\sqrt{n}}{2}\left(\psi - \frac{kT}{q} \ln \frac{n}{n_i} - \phi_n \right) = 0 \tag{10}$$

$$\nabla \cdot (b_p \nabla \sqrt{p}) + \frac{\sqrt{p}}{2}\left(\psi + \frac{kT}{q} \ln \frac{p}{n_i} - \phi_p \right) = 0 \tag{11}$$

$$\nabla \cdot (\mu_n n \nabla \phi_n) + \frac{\partial n}{\partial t} + r = 0 \quad (12)$$

$$\nabla \cdot (\mu_p p \nabla \phi_p) - \frac{\partial p}{\partial t} - r = 0 \quad (13)$$

where r is the net electron-hole pair recombination rate. Even though Maxwell-Boltzmann (MB) statistics are assumed here, Fermi-Dirac (FD) can easily be implemented. The boundary conditions for the above system of equations are the same as in DD except that there are explicit boundary conditions that are associated with Eqs. (10-11).

There are two special cases in which the above general equation set can be simplified:

1. Thermal equilibrium. Since the quasi-Fermi levels are known (being zero), Eqs. (12-13) can be dropped and the system reduces to a three-equation set to be solved for the carrier density and potential distributions.
2. Classical case. With $b_n = b_p = 0$ (or $\hbar \to 0$). Eqs. (10-13) can be combined to reduce to the electron/hole continuity equations of DD theory.

The above equation set is readily solved without stabilization by discrete methods, e.g. finite differences, finite volume, or finite element. A noticeable difference from the classical DD model, however, is that the pn product at the thermal equilibrium is no longer equal to n_i^2, i.e., $n_0 p_0 \neq n_i^2$ in the bulk of the semiconductor region. This requires the use of different models for recombination effects than those used with classical DD [8].

3.2 Equation set in barrier region for tunneling

If tunneling is permitted in barrier regions such as the gate oxide, then one must solve carrier equations in these regions too. We consider only elastic tunneling and therefore the mobility vanishes and the transport is ballistic. Elastic tunneling further implies that carriers (electrons or holes) entering from either end of the barrier do not mix and must therefore be treated separately as shown in Fig. 1 (a). There are thus four independent carrier concentrations to be solved for in the barrier region: two electron populations, n and u, and two hole populations, p and b. Ballistic transport implies that inertia is important. The variables of the problem are the four carrier densities, their respective quasi-Fermi levels and velocities, together with the electrostatic potential. Thus, the number of unknowns in the barrier region is thirteen in principle, for which we must have thirteen equations. But in the one-dimensional case, various approximations can vastly reduce the complexity of the system as well as the number of equations to solve. For details, refer to [8].

The electrostatics is governed by Poisson's equation (assuming no fixed charge in the oxide) which is

$$\nabla \cdot (-\epsilon \nabla \psi) = q(p + b - n - u) \quad (14)$$

where ψ is assumed continuous across the oxide/silicon interface.

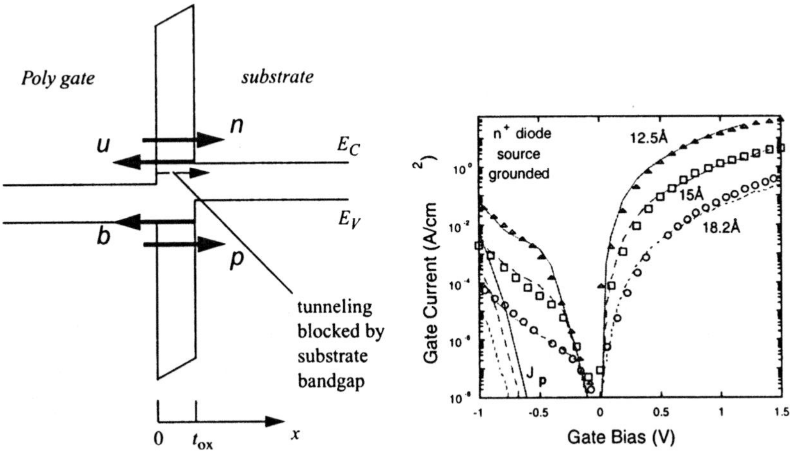

(a) Schematic for four tunneling carrier species

(b) Comparison of tunneling data and simulation

Fig. 1: Modeling and simulation of tunneling in DG framework.

The carrier flux in the barrier region is expressed as

$$\boldsymbol{F}_n = n\boldsymbol{v}_n \tag{15}$$

where $\boldsymbol{v}_n$ is the carrier velocity for forward electrons. The equations solved for n and $\boldsymbol{v}_n$ are: Eq. (10), with the density dependent term, $kT/q\ln(n/n_i)$, replaced by a general dependence of n, for n and the carrier continuity equation for $\boldsymbol{v}_n$,

$$\frac{\partial n}{\partial t} = -\nabla \cdot (n\boldsymbol{v}_n) \tag{16}$$

The equation for ϕ_n is the following equation of motion,

$$m_n^* \frac{d\boldsymbol{v}_n}{dt} \equiv m_n^* \left[\frac{\partial \boldsymbol{v}_n}{\partial t} + (\boldsymbol{v}_n \cdot \nabla)\boldsymbol{v}_n \right] = q\nabla\phi_n \tag{17}$$

Under steady state conditions the equations for the forward-going electrons are the following three equations to be solved for the three independent variables: n, ϕ_n, and $\boldsymbol{v}_n$,

$$\nabla \cdot (b_n \nabla\sqrt{n}) + \frac{\sqrt{n}}{2}\left[\psi + g(n) - \phi_n\right] = 0 \tag{18}$$
$$m_n^*(\boldsymbol{v}_n \cdot \nabla)\boldsymbol{v}_n = q\nabla\phi_n \tag{19}$$
$$\nabla \cdot (n\boldsymbol{v}_n) = 0 \tag{20}$$

where in Eq. (18) we have left the density-dependent part of the equation of state unspecified, using a generic function form $g(n)$ instead. It is probably the best choice to use an adiabatic expression (rather than the MB or FD forms).

To derive the boundary conditions for n and ϕ_n, we illustrate using the one-dimensional case with a barrier extending from 0 to t_{ox} and carriers launched into the barrier the directions shown in Fig. 1. The boundary conditions on these forward-electrons are

$$\begin{aligned} n(0^+) &= n(0^-) && (21)\\ \left.\frac{dn}{dx}\right|_{x=t_{ox}^-} &= 0 && (22)\\ \phi_n(0^+) &= \phi_n(0^-) && (23)\\ \left.\frac{d\phi_n}{dx}\right|_{x=t_{ox}^-} &= 0 && (24)\\ n(0^-)\mu_n(0^-)\left.\frac{d\phi_n}{dx}\right|_{x=0^-} &= n(0^+)v_n(0^+) + u(0^+)v_u(0^+) && (25)\\ n(t_{ox}^-)v_n(t_{ox}^-) &= n(t_{ox}^-)v_{n,TRV}(t_{ox}) && (26) \end{aligned}$$

where the subscript, *TRV*, indicates the tunneling recombination velocity. At the downstream end ($x = t_{ox}^-$), the normal component of the gradient of both carrier concentration and quasi-Fermi level is assumed to be zero [8].

Equation (26) involves the currents in the barrier region and as discussed in [5] these are modeled by tunneling recombination velocity conditions that describe how the carriers that are accelerated across the barrier by the electric field recombine with the equilibrium carrier population at the downstream end. In the above equations, Eq. (21) is simply a statement that carrier concentration is continuous across the material interface for carriers at the upstream end (i.e., $n(0^+)$). Eq. (23) states that the quasi-Fermi level is continuous across the upstream interface. Eq. (25) refers to the continuity condition of the carrier flux across the interface.

3.3 Interfacial boundary conditions in multi-layer structures

The proper specification of boundary conditions (BC) at material interfaces such as the Si/SiO_2 interface is critical to a robust numerical implementation of the DG model. For problems without tunneling, these interfacial BCs are

1. Continuity of ψ across the interface.
2. Vanishingly small carrier densities on the semiconductor side of the semiconductor/insulator interface (in practice a small but non-zero value is specified.)
3. Zero normal component of $\nabla\phi_n$ and $\nabla\phi_p$ across the semiconductor/insulator interface.

4 Implementation and examples

4.1 Model implementation

The above five-equation system has been implemented in PROPHET, a general PDE solver that provided user-specification of PDEs and boundary conditions via a high-

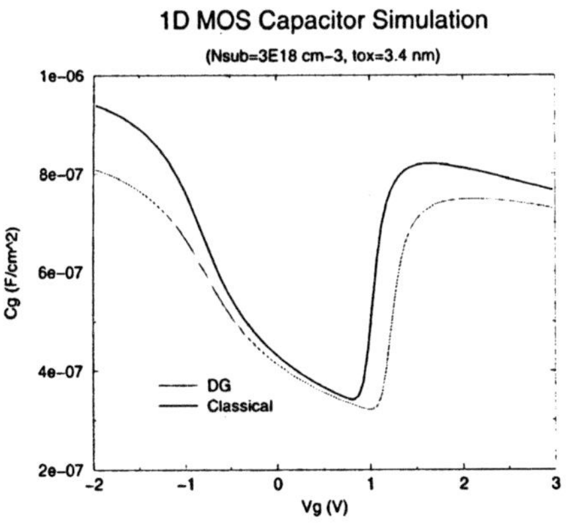

(a) Simulated *C-V* plot for a MOS capacitor using both DD and DG models.

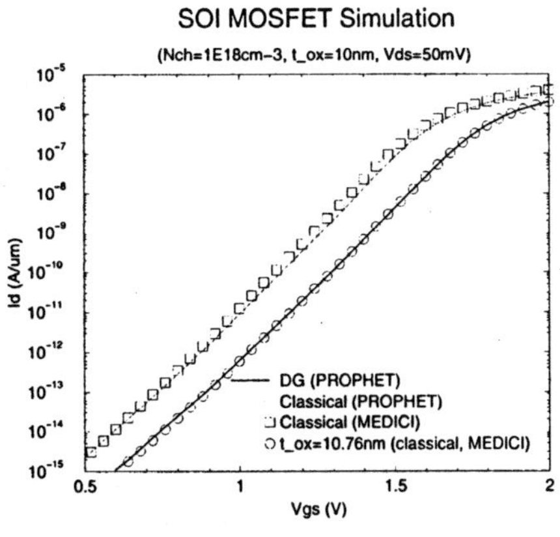

(b) Simulated *I-V* curve for an SOI MOSFET using both classical and DG models. They achieve the same *I-V* plot, the classical model has to use the "electrical" gate oxide which is 0.76 nm thicker than the physical thickness.

Fig. 2: Simulated *C-V* for a MOS capacitor and *I-V* for an SOI MOSFET.

level scripts language, and simulations have been performed for various 1/2/3D MOS structures. Full Newton-Raphson iterations are used to obtain the solution, providing the Jacobian necessary for frequency-dependent small signal *ac* analysis. The convergence rate is acceptable.

At the present time, tunneling current through a thin oxide has been demonstrated in a separate code [8] and the accuracy of the simulation is demonstrated in Fig. 1 (b).

4.2 Simulation examples

Two simulation examples are provided: quasi-static *ac* analysis of a 1D MOS capacitor and *dc* analysis of a 2D SOI MOSFET with the thickness of the silicon thin film varying from 3 to 20 nm and above. The latter simulation results are also compared to Poisson/Schrödinger equation solutions to establish their accuracy.

Fig. 2 (a) shows a simulated *C-V* curve for a MOS capacitor, comparing the classical DD model with the quantum corrected DG model. The DG model predicts a higher threshold voltage and a lower capacitance in both the accumulation and inversion regions. The *C-V* curve is smooth including in the flat band region where other simulations techniques often have difficulty in achieving convergence due to model singularities.

The second example analyzes an SOI MOSFET. Fig. 2 (b), shows the comparison of *I-V* characteristics. To achieve a similar drain current using conventional DD solutions, the "effective" gate oxide thickness would be 7.6 Å thicker than that of the physical one. This order of t_{ox} "correction" is in agreement with other observations, e.g. [9].

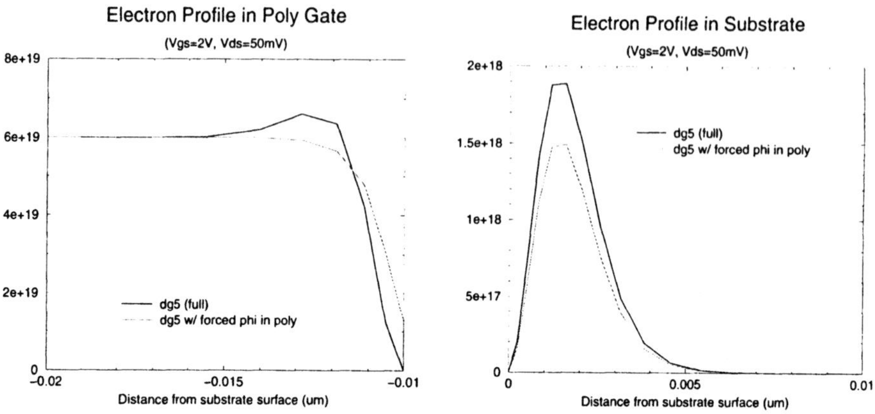

(a) Electron distribution in poly gate

(b) Electron distribution in substrate

Fig. 3: Simulated electron distribution in the poly gate and channel in the linear bias region ($V_{ds} = 50\,\text{mV}$ and $V_{gs} = 2\,\text{V}$) for an SOI MOSFET. Comparison is made between applying DG in all device regions and applying DG only to the silicon film/substrate.

Fig. 3 shows the electron distribution in the poly gate region and in the channel for the DG solutions. One distinctive feature of the simulation results is that there is a dipole in the poly gate region near the poly/oxide interface. This dipole is a result of the physical requirement of quasi-charge neutrality under flat band condition and the repulsion effect of the large SiO_2 barrier. The negative charge of the dipole is due to the pile up of electrons over the background doping. The existence of this dipole has been independently verified by NEMO [6], a program solving Schrödinger equation consistently. Finally, in Fig. 4 the simulated electron profile in the active region for silicon film thickness from 5 to 20 nm is compared with those from SCHRED, a Poisson and Schrödinger equation solver written by Professor D. Vasileska of Arizona State University. The agreement is excellent. It is noteworthy that in order to get a smooth distribution of carriers, SCHRED has to include more than thirty subbands.

Acknowledgment: Authors are grateful to contribution from Dan Connelly of Acorn Technologies. Software prototyping was provided for this research through consortium efforts supported by Texas Instruments and Agere.

References

[1] M.G. Ancona and G.J. Iafrate, "Quantum correction to the equation of state of an electron gas in a semiconductor," p. 9536, *Physical Review B*, vol. 39, no. 13, 1 May 1989.

[2] M.G. Ancona and H.F. Tiersten, "Macroscopic physics of the silicon inversion layer," p. 7959, *Physical Review B*, vol. 35, 1987.

[3] H. Tsuchiya and T. Miyoshi, "Quantum transport modeling of ultrasmall semiconductor devices," p. 880, *IEICE Tr. Electronics*, vol. E82-C, no. 6, June 1999.

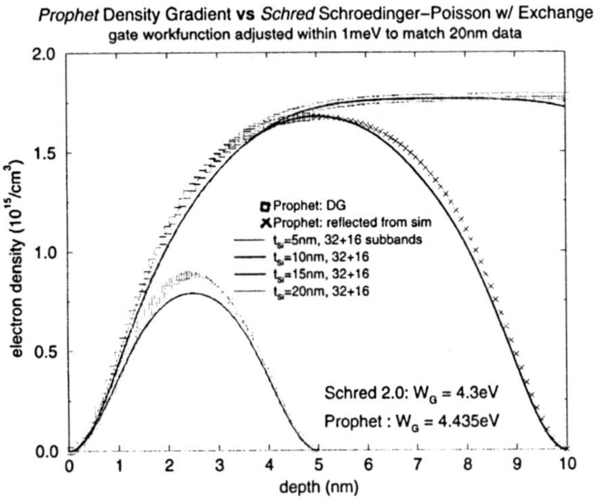

Fig. 4: Simulated electron profile in the active region of double-gate SOI MOSFETs with varying thin film thickness (5, 10, 20 nm). The comparison is made between DG model and the Schrödinger/Poisson's solver, SCHRED.

[4] D. Ferry, R. Akis, and D. Vasileska, "Quantum effects in MOSFETs: Use of an effective potential in 3D Monte Carlo simulation of ultra-short channel devices," p. 287, *IEDM*, Dec. 2000, San Francisco, CA.

[5] M.G. Ancona "Macroscopic description of quantum-mechanical tunneling," p. 1222, *Phys. Rev.*, B42, 1990.

[6] C. Bowen, *et al.*, "Physical oxide thickness extraction and verification using quantum mechanical simulation," p. 859, *IEDM Digest*, Dec. 1997, Washington, DC. Also see R. Lake, *et al.*, p. 7845, *J. Appl. Phys.* **81**, 1997.

[7] E.P. Wigner, "On the quantum correction for thermodynamic equilibrium," p. 749, *Physics Review*, vol. 40, June 1, 1932.

[8] M.G. Ancona, Z. Yu, R.W. Dutton, P.J. Vande Voorde, M. Cao, and D. Vook, "Density-gradient analysis of MOS tunneling," p. 2310, *IEEE T-ED* Dec. 2000.

[9] R. Chau, *et al.*, "30nm physical gate length CMOS transistors with 1.0 ps n-MOS and 1.7 ps p-MOS gate delays," p. 45, *IEDM*, Dec. 2000, San Francisco.

Atomistic Front-End Process Modelling: A Powerful Tool for Deep-Submicron Device Fabrication

M. Jaraiz*, P. Castrillo, R. Pinacho, I. Martin-Bragado, and J. Barbolla
Dept. de Electronica, Univ. de Valladolid, Spain
*Email: mjaraiz@ele.uva.es

Abstract

The complexity attained by current microelectronics process technology can hardly be handled with simulators based on the continuum approach. Over the last few years, atomistic Kinetic Monte Carlo has proven to be a new way to tackle the problems that arise as device dimension shrink into the deep submicron regime. We present some encouraging results of exploring the capabilities of this new process modelling approach.

1 Introduction

As a result of the current efforts to make progress into the deep submicron IC technology, silicon processing is facing an increasingly high level of complexity. In addition, this situation can only worsen as CMOS process technology is pushed closer to the limit of its possibilities. Front-end processing, in particular, is trying to extend the use of conventional, well established doping techniques (ion implantation plus furnace/RTP annealing) into the deep sub-micron size regime. However, many new effects show up when those techniques, well suited for larger feature sizes, are applied to sub-quarter micron processing. Since many of these effects are concurrent, the interpretation of experiments becomes ambiguous and the use of predictive process simulation becomes almost imperative. The continuum approach, based on solving partial differential equations (PDE), can barely handle the increasingly high number of different phenomena that take place, simultaneously, during deep sub-micron device fabrication. Over the last few years, we have been developing and testing the capabilities of Atomistic Front-End Process simulation based on the Kinetic Monte Carlo (KMC) technique. From our experience [1,2], Atomistic KMC process modeling seems especially suitable to fulfill these needs and is, thus, emerging as a most valuable simulation tool in the forefront of microelectronics device fabrication.

It is our purpose to show that atomistic process modeling, implemented in DADOS [3], can meet the challenge of handling, simultaneous and accurately, the wide variety of phenomena involved in current deep submicron front-end processing.

2 Setting the Problem: What do we want to model?

Figure 1 represents the formation of the source/drain (S/D) region of a 0.1um NMOSFET, after a short anneal. An amorphous region created by the As implant has recrystallize very quickly, leaving behind only the foreign impurity atoms.

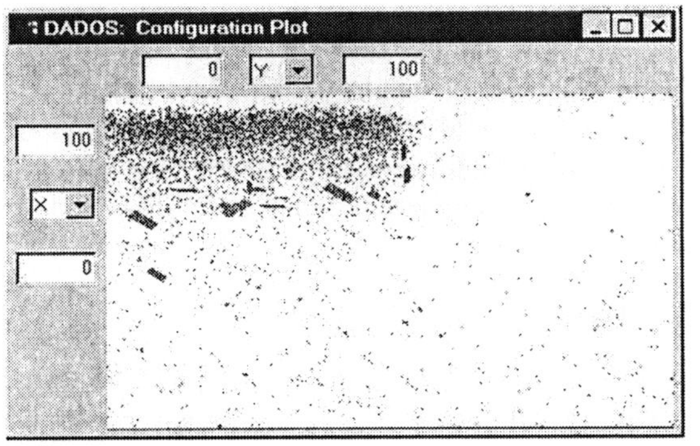

Figure 1. Typical view of a S/D extension implant. {311} defects can be seen beyond the a-c interface after recrystalization of the amorphised region.

Excess I beyond the a-c interface have formed small clusters and some {311} defects (and in some cases dislocation loops) that are now dissolving and producing TED, as well as forming I-B complexes with the B atoms present in the channel. The very high As concentration induces Fermi level effects on diffusion. Moreover, due to the dimensions involved, 2D or 3D simulation is needed to account for realistic effects. Finally, the discrete nature of the dopants and point defects together with their possible charge state can introduce particular correlations in their spatial distribution.

In what follows, we will briefly show how KMC Atomistic Process Simulation can implement these phenomena, in a way accurate and efficient enough to tackle the challenge of including *simultaneously*, in a *3D simulation* all of the mechanisms and highly *non-equilibrium* processes mentioned above. Since we have already presented some of the features of DADOS, we will describe here mainly mechanisms recently incorporated, namely Fermi level effects, and amorphization-recrystallization.

3 The KMC Concept

KMC, unlike Molecular Dynamics, is an event-driven technique, i.e., simulates events (e.g. diffusion hops) at random, with probabilities according to their respective event rates. In this way, it self-adjusts the timestep as the simulation proceeds, just to be able to account for the fastest event present at the current time.

Figure 2 illustrates the concept behind the KMC approach: Fig. 2a (from Ref. 4) shows a high resolution TEM view of a silicon sample with a {311} extended defect embedded in the silicon atom rows. In MD, all lattice and defect atoms are simulated. In KMC, only the atoms belonging to point or extended defects (represented as circles on the TEM view in Fig. 2b) are simulated. In the actual sample, all lattice atoms would be vibrating (with a period of about 10^{-13} s) and, from time to time (e.g. every 10^{-9} s), one of the

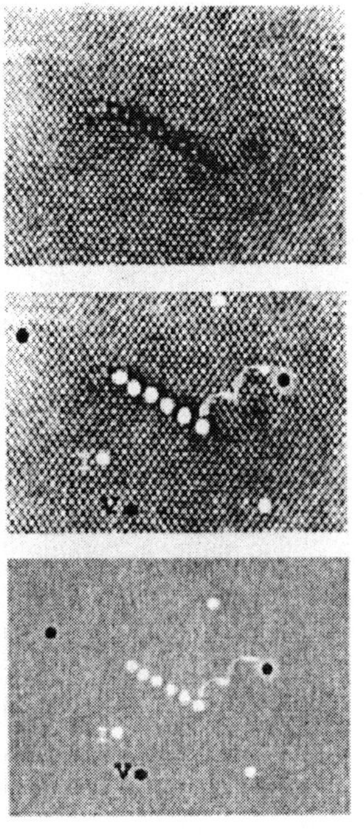

Figure 2. Sketch of the KMC concept.

isolated point defects would jump to a neighboring position, where it might be captured by an adjacent extended defect. At even longer time intervals, (e.g. every 10^{-3} s) a point defect would be emitted from the extended defect. To simulate this, KMC (Fig. 2c) would start out with timestep of about 10^{-9} s, since it only follows the defect atoms. In addition, the fast moving point defects disappear very quickly, leaving only the extended defect and allowing the timestep to automatically be raised to 10^{-3} s.

3.1 KMC Basic Components

A KMC simulator consists of a 3D Simulation Box, of dimensions ranging from tens of nanometers to a few microns, where point and extended defects can be created and allowed to perform events (point defect jump, emission from a cluster,) and interact with each other (annihilation, capture by a cluster,). Point defects can be created in the simulation box to simulate, for example, an ion implantation step. For this purpose, a BCA (Binary Collision Approximation) code can be used to calculate the coordinates of the I and V generated by each ion (cascade). Alternatively, V and I can be generated and recombined at one of the box surfaces to simulate a free surface

DADOS is a non-lattice KMC simulator. The main reason for this is that the processes and dimensions we want to be able to simulate do not need (i.e., are insensitive to using) a lattice. Besides, the computer requirements would be prohibitive for many practical situations. Lattice KMC is necessary, for example, to simulate grain boundaries in polycrystalline materials [5].

4 Atomistic KMC modeling of Diffusion and Defects in Si: a Brief Review

Ion implantation is routinely used in current silicon technology for different purposes like doping and amorphization. Ion implantation damages the crystal lattice. From a practical standpoint for microelectronics processing, implantation damage can be classified into two categories depending on whether or not amorphization is reached. Here we focus on the situation where the damage is not enough to amorphize the lattice. In that case, subsequent high temperature treatments transform the lattice damage into separate, well localized extended defects by either the agglomeration of point defects or by the anneal of high disorder regions.

The basic defects in Si are the intrinsic point defects: self-interstitials (I) and vacancies (V). In addition, there is always a certain amount of unwanted impurities, particularly carbon and oxygen, which form the native extrinsic defects. Finally, other impurities (boron, phosphorous, arsenic) are intentionally introduced to dope or otherwise modify the Si properties. Their presence adds new defect forms and interaction mechanisms. A generic picture of the basic interaction of a

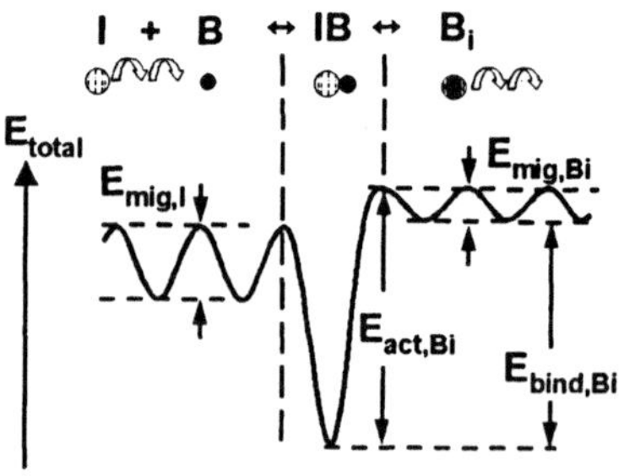

Figure 3. Generic interaction of a Si point defect (I) with an impurity atom (B). Energies involved in the process can be calculated by ab-initio methods.

point defect (I, V) and a foreign atom is shown in Figure 3. In this example, a migrating free I encounters a substitutional impurity atom B and binds to it forming an IB pair that may be mobile or not. The energies involved (migration, binding...) can often be calculated by ab-initio methods.

The agglomeration of V and I can lead to the formation of Si extended defects (like I {311} and dislocation loops, and V voids). Small I clusters are likely to have a non-monotonic behavior in terms of both shape and energy, as inferred from calculations and from inverse-model analysis of diffusion data (Figure 4a, from Ref. 6). Energy oscillations are important because they can act as a bottleneck during the Ostwald ripening process by which small, unstable clusters dissolve while the big, more stable ones grow. The {311} 'rod like' defect has been a subject of intense research over the last few years [7] because of their close association with transient enhanced diffusion (TED)[4]. If the dose and temperature are high enough another form of extended defect, the dislocation loop (DL), appears. Knowledge of geometry and energies (Figure 4b, from Ref. 8) permits direct implementation of these defects into an atomistic process simulator, as it has been recently done in DADOS.

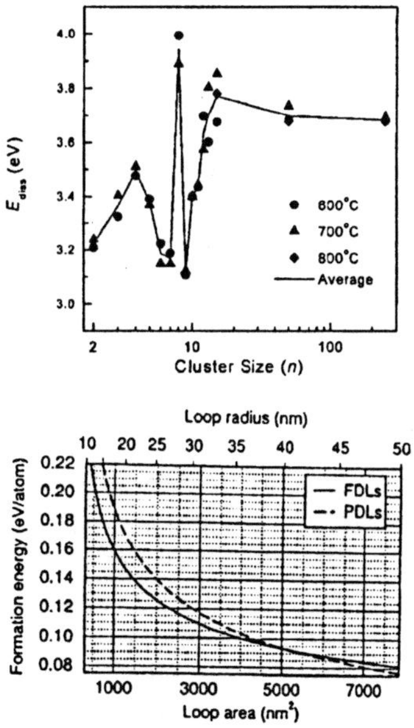

Figure 4. (a) Energies of {311} defects (from Cowern et al. Ref. 6) and (b) Energies of dislocation loops (from Cristiano et al. Ref. 8)

Overall, there is a continuous defect morphology evolution throughout a high temperature annealing stage. Mobile point defects very soon agglomerate forming small clusters. Then the Ostwald ripening process begins: less stable clusters (higher emission/capture ratio) dissolve while the more stable ones grow and evolve into {311}. If the free surface is far away and the dose is high enough the {311} can evolve into the more stable faulted DL and these, finally, into the perfect DL which have a lower formation energy per atom than the faulted DL beyond a threshold size.

Small vacancy defects have been experimentally identified by electron paramagnetic resonance. Vacancy clusters large enough to be seen by electron microscopy appear as spherical voids. Similar to the I clusters behavior just described, when a small V cluster grows beyond a given size it is re-shaped into a void defect. This is necessary to maintain the correct volume to surface ratio as the V cluster grows. A large cluster of N vacancies is re-shaped to be spherical, occupying the volume corresponding to the same number of Si lattice sites.

Under certain conditions, the mechanisms given so far are not enough to explain some experimental observations. For example, it is known that, above a certain concentration, dopants become electrically inactive. It has also been shown that a high I concentration can render electrically inactive a fraction of substitutional B even at a concentration below its solid solubility [9]. An I_nB_m clustering mechanism has been invoked to account for these facts [2]. These extended defects, can be modeled in a way parallel to the clusters, but with two lists of constituent atoms. An analogous modeling has also been successfully employed for carbon, as shown in Fig. 5 [10].

Electrical deactivation can also be due to precipitation. Small X-I clusters might be the embryos of SiX precipitates and clustering and precipitation occur together with one process dominating over the other depending on the experimental conditions.

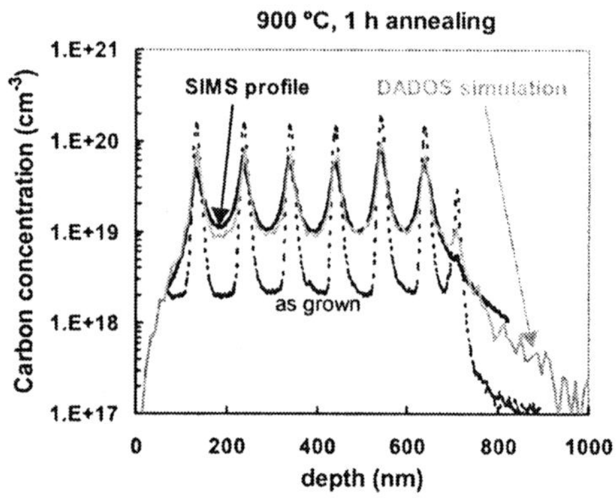

Figure 5. Experimental SIMS and DADOS-simulated carbon profiles in silicon after 1 hour annealing at 900C.

The models presented above can also account for some forms of gettering. Solid solubility effects have also been implemented in DADOS for Carbon, as an example. When C particles are captured by (or 'deposited at') the surface 'extended defect', it stores them and emits C_i point defects at a rate proportional to the C areal density up to a limit that is related to the solid solubility of C in Si.

5 Fermi Level Effects and Amorphization

Fermi level effects are implemented in DADOS as follows. The ratio between different charge states for a given species can be assumed to respond instantaneously to the local electron concentration. Every certain number of events or elapsed time, the atomistic configuration is analyzed to generate a set of data that includes the current 1D, 2D or 3D concentration profiles, cluster size histograms and so on. We shall refer to this set of data as a time frame. The time interval between time frames is chosen such that they change smoothly from one to the next. The electron concentration profile (1D, 2D or 3D) is updated every time frame and used to generate a corresponding transition probability profile for each species to each charge state. Once the time-frame data set is built, the simulation proceeds and, every time a point defect jumps, the charge state is updated. In addition, the effect of the electric field on charged point defects has to be taken into account. The presence of a macroscopic electric field (weaker than the local electric field) is assumed not to change the jump distance or jump rate, but to simply introduce a bias in the diffusion jump direction probabilities. Figure 6 shows a test example of charged V profiles.

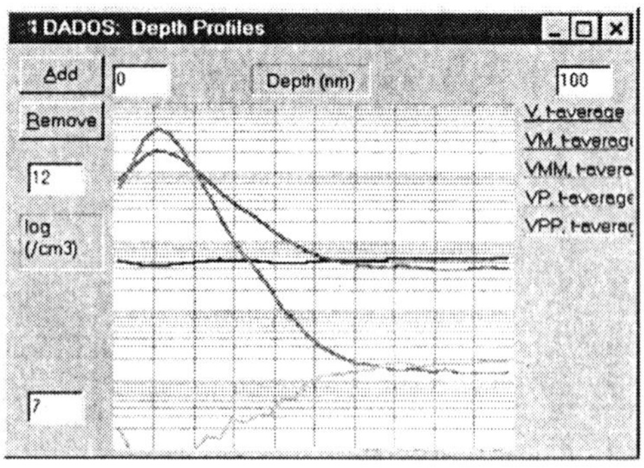

Figure 6. Depth profiles for different V charge states in an As doped region. The flat profile of neutral V indicates that the balance between electric field and charge population is correctly achieved by the atomistic model described in the text.

The global nature and behavior of implantation damage in silicon is a complex issue and is currently the subject of active research. For instance, as the temperature is raised clusters are evidenced experimentally [11] as V-voids and I-dislocations (Fig. 7a). This damage pattern (V clusters near the surface, I clusters deeper) can be readily understood taking into account the ion momentum. However, chemical effects must also have played a critical role, as it is obvious from the comparison of Fig. 7a and Fig. 7b, which include pairs of ions with almost the same mass but one is a dopant (As) while the other is isoelectric (Ge). Similar results have been obtained for P and Si [11].

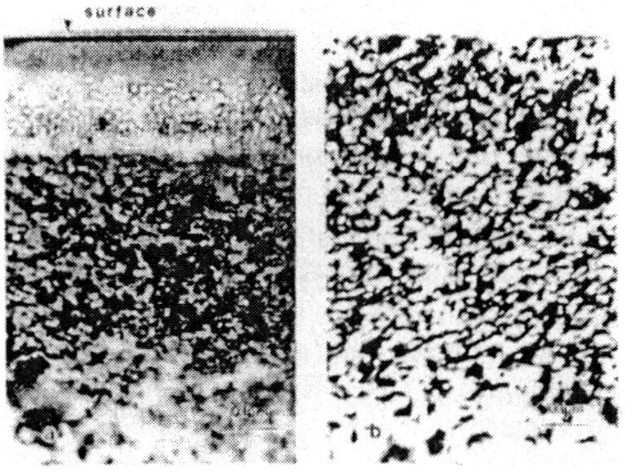

Figure 7. (From Holland et al. Ref. 11) Chemical effects have to be invoked to explain the different damage morphology of two atoms of almost the same ion mass, As (left) and Ge (right). In Atomistic modeling it is easy to specify interaction rules depending on the type of atom.

Amorphization, unlike the extended defects considered up to here, is a massive form of lattice disorder. As such, it is not amenable to the atomistic modeling employed so far. Yet, it has been recently incorporated within a KMC scheme in DADOS as follows.

For simplicity we shall refer to only one dimension histograms, but it is implemented in 3D with 2×2×2 nm^3 boxes. Figure 8 shows a damage depth profile at two consecutive time frames t1, t2, during an implant. The damage depth profiles correspond to the I and V profiles generated by the implanted ions. Normally, the I and the V profiles are indistinguishable from each other because their values are much larger than their difference. Indicated in the figure are the damage threshold for amorphization and a 'MaxStorage' level. The level for amorphization is typically between 20%-30% of the Si atomic density. The MaxStorage level is defined as the maximum allowable I and V concentration in the simulation. It is chosen much lower than the amorphization level, to keep the number of particles in the simulation manageable, but high enough to still reproduce dynamic annealing effects reasonably well. Typically, the MaxStorage level is two orders of magnitude lower than the amorphization level.

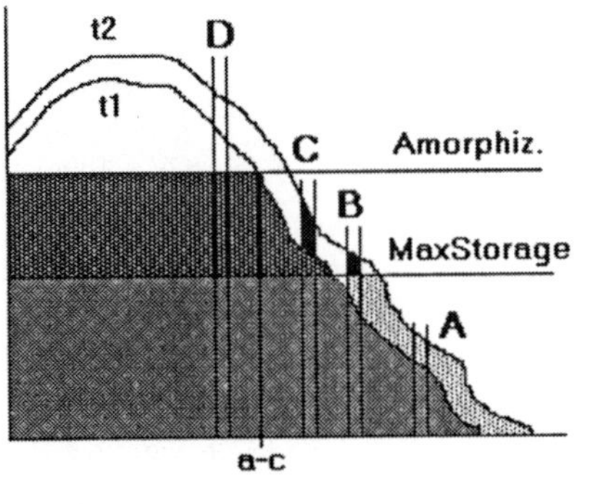

Figure 8. Schematics of the dynamic annealing, amorphization and recrystalization as implemented in the KMC simulator DADOS (see text).

At depth A, the damage level is low and all I and V particles generated by the cascades are incorporated into the atomistic simulation. At depth B and time t2, the damage has reached past the MaxStorage threshold: the number of I and V particles in excess are not inserted into the simulation box, but the damage histogram is updated. Likewise, at depth C, time t2, all I and V particles above the MaxStorage level are deleted, and the damage profile is updated. Finally, at depth D (already amorphised), a high concentration of I and V particles equal to MaxStorage is kept, but the region is labeled as 'amorphous'. Notice that, since there is dynamic anneal between t1 and t2, it is possible for regions B or C to go back to a damage level below the MaxStorage threshold. This, however, is not allowed to occur in region D that can only re-crystallize from the amorphous-crystalline (a/c) interface. The process of deleting I and V particles in excess of the MaxStorage limit always takes care of a particular requirement: the local imbalance between the I and V profiles, due to the ion momentum (see Fig. 8), has to be maintained. V and I particles generated by the cascades are inserted into the simulation as point defects. However, upon interaction, I and V point defects do not directly recombine but form an IV pair point defect with a barrier for recombination. In addition, I and V are assumed to initially form 'amorphous' agglomerates, similar to the clusters described above, but with a lower activation, in agreement with experimental observations [12,13]. Recrystalization is implemented as a special event type. The a/c interface advances into the amorphous regions at a temperature dependent velocity. The recrystalization procedure deletes all I and V particles, only impurity atoms are left behind.

The accuracy of the above described method is quite remarkable. An example can be seen in Fig. 9, in which DADOS correctly predicted that a 50 keV, 3.6 cm^{-2} Si implant will amorphize up to slightly beyond 100 nm, but not the 5 nm adjacent to the surface, in agreement with the experiment [14].

All of the mechanisms described above that are implemented in DADOS are active in all simulations without degrading computation speed noticeably. For example, on a 400 MHz Pentium-II it simulates more than 600 000 events per second, including charge update and for any number of different impurity atoms.

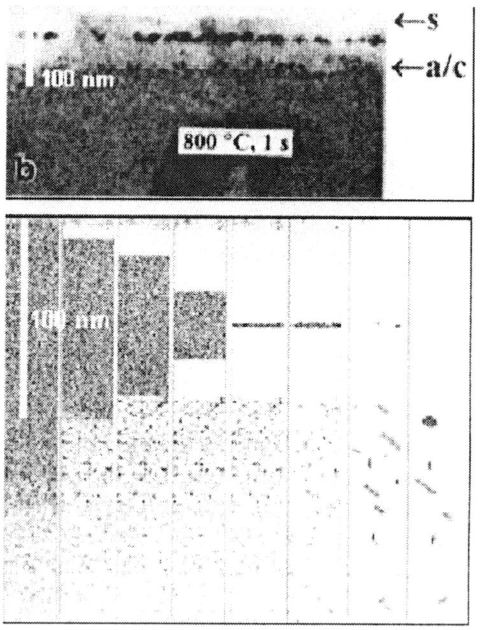

Figure 9. Top: (from Pan et al. Ref. 14). XTEM view showing recrystallized layer and defects at the center and beyond the a/c interface. Bottom: Sequence, from a DADOS simulation, predicting the correct amorphization and annealing behavior.

6 Conclusions

In addition to the accurate modeling of diffusion and defects, already demonstrated in atomistic KMC, we have recently included Fermi level effects and amorphization. This brings the KMC approach to a fully operational front-end process simulator stage, particularly efficient and realistic under extreme conditions such as deep submicron dimensions, complex processing scenarios and highly non-equilibrium situations. In summary, atomistic front-end process modeling has attained a level of maturity that enables it to advantageously simulate the processing steps of current deep submicron device technology.

Stimulating discussions with Drs. C. S. Rafferty, G. H. Gilmer and M. Hane are appreciated.

7 References

[1] M. Jaraiz, G. H. Gilmer, J. M. Poate, and T. D. de la Rubia, Appl. Phys. Lett. 68 (1997) p. 409.

[2] L. Pelaz, M. Jaraiz, G. H. Gilmer, H. J. Gossmann, C. S. Rafferty, D. J. Eaglesham, and J. M. Poate, Appl. Phys. Lett. 70 (1997) p. 2285.

[3] DADOS (Diffusion and Defects, Object-oriented Simulator), M. Jaraiz, L. Pelaz, E. Rubio, J. Barbolla, G. H. Gilmer, D. J. Eaglesham, H. J. Gossmann, and J. M. Poate, Mat. Res. Soc. Symp. Proc. 532 (1998) p.43.

[4] D. J. Eaglesham, P. A. Stolk, H. J. Gossmann, and J. M. Poate, Appl. Phys. Lett. 65 (1994) p. 2305.

[5] J. E. Rubio, M. Jaraiz, L. A. Bailon, J. Barbolla, M. J. Lopez, and G. H. Gilmer, Mat. Res. Soc. Symp. Proc. 514 (1998) p.127.

[6] N.E.B. Cowern, G. Mannino, P. A. Stolk, F. Roozeboom. H. G. A. Huizing, J. G. M. van Berkum, F. Cristiano, A. Claverie, and M. Jaraiz, Phys. Rev. Lett. 82, (1999) 4460.

[7] S. Takeda, Jap. J. Appl. Phys. 30 (1991) p. L639.

[8] F. Cristiano, J. Grisolia, B. Colombeau, M. Omri, B. de Mauduit, A. Claverie, L. F. Giles, and N. E. B. Cowern, J. Appl. Phys., 87 (2000) p. 8420.

[9] P. A. Stolk, H. J. Gossmann, D. J. Eaglesham, D.C. Jacobson, J.M. Poate, and H.S. Luftman, Appl. Phys. Lett. 66 (1995) p. 568.

[10] R. Pinacho, M. Jaraíz, H. J. Gossmann, G. H. Gilmer, and J. L. Benton, Mat. Res. Soc. Spring Meeting 2000. (In press).

[11] O. W. Holland, L. Xie, B. Nielsen, and D. S. Zhou, J. Electronics Mat., 25 (1996) p. 99.

[12] P. J. Schultz, C. Jagadish, M. C. Ridgway, R. G. Elliman, and J. S. Williams, Phys. Rev. B, 44 (1991) p.9118.

[13] R. G. Elliman, J. Linnros, and W. L. Brown, Mat. Res. Soc. Symp. Proc., 100 (1988) p. 363.

[14] G. Z. Pan, K. N. Tu, and S. Prussin, Appl. Phys. Lett., 71 (1997) p.659.

Monte Carlo Impurity Diffusion Simulation Considering Charged Species

Masami Hane, Takeo Ikezawa, and George H. Gilmer†
Silicon Systems Research Laboratories, NEC Corporation
1120 Shimokuzawa, Sagamihara, Kanagawa 229-1198, Japan
Email:hane@az.jp.nec.com
† Agere Systems, Murray Hill, NJ 07974, U.S.A.

Abstract

A Monte Carlo dopant diffusion simulation program has been developed which includes charged species, i.e. Fermi-level effects on drift-diffusion and reactions. In order to save computational time, an algorithm that determines variable time steps was improved to account for all the Fermi-level dependent quantities, such as different charge states of point-defects, pairs and complexes, and different diffusivities/reaction rates for them. Simulation of coupled arsenic and boron diffusion for a typical sub-100nm CMOS process has been demonstrated by using this MC program.

1 Introduction

Recently, several sub-50nm CMOS devices have been demonstrated [1][2]. Their extremely small features and extreme process conditions, e.g. 550℃ for 10 hour anneal or 1100℃ for spike anneal with a 300℃/s ramp-up rate, have challenged conventional process simulators.

In order to simulate such critical MOSFET characteristics, more precise impurity diffusion and activation models are required not only for accurate impurity profiles, but also for evaluating dopant distribution fluctuations which degrade some device characteristics, e.g. threshold voltage uniformity[3]. Some atomistic approaches for modeling dopant diffusion and activation kinetics have been reported over the last few years[4][5].

More recently, attempts were made to improve the predictive capability by incorporating *ab initio* calculations of the energetics of the key diffusing species in the model, e.g. boron clustering[6]. Even the full-CMOS process simulation may be feasible if such an atomistic approach could include some more practical aspects, such as more realistic boundary conditions, Fermi-level effects, and a more efficient computational task for 2-3 dimensional simulations. In this paper, a kinetic Monte Carlo diffusion simulation method is described with emphasis on these features.

2 Model

The basic model for the diffusion simulation is a random walk, where the hopping motion is off-lattice of crystal silicon, i.e. a kinetic Monte Carlo dopant diffusion simulation similar to DADOS[7] code.

Every jump and reaction event is calculated with the discretized variable time step which is determined as the inverse of the sum of every event frequency[7]. Our code considers more than one dopant atom, point defects, pairs, clusters and complexes. Ion implantation is simulated also by a separate MC code[8].

$I^- \; I^0 \; I^+, I^{++}, V^= \; V^- \; V^0 \; V^+,$
$B^- \; BI^- \; BI^0, BI^+,$
$As^+ \; AsV^+ \; AsV^0 \; AsV^-, As_2V^0, ...$

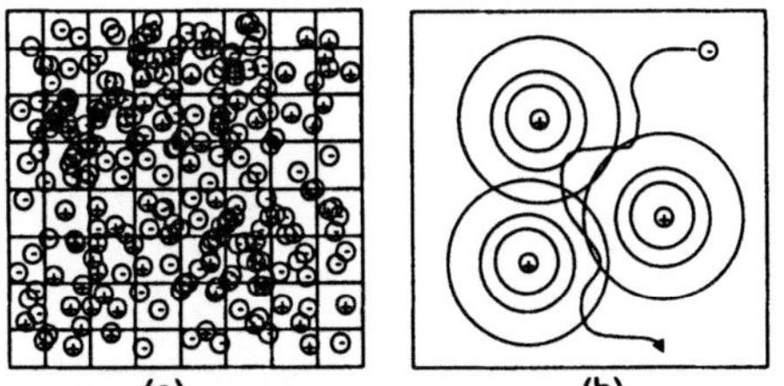

(a) (b)

Figure 1: Typical charged species considered in the model. Electrostatic potential is represented by (a) grid based Poisson equation, or (b) screened Coulomb potential.

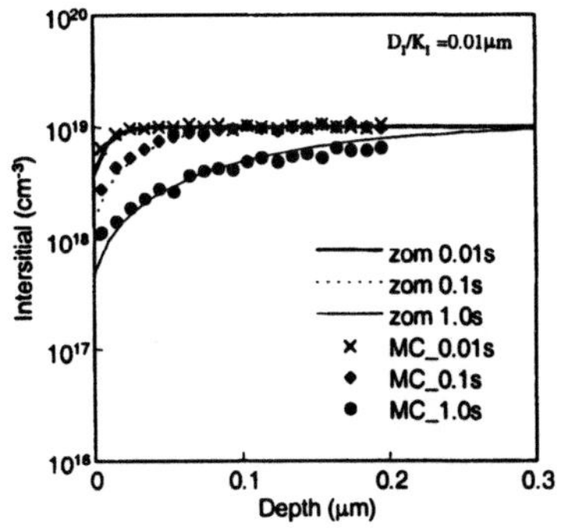

Figure 2: Calculation results for initially flat excess interstitial-Si diffusion behavior assuming finite surface recombination rate(K_I). Comparison is made to the continuum model solutions(solid lins).

Here, we have tested two features, one is the incorporation of all the charged-species, and the other is finite recombination rates for the point defects at the interface. Each substitutional dopant atom has its specific charge state. Monte Carlo particles represent the total number of point defects and pairs in the model, but the charge states are considered every time when they jump or react with other species according to the local electrostatic potential.

The local potential is determined by solving the Poisson equation on a discretized grid (Fig.1(a)) with Newton iteration, since the population of charged point-defects that is determined by the local Fermi-level. In most cases, however, the quasi-neutral approximation is applied instead of solving Poisson equation. The grid-based representation of the local potential implies some artificial averaging of potential fluctuations but these can be captured by considering individual screened Coulomb potentials from the charged species (Fig.1(b)). However, this grid-based representation is also useful for the efficient calculation of the variable time steps. Evaluation of the event is the most computationally intensive step, especially when the model includes disparate different diffusivities or reaction rates for each species.

The computational cost for re-calculation of the specific event frequency was significantly reduced by evaluating the local event or local potential only for the grid element where the particle population was changed. This method enables us to simulate practical process conditions.

All the boundary conditions are set to be periodic or mirror except for the interface and the bottom. The interface is not a perfect sink for point defects but can be described by the finite recombination length $D_I/K_I \simeq 0.01 \sim 0.1 \mu m$[9]. This boundary condition was successfully incorporated in our code (Fig.2).

3 Results and discussion

Almost all of the physical parameters in the model are taken from the literature, but some of them were modified slightly to account for the finite K_I and charge states assumed in this model.

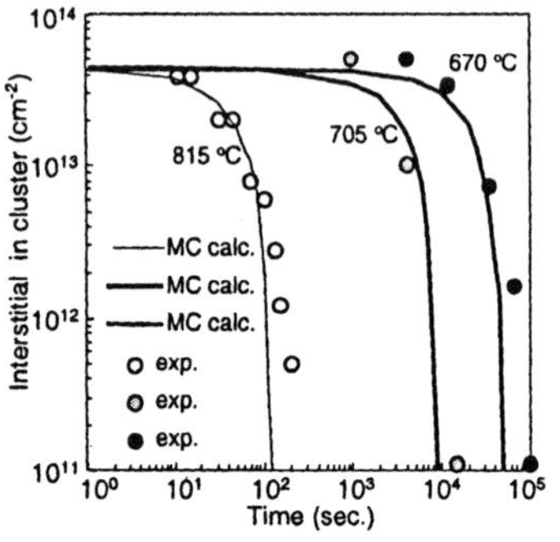

Figure 3: Calculation results for the {311} interstitial cluster dissolution. Experimental data are taken from [10].

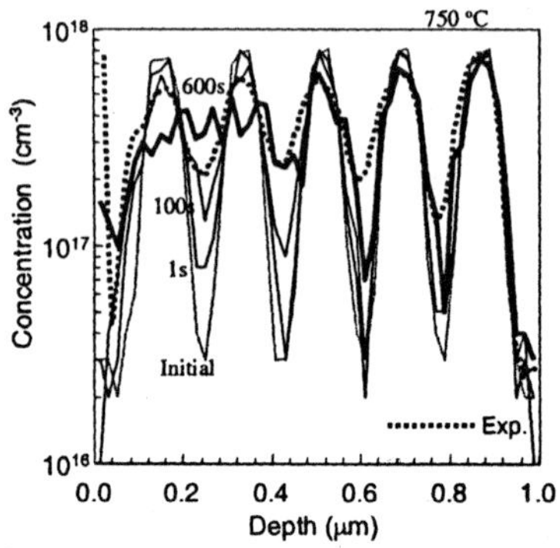

Figure 4: Calculation results compared with the data from the delta-doped boron markers with 40keV 5e13cm^{-2} Si implantation and annealing experiment[11].

The point defect cluster kinetics (Fig.3) and boron transient enhanced diffusion (Fig.4) can be reproduced by the single set of the parameters. Using this MC code, a coupled arsenic and boron diffusion was calculated considering typical process conditions for sub-100nm CMOS. The arsenic source/drain is not only shallow but also includes a high arsenic concentration surrounded by over $10^{18}cm^{-3}$ substrate boron doping(Fig.5). Boron redistribution occurs during the annealing due to the charged species population difference near the PN junction and the electric field effects. The MC calculation was done in a 3D computational cell of $0.1(L) \times 0.5(Depth) \times 0.2(W)[\mu m]$. The results are shown in 1D (Fig.6) and 2D (Fig.7 and Fig.8). A typical CPU-time for the annealing calculation is about 21 hours on a DEC-α 500MHz workstation.

The result shows boron pile-up in the high concentration arsenic source/drain region (Fig.6) and the resulting depletion near the source/drain (Fig.7) which would change the short channel effect.

Analyses of this Monte Carlo simulation results show that the boron redistribution is

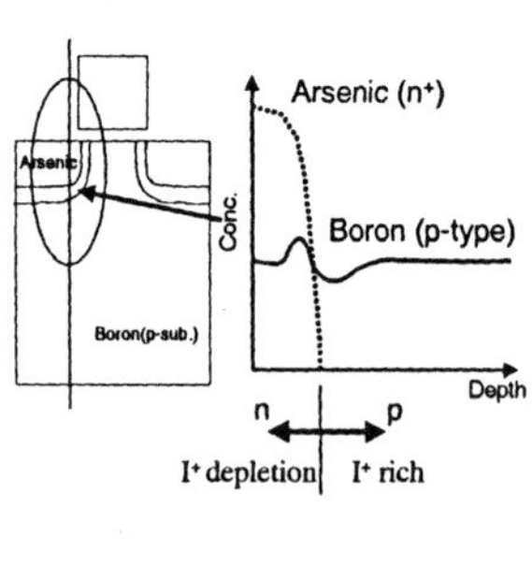

Figure 5: Schematic representation of the redistribution of the substrate boron near the arsenic source/drain.

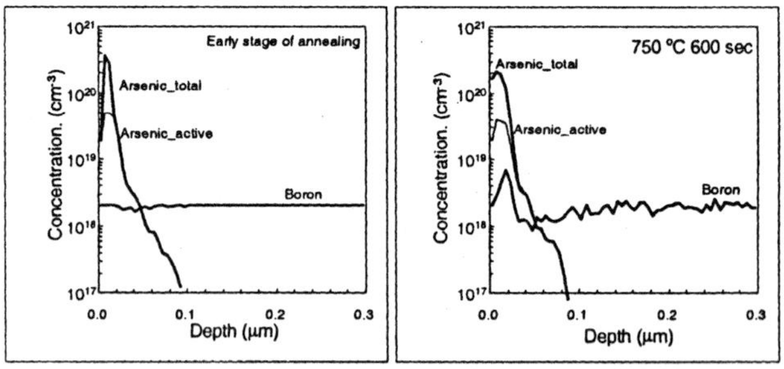

Figure 6: Calculation results for coupled arsenic and boron diffusion. Boron concentration was initially flat. Arsenic was implanted at 10keV, $5e14cm^{-2}$. Calculation region is $0.1(L) \times 0.5(Depth) \times 0.2(W)\mu m$.

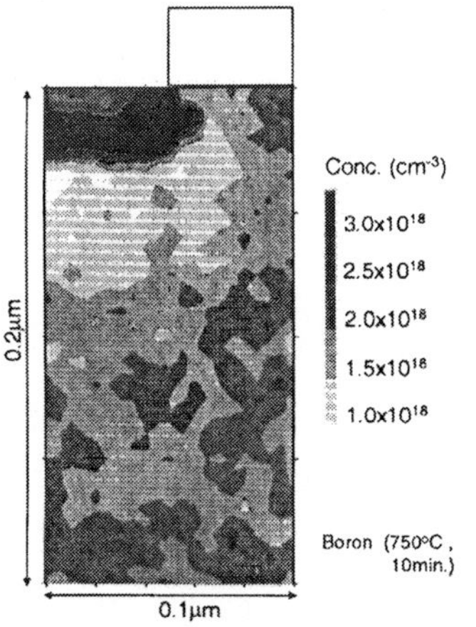

Figure 7: Two-dimensional contour representation of the calculation result for boron shown in Fig.6. (Concentrations were averaged over the W-direction.)

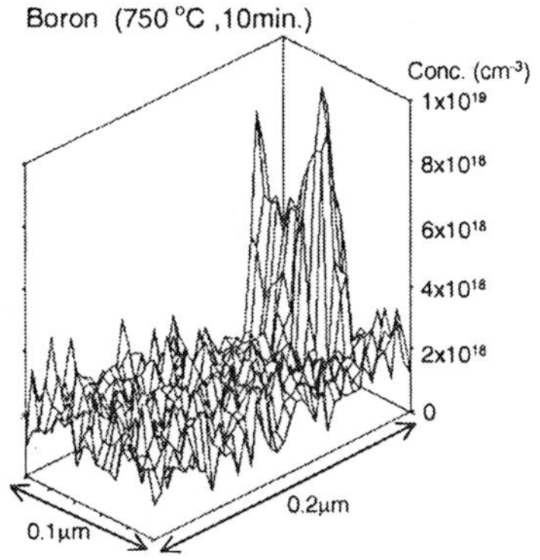

Figure 8: Same as Fig.7 but with boron concentration plotted on the vertical axis.

mostly caused by the population difference of BI pairs in different charge states between the highly doped n^+type region and the p^-type substrate region.

4 Summary

The Monte Carlo diffusion (and implantation) simulator described here has been shown to be suitable for application to sub-100nm CMOS, where the extremely small features require advanced, physically correct (atomic-level) models. Such models can incorporated in our MC code, and simulations of actual process conditions can be accomplished in a reasonable amount of computer time.

Acknowledgment

Authors wish to thank Prof. M. Jaraiz in Valladolid University and Dr. C. S. Rafferty in Agere systems for useful comments and discussion.

References

[1] K. Tsuji *et al.*, VLSI Symp., p.9 (1999)
[2] H. Wakabayashi *et al.*, IEDM Tech. Dig.; p.441 (2000)
[3] P. Stalk *et al.*, SISPAD '97, pp.153-156 (1997)
[4] M. Jaraiz *et al.*, Appl. Phys. Lett., **68**, pp.409-501 (1996)
[5] M.J. Caturla, Comp. Mat. Sci., **12**, pp.319-332 (1998)
[6] S.K. Theiss *et al.*, SISPAD '00, pp.18-22 (2000)
[7] M. Jaraiz *et al.*, MRS '98 (1998)
[8] M. Hane *et al.*, NEC R&D J., **37**(2), pp.170-178(1996)
[9] D.R. Lin *et al.*, Appl. Phys. Lett., vol.67, no.16, pp.2302-2304 (1995)
[10] J.M. Poate, *et al.*, IEDM Tech. Dig., pp.77-80 (1995)
[11] M. Hiroi, T. Ikezawa, M. Hane and A. Furukawa, SISPAD '99, p.63-66 (1999)

A Novel Model for Boron Diffusion in SiGe Strained Layers Based on a Kinetics Driven Ge-B Pairing Mechanism

D. Villanueva, P. Moens
Alcatel Microelectronics, Westerring 15, B-9700 Oudenaarde, Belgium
davy.villanueva@mie.alcatel.be

K. Rajendran, W. Schoenmaker
IMEC, Kapeldreef 75, B-3001 Leuven, Belgium

Abstract

A new temperature dependant model for the Ge-B pairing mechanism has been established and implemented in a process simulator. The combination of both lattice strain effects and GeB clustering has been successfully applied to various anneals. The match with experimental SIMS profiles is excellent.

1 Introduction

Controlling the B outdiffusion in SiGe strained layers is a major concern since it rules the electrical properties of high frequency bipolar transistors. Studies have been carried out describing how Ge lowers the B diffusivity [1-3]. It appears that the effect of strain in the lattice and a sort of clustering reaction between Ge and B are the most relevant mechanisms to investigate. In this paper, results from simulations featuring both phenomena are presented. The novelty is that the clustering reaction is no longer rendered by an additional term to Fick's law [1] but is now treated as a chemical reaction containing the models for kinetics and thermodynamics. Appropriate SIMS profiles after various relevant RTP and furnace anneals have allowed to precise the reaction equation for the Ge-B clustering.

2 Experimental Data

The calibration of the presented model is based on SIMS profiles analysis. Wafers have been processed with a B box shaped profile (8e18/cm3) partly included in a Ge graded profile (up to 7.5e21/cm3). In order to emphasise chemical reactions, the samples have been submitted to different thermal budgets, ranging from 1000 to 1100°C. Another set of SIMS profiles based on a three B spikes profile in a Ge box profile and submitted to typical furnace anneals, has been used to check the validity of the presented model.

3 Diffusion model & Clustering Reaction

The diffusion model is largely based on the one presented by Cowern et al. [3], including the effect of lattice strain on the B diffusivity. It also features the dependency of the intrinsic carrier concentration on Ge as presented by Rajendran et al. [1].

The main focus of this work is the mathematical treatment of the pairing mechanism between Ge and B. When including large Ge atoms in the Si, the lattice is compressed in the growth plane and expanded in the growth direction, which leads to a strained structure. Due to the smaller size of B atoms compared to Si atoms, a local tensile strain exists around the B atoms. On the contrary Ge atoms will generate local compressive strain points. In order to realise some stress energy, B and Ge atoms will attract each other and thus create GeB clusters. This mechanism can be modelled as follows:

- The first solution is to include the trapping mechanism as an additional term to Fick's law [1]:

$$\frac{\partial C_B}{\partial t} = -\nabla J_B - \frac{C_B}{\tau_{GeB}} \qquad [1]$$

In this way, very good results can be achieved for temperatures in the range between 800 and 1050°C, but it fails for higher temperatures, as shown in Figure 1.

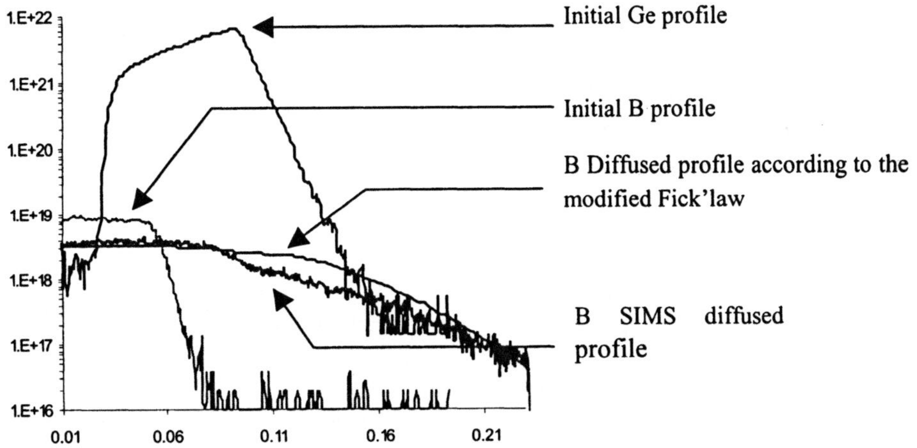

Fig.1. Failure of the modified Fick's law model at 1100 °C during 40sec

- The second option is to go for a chemical reaction between Ge and B as defined by thermodynamics and kinetics. This solution suggests that, since the clustering mechanism is seen as a chemical reaction, it should be thermally enhanced and consequently the study will be mainly focussed on the samples with the highest temperatures. In general, the reaction term will be defined as follows:

$$R = \left[a \exp\left(\frac{-E}{kT} \right) \cdot \sum_{m,n} a_{m,n} [Ge]^m [B]^n \right] \cdot K_0 \qquad [2]$$

Basically, the resultant term consists of a kinetic part, which is weighted by the thermodynamic factor K_0.

As a starting point, it will be assumed that the reaction is not limited by thermodynamics, and hence the main focus is on kinetic considerations. Such assumptions may be justified by the fact that the thermodynamic equilibrium is not very likely to be reached within a few seconds, i.e. during most RTP steps.

From a large number of simulations and from the comparison of the simulated data with the experimental SIMS profiles, it was found that at least two terms are required in order to reproduce the observed features. The most simple reaction rate equation is:

$$Rate = \alpha[Ge] \bullet [B] + \beta[Ge]^{m \geq 2} \qquad [3]$$

The first term is responsible for the GeB cluster formation at medium Ge concentrations, whereas the second term is required to simulate the cluster formation at higher Ge concentrations. In this way, the enhanced B pile-up at high Ge concentrations is well reproduced. The parameters of equation [3] have been determined by fitting various experimental SIMS profiles.

4 Results & Discussions

The simulations were performed by implementing equation [3] in the process simulator DIOS included in the ISE TCAD tools package [(4)]. Figure 2 shows the simulated B profile after a 40sec anneal at 1100°C.

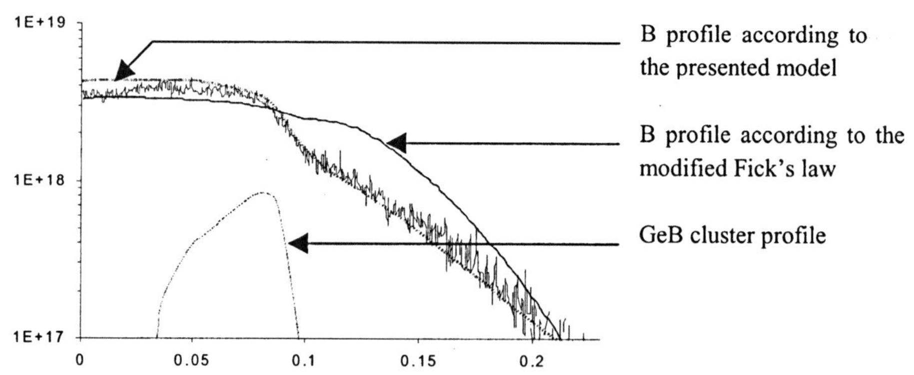

Fig.2. B profiles after 40sec@1100°C

A perfect match with the experimental profile is obtained only with the new model. It follows that up to 30% of the B atoms are trapped into GeB clusters, resulting in a significantly retarded diffusion for high Ge doping level. The GeB cluster profile exhibits two peaks: one at 0.5μm depth and another at 0.8μm depth, due to the first and second terms in the reaction equation respectively. The presented model has also been applied to the "three spikes" B profile of reference [1] annealed at 900°C during 12 min. For the 12min anneal, the simulated profile follows very closely the experimental one, whereas only including strain cannot mimic the B peak, as shown in Figure 3.

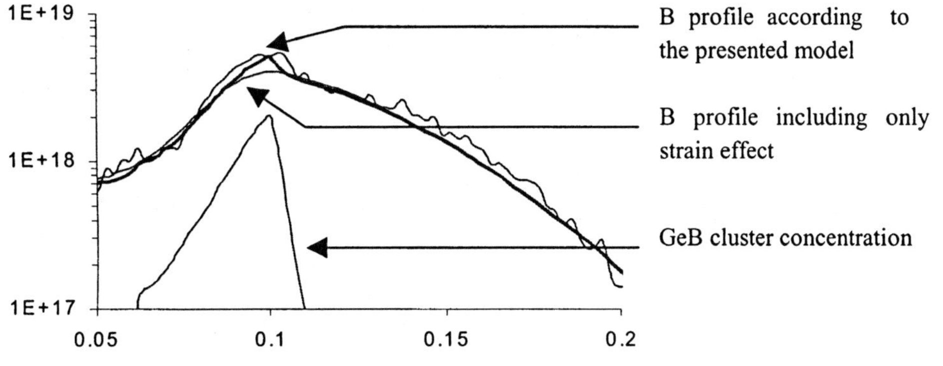

Fig.3. B profiles from the three spikes profile after a 12min anneal at 900°C

Even at such temperature, it is interesting to notice that up to 40% of the B concentration may be trapped by Ge. The efficiency of the presented model for this anneal largely overshoots the limits defined by the thermodynamic assumptions, and thus extends its applicable range to most of the usual thermal budgets for processing.

5 Conclusion

It has been shown that modelling the B retarded diffusion requires the use of a temperature dependent GeB clustering model. Therefore the chemical approach has proven to be the most adequate and flexible solution over a wide temperature range.

Acknowledgements

The authors wish to thank H. Ziad and P. Chevalier for their contribution. Part of this work was performed in the frame of the MEDEA T555 project.

References

[1] K. Rajendran et al., Simulation of B Diffusion in SiGe layers. SISPAD2000, pp. 206-209

[2] R.F. Lever et al., Boron diffusion across silicon-silicon germanium boundaries. J. Appl. Phys., vol.83, pp.1988-1994, 1998.

[3] N. Cowern et al., Diffusion in strained (SiGe), Phys. Rev. Lett., vol.72, pp. 2585, 1994.

[4] DIOS, process simulator, Ver 6.1, ISE AG, Zurich, Switzerland.

The Role of Incomplete Interstitial-Vacancy Recombination on Silicon Amorphization

Luis A. Marqués, Lourdes Pelaz, Jesús Hernández and Juan Barbolla
Departamento de Electrónica, Universidad de Valladolid
E.T.S.I. Telecomunicación, Campus Miguel Delibes, 47011 SPAIN
lmarques@ele.uva.es

Abstract

We investigate the role that point defects and interstitial-vacancy pairs have on the Si amorphization process using molecular dynamics techniques. We show that accumulation of interstitial-vacancy pairs in concentrations of 25% and above lead to homogeneous amorphization. We identify very stable defect structures, consisting of the combination of the pair and Si self-interstitials, which form when there is an excess of interstitials or by incomplete interstitial-vacancy recombination in a highly damaged lattice. These defects could survive long enough at room temperature to act as embryos for the formation of extended amorphous zones and/or point defect clusters.

1 Introduction

The study of the ion irradiation induced amorphization of Si is of particular interest due to the use of increasingly high ion implantation doses in the microelectronics industry. Experiments show that amorphization starts near the ion end-of-range [1] where lattice damage and a net excess of Si interstitials coexist [2]. It has been proposed that the excess interstitials could have an influence on the amorphization process, perhaps through their interaction with the lattice defects created by the irradiation or by forming interstitial clusters that would act as amorphous embryos [3]. This idea has been also suggested to explain the recently achieved amorphization of Si by electron beams [4]. On the other hand, Tang *et al.* proposed that the lattice defect known as *IV pair* could be responsible for the silicon amorphization process [5]. Using molecular dynamics techniques, we have studied the amorphization produced by IV pair accumulation in an attempt to develop a microscopic model to be included in the atomistic diffusion code DADOS [6].

2 The IV pair

The IV pair consists of the atomic bond rearrangement displayed in Fig. 1. The bond lengths obtained from our simulations using the Tersoff 3 potential [7] are in very good agreement with Tight Binding [5] and *ab initio* [8] results, as can be deduced from inspection of Table 1. The obtained formation energy, 3.01 eV, compares with the 3.51 eV from Tight Binding [5] and the 3.26 eV from *ab initio* calculations.

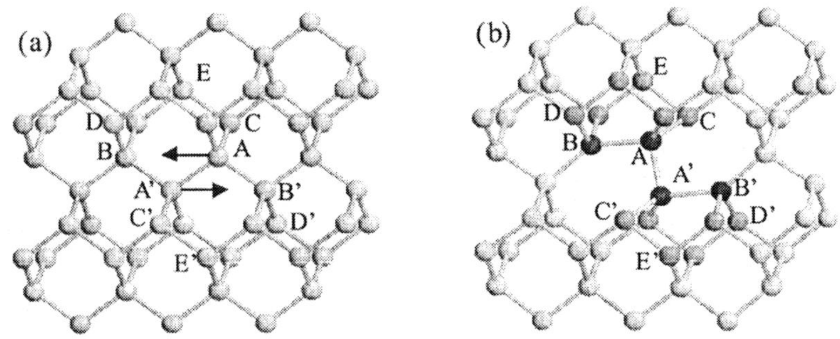

Fig. 1. Formation scheme of the IV pair. Atoms A and A' switch their bonds with atoms B and B'. Darker tones correspond to higher potential energies.

Table 1. Lengths of the bonds involved in the IV pair, expressed in Å. Letters refer to the corresponding atoms in Fig. 1(b).

Bond	This work (T3)	TBMD [5]	*Ab initio* [8]
A-A′	2.28	2.28	2.27
A-B	2.51	2.46	2.46
A-C	2.35	2.38	2.39

We have also studied the stability of the IV pair by computing its time for recombination at different temperatures. The results are shown in the Arrenhius plot of Fig. 2. Extrapolating at room temperature we find a lifetime of only 3 μs, which is too short for the IV pair to survive between successive cascades and thus act as the embryo for Si amorphization, unlike it was proposed previously [5,8].

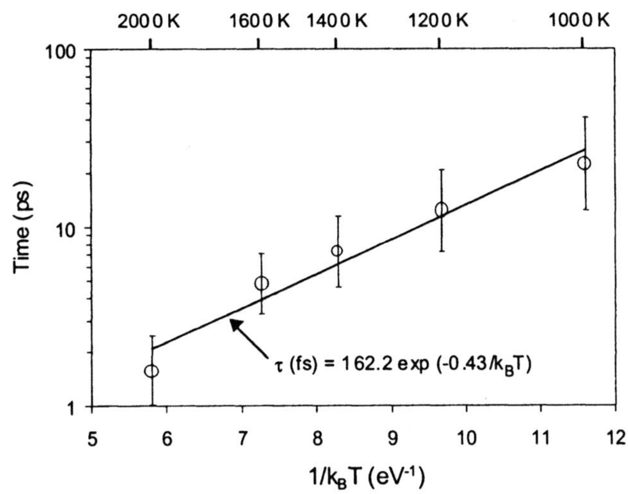

Fig. 2. Arrenhius plot of the time needed for I-V recombination. The line is the best fit to the data, which determines an activation energy of 0.43 eV.

3 Amorphization simulations

In Fig. 3 we display the evolution of the potential energy per atom in samples with different concentrations of IV pairs at several temperatures. As can be seen, samples with concentrations of 10 and 20% recrystallized for all the simulated temperatures. However, samples with 30% of IV pairs remained amorphous. Interestingly enough, the behavior of the sample with a concentration of 25% depends on temperature. After introducing 25% of IV pairs some traces of crystallinity are left. These crystal seeds grow at lower temperatures, but thermal agitation at higher temperatures dissolve them. Consequently, for each temperature there exists a critical damage concentration that prevents recrystallization.

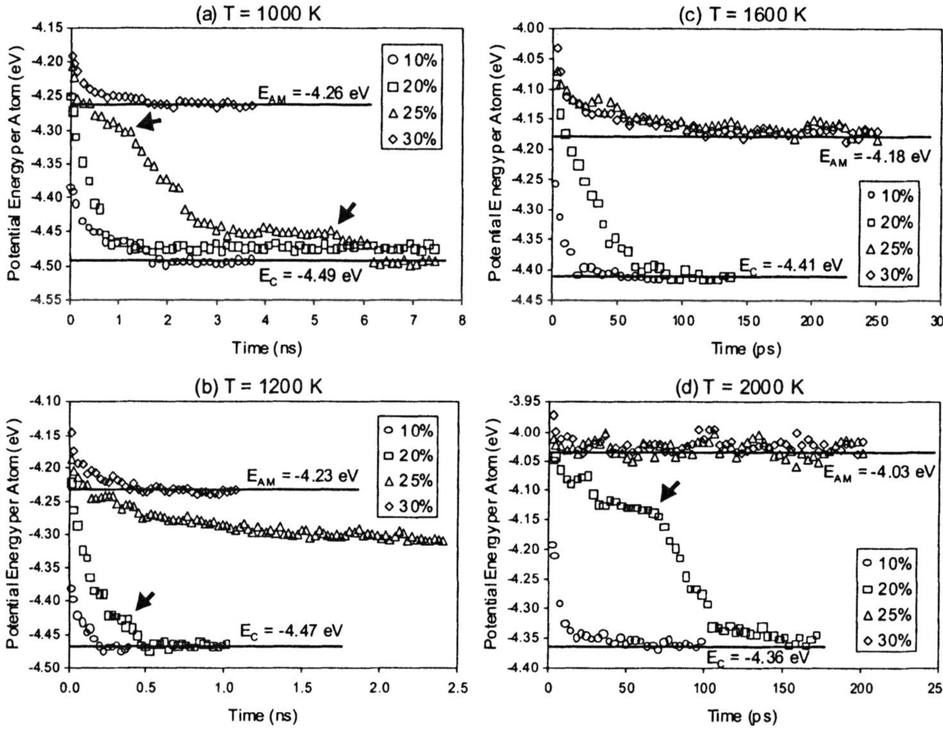

Fig. 3. Time evolution of the potential energy per atom in samples with different IV pair contents during the annealing at several temperatures. Solid lines indicate the mean potential energy per atom in amorphous and crystal Si at each temperature, E_{AM} and E_C, respectively.

For a concentration of 10% energy decay is exponential, and corresponds to an activation energy of 0.45 eV, very close to the barrier obtained for IV pair recombination. This means that 10% is a concentration so low that IV pairs do not interact with each other, and thus the overall crystallization behavior is the same that when there is just one IV pair. However, for higher concentrations the evolution of the potential energy per atom shows plateaus followed by steep decreases, indicated by arrows in Fig. 3. In those cases, IV pairs interact with each other and form more stable structures. The crystallization then implies the collective movement of several atoms which produces a sudden decrease in the potential energy per atom.

4 The IV+2I complex

In our simulations we found very stable defects formed by the combination of an IV pair and two interstitials, shown in Fig. 4(a). We observed that this defect, that we will refer to as *IV+2I complex*, is mobile. Its displacement takes place following the scheme depicted in Fig. 4. This defect can remain stable in the lattice after more than 10 ns at 1000 K, which implies a lifetime at least two orders of magnitude longer that that of the IV pair. This fact means that self-interstitials can stabilize the lattice disorder represented by the IV pair. If the disorder survives between successive cascades, it could accumulate and eventually lead to amorphization. The formation of these complexes is favored when excess of interstitials coexists with the lattice damage, as in the case of ion irradiation. They can also be formed with no net excess of Si self-interstitials if high enough density of IV pairs is present, as in the case of electron irradiation at high doses and low temperatures.

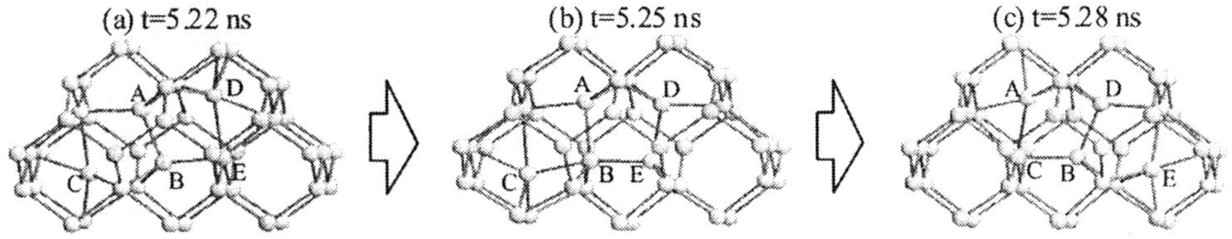

Fig. 4. Atomic pictures showing the time evolution of the IV+2I complex formed during the annealing at 1000 K of the sample with a initial IV pair concentration of 20%.

References

[1] Maszara, W.P., Rozgonyi, G.A. (1986): Kinetics of damage production in silicon during self-implantation. J. Appl. Phys. 60: 2310-2315

[2] Holland, O.W., White, C.W. (1991): Ion-induced damage and amorphization in silicon. Nucl. Instrum. Methods Phys. Res. B 59/60: 353-362

[3] Seidman, D.N., Averback, R.S., Okamoto, P.R., Baily, A.C. (1987): Amorphization processes in electron- and/or ion-irradiated silicon. Phys. Rev. Lett. 58: 900-903

[4] Takeda, S., Yamasaki, J. (1999): Amorphization in silicon by electron irradiation. Phys. Rev. Lett. 83: 320-323

[5] Tang, M., Colombo, L., Zhu, J., Diaz de la Rubia, T. (1997): Intrinsic point defects in crystalline Si: Tight-binding molecular dynamics studies of self-diffusion, interstitial-vacancy recombination, and formation volumes. Phys. Rev. B 55: 14279-14289

[6] Jaraíz, M., Pelaz, L., Rubio, E., Barbolla, J., Gilmer, G.H., Eaglesham, D.J., Gossmann, H.J., Poate, J.M. (1998): Atomistic modeling of point and extended defects in crystalline materials. MRS Symp. Proc. 532, MRS Spring Meeting, San Francisco, pp. 43-53

[7] Tersoff, J. (1988): Empirical interatomic potential for silicon with improved elastic properties. Phys. Rev B 38: 9902-9905

[8] Cargnoni, F., Gatti, C., Colombo, L. (1998): Formation and annihilation of a bond defect in silicon: An *ab initio* quantum-mechanical characterization. Phys. Rev. B 57: 170-177

Atomistic simulations of extrinsic defects evolution and transient enhanced diffusion in silicon

F. Cristiano[1], B.Colombeau[2], C.Bonafos[2], J.Aussoleil[2], G.Ben Assayag[2] and A.Claverie[2]
[1]LAAS/CNRS, 7 av. du Col. Roche, 31077 Toulouse, France
[2]CEMES/CNRS, 29 rue J. Marvig, 31055 Toulouse, France

Abstract

We present a "unified" model for the simulation of boron transient enhanced diffusion in both crystalline and preamorphised structures. The model describes the thermal evolution of a supersaturation of Si interstitial atoms in dynamical equilibrium with all types of extrinsic defects. We show some successful applications of our model to a variety of experimental conditions and give an example of its predictive capabilities at ultra low implantion energies.

1 Introducion

The transient enhanced diffusion (TED) of ion implanted boron in silicon is a major problem in advanced IC fabrication. It is driven by the excess Si interstitial atoms ($Si_{int}s$) generated by the ion implantation process. Early models aimed at the simulation of the basic experimental case of boron implanted into a crystalline substrate, where the dominant defects are typically the {113}s. These models assumed that a flux of $Si_{int}s$ (responsible for TED) originates from the continuous dissolution of the {113}s defects (Chao et al. 1996). More recently (Bonafos et al. 1997, Claverie et al. 2000), it has been proposed that TED is governed by the thermal evolution of a supersaturation of $Si_{int}s$ in dynamical equilibrium with all types of extrinsic defects formed after implantation (clusters, {113}s and dislocation loops (DLs)). Based on this description (Ostwald ripening), new models were proposed (Gencer and Dunham 1997) but often rely on many fitting parameters. While successfully fitting particular sets of experiments, they are not sufficiently accurate for predictive TCAD applications. Moreover, the increasing need to achieve ultra shallow junctions, has resulted today in the extensive use of preamorphised structures, where most of the extended defects formed after the implant are the DLs. Very few models have been proposed to simulate TED in presence of DLs (Chao et al. 1996). In addition, these models assume that DLs are perfect sinks for the interstitials, i.e. they behave in an opposite way to the {113}s.
In this paper, we present a "unified" model we have developed which describes the concomitant evolution of the defects and of the supersaturation of $Si_{int}s$ in both crystalline and preamorphised structures. Great effort has been put into the "physical" definition of the various parameters needed for the calculations, with the overall aim of reducing to a minimum the number of fitting parameters. We will present some applications of our model to a variety of experimental conditions and we will give an example of its predictive capabilities.

2 Simulation of defect growth and TED

Our model describes the concomitant time evolution of extrinsic defects and of the supersaturation of $Si_{int}s$ in the defect region, according to the Ostwald ripening theory. Each defect (of size n from 2 up to thousands of atoms) is characterised by its formation energy $E_f(n)$ (the energy required to add one extra atom to the defect) and capture efficiency A_n/R_{eff}. Both $E_f(n)$ and A_n/R_{eff} depend on the geometry of the defect (type and size), as recently described by Claverie et al. (2001). The growth of a given defect is calculated by the difference between its capture rate, F_n, and its emission rate, R_n. The capture rate F_n depends on the defect environment and is directly proportional to the mean interstitial supersaturation, S, in the defect region, while the emission rate R_n depends only on the defect formation energy $E_f(n)$. The evolution of the whole defect population is mathematically expressed as follows:

$$\frac{dN_n}{dt} = F_{n-1}N_{n-1} - F_nN_n + R_{n+1}N_{n+1} - R_nN_n \tag{1}$$

where N_n is the number of defects of size n, while the mean interstitial supersaturation (free $Si_{int}s$ in dynamical equilibrium with the defects) is given by

$$S = \frac{C_i}{C_i^*} = \frac{\sum_{n=2}^{\infty}\beta_n R_n N_n}{D_i C_i^* \left(\left(\sum_{n=2}^{\infty} \frac{A_n}{R_{eff}} N_n \right) + \frac{1}{L_{surf} + R_p} \right)} \tag{2}$$

β is the number of Si atoms released by the break-up of a cluster ($\beta = 2$ for $n=2$, $\beta = 1$ otherwise). R_p is the distance between the defect region and the surface, while the surface recombination length, L_{surf} is the only fitting parameter of the model.
The presence of an interstitial sink at the wafer surface (term $(L_{surf}+R_p)$ in eq. 2) imposes a "pinning" of the $Si_{int}s$ supersaturation that linearly decreases from the defect region towards the surface, as schematically shown in fig. 1a. This result implies that our model is equally applicable to both cases of crystalline and preamorphised substrates. In the first case, the defect region is generated by the boron implant itself and is located around the boron concentration peak. The boron diffusivity enhancement is therefore directly given by eq. 2. In the second case, the defect region is generated by the preamorphisation implant and can be much deeper than the boron profile. The boron diffusivity enhancement is therefore calculated by

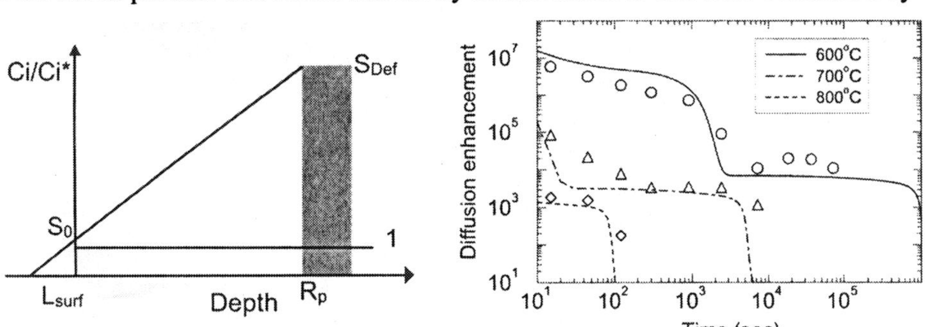

Fig. 1. *(a)* Depth variation of the amplitude of the $Si_{int}s$ supersaturation in the surface region. *(b)* Model validity test. TED of boron delta layers: comparison of measured diffusion enhancement (symbols) and simulated values (lines).

interpolating the $Si_{int}s$ supersaturation amplitude between the defect region (eq. 2) and the silicon surface, at the depth corresponding to the boron profile.

3 Results and Discussion

Typical results of our simulations are given in fig. 1b, where we have tested our model against the experimental study by Cowern et al. (1999), concerning the TED of CVD-grown boron delta layers after low dose Si^+ implantation. The measured evolution of the boron diffusivity enhancement, i.e. of the $Si_{int}s$ supersaturation, is well reproduced by the simulated curves, including the two "plateaus" associated to small stable clusters (size 4 and 8) and {113} defects observed at 600 °C. We have also been able to perfectly fit published data concerning the dissolution of {113} defects (Colombeau et al. 2001).
We now consider the case of direct boron implantation into a crystalline substrate. We have studied through computer simulations the effect of the implant energy in the range 1-25 keV, for a 10^{14} cm^{-2} boron implant. Fig. 2 shows the predicted evolution of the $Si_{int}s$ supersaturation in the defect region during a 950°C anneal. All implants show a sharp "peak" at very short times (when only small clusters exist), followed by a "plateau" associated with the ripening of {113} defects whose lifetime increases with increasing energy. This result demonstrates how the vicinity of the surface to the boron profile plays a crucial role in determining the TED decay time, especially at ultra low implantation energies. Indeed, these predictions are confirmed by the successful simulation of TED experiments of 2 keV implants (not shown).
We now present the application of our model to a typical case of boron TED in a preamorphised structure. Fig. 3a shows the SIMS profiles (symbols) measured after annealing at 950, 1000 and 1050°C for 10 sec of a series of samples implanted with 1 keV B^+ ions to a dose of 1×10^{14} ions/cm^2 into preamorphised silicon (150 keV Ge^+, 2×10^{15} ions/cm^2, a/c interface at 175 nm). The values of the average boron diffusion acceleration $<D_B/D_B^*>$ for the three samples are reported in the figure inset. They have been extracted by fitting the measured profiles using a commercially available process simulator and taking the boron diffusion coefficient as the fitting parameter (lines in fig. 3a). We have then used our model to calculate the average $Si_{int}s$ supersaturation in equilibrium with the defects (dislocation loops) for each of the three temperatures and the corresponding values in the region containing the boron

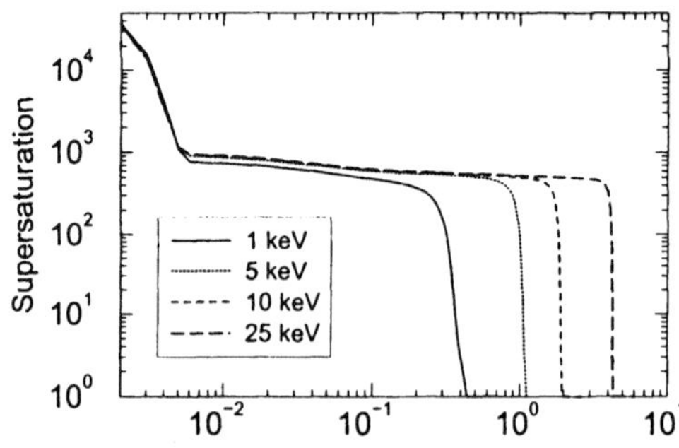

Fig. 2. Simulation of the $Si_{int}s$ supersaturation evolution for 1×10^{14} cm^{-2} boron implants in the 1-25 keV rage at an annealing temperature of 950°C.

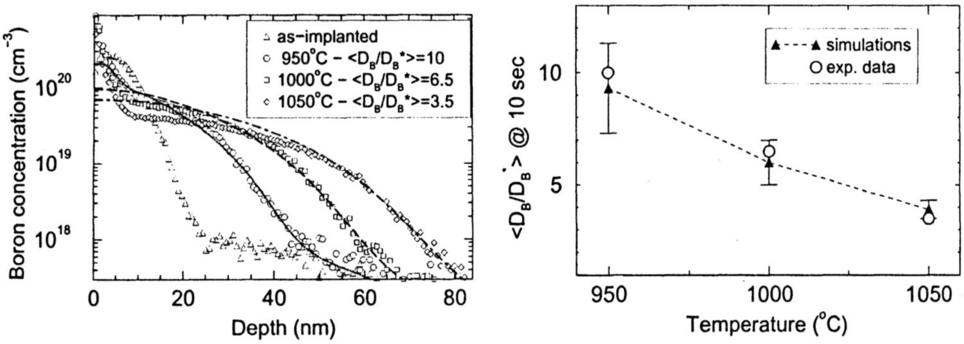

Fig. 3. *(a)* SIMS profiles (symbols) of 1 keV B^+ implants and acceleration coefficients (inset) extracted by a fitting procedure (lines). *(b)* Comparison between the simulated and measured values of the average acceleration coefficients for a 10 sec anneal.

profile, following the schematic description of fig. 1a. The calculated values are reported in fig. 3b (triangles) together with the measured ones (circles). An excellent agreement is obtained for all cases without changing any of the input parameters, except for the annealing temperature.

It is interesting to note that the low diffusivity enhancement in the preamorphised region is a "simple" consequence of the pinning of the $Si_{int}s$ supersaturation at the surface. There is therefore no need to describe the dislocation loops as a more or less efficient barrier against the the $Si_{int}s$ flux towards the surface.

4 Conclusions

We have developed an atomistic simulation of boron TED based on the Ostwald ripening of extrinsic defects occurring during annealing. The model applies to both crystalline and preamorphised structures. We have tested the model against some notorious experimental studies of TED and defect evolution. We have successfully applied our model to a variety of experimental conditions including ultra low energy implants. The strength of the model relies on the accurate physical description of the defects formation and evolution during annealing.

References

[1] C.Bonafos, M.Omri, B.de Mauduit, G.Benassayag, A.Claverie, D.Alquier, A.Martinez and D. Mathiot, J. Appl. Phys. 82 (1997) 2855.

[2] H..S.Chao, S.W.Crowder, P.B.Griffin and J.D.Plummer, J. Appl. Phys., 79 (1996) 2352.

[3] A.Claverie, F.Cristiano, B.Colombeau and N.Cowern, MRS Proceedings vol 610 (2000).

[4] A.Claverie, B.Colombeau, F.Cristiano, A.Altibelli and C.Bonafos, MRS Proc. 2001.

[5] B.Colombeau, F.Cristiano, A.Altibelli, C.Bonafos, G.Benassayag and A.Claverie, Appl. Phys. Lett., 78 (2001) 940.

[6] N.Cowern et al., Phys. Rev. Lett., 82 (1999) 4460.

[7] A.H. Gencer and S.T. Dunham, J. Appl. Phys., 81 (1997) 631.

Initial Conditions for Transient Enhanced Diffusion: Beyond the Plus-Factor Approach

G. Hobler
Univ. of Technology Vienna
A-1040 Vienna, Austria
gerhard.hobler@tuwien.ac.at

V. Moroz
Avant! Corporation,
Fremont, CA 94538, USA
victorm@avanticorp.com

Abstract

A new model is proposed for interstitial and vacancy profiles due to ion implantation. When used in TED simulations, more accurate results are obtained than with the "plus-factor" model. The user is not required to run Monte Carlo simulations nor to explicitly take spatial correlations between interstitials and vacancies into account. The model is formulated in terms of effective Frenkel pair numbers and vertical and lateral shift distances between interstitial and vacancy profiles. It is well suited for 1D/2D/3D process simulation.

1 Introduction

According to the "plus-factor model" the initial interstitial profile for transient enhanced diffusion (TED) simulations is assumed to be the as-implanted dopant profile multiplied with the "plus-factor". The plus-factor has been determined in [1] as to yield the same amount of diffusion in the bulk as simulations using the full point defect distributions. Values as large as 4 have been found for practical implant conditions (a value of unity corresponds to the "+1" model [2]). The origin of plus-factors different from unity are deviations between the interstitial (I) and vacancy (V) profiles, with net-excess interstitials (netI) deeper in the bulk and net-excess vacancies (netV) near the surface, and the out-diffusion of vacancies during Frenkel pair recombination.

While the plus-factor model is a major improvement over the "+1" model, it still has limitations: (i) The depth dependence of TED is not accurately described. When the plus-factor is calibrated in the bulk, errors up to 20% may occur at $x = R_p$ and even larger errors near the surface [3]. (ii) There is no information about the number of Frenkel pairs. Therefore, it cannot be used to simulate processes taking place during Frenkel pair recombination such as the competition between dopant-interstitial and interstitial-interstitial clustering [4]. (iii) While the plus-factor model can be applied to amorphizing implants, the plus-factors assume high values and depend on both the implant energy and the dose, as will be shown. This complicates finding an analytical expression for the plus-factor as a function of the implant parameters [3].

In order to overcome these problems, we propose alternative ways of constructing I and V profiles as initial conditions for TED simulations. They are more accurate than the plus-factor model but avoid the complexity of the original model [1, 5]. In the following we present models for non-amorphizing and amorphizing implants (Sections 2 and 3, respectively), and extend these models to two and three spatial dimensions (Section 4).

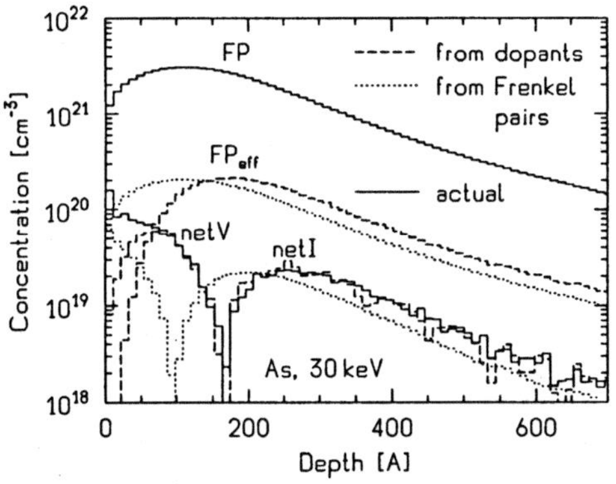

Fig. 1: Construction of effective point defect profiles after a non-amorphizing implant by scaling and shifting the dopant profile (dashed lines) and the Frenkel pair (FP) profile (dotted lines). netI=max(I-V,0), netV=max(V-I,0).

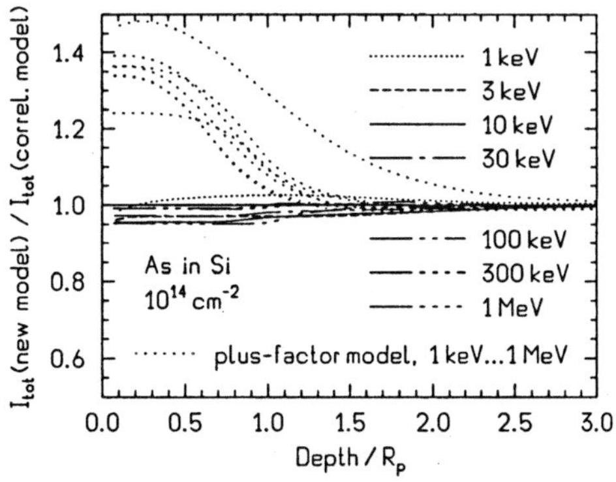

Fig. 2: Ratio of the time integrated interstitial supersaturations calculated with the proposed model and with the more sophisticated model of [1, 5]. For comparison the corresponding ratio is shown for the plus-factor approach by the wide-dotted lines.

2 Non-amorphizing implants

The first feature to be eliminated from the original model [1, 5] are the spatial correlations between I's and V's. We do this by scaling the I and V profiles such that the scaled profiles together with the standard recombination rate $4\pi r_0 D C_\mathrm{I} C_\mathrm{V}$ yield the same amount of TED as the actual profiles together with the more sophisticated recombination model [1, 5]. The scaling factor, which can be written as the ratio of the corresponding numbers of Frenkel pairs per ion, may be estimated as

$$n_\mathrm{FPeff}/n_\mathrm{FP} = \left(1 + 4\pi r_0 D \int_0^\infty w(t)\,\mathrm{d}t\right)^{-1} \tag{1}$$

using the theory developed in [1] and assuming that correlations disappear quickly from the distribution. r_0 denotes the capture radius, D the sum of the I and V diffusivities, and $w(t)$ a function defined in [1] that contains information about the spatial correlations. Alternatively, $n_\mathrm{FPeff}/n_\mathrm{FP}$ can be determined by comparing TED simulations with and without correlations. The two approaches agree reasonably well.

The second task is to construct the scaled vacancy profile $C_\mathrm{Veff}(x)$ from the scaled interstitial profile $C_\mathrm{Ieff}(x)$. This allows to use analytical expressions for $C_\mathrm{Ieff}(x)$ [6], to determine $C_\mathrm{Ieff}(x)$ from Monte Carlo simulations with poorer statistics, or to approximate $C_\mathrm{Ieff}(x)$ by the scaled dopant profile $C_\mathrm{A}(x)$. Following [7], we have tried to construct $C_\mathrm{Veff}(x)$ by shifting $C_\mathrm{Ieff}(x)$ towards the surface by some distance Δx. Fig. 1 shows that the resulting netI and netV profiles (dotted lines) do not agree well with the corresponding profiles determined from the actual I and V distributions (solid lines). Much better results are obtained by using the scaled dopant profile (dashed lines):

$$C_\mathrm{Ieff}(x) = n_\mathrm{FPeff} C_\mathrm{A}(x), \qquad C_\mathrm{Veff}(x) = C_\mathrm{Ieff}(x + \Delta x). \tag{2}$$

With properly chosen parameters n_FPeff and Δx the depth dependence of TED is much better described than by the plus-factor model. In the example shown in Fig. 2 the maximum error is reduced from 50% to 5%.

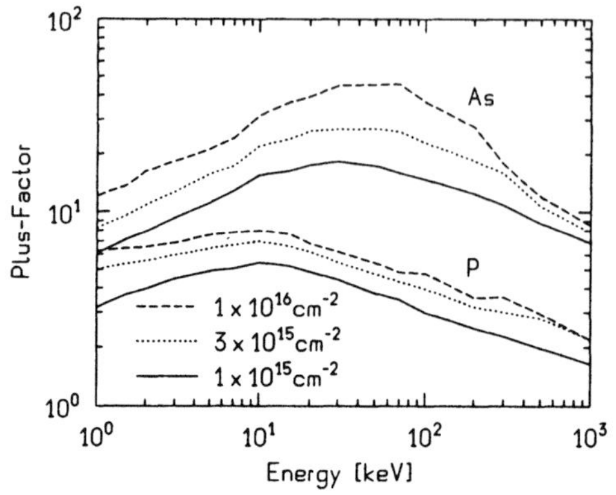

Fig. 3: Plus-factor as a function of energy for P and As ions at three amorphizing implantation doses. Note the dose dependence.

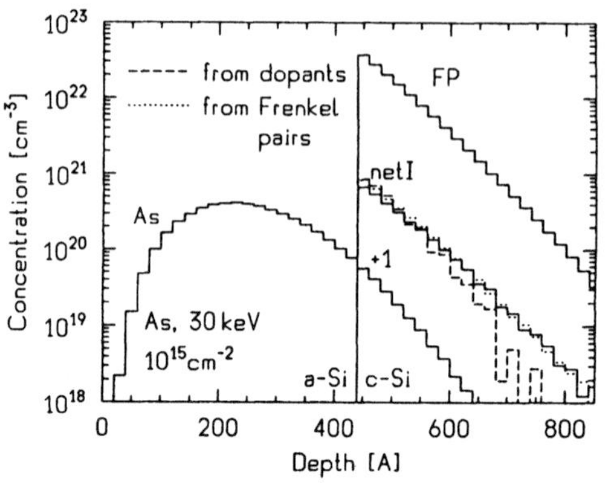

Fig. 4: Construction of net-excess interstitial (netI) profiles after an amorphizing implant by shifting the Frenkel pair (FP) profile (dotted lines) and the scaled dopant profile (dashed lines).

3 Amorphizing implants

For amorphizing implants it is assumed that recrystallization sweeps away all point defects within the amorphized zone. Because of the high concentrations in the remaining profile, recombination is fast and all that matters is the netI distribution [8]. Therefore, I–V correlations are not an issue. Plus-factors can be determined in the same way as in the case of non-amorphizing implants. Fig. 3 shows that large values are obtained, particularly for the heavy ion species As. Moreover, the plus-factor significantly depends on the dose. This is explained by the flatter tails of the point defect distributions compared to the dopant profile. Thus, when the amorphous zone increases, larger scaling factors are needed to obtain approximate netI distributions from the dopant profiles.

For the same reason it is better to use the actual interstitial profile than the scaled dopant profile in a model similar to that proposed in Section 2. An example illustrating this is given in Fig. 4. The netI profile is obtained by

$$C_{\mathrm{netI}}(x) = C_{\mathrm{I}}(x) - C_{\mathrm{V}}(x) \approx C_{\mathrm{I}}(x) - C_{\mathrm{I}}(x+\Delta x) \approx -\frac{\partial C_{\mathrm{I}}}{\partial x}\Delta x. \tag{3}$$

The interstitial profiles can be approximated by the analytical functions published previously [4]. The shift distance Δx is independent of the implant dose in contrast to the plus-factor in the plus-factor model (Fig. 3).

4 Two- and three-dimensional model

Fig. 5 shows the two-dimensional distributions of netI's and netV's after implantation by a mask edge, calculated by shifting the vertical distribution function as described above and using the same lateral distribution function for I's and V's. With this approach netV's appear near the surface, which clearly disagrees with the actual 2-D Monte Carlo results (solid lines in Fig. 6). The deficiency can be repaired by shifting the lateral I distribution $f_{\mathrm{I}}^{\mathrm{lat}}(y)$ away from the center by a distance Δy. The dose added

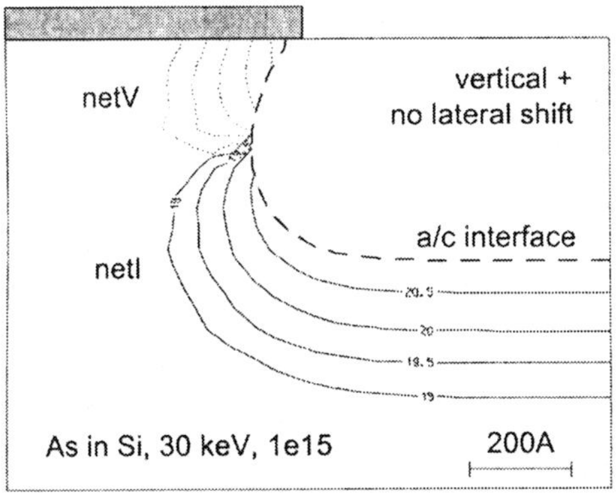

Fig. 5: 2-D net-excess interstitial (netI) and vacancy (netV) distribution calculated from shifted vertical and identical lateral I and V distribution functions. The presence of netV's near the surface is in disagreement with the full Monte Carlo simulations (see Fig. 6).

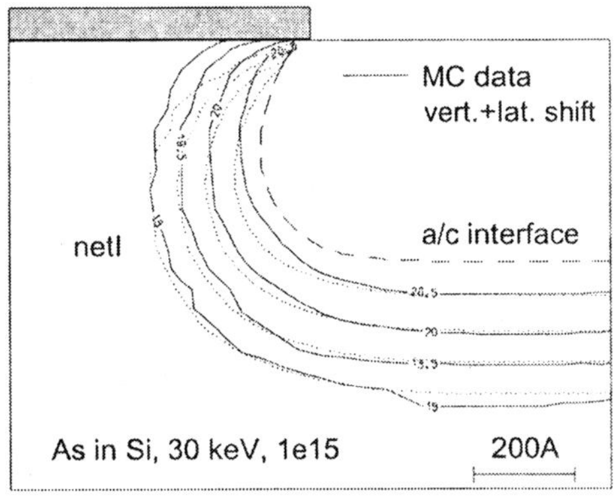

Fig. 6: 2-D net-excess interstitial (netI) and vacancy (netV) distributions as obtained from Monte Carlo simulations (solid lines) and by shifting the vertical and the lateral distribution function (dotted lines). In both models there are no netV's near the surface.

in the center can be compensated for by subtracting a Gaussian function with a standard deviation σ, which has to obey $\Delta y \ll \sigma \ll \sigma_{\text{lat}}$ with σ_{lat} the standard deviation of the lateral V distribution:

$$f_{\text{I}}^{\text{lat}}(y) = f_{\text{V}}^{\text{lat}}(\max(|y| - \Delta y, 0)) - \frac{2\Delta y f_{\text{V}}^{\text{lat}}(0)}{\sqrt{2\pi}\sigma} \exp\left(-\frac{y^2}{2\sigma^2}\right). \tag{4}$$

This model has Δy and σ as parameters where the value of σ usually is uncritical. Similarly, in 3-D such a shift can be performed in radial direction.

Thus, we have proposed a fairly simple and complete model of initial conditions for TED which can be used for 1D/2D/3D process simulation.

References

[1] G. Hobler, L. Pelaz, and C. S. Rafferty, *J. Electrochem. Soc.*, 147(9), 3494–3501, 2000.

[2] M. D. Giles, *J. Electrochem. Soc.*, 138(4), 1160–1165, 1991.

[3] G. Hobler and V. Moroz, In *ESSDERC 2000*, pp. 168–171. Frontier Group, 2000.

[4] L. Pelaz, G. H. Gilmer, H.-J. Gossmann, C. S. Rafferty, M. Jaraiz, and J. Barbolla, *Appl. Phys. Lett.*, 74(24), 3657–3659, 1999.

[5] G. Hobler, L. Pelaz, and C. S. Rafferty, *Nucl. Instr. Meth.*, B 153, 172–176, 1999.

[6] G. Hobler and S. Selberherr, *IEEE Trans. Comp.-Aided Des.*, 7(2), 174–180, 1988.

[7] R. Kalyanaraman, T. E. Haynes, D. C. Jacobson, H.-J. Gossmann, and C. S. Rafferty, In *Mat. Res. Soc. Symp. Proc.*, vol. 610, B9.4. MRS, Warrendale, 2000.

[8] L. Laanab, C. Bergaud, M. M. Faye, J. Faure, A. Martinez, and A. Claverie, In *Mat. Res. Soc. Symp. Proc.*, vol. 279, pp. 381–386. MRS, Pittsburgh, 1993.

Local Iterative Monte Carlo investigation of the influence of electron-electron scattering on short channel Si-MOSFETs

J. Jakumeit
Institute for Algorithm and Scientific Computing (SCAI),
GMD-German National Research Center for Information Technology,
Schloß Birlinghoven, 53754 Sankt Augustin, Germany,
Phone: +49 2241 14-2746, Fax: +49 2241 14-2181, email: jakumeit@gmd.de

U. Ravaioli
Beckman Institute, University of Illinois at UC, Urbana, IL 61801, USA

Abstract

The Local Iterative Monte Carlo (LIMO) technique is used to investigate the effect of electron-electron scattering on hot electron effects in Si-MOSFETS with channel length as short as 25 nm. The results indicate that electron-electron scattering might be an important source for hot electrons in the next generations of Si-MOSFETs. But the effect decreases if the channel length comes close to 25 nm.

1 Introduction

The correct treatment of electron-electron and electron-impurity scattering is still a challenge for modern semiconductor device simulation [1-3]. A two particle model must be used, making the simulation of the short range electron-electron interaction very expensive within the Monte Carlo approach. On the other hand, electron-electron interaction may be the single most important source of hot electrons in modern short channel Si-MOSFETs with implications for the life-time of these devices.

We introduced the local iterative Monte Carlo (LIMO) algorithm in order to reduce the computation time of Monte Carlo (MC) simulations of rare events like hot electron effects by a more efficient use of the computational resources [4-5]. The LIMO approach gives new possibilities for an effective treatment of electron-electron scattering. Comparison with experimental findings for a 90 nm MOSFET show an excellent agreement when electron-electron scattering is included in the LIMO calculation. We use this new approach to investigate MOSFET scaling trends down to 25 nm channel length.

2 Theory

Two properties of the LIMO algorithm simplify the simulation of electron-electron interaction effects compared to standard MC algorithms. First, the energy distribution of electrons is known at each iteration of the LIMO algorithm. LIMO does not simulate the flight of an ensemble of electrons through the device like standard MC algorithms but changes the distribution iteratively by using information from MC simulations of relatively short electron trajectories ($\sim 10^{-14}$ sec). The electron distribution is always accessible and the self-consistent long range electron-plasmon and short range electron-electron scattering rates can be calculated directly using Fermi's Golden Rule. It is not necessary to perform a time average in order to gain an electron distribution.

Second, the partner electron for a short range electron-electron scattering can be created out of the distribution and then returned to the distribution for each scattering process. The computationally expensive search for a partner electron from the ensemble of simulated electron is therefore not necessary in LIMO. Since the distribution memorizes only the energy and not the complete momentum of the electron, the momentum direction of the partner electron can be chosen arbitrarily and the final direction need not be calculated. This simplifies a correct treatment of energy and momentum of the interacting electrons, compared to standard MC algorithms.

3 Results for short channel MOSFET

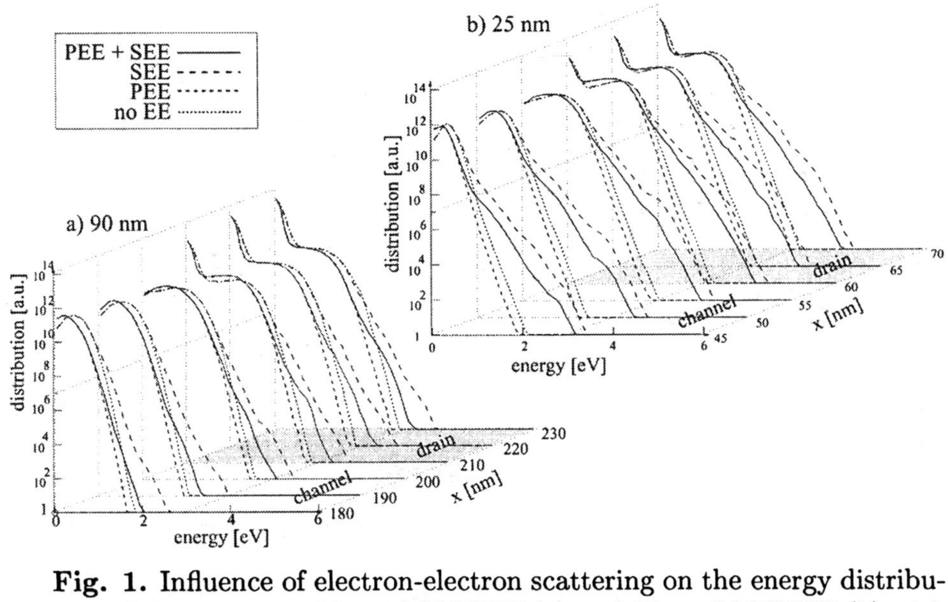

Fig. 1. Influence of electron-electron scattering on the energy distribution of electrons in a 90 nm MOSFET (a) and 25 nm MOSFET (b) with 1.5 V applied to drain and gate. Distributions obtained including no electron-electron scattering (no EE), only electron-plasmon scattering (PEE), only the short range direct interaction (SEE) and the full model (PEE+SEE) are displayed.

The structures from the "well-tempered MOSFET" project (http://www-mtl.mit.edu:80/Well) with channel length 90 nm, 50 nm and 25 nm were used for the investigation of hot electron effects in short channel Si-MOSFETs. Figure 1 shows the influence of electron-electron scattering on the energy distribution of electrons in: (a) 90 nm MOSFET (a), (b) and 25 nm MOSFET, with 1.5 V applied to drain and gate. In both MOSFETs electron-plasmon scattering (PEE) as a relaxation mechanism leads to a reduction of the distribution spread in energy, while the direct short range interaction (SEE) increases the high energy tail. While the 90 nm MOSFET shows only a moderate tail increase due to the short range interaction, the effect simulated in the 25 nm MOSFET is much stronger. Substrate and gate current were calculated from the simulated electron distributions for the analysis of hot electron effects (see [6] for a description of the procedure). Figure 2 compares the resulting currents for the 90 nm MOSFET with measurements. With electron-electron interaction, an excellent agreement could be achieved.

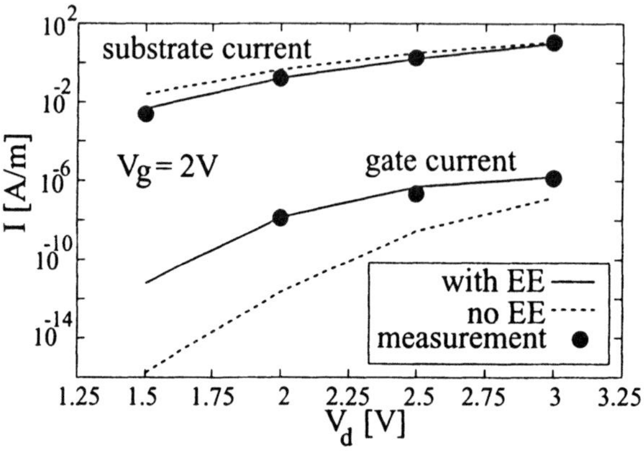

Fig. 2 Substrate current and gate current as function of the drain voltage for the 90 nm Si-MOSFET with a gate voltage of 2 V. Measurement results are compared with the LIMO results calculated with and without electron-electron scattering (EE).

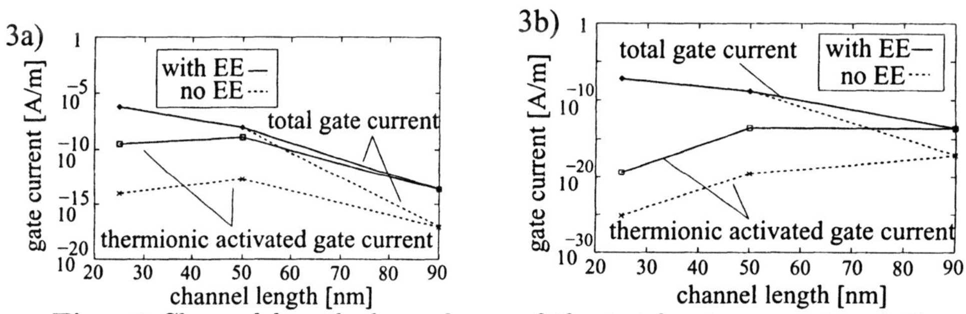

Fig. 3 Channel length dependence of the total gate current and the thermionic emission component of the gate current calculated with and without electron-electron scattering (EE). (3a) shows the results using constant bias scaling with 1.5 V at drain and gate and (3b) the same with quasi constant bias scaling with 1.5 V, 1.1 V and 0.8 V for the 90 nm, 50 nm and 25 nm MOSFET, respectively.

The channel length dependence of the gate current, as a measure for degradation effects, is investigated in Fig. 3 for constant bias scaling (3a) with 1.5 V at drain and gate and quasi constant bias scaling (3b) with 1.5 V, 1.1 V and 0.8 V for the 90 nm, 50 nm and 25 nm MOSFET, respectively. The 90 nm device shows a significant increase of the gate current due to electron-electron scattering. The smaller devices show no effect for both types of scaling, because tunneling of cold electrons at the source is the main contribution to the gate current. Since degradation effects in MOSFETs may be related to hot electrons and not tunneling cold electrons, we plotted the thermionic emission component of the gate current separately. The amount of hot electrons which are heated over the 3.2 eV barrier of the SiO_2 increases significantly, when the channel length is reduced from 90 nm to 50 nm with constant voltage but decreases slightly with a further channel length reduction to 25 nm. In case of the quasi constant bias scaling the amount of hot electrons remains almost the same for a channel length of 50 nm and a voltage of 1.1 V. The amount is significantly reduced by a further decrease of the channel length to 25 nm with 0.8 V at drain and gate. Thus, device degradation due to hot electron effects may remain an important issue for the next generations of MOSFETs but will become less important for ultra small devices with channel length in the range of 25 nm.

4 Conclusion

The LIMO technique was used to investigate the effect of electron-electron scattering on hot electron effects in down-scaled Si-MOSFETs. This approach has some advantages compared to the way electron-electron scattering is treated in standard Monte Carlo algorithms. The results for Si-MOSFETs with channel length down to 25 nm indicate that electron-electron scattering is an important source for hot electrons in small MOSFETs. It may play an important role for the next generations of Si-MOSFETS but we expect less pronounced effects for ultra small MOSFETs with channel length in the 25 nm range.

5 Acknowledgements

This work was partially supported by the Semiconductor Research Corporation, contract 99-NJ-726, by the Distributed Center for Advanced Electronics Simulation (DesCArtES) through National Science Foundation grant ECS 98-02730, and the Deutsche Forschungsgemeinschaft GK *Scientific Computing.*

References

[1] Fischetti, M. V., Laux, S. E. and E. Crabbé, E. (1995):Understanding hot-electron transport in silicon devices: Is there a shortcut?, J. Appl. Phys. 78: 1058-1087

[2] Gross, W. J., Vasileska, D., and Ferry, D. K. (1999): A Novel Approach for Introducing the Electron-Electron and Electron-Impurity Interactions in Particle-Based Simulations, IEEE Electron Dev. Lett. 20: 463-465

[3] Fischetti, M. V. and Laux, S. E. (2000): Performance degradation of small silicon devices caused by long-range Coulomb interactions, Appl. Phys. Lett. 76: 2277-2279

[4] Jakumeit, J. and Ravaioli, U. (2001), Semiconductor transport simulation with the Local Iterative Monte Carlo Technique, accepted for the April 2001 issue of IEEE Trans. Electron Devices

[5] Jakumeit, J. (2000) Computational Aspects of the Local Iterative Monte Carlo Technique, Int. J. Mod. Phys. C 11: 665-673

[6] Duncan, A., Ravaioli, U. and Jakumeit, J. (1998): Full-Band Monte Carlo Investigation of Hot Carrier Trends in the Scaling of Metal-Oxide-Semiconductor Field-Effect Transistors. IEEE Trans. Electron Devices 45: 867-876

Simplified Inelastic Acoustic–Phonon Hole Scattering Model for Silicon

F. M. Bufler, A. Schenk, and W. Fichtner

Institut für Integrierte Systeme, ETH Zürich, CH-8092 Zürich, Switzerland

Abstract

A simplified model for inelastic acoustic phonon scattering of holes in silicon is developed. It consists in approximating both the acoustic phonon energy and the square of the phonon wave vector by lattice–temperature dependent constants. The resulting scattering rate depends only on energy and thus facilitates the search of after–scattering–states during full–band Monte Carlo simulation. The simulation results for the velocity–field characteristics accurately agree with the experimental data at different lattice temperatures, while the population of hot–hole states is significantly enhanced compared to the elastic equipartition approximation. The value of the energy relaxation time to be used in hydrodynamic device simulations is roughly 0.1 ps.

1. Introduction

Modeling for TCAD applications is always a trade–off between computational speed and the accuracy of the models employed. In the context of full–band Monte Carlo (FBMC) device simulation, the elastic equipartition approximation [8] for acoustic phonon scattering has proved very useful. On the one hand, the scattering rate then depends on energy only and not on the wave vector, thus facilitating the search of the state after scattering. In addition, this scattering model allows for an exact computation of the Ohmic drift mobility via the microscopic relaxation time which requires only a one–dimensional integration over energy [2]. On the other hand, this elastic model is also able to accurately reproduce experimental drift mobility or velocity data both in the linear [2, 7] and in the nonlinear regime [1, 5] over a wide range of lattice temperatures. Nevertheless, the elastic approximation significantly underestimates the number of hot holes as has recently been shown [6] by a comparison with a model which includes inelastic acoustic phonons and yields — after recalibration of the acoustic and the optical coupling constants — similar good agreement with the experimental velocity data as the elastic model. It is therefore the aim of this paper to propose a new model for acoustic phonon scattering which combines the energy dissipation of the wave–vector dependent inelastic acoustic model with the simplicity of the elastic model while preserving the good agreement with the experimental drift velocity data.

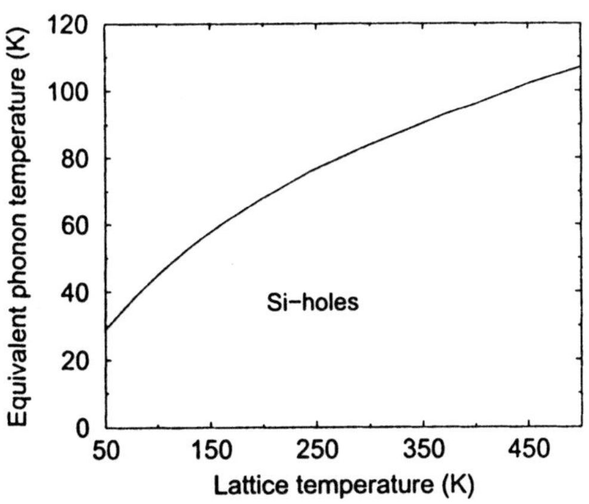

Fig 1: Mean acoustic phonon energy expressed in terms of the equivalent phonon temperature defined via $k_B \theta \equiv \langle \hbar\omega(q) \rangle$.

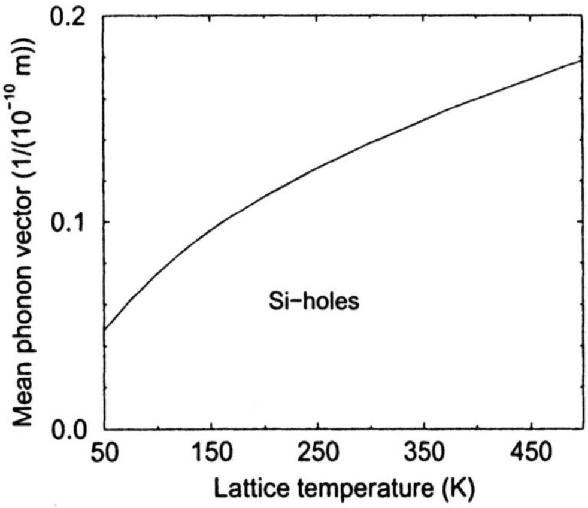

Fig 2: Modulus of the phonon wave vector transferred during acoustic phonon-hole scattering determined via $\sqrt{\langle q^2 \rangle}$.

2. Model

The starting point is the scattering rate which can be written in the form (e.g. [6])

$$S(\mathbf{k}) = \sum_{\mathbf{k}'} \frac{\pi}{\rho\,\omega(q)} \Delta^2(q) \left(N_q + \frac{1}{2} \mp \frac{1}{2} \right) \frac{1}{V} \delta\left(\epsilon(\mathbf{k}') - \epsilon(\mathbf{k}) \mp \hbar\omega(q)\right) \qquad (1)$$

where the overlap integral has been set to unity and the phonon wave vector $\mathbf{q}$ results from wave vector conservation. For optical phonons, the coupling constant $\Delta(q) = D_t K$ and the phonon energy $\hbar\,\omega(q) = \hbar\,\omega_{\text{opt}} \equiv k_B\,\theta_{\text{opt}}$ can be approximated by constants with an equivalent phonon temperature of $\theta_{\text{opt}} = 731$ K. The inelastic acoustic phonons are characterized in the isotropic approximation via $\Delta(q) = \mathcal{E}\,q$ and an analytical formula for the phonon dispersion $\omega(q)$ as given in Ref. [6]. Instead of keeping the wave–vector dependent acoustic phonon model or applying the elastic equipartition approximation, our new approach consists of replacing in the above scattering rate both $\omega(q)$ and q^2 by constants which are determined by an adequate averaging procedure. For this, first the modulus of the phonon wave vector, $q = ||\mathbf{k}' - \mathbf{k}||$, is averaged over a sphere in the spirit of an isotropic, parabolic band structure. This yields $q = \frac{4}{3}k$. Then the mean phonon energy, which is absorbed or emitted by a hole with wave vector $\mathbf{k}$, is obtained via an average with the hole distribution function. Approximating further the distribution function by the Maxwell–Boltzmann distribution leads with $\beta = 1/(k_B T)$ to

$$k_B\,\theta_{\text{ac}} \equiv \hbar\,\omega_{\text{ac}} = \frac{\int d\epsilon\,\hbar\,\omega(\frac{4}{3}k(\epsilon))\,\mathcal{D}(\epsilon)\,e^{-\beta\epsilon}}{\int d\epsilon\,\mathcal{D}(\epsilon)\,e^{-\beta\epsilon}} \qquad (2)$$

with the density of states $\mathcal{D}(\epsilon)$ and $k(\epsilon) = \sqrt{2\,m(\epsilon)\,\epsilon}/\hbar$. The effective mass $m(\epsilon)$ is taken to be the energy–dependent effective density–of–states mass and thereby the main full–band–structure dependence is retained. As a consequence of this procedure, the mean acoustic phonon energy depends on the lattice temperature T which is shown in Fig. 1. The same averaging procedure is applied to q^2 and the result is displayed in Fig. 2. The similar lattice–temperature dependence of $\sqrt{\langle q^2 \rangle}$ indicates that the nonlinear part of the phonon dispersion $\omega(q)$ is negligible.

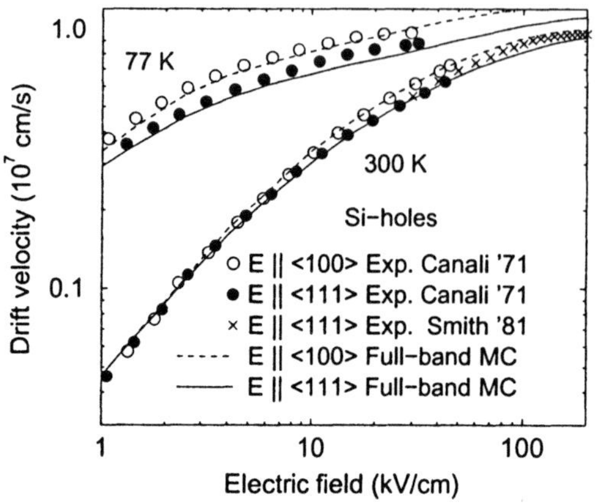

Fig 3: Results of the simplified inelastic model for the velocity–field characteristics of holes in Si at 77 and 300 K in comparison to experimental results [4, 10]

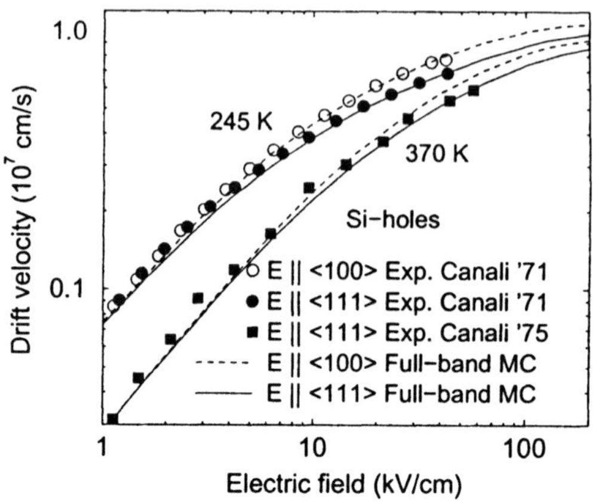

Fig 4: Results of the simplified inelastic model for the velocity–field characteristics of holes in Si at 245 and 370 K in comparison to experimental results [4, 3]

3. Results

For this inelastic acoustic phonon model, the two coupling constants have been recalibrated with the result $\mathcal{E} = 5.03$ eV and $D_tK = 8.7$ eV/Å (The equidistant mesh in **k**–space for the full-band structure is $1/96\ 2\pi/a$ where a is the lattice constant). Note that the *optical* coupling constant is, in analogy to Refs. [7] and [6], smaller than in the corresponding model with elastic acoustic phonons [1, 2]. Figures 3 and 4 demonstrate that the inelastic model with the new values for the coupling constants leads to a similar good agreement with the experimental velocity–field characteristics [4, 3, 10] as the elastic model [1]. However, due to the smaller optical coupling constant, the population of high–energetic states by holes is significantly increased in the new model as is illustrated by the energy distributions in Fig. 5. This has, of course, corresponding consequences for the impact ionization coefficient and the substrate current in p–MOSFETs. It should be noticed that the number of high–energetic holes can still be increased by simply further reducing optical phonon scattering and readjusting the acoustic scattering rate. This would result, however, in a significant underestimation of both the experimental saturation velocity at high lattice temperatures and the experimental drift velocities in the low–field regime at low lattice temperatures [9]. Figure 6 finally reports the energy relaxation time for hydrodynamic device simulations where an increase from roughly 0.08 ps to 0.10 ps can be observed.

4. Conclusions

A new inelastic acoustic–phonon scattering model has been developed which (i) has only an energy–dependent scattering rate thus enabling an efficient search of after–scattering states, (ii) recovers the hot–hole tail of the energy distribution of more complicated inelastic models and (iii) accurately reproduces the experimental velocity–field characteristics over a wide range of lattice temperatures.

Acknowledgments This work was in part supported by the Kommission für Technologie und Innovation under Contract 4082.2.

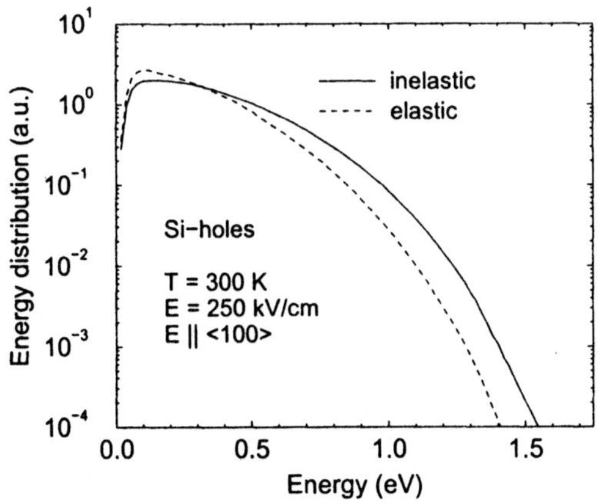

Fig 5: Energy distribution as a function of the hole energy $\epsilon(\mathbf{k})$ at 300 K for a field of 250 kV/cm in $\langle 100 \rangle$ direction.

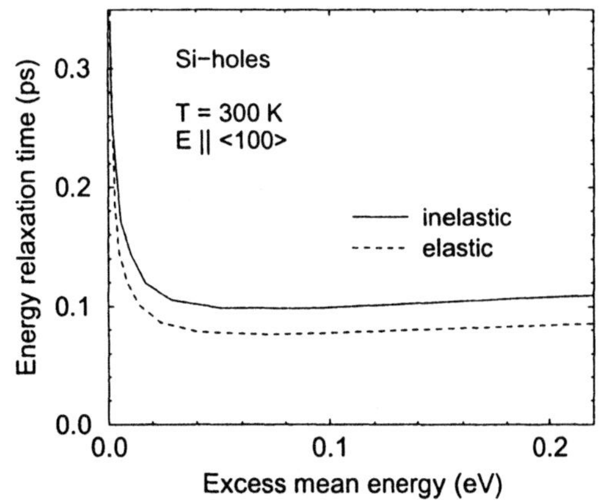

Fig 6: Energy relaxation time as a function of the excess mean energy $\langle \epsilon(\mathbf{k}) \rangle - \langle \epsilon(\mathbf{k}) \rangle_{\text{eq}}$ at 300 K in $\langle 100 \rangle$ direction.

References

[1] F. M. Bufler, P. Graf, and B. Meinerzhagen. High-field hole transport in strained Si and SiGe by Monte Carlo simulation: Full band versus analytic band models. *VLSI Design*, 8:41–45, 1998.

[2] F. M. Bufler and B. Meinerzhagen. Hole transport in strained $\text{Si}_{1-x}\text{Ge}_x$ alloys on $\text{Si}_{1-y}\text{Ge}_y$ substrates. *J. Appl. Phys.*, 84:5597–5602, 1998.

[3] C. Canali, G. Majni, R. Minder, and G. Ottaviani. Electron and hole drift velocity measurements in silicon and their empirical relation to electric field and temperature. *IEEE Trans. Electron Devices*, 22:1045–1047, 1975.

[4] C. Canali, G. Ottaviani, and A. Alberigi-Quaranta. Drift velocity of electrons and holes and associated anisotropic effects in silicon. *J. Phys. Chem. Solids*, 32:1707–1720, 1971.

[5] B. Fischer and K. R. Hofmann. Discretization of the Brillouin zone by an octree/Delaunay method with application to full–band Monte Carlo transport simulation. In *Proc. SISPAD*, pages 181–184, Leuven (Belgium), September 1998.

[6] B. Fischer and K. R. Hofmann. A full–band Monte Carlo model for the temperature dependence of electron and hole transport in silicon. *Appl. Phys. Lett.*, 76:583–585, 2000.

[7] M. V. Fischetti and S. E. Laux. Band structure, deformation potentials, and carrier mobility in strained Si, Ge and SiGe alloys. *J. Appl. Phys.*, 80:2234–2252, 1996.

[8] C. Jacoboni and L. Reggiani. The Monte Carlo method for the solution of charge transport in semiconductors with application to covalent materials. *Rev. Mod. Phys.*, 55:645–705, 1983.

[9] Y. Kamakura, I. Kawashima, K. Deguchi, and K. Taniguchi. Verification of hot hole scattering rates in silicon by quantum–yield experiment. *J. Appl. Phys.*, 88:5802–5809, 2000.

[10] P. M. Smith and J. Frey. High–field transport of holes in silicon. *Appl. Phys. Lett.*, 39:332–333, 1981.

An Impact Ionization Model Including Non-Maxwellian and Non-Parabolicity Effects

T. Grasser, H. Kosina, and S. Selberherr
Institute for Microelectronics, TU Vienna, Gusshausstrasse 27–29, A-1040 Vienna, Austria
Phone: +43-1-58801/36023, Fax: +43-1-58801/36099, Email: Grasser@iue.tuwien.ac.at

Abstract

Accurate modeling of impact-ionization is a critical issue for submicron devices. It is well known that models based on the electric field or on the average carrier energy give a rather poor description of the problem. We show that by accounting for the average square energy an accurate analytical description of the distribution function can be given which can then be used to evaluate microscopic models in a macroscopic device simulator. The new model is accurate for both bulk and submicron devices and involves only local quantities.

1 Introduction

For fast and accurate evaluation of CMOS reliability issues in sub-100 nm technology very precise high field transport models are needed. In particular, hot-carrier effects like impact ionization are poorly described by models using the local average carrier energy as parameter. To obtain a more accurate description, equation sets based on six moments of Boltzmann's equation have been considered where in addition to the carrier concentration and temperature the next higher even moment, the average square energy of the distribution function is taken into account. First promising results were obtained by Sonoda *et al.* [1] who proposed an analytical expression for the impact ionization rate as a function of the average energy and the average square energy by fitting Monte Carlo (MC) results. Being a fit-formula, the model lacks theoretical background and we obtained poor accuracy for other channel lengths and biases than those used in [1]. Therefore we developed an analytical model for the symmetric part of the distribution function which goes beyond the Maxwellian approximation. This model is then used to derive a macroscopic impact ionization model based on microscopic descriptions used in MC simulators. Impact ionization models based on six moments of the distribution function (DF) nicely fit into the concept of conventional device simulators and do not require e.g., integration along current paths, as only local quantities are involved.

2 The new model

It has been frequently shown that the shape of the DF depends on whether the gradient of the field is positive or negative with respect to the current density. Several cases have been considered and evaluated and we found that

$$f(\mathcal{E}) = A \exp\Big[-\Big(\frac{\mathcal{E}}{a}\Big)^b\Big] \tag{1}$$

which is a generalization of [2] and [3], neatly captures the main features of the distribution function throughout the whole device. The parameters $a = a(T_n, \beta_n)$ and $b = b(T_n, \beta_n)$ are functions of the local temperature and kurtosis. Instead of using the average square energy directly, we define the kurtosis of the distribution function as $\beta_n = (3/5)\langle\mathcal{E}^2\rangle/\langle\mathcal{E}\rangle^2$. We now take the moments of order l of the analytical energy distribution function (1)

$$m_l = \int \mathcal{E}^l g(\mathcal{E}) f(\mathcal{E})\ \mathrm{d}\mathcal{E} \tag{2}$$

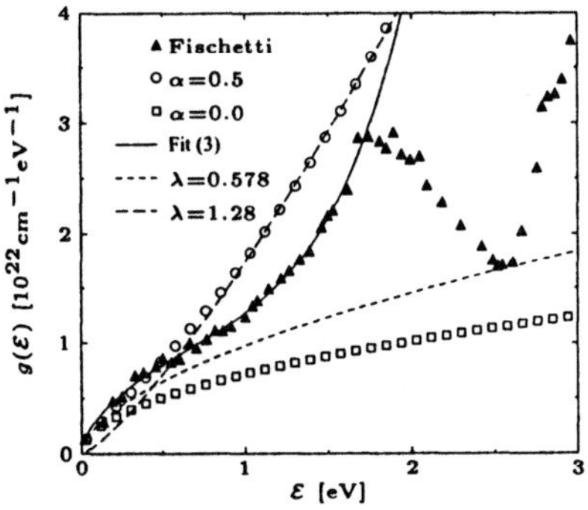

Fig. 1: Comparison of different expressions for the density of states

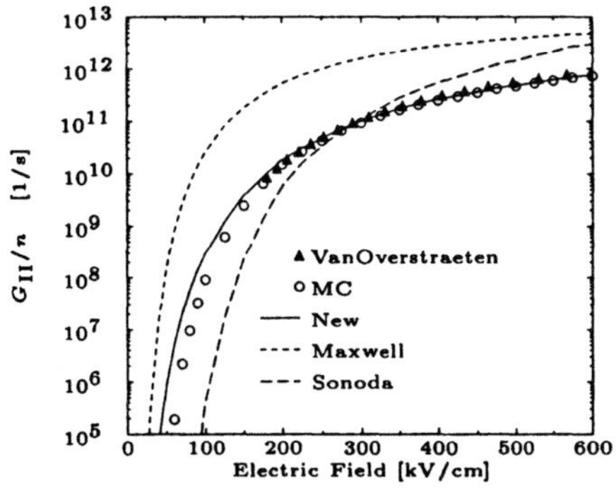

Fig. 2: Comparison of the new expressions with other models for the bulk case

To evaluate (2) an expression $g(\mathcal{E})$ for the density of states is needed. Results based Kane's first-order non-parabolicity correction [4] cannot be handled analytically. As can be shown, a power-law approximation $g(\mathcal{E}) = g_0\,\mathcal{E}_C(\mathcal{E}/\mathcal{E}_C)^\lambda$ similar to the one given by Cassi and Riccò [2] cannot capture the non-parabolic nature of the bands, because only the low- or the high-energy part can be fit accurately, but not both simultaneously as shown in Fig. 1. Instead, we use the expression

$$g(\mathcal{E}) = g_0\sqrt{\mathcal{E}}\Big(1 + (\eta\mathcal{E})^\zeta\Big) \tag{3}$$

with $\eta = 1.4\,\mathrm{eV}^{-1}$ and $\zeta = 1.08$ which were obtained by a fit to Kane's expression. In addition (3) can be nicely fit to the pseudo-potential data obtained by Fischetti and Laux [5] using the parameter values $\eta = 0.9\,\mathrm{eV}^{-1}$ and $\zeta = 1.4$ (cf. Fig. 1).

With this approximation for $g(\mathcal{E})$ expressions for $T_n(a,b)$ and $\beta_n(a,b)$ can be found

$$T_n = \frac{2}{3}\frac{a}{\mathrm{k_B}}\frac{m_1}{m_0} \qquad\qquad \beta_n = \frac{3}{5}\frac{m_0\,m_2}{m_1^2} \tag{4}$$

$$m_l = Ag_0\frac{a^{l+3/2}}{b}\Big[\Gamma\Big(\frac{2l+3}{2b}\Big) + (\eta a)^\zeta\Gamma\Big(\frac{2l+2\zeta+3}{2b}\Big)\Big] = Ag_0\frac{a^{l+3/2}}{b}F_l(a,b) \tag{5}$$

As there is no analytical inversion of (4) with respect to a and b, the solution has been obtained by a two-dimensional Newton procedure. A comparison of DFs obtained by Monte Carlo (MC) simulation and their analytical counterparts is given in Fig. 3 and Fig. 4 for the bulk case and the short channel device, respectively. Also shown are the results based on a parabolic band approximation ($\eta = 0$). The analytical expression slightly overestimates the high-energy tail when the curvature of the DF is too strong. However, with a more realistic MC model employing a fullband structure and including electron-electron scattering, the temperature of the high-energy tail is increased and the curvature of the DF diminished which in turn improves the quality of the analytical expression.

With this analytical description for the distribution function, microscopic scattering rates can be evaluated to yield models suitable for macroscopic transport models. We considered the anisotropic model of Quade *et al.* [6], the isotropic expressions given by Keldysh [7], and the extension used by Fischetti *et al.* [8]. In the following we

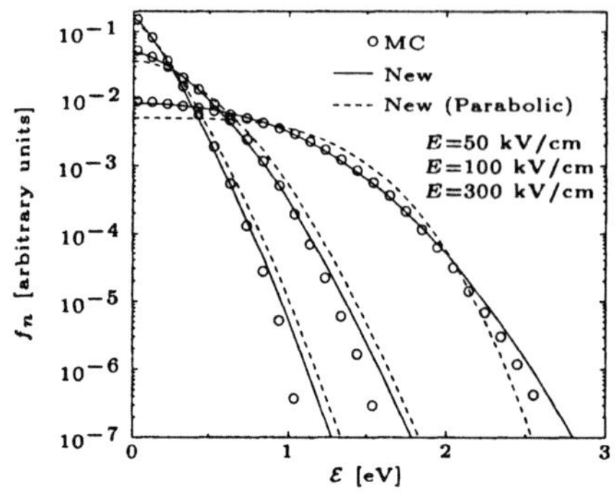

Fig. 3: Comparison of analytical expressions for the DF with MC results for the bulk case

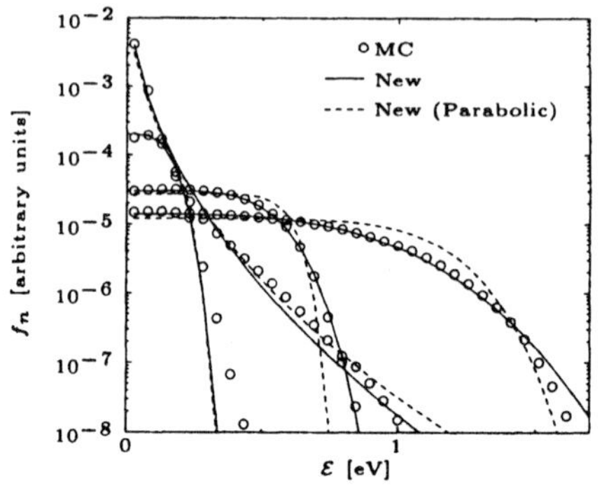

Fig. 4: Comparison of analytical expressions with MC results for the short channel device

restrict ourselves to the Keldysh formula $P_{\mathrm{II}}(\mathcal{E}) = P_0(\mathcal{E} - \mathcal{E}_{\mathrm{th}})^2/{\mathcal{E}_{\mathrm{th}}}^2$, because it gives rather compact results and can be easily extended to Fischetti's expression. For the six moments model [9] the additional terms to the right hand side of the balance equations are required which can be obtained via

$$G_{\mathrm{II},l} = \int\limits_{\mathcal{E}_{\mathrm{th}}}^{\infty} \mathcal{E}^l\, P_{\mathrm{II}}(\mathcal{E})\, f(\mathcal{E})\, g(\mathcal{E})\, \mathrm{d}\mathcal{E} \tag{6}$$

Here $l = 0, 1, 2$ denote the entries for the continuity, energy balance, and kurtosis balance equations, respectively. As we are only interested in the high energy part of the density of states, Cassi's expression can be savely used to give the final expression

$$G_{\mathrm{II},l} = n\, P_0\, {\mathcal{E}_{\mathrm{C}}}^{1-\lambda}\, \frac{a^{l+\lambda-1/2}}{F_l(a,b)} \left(\Gamma_{1,l} - 2{z_{\mathrm{th}}}^{-1/b}\Gamma_{2,l} + {z_{\mathrm{th}}}^{-2/b}\Gamma_{3,l}\right) \tag{7}$$

with $\lambda = 1.32$, $\mathcal{E}_{\mathrm{C}} = 0.07$ eV, $\Gamma_{j,l} = \Gamma[(j + l + \lambda)/b, z_{\mathrm{th}}]$ where $\Gamma[a, x]$ is the incomplete Gamma function and $z_{\mathrm{th}} = (\mathcal{E}_{\mathrm{th}}/a)^b$. We used $P_0 = 4.18 \cdot 10^{12}\ \mathrm{s}^{-1}$ and $\mathcal{E}_{\mathrm{th}} = 1.12$ eV consistently with the MC simulation, values which fit available experimental data as shown in Fig. 2.

A comparison of $G_{\mathrm{II}} = G_{\mathrm{II},0}$ from (7) with the MC results and the model proposed by Sonoda *et al.* is given in Fig. 5 and Fig. 6 for n^+-n-n^+ test-structures with channel lengths of 1.5 μm and 0.2 μm, respectively. The short channel device was taken from [1] and the channel length was increased to give the long channel version. The maximum electric fields where 120 kV/cm and 320 kV/cm for the long and short channel devices, respectively. Note that the models were evaluated using T_n and β_n from the MC simulation. To fit the short channel device the parameters of Sonoda's model had to be adjusted whereas the new model uses the same parameters as in the MC simulation. For the long channel device, the new model slightly overestimates the ionization rates because the high energy tail in the DF is overestimated by the analytical model. For the short channel device, the new model predicts the tail in the drain region. Unfortunately, the high energy tail is overestimated because the analytical DF cannot capture the turning point.

(7) can also be used to demonstrate the inaccuracies introduced by assuming a heated Maxwellian distribution when deriving macroscopic impact-ionization rates as done

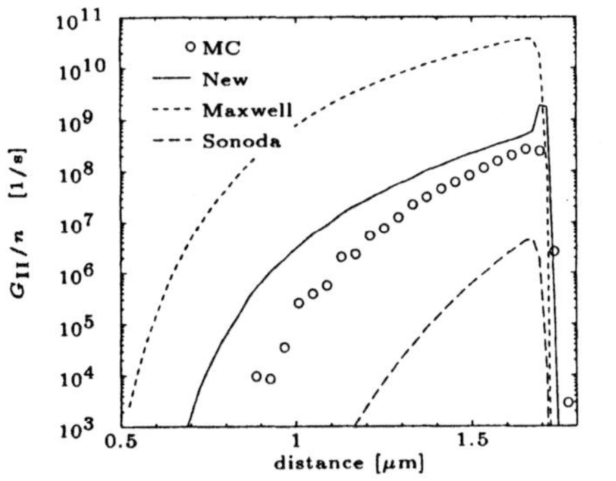

Fig. 5: Comparison of the new expressions with other models for a long channel device

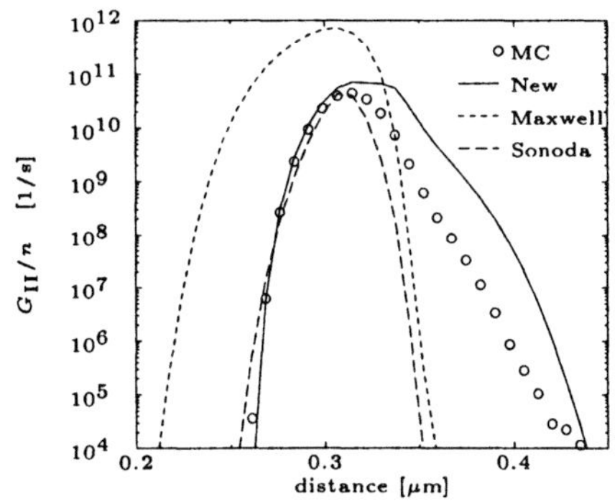

Fig. 6: Comparison of the new expressions with other models for a short channel device

e.g. in [6]. For a heated Maxwellian distribution function which is characterized by $a = \mathrm{k_B}T_n$ and $b = 1$, (7) delivers results similar to local energy models [10] which considerably overestimate the ionization rates as also shown in Fig. 2, Fig. 5, and Fig. 6. This overestimation leads to non-physical calibrations of the model parameters which are only valid for a very narrow range of devices. Furthermore, impact ionization caused by hot electrons in the drain cannot be captured because the cold carriers dominate the average energy which is close to the lattice temperature in the drain region.

3 Conclusion

We have demonstrated that the average square energy provides the necessary information to develop an analyical expression for the distribution function which goes beyong the heated Maxwellian assumption. Replacing the heated Maxwellian DF with the new expression allows for accurate evaluation of microscopic impact ionization rates for bulk and submicron devices. The new model nicely fits into the concept of macroscopic device simulators as it only employs local quantities.

[1] K. Sonoda, M. Yamaji, K. Taniguchi, C. Hamaguchi, and S. Dunham, "Moment Expansion Approach to Calculate Impact Ionization Rate in Submicron Silicon Devices," *J.Appl.Phys.*, vol. 80, no. 9, pp. 5444–5448, 1996.

[2] D. Cassi and B. Riccò, "An Analytical Model of the Energy Distribution of Hot Electrons," *IEEE Trans.Electron Devices*, vol. 37, pp. 1514–1521, June 1990.

[3] K. Hasnat, C.-F. Yeap, S. Jallepalli, S. Hareland, W.-K. Shih, V. Agostinelli, A. Tasch, and C. Maziar, "Thermionic Emission Model of Electron Gate Current in Submicron NMOSFETs," *IEEE Trans.Electron Devices*, vol. ED-44, no. 1, pp. 129–138, 1997.

[4] E. Kane, "Band Structure of Indium Antimonide," *J.Phys.Chem.Solids*, vol. 1, pp. 249–261, 1957.

[5] M. Fischetti and S. Laux, "Monte Carlo Analysis of Electron Transport in Small Semiconductor Devices Including Band-Structure and Space-Charge Effects," *Physical Review B*, vol. 38, no. 14, pp. 9721–9745, 1988.

[6] W. Quade, E. Schöll, and M. Rudan, "Impact Ionization within the Hydrodynamic Approach to Semiconductor Transport," *Solid-State Electron.*, vol. 36, no. 10, pp. 1493–1505, 1993.

[7] L. Keldysh, "Concerning the Theory of Impact Ionization in Semiconductors," *Sov.Phys.JETP*, vol. 21, pp. 1135–1144, 1965.

[8] M. Fischetti, S. Laux, and E. Crabbé, "Understanding Hot-Electron Transport in Silicon Devices: Is There a Shortcut?," *J.Appl.Phys.*, vol. 78, no. 2, pp. 1058–1085, 1995.

[9] T. Grasser, H. Kosina, M. Gritsch, and S. Selberherr, "Using Six Moments of Boltzmann's Transport Equation for Device Simulation," 2001. to appear in *J.Appl.Phys.*

[10] K. Souissi, F. Odeh, H. Tang, A. Gnudi, and P. Lu, "Investigation of the Impact Ionization in the Hydrodynamic Model," *IEEE Trans.Electron Devices*, vol. 40, pp. 1501–1507, Aug. 1993.

Density of States and Group Velocity Calculations for SiO_2

E. Gnani, S. Reggiani, and *M. Rudan*

Dipartimento di Elettronica, Università di Bologna,
viale Risorgimento 2, 40136 Bologna, Italy
mrudan@deis.unibo.it

Abstract

Ab initio calculations of the electron group velocity for SiO_2 are worked out. The conduction bands are calculated by means of two different techniques: Hartree-Fock (HF) and Density-Functional Theory (DFT). Eight energy bands have been used to calculate the density of states and group velocity for the energies of interest. Two different crystal structures of SiO_2, built-up by the same fundamental unit, namely, the SiO_4 tetrahedron, are investigated: they are the α-quartz, and the β-cristobalite.

1 Introduction

The interest in silicon dioxide as material for technological applications in electronic devices is growing steadily. In modern semiconductor technology, SiO_2 films grown on silicon have been employed in the construction of integrated circuits in various devices. Thus, especially in recent years, experimental and theoretical investigations focused on microscopic properties of this material related with reliability problems. Electron transport in silicon dioxide has already been tackled in the frame of the spherical-harmonics expansion (SHE) method applied to the solution of the Boltzmann transport equation (BTE), using the parabolic-band approximation [6]. In particular, the first conduction band was assumed to be a spherical and parabolic function terminating at 6 eV, this limiting the validity of the analysis when high electric fields are considered. A more accurate description of the bands at higher energies has already been achieved by tackling the calculation of the full-band structure of SiO_2. In [5] the solution of the Schrödinger equation has been carried out by means of two different *ab initio* techniques (Hartree-Fock (HF) and Density-Functional Theory (DFT)). In this paper, the same techniques are exploited to determine the coefficients of the SHE method, namely the density of states (DOS) and group velocity (GV) as a function of energy, along the same line already worked out for silicon [7]. As in the SHE scheme the band structure of the material appears through these electronic properties, the calculations reported in this work constitute the basis for an accurate description of the transport properties for SiO_2 at higher energies. In the authors' knowledge, such calculations have not been carried out before.

2 Theory and Implementation

In VLSI-CMOS technology, the gate oxide is grown on a silicon crystal in its amorphous phase. Early investigations [1] showed that the amorphous SiO_2 is constituted by SiO_4 tetrahedra with bond-length and bond-angle distortions; as a consequence, the calculation of the transport properties of the crystalline phases of SiO_2 can be exploited to understand the main aspects of the amorphous phase as well. The analysed systems are polymorphs of silica, characterized by an electron density similar to that of amorphous SiO_2. In this work, the α-quartz and β-cristobalite are addressed.
Although the calculation at hand could in principle be restricted to the irreducible Brillouin zone, for programming simplicity the band structure has been tabulated over the first octant of the Brillouin zone (BZ). The band structure has

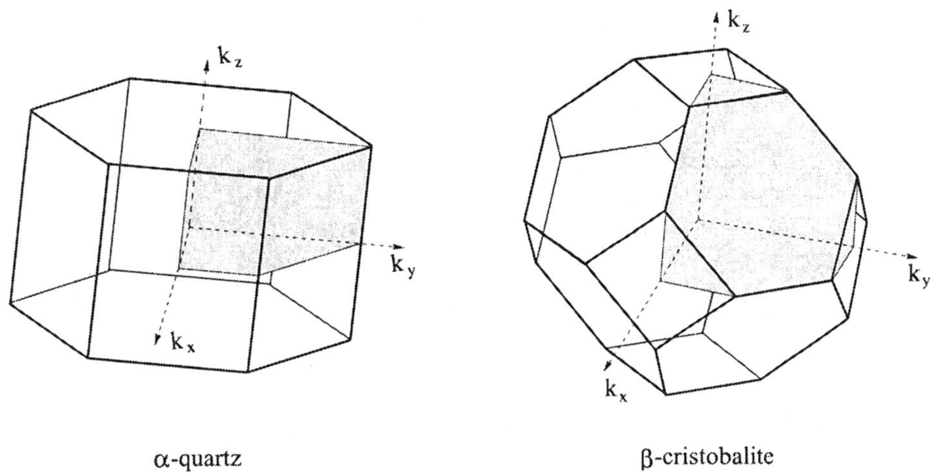

Fig. 1: First Brillouin zone of α-quartz and β-cristobalite.

been calculated in the reciprocal-space domain, by adopting a spatial discretization. More specifically, a $\Delta = 0.05$ mesh step has been adopted to discretize the k_x, k_y and k_z axes with respect to normalized boundaries, this defining an elementary cube associated to each discretization point. According to the crystalline structure, the total number of mesh nodes depends on the shape of the first BZ, shown in gray in Fig. 1. In particular, a mesh tabulation of 5775 and 6951 points has been obtained for the β-cristobalite and α-quartz, respectively. For each point, the eigenvalues of the eight lowest conduction bands have been computed using the CRYSTAL98© code [2]. In order to determine the density of states (DOS) and the group velocity (GV) at a given energy E, and to finally obtain the tabulation of DOS and GV over a wide range of energies, a linearization algorithm has been implemented, based on the Gilat-Raubenheimer method for $\boldsymbol{k}$-space integration [4]. Both DOS and GV have been tabulated on a range of energies up to 6 eV.

3 Results and Conclusions

The numerical accuracy in the description of the DOS and GV functions demanded by the SHE method is very high. Thus, special care has been devoted to the definition of the 3-D mesh in the $\boldsymbol{k}$ space and to the algorithm for calculating the two functions. As consistent data for DOS and GV in SiO_2 are not available elsewhere, the procedure has been validated first by calculating the same functions for Si and comparing the results with available data [3]. Good agreement has been obtained, especially for DFT which is known to provide a more realistic description of the band structure. Then, the procedure has been applied

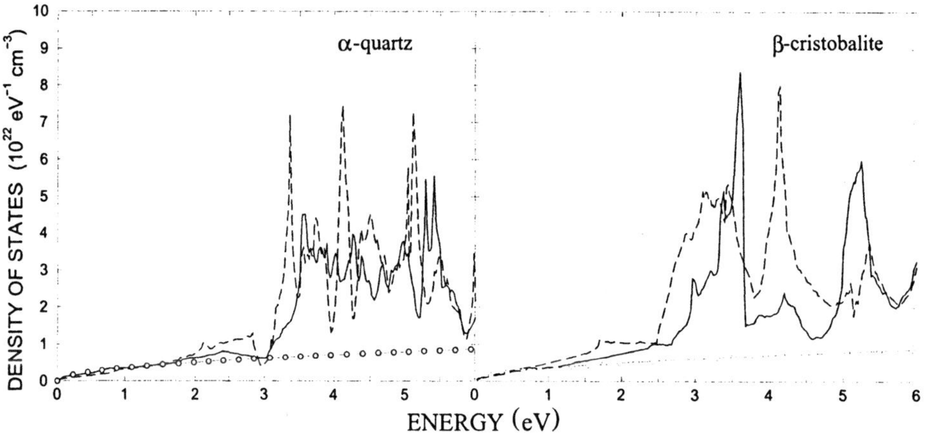

Fig. 2: DOS vs. energy for α-quartz and β-cristobalite. Solid lines: HF; dashed lines: DFT; circles: parabolic-band approximation (the effective masses are $m^* = 0.55\,m_0$ and $m^* = 0.68\,m_0$, respectively).

to the full-band structures of the polymorphs under study. In Figs. 2 and 3, the densities of states and group velocities for the two structures are reported. Both the DFT and HF data are shown, along with the parabolic-band approximation. The latter is seen to hold up to about 1 eV only. The difference between HF and DFT results is due to the intrinsic aspects of the calculation methods [5].

Two computational methods have been applied to determine the density of states and group velocity as functions of energy, with the purpose to be used for solving the BTE by the SHE method within the SiO_2 domain. This investigation is part of a larger project aiming at solving the carrier transport in semiconductor devices, in the frame of the semiclassical Boltzmann Transport Equation in real space and energy. Eight energy bands have been used to calculate both the electronic properties, in the energy range of interest for two different polymorphs of SiO_2. Such information, coupled with those already available for silicon, builds up the physical basis for the full-band calculation of the distribution function in the whole MOS structure.

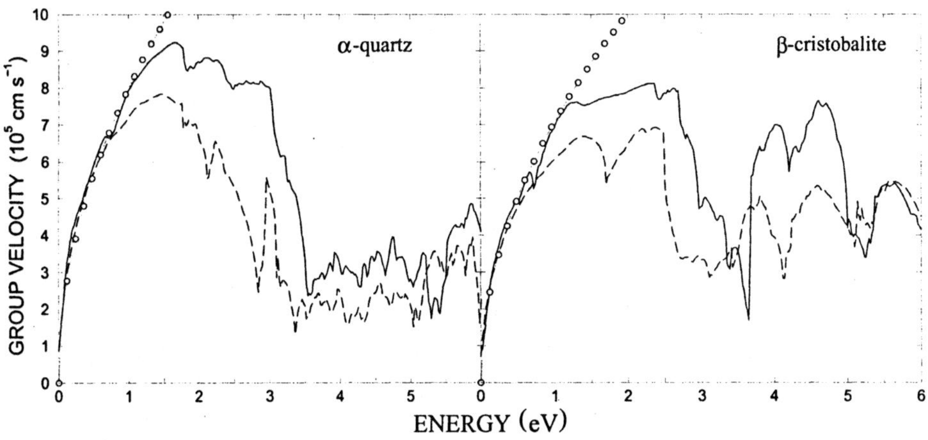

Fig. 3: GV vs. energy for α-quartz and β-cristobalite. Solid lines: HF; dashed lines: DFT; circles: parabolic-band approximation.

References

[1] Di Pomponio A., Continenza A., Lozzi L., Passacantando M., Santucci S., Picozzi P. (1995): Electronic Properties of Crystalline and Amorphous SiO_2 Investigated Via All-Electron Calculations and Photoemission Spettroscopy, Sol. State Comm. 95, 1995, pp. 313–317.

[2] Dovesi R., Roetti C., Causà M., Saunders V. R., Harrison N. M. (1999): CRYSTAL98–User's Manual.

[3] Fischetti M. V. (1991): Monte Carlo Simulation of Transport in Technologically Significant Semiconductors of the Diamond and Zinc-Blende Structures-Part I: Homogeneous Transport, IEEE Trans. on ED 39, pp. 634.

[4] Gilat G., Raubenheimer L. J. (1980), Computer Phys. Comm. 349.

[5] Gnani E., Reggiani S., Colle R., Rudan M. (2000): Band-Structure Calculations of SiO_2 by Means of Hartree-Fock and Density-Functional Techniques, IEEE Trans. on Electron Devices 47, pp. 1795–1803.

[6] Scozzoli L., Reggiani S., Rudan M. (2000): Homogeneous Transport in Silicon Dioxide Using the Spherical-Harmonics Expansion of the BTE, IEICE-TE E83-C, pp. 1183–1188.

[7] Vecchi M. C., Rudan M. (1998):Modeling Electron and Hole Transport with Full-Band Structure Effects by Means of the Spherical-Harmonics Expansion of the BTE , IEEE Trans. on Electron Devices 45, pp. 230–238.

Investigation of Spurious Velocity Overshoot Using Monte Carlo Data

T. Grasser, H. Kosina, and S. Selberherr

Institute for Microelectronics, TU Vienna, Gusshausstrasse 27–29, A-1040 Vienna, Austria
Phone: +43-1-58801/36023, Fax: +43-1-58801/36099, Email: Grasser@iue.tuwien.ac.at

Abstract

For the simulation of state-of-the-art devices hydrodynamic and energy transport models allow to account for non-local effects which cannot be captured by the drift-diffusion model. Although these models have been available for several decades, there are still unresolved issues. One of these issues is the occurrence of spurious peaks in the velocity profile which have originally been related to Bløtekjær's model. Recent research, however, showed that these peaks are inherent to both Stratton's and Bløtekjær's model. We investigate the origin of these peaks by introducing relaxation times, mobilities, and closure relations directly from a coupled Monte Carlo simulator. Although accurate modeling of the relaxation times and mobilities is important, it appears that the origin of the spurious peaks lies in the truncation of the infinite series of moments.

1 Introduction

In the traditional drift-diffusion (DD) approach the carrier energy is assumed to be in equilibrium with the electric field. Therefore, the DD model cannot predict non-local effects which occur in modern devices. Especially the overshoot in velocity is important to capture. Monte Carlo (MC) simulations predict an overshoot in the velocity when the electric field increases rapidly. This velocity overshoot can be qualitatively captured by higher order moment based models which, however, tend to overestimate this effect. Interestingly, these models also predict a velocity overshoot when the electric field decreases rapidly, e.g., at the end of a channel in a MOS transistor. This velocity overshoot is not observed in MC simulations and thus termed spurious velocity overshoot (SVO). Although the influence of this effect on device characteristics is probably not very dramatic, we feel that a thorough investigation of its cause is indeed important for the basic understanding of moment based models.

There has been an ongoing discussion in the literature on the cause of this effect. One of the first speculations was a weakness in the energy-transport (ET) model which is based on Bløtekjær's hydrodynamic model [1]. ET models and energy-balance (EB) models based on Stratton's approach [2] are closely related [1, 3], with the exception that the mobilities are defined in a different way. This leads to an additional driving term in the current relation of the EB model which is proportional to the spatial gradient of the mobility. This driving term can be reformulated and combined with the driving term of the temperature gradient yielding a modified prefactor of the temperature gradient. The modified prefactor is traditionally modeled as a constant and adjusted in such a way that the SVO is minimized [4]. It has already been pointed out that this approach is questionable [5], because the prefactor depends on the dominant scattering process which in turn depends on the doping and the applied voltages, and, therefore, no unique value can be found. Another important factor that has been attributed to SVO is the modeling of the mobilities which are not single valued functions of the energy [6]. As already argued by [7], SVO is not likely to be caused by the mobility alone.

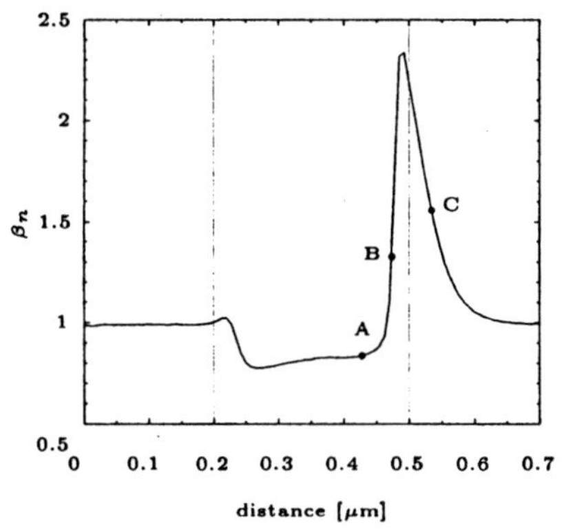

Fig. 1: Fourth order moment β_n inside a n^+-n-n^+ test-structure

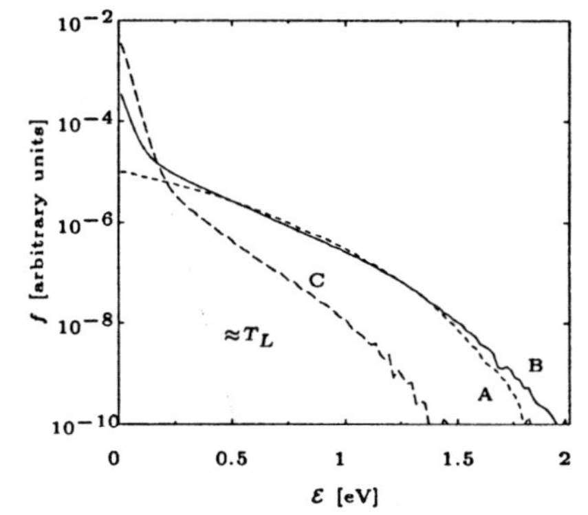

Fig. 2: Electron distribution functions inside a n^+-n-n^+ test-structure

2 Transport Models Coupled to Monte Carlo Simulator

A characteristic feature of moment based models is that the infinite hierarchy of moment equations has to be truncated to yield a finite number of equations. The highest order equation now contains a next higher moment which has to be modeled as a function of the lower order moments. Normally a heated Maxwellian shape is assumed for the distribution function (DF) which has the property that the fourth order moment $\langle \mathcal{E}^2 \rangle$ is proportional to the square of the average energy $\langle \mathcal{E} \rangle^2$. The accuracy of this relationship depends on the position in the device. To quantify the accuracy of this expression we define the kurtosis as [8]

$$\beta_n = \frac{3}{5} \frac{\langle \mathcal{E}^2 \rangle}{\langle \mathcal{E} \rangle^2} \tag{1}$$

which equals unity for a Maxwellian distribution and parabolic bands. A plot of β_n inside an n^+-n-n^+ test-structure is given in Fig. 1 for a parabolic MC simulation. In the channel β_n is in the range [0.8, 1] and the error assumed by a Maxwellian closure is probably not too dramatic. However, when the hot channel carriers meet the large pool of cold carriers in the drain, β_n reaches values above 2. Interestingly, this is exactly the region where the spurious velocity overshoot occurs. This is confirmed when we look at the distribution function at the points A, B, and C in Fig. 2 . For the channel point A, the curvature of the DF is smooth, reflected by a small value of β_n, whereas between the points B and C the contribution of the cold drain carriers increases while the high-energy tail relaxes slowly. Especially in these regions the Maxwellian shape which gives a straight line, is definitely a rather poor approximation.

To eliminate as many uncertainties as possible we use a parabolic ET model because non-parabolicity effects can only be approximately captured in the driving terms [9]. Furthermore we assume the validity of the diffusion approximation which allows to write the tensors as scalars and to neglect the contribution of the kinetic energy to the average carrier energy. These approximations are very common and will be reexamined later. In addition we restrict ourselves to the one-dimensional case, mainly because SVO is present in one-dimensional simulations and two-dimensional effects like real-space transfer further complicate any discussion. To remove the uncertainties in the relaxation time and mobility models we coupled our device simulator MINIMOS-NT to a MC simulator in a self-consistent manner where the potential distribution for the MC simulation was taken from the device simulator and the relaxation times for the device simulator from the last MC solution. A large number of scattering events had to be processed to achieve smooth relaxation times with small variance to obtain convergence within the moments models. In addition to the ET model we use our six moments model (SM) [8] for comparison which retains β_n as a solution variable and introduces a closure for $\langle\mathcal{E}^3\rangle$. Due to the existence of two carrier populations, the empirical closure we employed for $\langle\mathcal{E}^3\rangle$ is most critical in the very same region as the closure for the four moments ET model [8].

3 Comparison

A comparison of simulated velocity profiles obtained from the ET and SM model with MC results is given in Fig. 3 which confirms that SVO is not caused by the mobility alone as the SVO spike is still observed using MC data for all relaxation times. However, the SM model improves the closure of the equation system which may be the reason for the reduced spike observed within the SM model.

To further investigate this effect, another set of simulations was performed using β_n from the MC simulation to close the ET model, thus resulting in a correct closure of the four moments model. Furthermore, in one simulation the MC relaxation times were replaced by standard models, namely a constant energy relaxation time (0.33 ps), a simplified Hänsch mobility model [10]

$$\mu(T_n) = \frac{\mu_0}{1 + a(T_n - T_L)} \tag{2}$$

and a constant mobility ratio $\mu/\mu_S = 0.8$ where μ_S is the mobility of the energy flux [5].

$$\mathbf{S}_n = -\frac{5}{2}\frac{\mathrm{k_B^2}}{\mathrm{q}}\frac{\mu}{\mu_S}\left(\nabla(nT_n^2\beta_n) + \frac{\mathrm{q}}{\mathrm{k_B}}\mathbf{E}nT_n\right) \tag{3}$$

A comparison of these simulations with the MC values is given in Fig. 4. With the MC closure *and* the MC relaxation times the spike is completely removed whereas with the standard models it is quite pronounced, despite the correct closure. This leads us to the conclusion that SVO is a result of *both* the closure *and* the hysteresis in the relaxation times and can probably never be completely eliminated using a finite number of moment equations. However, as can be seen in Fig. 3, the error introduced by the closure is reduced when the number of moments is increased from four to six.

Fig. 4 serves as a good test for the approximations introduced in the derivation of the ET and SM model. Mainly, by employing the diffusion approximation, terms of the form $\langle\mathbf{u}\rangle \otimes \langle\mathbf{u}\rangle$ are neglected against $\langle\mathbf{u}\otimes\mathbf{u}\rangle$, one of the consequences being that the drift kinetic energy $m_\nu\langle\mathbf{u}\rangle^2/2$ is neglected against $\mathrm{k_B}T_n$. This contribution is only relevant as long as the carrier temperature is low and the velocity is high which is at the beginning of the channel in the velocity overshoot region. Indeed, a very small overestimation of the carrier temperature is observed in this region (Fig. 4). This contribution will gain more importance when the channel length is reduced. Fig. 4 stresses the importance of proper relaxation time and mobility models which influence the solution inside the whole device whereas the influence of the closure relation is strongest at the end of the channel and restricted to a relatively small region.

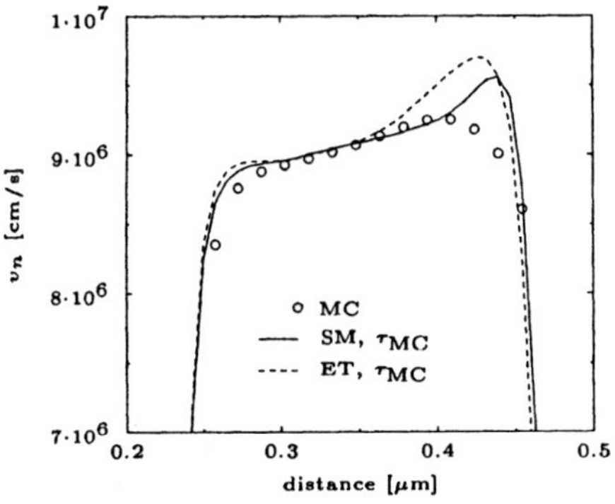

Fig. 3: Velocity profiles with MC relaxation times for the two transport models

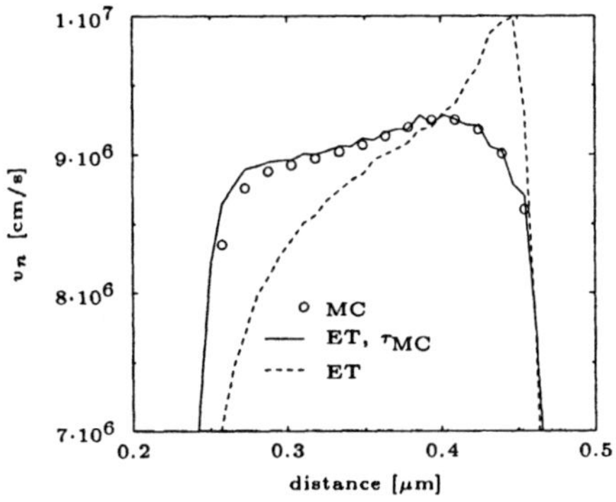

Fig. 4: Velocity profiles with MC relaxation times and closure

4 Conclusion

Spurious velocity overshoot has been attributed to several approximations introduced in the derivation of energy-transport models. We investigated the two most dominant effects which are caused by inaccuracies in the mobility and relaxation time models and by the truncation of the infinite series of moments. It has been demonstrated that accurate models for the mobilities and relaxation times are of utmost importance to reproduce velocity profiles inside the device. However, even with parameter values taken directly from Monte Carlo simulations, the spurious velocity overshoot remains in the velocity profiles but can be reduced by increasing the number of moments. This leads us to the conclusion that SVO is a result of both the closure and the inaccuracies in the physical models and can probably never be completely eliminated using a finite number of moments.

[1] K. Bløtekjær, "Transport Equations for Electrons in Two-Valley Semiconductors," *IEEE Trans.Electron Devices*, vol. ED-17, no. 1, pp. 38–47, 1970.

[2] R. Stratton, "Diffusion of Hot and Cold Electrons in Semiconductor Barriers," *Physical Review*, vol. 126, no. 6, pp. 2002–2014, 1962.

[3] T.-W. Tang and H. Gan, "Two Formulations of Semiconductor Transport Equations Based on Spherical Harmonic Expansion of the Boltzmann Transport Equation," *IEEE Trans.Electron Devices*, vol. ED-47, no. 9, pp. 1726–1732, 2000.

[4] D. Chen, E. Sangiorgi, M. Pinto, E. Kan, U. Ravaioli, and R. Dutton, "Analysis of Spurious Velocity Overshoot in Hydrodynamic Simulations," in *Intl. Workshop on Numerical Modeling of Processes and Devices for Integrated Circuits NUPAD IV*, (Seattle, Washington), pp. 109–114, May 1992.

[5] T. Tang, S. Ramaswamy, and J. Nam, "An Improved Hydrodynamic Transport Model for Silicon," *IEEE Trans.Electron Devices*, vol. 40, no. 8, pp. 1469–1476, 1993.

[6] G. Wolokin and J. Frey, "Overshoot Effects in the Relaxation Time Approximation," in *Proc.NASECODE VIII*, (Vienna), pp. 107–108, 1992.

[7] M. Stettler, M. Alam, and M. Lundstrom, "A Critical Examination of the Assumptions Underlying Macroscopic Transport Equations for Silicon Devices," *IEEE Trans.Electron Devices*, vol. 40, pp. 733–740, Apr. 1993.

[8] T. Grasser, H. Kosina, M. Gritsch, and S. Selberherr, "Using Six Moments of Boltzmann's Transport Equation for Device Simulation," 2001. to appear in *J.Appl.Phys.*

[9] T. Bordelon, X.-L. Wang, C. Maziar, and A. Tasch, "Accounting for Bandstructure Effects in the Hydrodynamic Model: A First-Order Approach for Silicon Device Simulation," *Solid-State Electron.*, vol. 35, no. 2, pp. 131–139, 1992.

[10] S.-C. Lee and T.-W. Tang, "Transport Coefficients for a Silicon Hydrodynamic Model Extracted from Inhomegenous Monte-Carlo Calculations," *Solid-State Electron.*, vol. 35, no. 4, pp. 561–569, 1992.

Elasto-Plastic Modeling of Microelectronics Materials for Accurate Prediction of the Mechanical Stresses in Advanced Silicon Technologies

Vincent Senez and Thomas Hoffmann
IEMN-ISEN, UMR CNRS 8520, 59652 Villeneuve d'Ascq Cédex, FRANCE
vincent.senez@isen.iemn.univ-lille1.fr

Abstract

This paper reports the implementation of an elasto-plastic model in advanced stress simulation system within the process simulator IMPACT to predict the mechanical behavior of (poly)crystalline materials used in the microelectronic technologies. The benefits of this new model are a better prediction of the stresses magnitude in elasto-plastic materials, the capability to calculate the location and size of plastic area in silicon substrate and the possibility to analyze very accurately stresses evolution in (un)passivated metal interconnects networks, a serious reliability issue for the IC industry in the recent years.

This work has been partly carried out within the STREAM project funded by the EC as IST-1999-10341.

1. Introduction

Stresses develop in integrated circuit (IC) structure during processing as a result of a variety of different effects. These stresses can lead to process modifications as well as change in the electrical behavior of the final device. More seriously, they can induce plastic flow, cracking, void formation and mechanical failure. That is why the knowledge of both the locations and magnitudes of such stresses is important. The finite element method is a powerful tool for solving problems of this nature. However, an accurate simulation of the stresses evolution requires informations on the materials, on the geometries of the structures and on the stresses sources. Only process simulation can provide all of these data at the same time and that is why we started few years ago the development of an homogenous and advanced stress simulation system in IMPACT, allowing the evaluation of the cumulative mechanical stresses at each process steps [1]. Our last work concerns the introduction of a complex rheological behavior for (poly)crystalline microelectronics materials which exhibit elasto-plastic properties under large stress sollicitations. To our knowledge, there is very few works reporting the implementation of an elasto-plastic algorithm in a process simulator [2]. Furthermore, they only focuse on the silicon behavior and do not take into account several complex mechanical behaviors such as stress-dependent viscoelasticity and elastoplasticity in the same numerical calculations. The benefits of this new model are a better prediction of the stresses magnitude in elasto-plastic materials, the capability to calculate the location and size of plastic area in silicon substrate and the possibility to analyze very accurately stresses evolution in (un)passivated metal interconnects networks, a serious reliability issue for the IC industry in the recent years.

2. Numerical implementation

The numerical "Return Mapping" algorithm is based on the works of Zienkiewicz [3] and Simo [4] and corresponds to an elastic perfectly plastic behavior. It is schematically summarized in Fig.

1a: from a residual stress σ_0 in the elastic domain, a new elastic trial stress σ'_1 is computed. If the resulting state defined by σ'_1 lies outside of the elastic region enclosed by the yield surface (F=0), one defines the final state as the "closest-point-projection" of σ'_1 onto the yield surface. The yield surface is defined with the von-Mises stress at the end of the time step. Then the new elastic residual stress σ_1 is calculated and $\Delta\sigma_1$ is converted into an un-balanced deviatoric stress, which by definition, is tangent to the yield surface. A new mechanical equilibrium is computed which leads to the residual stress σ'_2. This non-linear algorithm is accomplished by a Newton-Raphson procedure and is considered converged when all displacements are smaller than 10^{-9}cm. For problems requiring to solve both non-linear viscoelasticity and elastoplasticity, we use a decoupled approach given in Fig. 1b. First, we solve the non-linear (Eyring) viscoelasticity using SOR technique where poly(crystalline) materials are considered elastic. If, the elastic limit is reached,

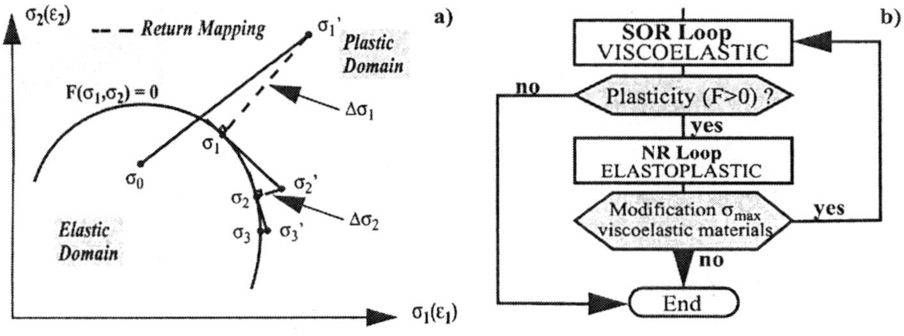

Fig. 1. a) Graphic interpretation of the numerical treatment of elasto-plasticity. b) Schematic algorithm of the decoupled treatment of elasto-plasticity and of viscoelasticity.

we treat the elastoplasticity considering that the viscosity of viscoelastic materials is constant. When we have converged on the elastoplastic problem, if the maximum shear stress σ_{max} of each viscoelastic materials has been modified by more than 1% after the NR loops, we go back to the viscoelastic problem. To evaluate the performances of this algorithm, we have simulated the "fictive" annealing from 25 to 1200°C of an AlCu line patterned on an oxidized silicon wafer (Fig. 2a). On can see that the plasticity of AlCu is activated around 100°C and the viscoelasticity of SiO_2 around 975°C. During the simulation, there is no iterations between the elastoplastic and the viscoelastic problems, the higher variation of σ_{max} being 0.6 MPa at 1100°C. If we now analyse

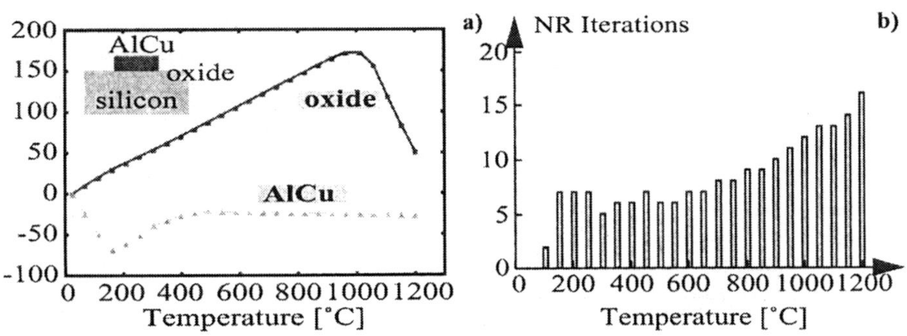

Fig. 2. a) Average stress in AlCu and oxide versus annealing temperature for a structure given in the insert. b) Histogram of the number of Newton-Raphson iterations versus temperature for the mechanical simulation reported in fig. 2a.

the elastoplastic treatment (Fig. 2b), a very good convergence rate is reached which slighlty decreases at high temperatures due to the purely viscous behavior of SiO_2. To validate this modeling, we have chosen the experiment of Vanhellemont [5] where a very thick LOCOS is grown in wet ambient. It leads to the creation of a dislocation in silicon in the bird's beak region (Fig. 3c). The simulations of this process, using either an elastic behavior for silicon (Fig. 3a) or an elastoplastic behavior (Fig. 3b) have been performed considering a rather high elastic limit (σ_Y) equal to 1GPa. Figs 3a-b gives the 2D distributions of the von-Mises stress and show the appearance of a plastic area in the bird's beak region in agreement with the experiment. The use of the elastoplastic algorithm increases the computing time by 6% only.

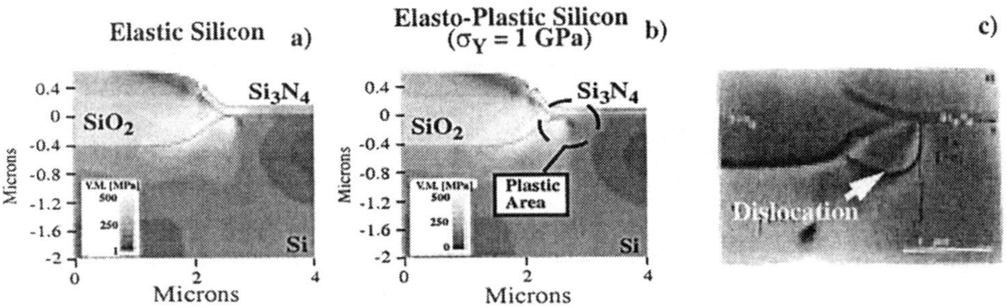

Fig. 3. Growth of a LOCOS isolation considering either a) an elastic or b) an elasto-plastic silicon, c) corresponding SEM picture showing the presence of a dislocation [5].

3. Applications to the behavior of aluminum-copper interconnects

For this material, it is known that σ_Y is function of the thickness of the film as well as the temperature. The experimental variations of the average stress in different films versus the temperature are given in Fig. 4 [6-7]. They reveal a complex mechanical elasto-plastic behavior which can be reproduced by the calibrated model. This calibration, performed for seven temperatures, is compared (Fig. 5a) with other works [8-9] and shows a good agreement. The full set of calibrated data is plotted in Fig. 5b and one can observe that at low temperature σ_Y is less sensitive to the film thickness than at high temperature. To validate this modeling, we have chosen the experiments of Yeo [10] where a network of AlCu lines (0.8μm thick) are patterned with different widths on an

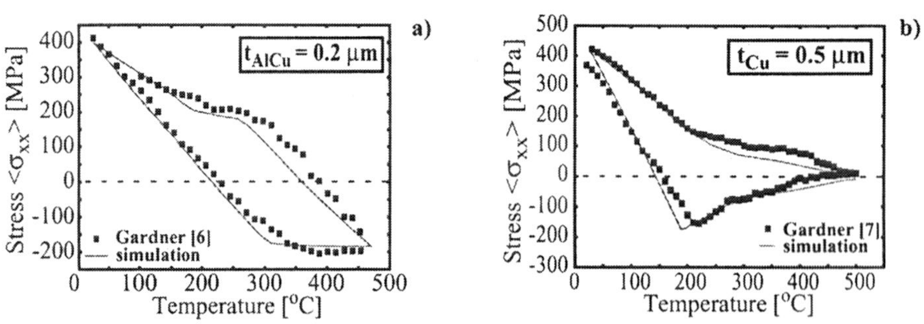

Fig. 4. Validation of the elasto-plastic modeling of aluminum-copper and of copper films for various thicknesses: a) 0.2μm, b) 0.5μm.

oxidized silicon wafer. The variations of the average stress versus temperature are given in Fig. 6a and, without any additional calibration, one can see they are well predicted by the model (Fig. 6b). One can notice that the magnitude of the hysteresis becomes smaller when the width is smaller. If we now analyze the effect of the passivation (Fig. 6c), one can see that the confinement increases the magnitude of the stress and that the aspect ratio of the line is an important factor in the activation of plasticity. If we study more specifically this parameter, one can show that the optimization of the geometry of the line is not only dependent of its aspect ratio as concluded by Yeo but also of its thickness. This later aspect will be developed in details in our oral presentation.

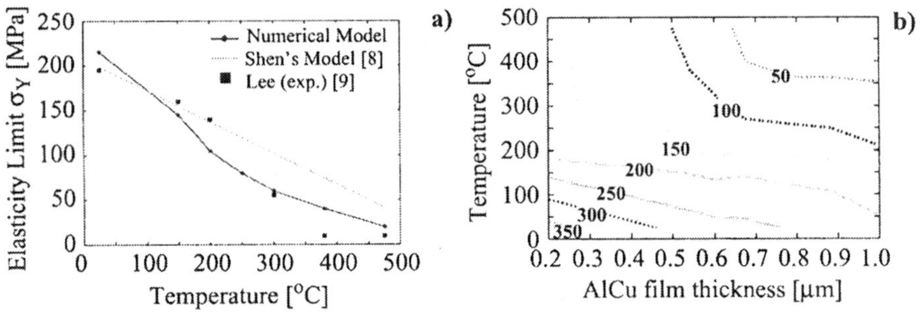

Fig. 5. Calibration of the elasto-plastic model for AlCu films: a) comparison of the elasticity limits for a 1μm thick film, b) variation of the elasticity limit versus temperature and thickness.

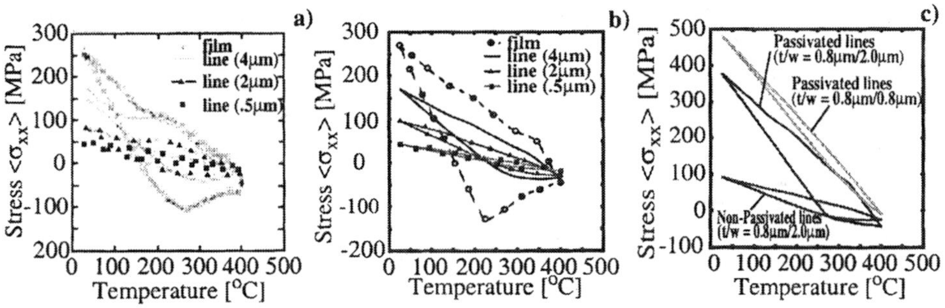

Fig. 6. a) Experimental variations of the average stress in AlCu lines (0.8μm thick) having different widths versus temperature [10], b) corresponding simulation results, c) influence of the passivation on the stresses in the AlCu lines.

4. References

[1] Ferreira, P., Senez, V., Baccus, B., Varon, J., Lebailly, J. (1995), IEDM'95, pp. 503-506.
[2] Uchida, T., Kotani, N., Kobayashi, K., Mashiko, Y., Tsubouchi, N. (1993), IEDM'93, pp. 315-318.
[3] Zienkiewicz, O.C., Valliapan, S., King, I.P. (1969). Int. J. of Numerical Methods 1: 75-100.
[4] Simo, J.C., Taylor, R.L. (1985). Comp. Meth. Appl. Mech. & Eng. 48, 101-118.
[5] Vanhellemont, J., Amelinckx, S., Claeys, C., (1987). J. Appl. Phys. 61, 2176-2188.
[6] Gardner, D.S., Flinn, P.A., (1988). I3E Trans. Elect. Dev. 35, 2160-2169.
[7] Gardner, D.S., Longworth, H.P., Flinn, P.A., (1992). J. Vac. Sc. & Tech. A, 1426-1441.
[8] Shen, Y.L., (1997). J. Appl. Phys. 82, 1578-1581.
[9] Lee, J., Sauter Mack, A.I. (1998). I3ETrans. Semicon. Manuf. 11, 458-464.
[10] Yeo, I.S., Ho, P.S., Anderson, S.H., (1995). J. Appl. Phys. 78: 945-961.

A Unified Model of Dopant Diffusion in SiGe.

Ardechir Pakfar (1,2), A. Poncet (2), T. Schwartzmann (1) and H. Jaouen (1).
(1) STMicroelectronics, 850 rue Jean Monnet. F-38926 Crolles cedex, France.
(2) LPM - INSA de Lyon, F-69621 Villeurbanne cedex, France.
Email: ardechir.pakfar@st.com

Abstract

The understanding of the effect of each physical mechanism driving dopant and point defect diffusion due to Ge leads to a unified formulation of diffusion for the usual dopants in SiGe material. The model calibration is deduced from a critical synthesis of the theoretical and experimental published studies.

1 Introduction

The control of dopant profiles in SiGe has a direct influence on the electrical characteristics of HBT devices. Furthermore, Ge can be considered as an additional "degree of freedom" for the analysis of the different material parameters on dopant diffusion.

2 Modeling of physical mechanisms

The presence of Ge in Si modifies some of the material physical characteristics with respect to pure Si. The study of the effect of physical phenomena on the equilibrium concentration of point defects is developed, allowing to express the modification of dopant diffusivity as it is proportional to the point defect concentration.
Other phenomena being supposed identical in Si and SiGe, our model considers each physical mechanism acting on dopant diffusion in SiGe as a perturbation of the standard Si diffusion.

2.1 Layer strain effect on diffusion

The synthesis by Zhao et al. [10] on dopant diffusion in strained Si leads to the quantification of the dual effect of strain on the formation energy and thus on the equilibrium concentration of point defects.
A biaxial compression will increase the Vacancies equilibrium concentration (V) and decrease the Interstitials concentration (I). In a SiGe layer, V-mediated dopants like Antimony will diffuse faster in strained than in relaxed SiGe [7] as illustrated in Fig. 1. On the other hand, I-mediated dopants, are believed to have a lower diffusivity in strained than in relaxed SiGe layers as shown in Fig. 2.

2.2 Ge chemical effect

The point defect equilibrium concentrations are closely related to the bonding energy of atoms. In purely covalent crystals, the heat of sublimation corresponds to the rupture of half of the atomic bonds.

Table 1. Material characteristics of pure Si and Ge

Material	Heat of Sublimation ($kJ.mol^{-1}$)	Bonding Energy (eV)
Si	455.6	2.36
Ge	374.5	1.94

In Table 1, we report the lower bonding energy of Ge as compared to Si. As a consequence, the atomic bonding energy in relaxed SiGe is lowered by the presence of Ge, so the formation energy of point defects is decreased with Ge concentration. It explains that both I and V point defects concentration are increased by the Ge chemical effect as suggested by Christensen et al. [2].

2.3 Band gap modification

The strain value is of prime importance on the formulation of the band gap change. The two stress limit values are considered for SiGe layers:

Relaxed (R): $\Delta Eg_R = -0.55\, x_{Ge} + 0.33\, x_{Ge}^2$ from [1]

Strained (S): $\Delta Eg_S = -1.01\, x_{Ge} + 0.835\, x_{Ge}^2$ from [3]

The band gap of Si is lowered by the presence of Ge atoms in the matrix. The equilibrium of charged point defects is modified by this effect and the dopant-defect interactions are reduced, lowering dopant diffusivity.

2.4 B-Ge pairing

For Boron, an other physical mechanism, verified by the dedicated experimentations of Lever et al. [6], has to be considered: a mechanical pairing between Boron and Germanium atoms reducing the number of mobile B atoms.
In a first approximation, we consider the B-Ge pairing parameter as independent of Ge concentration. Indeed, this has to be moderated due to the experiments on B diffusivity in $Si_{0.99}Ge_{0.01}$ of Zangenberg et al. [9] and by the fact that, for a Ge content higher than 50%, no local atomic strain around Ge atoms has to be supposed.

2.5 Formulation

Considering their diffusion mechanism, the diffusivity ratio for all dopants between SiGe and pure Si is expressed in our model with the following formulation:

$$\frac{D(SiGe)}{D(Si)} = \left(\frac{1}{1+K_2 X_{Ge}}\right)\exp\left(\frac{(Q'+Q_{Ge})\;0.042 X_{Ge} + K_1 \Delta Eg}{k_b T}\right)$$

Where X_{Ge} is the Ge atomic content of the SiGe layer.
Each basic driving force is introduced through the parameters: K_1 for the effect of the modification of the band gap, Q' for the effect of strain, Q_{Ge} for the Ge chemical effect and finally K_2 for the B-Ge pairing effect in case of Boron diffusion.

3 Discussion

In Fig. 1 and 2, we have gathered all the reliable published data on Antimony in [8] and [4], on Boron in [7], [8] and [9] and finally on Phosphorus diffusion in SiGe strained and relaxed layers in [2] and [5].
In our model, Antimony diffusivity is found to increase rapidly with the Ge content of the layer as both stress and Ge chemical effects enhance V-mediated dopant diffusion.

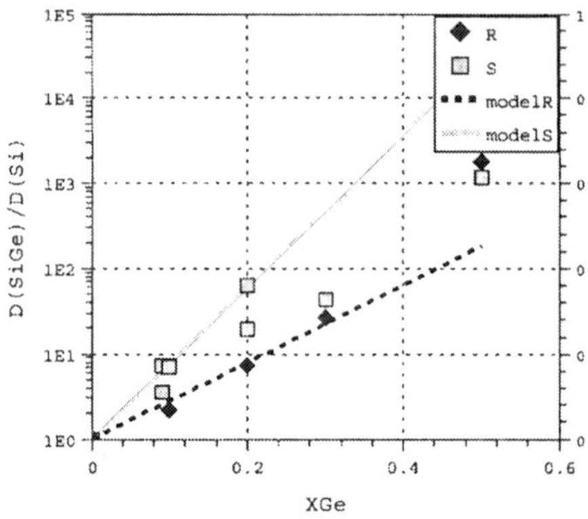

Fig. 1. Antimony diffusivity ratio between SiGe and Si material at 800°C. Squares are experimental data in coherently Strained (S) and diamonds are data in fully Relaxed (R) SiGe. The curves are the model prediction for Strained layer in solid and Relaxed layer in dashed.

The B-Ge pairing is the main effect to consider for Boron diffusion as the effect of the other phenomena on Boron diffusion is weak. Q' and k_2 are on the same order of magnitude than previously published values of [10] and [6].

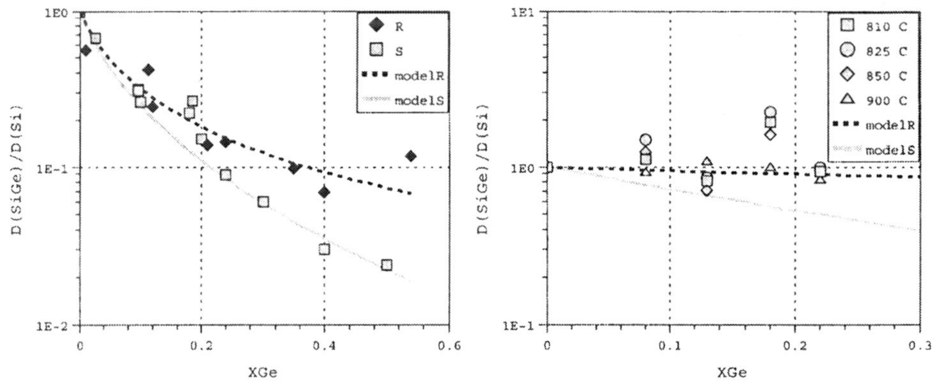

Fig. 2. Boron (left) and Phosphorus (right) diffusivity ratio between SiGe and Si material at 800°C. Legend similar to Fig. 1.

The Phosphorus diffusion in strained and relaxed SiGe, not affected by the pairing effect, is very useful for the study of the different mechanisms acting on Interstitial point defect. P diffusivity is found to be on the same order in Si and SiGe [2] and higher for relaxed layers than for strained ones.

Even if no experimental data of Arsenic diffusion have been reported, our model predicts an enhancement of As diffusivity in Relaxed and Strained SiGe.

4 Conclusion

The understanding of each physical phenomenon involved in the impurity diffusion leads to a unified model for all usual dopants in Strained and Relaxed SiGe in accordance with physical considerations. Despite the dispersion of the experimental data, the diffusion behavior of Sb, P and B in SiGe is well described by our model. Nevertheless, a detailed atomistic study is needed relative to the behavior of dopant point defects formation energies, and to the B-Ge pairing with the Ge content.

References.

[1] Braunstein R., Moore A.R. and Herman F. (1958): Intrinsic optical absorption in Germanium-Silicon alloys. Phys. Rev. 109(3), pp 695-710.

[2] Christensen J.S., Kuznetsov A.Y., Radamson H.H. and Svensson B.G. (2000): Phosphorus diffusion in SiGe. Fifth International conference on DIffusion in MATerials, DIMAT 2000, Paris (France), July 2000 (to be published).

[3] Dutartre D., Bremond G., Souifi A. and Benyattou T. (1991): Excitonic photo-luminescence from Si-capped strained SiGe layers. Phys. Rev. B 44(20), pp 11525-11527.

[4] Kuznetsov A.Y., Cardenas J., Schmidt D.C., Svensson B.G. and Nylandsted Larsen A. (1999): Sb-enhanced diffusion in strained SiGe. Phys. Rev. B 59(11), pp 7274-7277.

[5] Kuznetsov A.Y., Christensen J.S., Linnarsson M.K., Svensson B.G., Radamson H.H., Grahn J. and Landgren G. (1999): Diffusion of Phosphorus in strained Si/SiGe/Si heterostructures. MRS Symp. Proc. No. 568: "Silicon Front-End Processing-Physics and Technology of Dopant-Defect Interactions." Edited by Gossmann H-J.L., Haynes T.E., Law M.E., Nylandsted Larsen A. and Odanaka S., pp 271-276 and references.

[6] Lever R.F, Bonar J. and Willoughby A.F.W. (1998): Boron diffusion across Silicon - Silicon Germanium boundaries. J. Appl. Phys. 83(4), pp 1988-1994 and references.

[7] Nylandsted Larsen A. (1998): Impurity Diffusion in SiGe Alloys: Strain and Composition Effects. MRS Symp. Proc. No. 532: "Silicon Front-End Technology-Materials Processing and Modeling." Edited by Cowern N.E.B., Jacobson D.C., Griffin P.B., Packan P.A. and Webb R.P., pp 187-198 and references.

[8] Willoughby A.F.W., Bonar J. and Paine A.D.N. (1999): Diffusion Mechanisms in SiGe Alloys. MRS Symp. Proc. No. 568: "Silicon Front-End Processing-Physics and Technology of Dopant-Defect Interactions." Edited by Gossmann H-J.L., Haynes T.E., Law M.E., Nylandsted Larsen A. and Odanaka S., pp 253-264 and references.

[9] Zangenberg N.R., Fage-Pedersen J., Lundsgaard Hansen J. and Nylandsted Larsen A. (2000): Boron diffusion in strained and relaxed SiGe. Fifth International conference on DIffusion in MATerials, DIMAT 2000, Paris (France), July 2000 (to be published).

[10] Zhao Y., Aziz M.J., Gossmann H-J., Mitha S. and Schiferl D. (1999): Activation volume for Boron diffusion in Silicon and implications for strained films. Appl. Phys. Letters 74(1), pp 31-33 and references.

A Simple Modeling and Simulation of Complete Suppression of Boron Out-Diffusion in $Si_{1-x}Ge_x$ by Carbon Insertion

K. Rajendran[1] and W. Schoenmaker[2]
STDI/TCAD Division, IMEC, Kapeldreef 75, B-3001 Leuven, **Belgium**.
1.rajendra@imec.be. 2. Schoen@imec.be

Abstract

We present a simple modeling of boron diffusion in $Si_{1-x-y}Ge_xC_y$ by manipulating the strain and the intrinsic carrier concentration. We show that the diffusion of boron is strongly suppressed by a moderate concentration of substitutional C in $Si_{1-x}Ge_x$. This suppression is due to an under saturation of Si self-interstitials in the C-rich region. The results obtained from the proposed model are in good agreement with the measured values.

1. Introduction

The integration of SiGe alloy layers has opened the possibility of combining advanced Si microelectronics technology with the unique advantages of heterostructure devices. A prominent example for this exploitation of bandgap engineering is the heterojunction bipolar transistor (HBT). A key problem in npn SiGe HBT technology is retaining the narrow as-grown boron profile within the SiGe layer during post-epitaxial processing. However, boron diffusion into the Si region during device processing causes undesirable conduction band barriers, and thus significantly degrades device performance. Boron out-diffusion can be caused by: (1) thermal annealing, (2) transient enhanced diffusion (TED) due to an extrinsic boron implantation and anneals etc., An additional degree of freedom of the fabrication process is given by adding substitutional C in SiGe HBTs as a further diffusion suppressing agent [1]. The suppression of B diffusion due to Ge and C allows the control of steep doping profiles during device processing. Addition of C has also the advantage that it reduces the misfit strain and eliminates secondary defects at the amorphous/crystal (a/c) interface in ion-implanted Si was proposed by Nishikawa et al. [2]. The present modeling and simulation suggests that substantially higher thermal budgets can be tolerated when substitutional carbon in concentrations of the order of 0.1 atomic % is added to the SiGe layer [1]. We show that the strong suppression of B diffusion can be achieved by a small amount ($\sim 10^{20}$ cm^{-3}) of substitutional C in SiGe.

2. Model

Diffusion of C out of regions with higher C concentrations can cause localized under saturation of Si self-interstitials in the C rich region, which, in turn, results in suppressed diffusion of boron. Mobile interstitial carbon atoms (C_I) are created through the reaction of Si self-interstitials (I) with immobile substitutional carbon (C_s):

$$C_s + I \Leftrightarrow C_I \qquad (1)$$

The formation of C_I pairs through the replacement reaction of Eq. (1) reduces the amount of interstitials available for pairing with substitutional B, thereby lowering the B diffusivity. The boron impurity flux is given as [3-4]:

$$J_B = -D_B \Psi\left(\frac{\partial C_B}{\partial x} + \frac{C_B}{Q}\frac{\partial N_T}{\partial x} + \frac{2C_B n}{Q}\frac{\partial \ln n_i}{\partial x}\right) \quad (2)$$

Ψ allows for the concentration dependence of D_B.
The boron diffusivity is modeled in a simplified form as follows:

$$D_B = \left(\frac{C_I}{C_I^*}\right)\Psi D_{B0}\exp(-E_A / kT) \quad (3)$$

where D_{B0} is the pre-exponential factor for the diffusivity. Here, the boron diffusivity includes the effect of Si interstitials, because it is proportional to the ratio between the concentration of Si interstitial C_I and the equilibrium concentration of Si interstitial C_I^*. E_A is the activation energy of B in Si, which is modified to fit for $Si_{1-x-y}Ge_xC_y$ is: $E_A = 3.62 + (0.4.x.y) \pm 0.0154$, where x and y are at. % fractional values of germanium and carbon respectively. According to the Eqs. (2) and (3), the retardation of B diffusion in the C implanted sample may be attributed to the deactivation of B, which leads to a lower hole concentration p. The addition of C in $Si_{1-x}Ge_x$ has been used for strain compensation [1]. At certain Ge and C levels, the strain due to Ge and C atoms becomes important for the boron diffusivity in $Si_{1-x-y}Ge_xC_y$. So, the boron diffusivity, given in Eq. (3) can be modified by multiplying with exp(-*f*S/kT). Here S is the local strain due to the presence of Ge and C and defined as: $S = |S_{Ge} - S_C|$; $S_{Ge} = \delta \cdot x$; $S_C = R\delta \cdot y$, where δ is the moderate lattice mismatch between Si and Ge (~0.042), R is the strain compensation ratio (~10.71) [1], and f is the activation energy per unit strain, which is modified as: $f \rightarrow (f' - 0.4 * x * 100)$. The bandgap difference (ΔE_g) between $Si_{1-x}Ge_x$ and $Si_{1-x-y}Ge_xC_y$ is supposed to affect intrinsic carrier concentration n_i and results in the difference in D_B as well as in boron flux in Eq. (2). The intrinsic carrier concentration as a function of temperature is given by: $n_i^{SiGeC} = \sqrt{(N_C N_V)^{SiGeC}}\exp(-(E_g + \Delta E_g)/2kT)$ where N_C and N_V are the effective density of states in the conduction and valence bands for SiGeC and
$\Delta E_g = E_{gsT} + (E_{grT} - E_{gsT})(1 - F_{pm})$. Here E_{gsT} is the energy gap due to strain, is approximated by $E_{gsT} = x(0.767 + (1.71/R))$.

Since B diffusion is primarily attributed to an interstitials mechanism, it is retarded if the concentration of interstitials C_I is reduced as compared to the equilibrium value C_I^*. To examine this point, we simulated the B diffusion in $Si_{1-x-y}Ge_xC_y$ by using the process and device simulator TAURUS through Physical Model and Equation Interface (PMEI) [4]. Since we assumed that C_I is equal to C_I^*, the change in boron diffusivity defined in the Eqs. 2-3 should be a measure of C_I averaged around the boron diffusion region. We changed the value of pre-factor D_{Bo} and by the inclusion of models for the modified activation energy of B (which includes the strain with f factor) and intrinsic carrier concentration for $Si_{1-x-y}Ge_xC_y$ to fit the simulated profile to the experimental one as discussed below.

3. Results and discussion

Fig. 1 shows the SIMS as-grown and annealed profiles of boron spikes and C with peak concentration of 5×10^{19} cm^{-3} (0.1 at. %), as well as the profiles from the present model. Nearly four peaks of B spikes were incorporated in the C-rich regions, and in the adjacent region without interstitial C doping. We found strongly enhanced diffusion of B spikes in the C- poor region. The suppression of B diffusion at high C

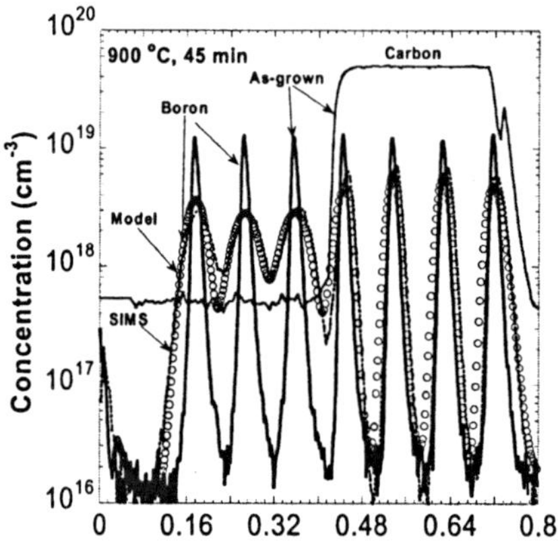

Fig.1 Comparison of SIMS with simulation for C: 5x10^{19} cm^{-3}. Open circules and dotted lines are after anneal from simulation and SIMS. SIMS profile are from Rucker et al. [5].

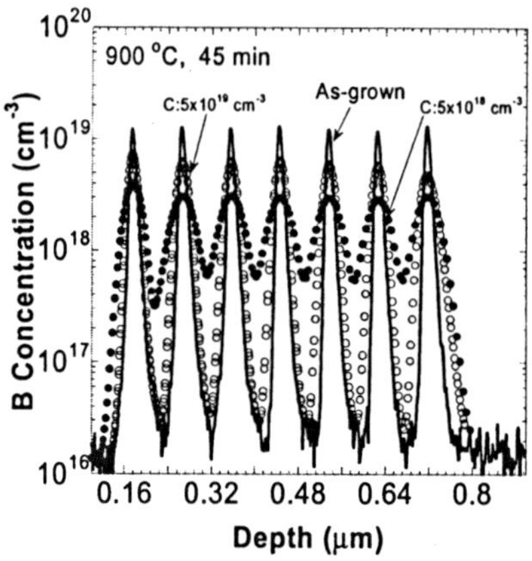

Fig.2 Depth profiles of B with different background C conc. Solid line is the as-grown and symbols are the annealed samples after 900 °C, 45 min.

concentrations are consistent with the under-saturation of Si self-interstitials in the C-rich region predicted by theory from Rucker et al. [5]. The profiles obtained from the present model agree very well with the experimental profiles in both C-poor and rich regions. The deduced D_{B0} for the sample with a C concentration of 5x10^{19} cm^{-3} was 0.159 cm^2/s, which is obviously smaller than the reported value for Si [2] as well as for $Si_{1-x-y}Ge_xC_y$ [1] due to the inclusion of models for strain and n_i. Fig.2 shows the dependence of TED suppression by various C concentrations (at the background) for three B doping super-lattices. It has been reported that TED of boron is strongly suppressed in C-rich Si [1]. The calculated B profile in the C-poor sample shows strongly enhanced diffusion of B. The calculated results agree well with the measurement, as well as with those reported by Rucker et al. [5]. We conclude from the present simulation and measurement [5] that TED of B is suppressed in adjacent regions of low C concentrations.

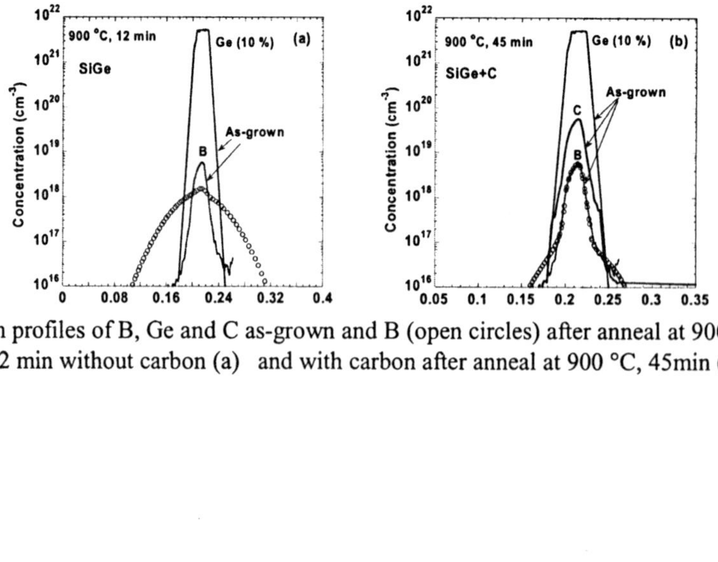

Fig.3 Depth profiles of B, Ge and C as-grown and B (open circles) after anneal at 900 °C, 12 min without carbon (a) and with carbon after anneal at 900 °C, 45min (b).

Figs. 3(a) and (b) show the measured as-grown and the annealed simulated results without and with C respectively, which demonstrate that the boron is confined well within the SiGe (or SiGeC) layers. The layered structure without C exhibits strong boron out-diffusion. Fig. 4 shows the intrinsic boron diffusivity D_B that was

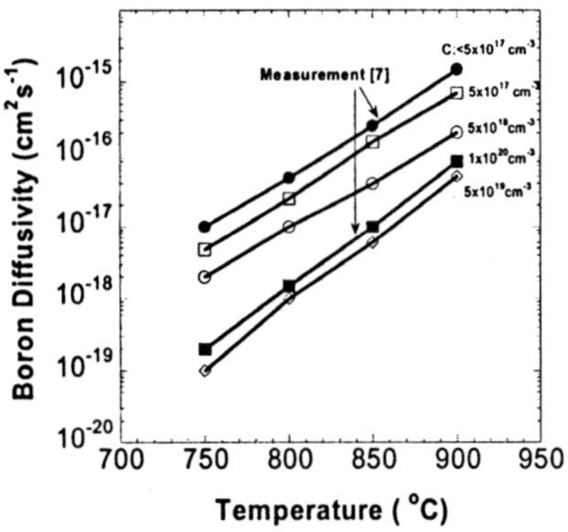

Fig.4 Intrinsic B diffusivity as a function of annealing temperature for three C concentration. Measurement values are also given in the plot.

determined as a function of C concentration and annealing temperatures from the measurement [5], and from the present simulation. For increasing C concentrations, we find a strong suppression of the B diffusivity by a factor, which is almost independent of temperature.

4. Conclusion

We have developed a model for the diffusion of boron in $Si_{1-x-y}Ge_xC_y$ that is based on the modification of the strain, intrinsic carrier concentration and band gap narrowing. We have demonstrated that the diffusivity of boron is strongly suppressed by high concentrations of substitutional C. This observation is interpreted in terms of an under-saturation of Si self-interstitials in the C-rich region. Finally, we have shown that the results from the present simulation agree well with the measurement values.

References

[1] Osten, HJ, Lippert G, Knoll D, Barth R, Heinemann B, Rucker H, and Schley P (1997):The effect of carbon incorporation on SiGe heterobipolar transistor performance and process margin. Tech. Dig. IEDM, pp. 803-06.

[2] Nishikawa, S., Takeda, A., and Yamaji, T (1992): Reduction of transient boron diffusion in preamorphized Si by carbon implantation. Appl. Phys. Lett. **60:** 2270-72

[3] Lever RF, Bonar JM, and Willoughby AFW (1998): Boron diffusion across silicon-silicon germanium boundaries. J. Appl. Phys. **83**: 1998-04

[4] Rajendran K and Schoenmaker W (2001): Studies of boron diffusivity in strained SiGe epitaxial layers. J. Appl. Phys. **89**: 980-86

[5] Rucker H, Heinemann B, Bolze D, Kurps R, Kruger D, Lippert G, Osten HJ (1999): The impact of supersaturated carbon on transient enhanced diffusion. Appl. Phys. Lett.**74:** 1682-4.

On the Effect of Local Electronic Stopping on Ion Implantation Profiles in Non-Crystalline Targets

A. Burenkov[a], Y. Mu[ab], H. Ryssel[a]

[a] Fraunhofer Institute of Integrated Circuits
Schottkystrasse 10, 91058, Erlangen, Germany
e-mail: burenkov@iis-b.fhg.de

[b] Physics Department, Shandong University, Jinan, 250100, P. R. China,
presently at the Institute of Physical and Theoretical Chemistry,
Goethe University Frankfurt/Main
Marie Curie Strasse 11, 60439 Frankfurt/Main, Germany

1 Introduction

The standard model for simulation of ion implantation into amorphous materials based on the LSS concept of ion penetration [1] remained a reliable work horse for the simulation of ion implantation for a long time, but recent accurate experimental measurements of the implantation profiles in non-crystalline materials (photoresist, amorphous carbon, pre-amorphized silicon) indicated systematic deviations of the measured profiles from the predictions of that standard model. The experimentally measured implantation profiles are usually broader than predictions of simulations based on the amorphous material model, and the relative profile broadening is especially significant for heavy ions at elevated implantation energies. Several physical effects may come in question to explain the broadening of the implantation profiles: experimental uncertainty during the profile analysis, radiation enhanced diffusion during ion implantation, deficiency of the physical model for ion-atomic interaction. The high accuracy of the measurements and the fact that the observed profile broadening is independent of the chemical nature of the target atoms and ions speaks in favour of the last assumption about the physical background of these deviations between the model and experiment.

2 Models of electronic stopping in non-crystalline materials

The standard model for simulation of ion implantation in amorphous material assumes that nuclear collisions and electronic stopping are independent, with electronic stopping being a function of ion energy and being position independent. The range straggling of the implanted ions in this model is mainly determined by nuclear collisions, the electronic stopping acts merely as a continuous friction force. From such a model it follows that the relative projected range straggling $\Delta R_p/R_p$ remains unchanged when electronic stopping is adjusted to reproduce the mean projected range R_p in a simulation. Moreover, the enlarged values of $\Delta R_p/R_p$ observed experimentally could not be reproduced in the simulation when different but realistic models for the nuclear scattering were tested.

The position dependent local electronic energy loss is known to play an important role in channeling phenomena increasing the range of channeled ions. A similar increase of the ion ranges can also take place in amorphous materials if one considers that the electronic stopping in an amorphous material is strongly impact parameter dependent and therefore is strongly correlated with nuclear energy loss. In fact, the average electronic stopping of ions which have a maximum possible range, i. e. which predominantly participate in distant collisions, is then expected to be lower than for the mean stopping of the ions and such ions therefore have a larger projected range than predicted by standard models leading to an overall increase of the profile width.

An attempt to quantitatively describe the effect of the widening of ion implantation profiles in non-crystalline materials used in semiconductor technology due to position dependent local electronic energy loss was done in this work. To describe channeling phenomena, an exponential dependence of the electronic stopping on the minimum approach distance between ion and the target atom was suggested by Oen and Robinson [2]. The energy loss in a single collision, $T_e(E,p)$, as a function of the ion kinetic energy, E, and the impact parameter, p, is described as:

$$T_e(E,p) = A\, S_e \exp\left[-B\, r_0(E,p)/a\right], \qquad (1)$$

where A is the normalisation coefficient to ensure the proper value of the total electronic stopping cross section S_e which is the measure of the average electronic stopping, B is the model coefficient which determines how sharply the energy loss T_e decreases if the impact parameter p increases, a is the inter-atomic screening length, and r_0 is the distance of closest approach in a collision at ion energy E and impact parameter p. A value of *0.3* was suggested originally by Oen and Robinson for the constant B to describe the channeling phenomena. Later, Xia et al. [3] applied the impact parameter dependent energy loss of Eq. (1) to describe ion implantation into crystalline silicon in non-channeled directions and for ion implantation into amorphous carbon. Larger values of B equal to 0.45 and 0.76 were found to describe ion implantation of boron and phosphorus, respectively, into silicon and B=0.4 was found for argon implantation into carbon.

3 Simulation results, comparison to experiment

To simulate the implantation profiles, a modified Monte-Carlo code of the TRIM program [4] was used in this work. The standard model for amorphous materials included a position independent electronic stopping according to Ziegler-Biersack-Littmark (ZBL) tabulation [4], and the maximum impact parameter in the simulation was energy dependent so that only a part of all possible impact parameters were considered at elevated implantation energies. The model suggested in this work used an impact parameter dependent local electronic stopping according to Eq. (1) with B=1, the average electronic energy loss was equal to the ZBL tabulation or in some cases corrected with a constant energy independent scaling factor to reproduce the average projected range in the simulation, and the maximum impact parameter was energy independent and equal to the half distance between next neighbours in the amorphized target material. The choice of the constant maximum impact parameter in the present work prevents the unwanted limitation to small impact parameters at higher energies in TRIM program and distinguish our model from the model of Xia et al. [3].

Figures 1 to 4 show some of the results of simulations using the standard and new simulation models in comparison to the experimental results of Glawischnig and Parks [5] obtained for ion implantation of arsenic and phosphorus ions into photoresist. The best reproduction of the experimentally observed profiles in the simulation has been achieved with the model suggested in this work using the impact parameter dependent model for electronic stopping with a screening length equal to the screening length of inter-atomic interaction, i.e. B=1 in Equation (1).

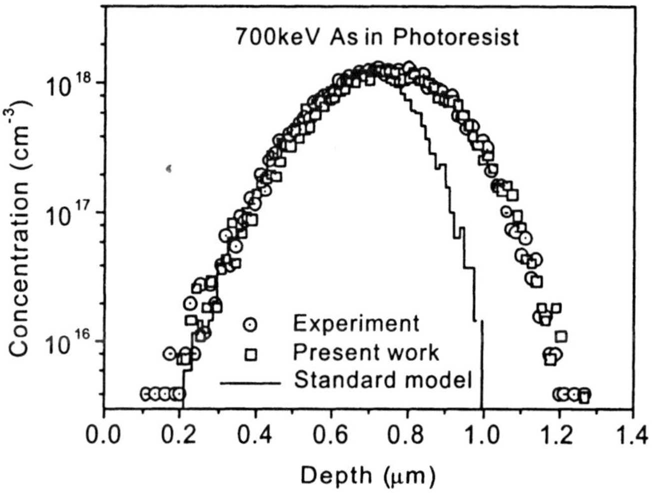

Fig. 1. Implantation profile of arsenic in photoresist

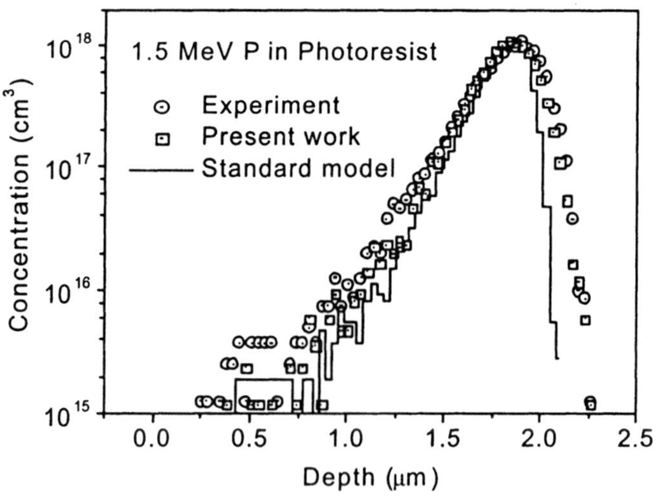

Fig. 2. Implantation profile of phosphorus in photoresist

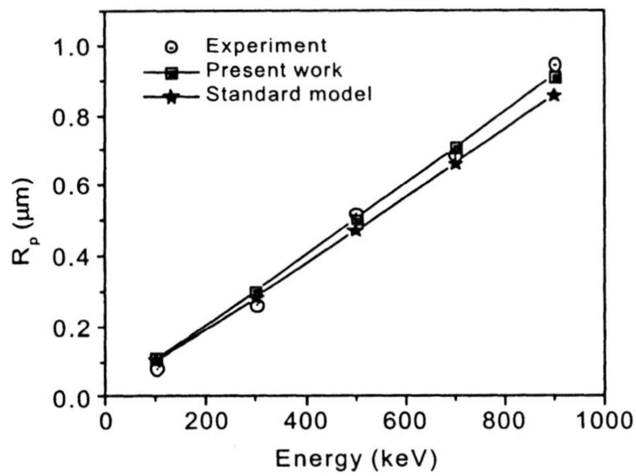

Fig. 3. Mean projected range of arsenic ions in photoresist as a function of implantation energy

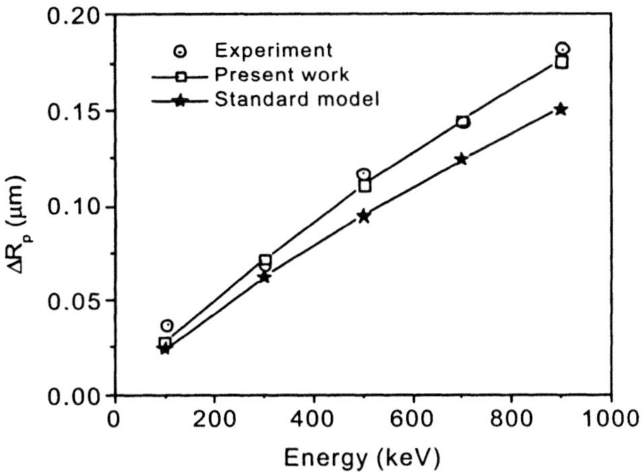

Fig. 4. Mean projected range straggling of arsenic ions in photoresist as a function of implantation energy

References

[1] Lindhard J, Scharff M, Schiøtt HE. Range concepts and heavy ion range. Mat. Fys. Medd. Dan. Vid. Selskab. 1963, 33, No. 14: 1-36

[2] Oen OS, Robinson MT. Computer studyes of the reflection of light ions from solids. Nucl. Instr. and Meth. 1976, 132: 647-653

[3] Xia Y, Lau WM, Lennard WN. Effects of electronic energy loss straggling on range distributions. Nucl. Instr. and Meth. in Phys. Res. 1992, B71: 165-170

[4] Ziegler JF, Biersack JP, Littmark U. The Stopping and range of ions in solids, vol. 1. Pergamon Press, New York, Oxford, Toronto, Sydney, Frankfurt, Tokio, 1985

[5] Glawischnig H, Parks CC. SIMS and modeling of ion implants into photoresist. In: Ishida E, Barusjee S, et al. (eds): Ion Implantation Technology - 96, IEEE, New York, 1997, pp. 579-582 (IEEE Proceedings, v. 579)

Dynamics of p^+ polysilicon gate depletion due to the formation of boron compounds in $TiSi_2$

F.G. Lau and W. Molzer
Infineon Technologies AG, CFE TD SIM, Otto-Hahn-Ring 6
81730 Munich, Germay
frank.lau@infineon.com, wolfgang.molzer@infineon.com

Abstract

In dual workfunction gate technologies it can be observed, that p^+ poly gates of pMOSFETs tend to lose boron doping. This work presents a model for the transport of Si and B in the $TiSi_2$/polysilicon bilayer system that can explain the saturation of the B dose loss.

1 Introduction

Many CMOS applications are driven by low power and high performance issues. Both demands can be met by the dual workfunction gate technology (p^+ poly gates for pMOSFETs and n^+ poly gates for nMOSFETs). It is observed, that p^+ poly gates of pMOSFETs tend to lose boron doping [1]. Depletion of gate degrades the performance of pMOSFETs. On the other hand for metallization titanium disilicide $TiSi_2$ was shown to have low sheet resistance and low contact resistance on p and n type silicon. The focus of this work is to provide insight to the boron transport in the $TiSi_2$/polysilicon bilayer system.

2 Experimental

Fig. 1 shows boron profiles from secondary ion mass spectrometry (SIMS) after anneals at 900 °C for 1, 10 and 60 minutes. Fig. 2 shows corresponding boron doses in polysilicon (from Fig. 1, thickness of the polysilicon layer: $0.170\mu m$) as a function of the anneal time. A pronounced dose loss is observed after 10min. For larger times the boron dose saturates.

3 Theory

The boron kinetics in $TiSi_2$ is dominated by the formation of titanium diborides TiB_2 [2], i.e. by

$$TiSi_2 + 2B \leftrightarrow TiB_2 + 2Si. \tag{1}$$

The corresponding law of mass action is

$$C_{TiSi_2}(C_B^t)^2 = K_1 C_{TiB_2}(C_I^t)^2, \tag{2}$$

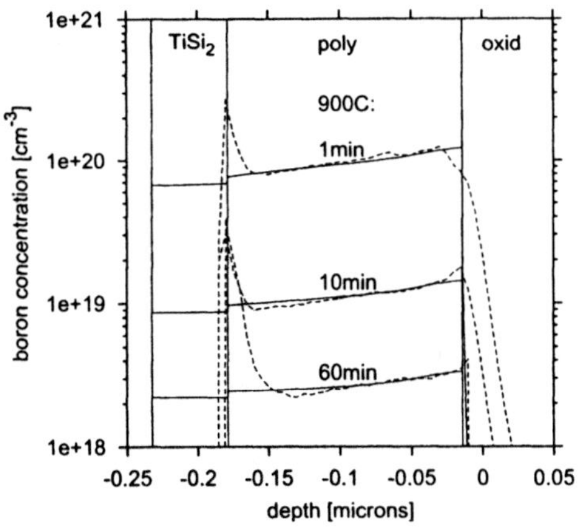

Figure 1: Boron distributions from SIMS (dashed lines) and from simulation (solid lines) after annealing at 900 °C for 1, 10 and 60 minutes.

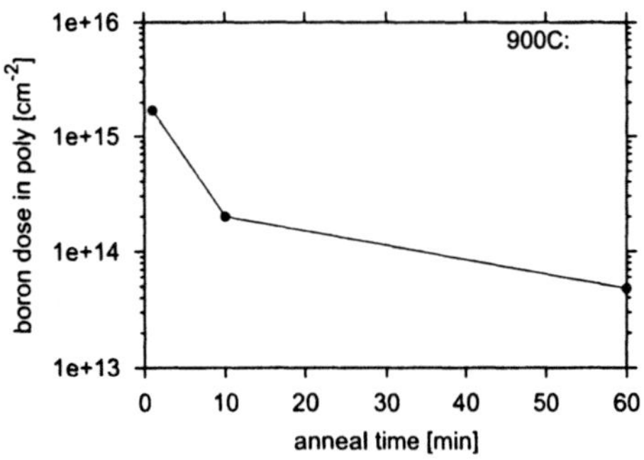

Figure 2: Boron dose in polysilicon (from Fig. 1) as a function of the anneal time at 900 °C.

where C_{TiSi_2}, C_B^t, C_{TiB_2} and C_I^t are the volume concentrations of $TiSi_2$, mobile B, TiB_2 and of mobile silicon atoms respectively. The superscript 't' indicates mobile species in the $TiSi_2$ matrix, K_1 is a mass action constant. The concentration of molecules in the host material C_{TiSi_2} is assumed to be constant ($K = C_{TiSi_2}/K_1$), leading to

$$C_{TiB_2} = K(C_B^t)^2/(C_I^t)^2. \tag{3}$$

The reaction (1) is assumed to be in equilibrium on the time scale of dopant diffusion and other reactions. Thus the concentrations C_B^t, C_{TiB_2} and C_I^t cannot change independently over time, but satisfy

$$\frac{\frac{d}{dt}C_{TiB_2}}{C_{TiB_2}} = \frac{\frac{d}{dt}K}{K} + 2\frac{\frac{d}{dt}C_B^t}{C_B^t} - 2\frac{\frac{d}{dt}C_I^t}{C_I^t}. \tag{4}$$

The first term shows up because K implicitly depends on time by the temperature dependence of the mass action constant. It can be neglected when only constant temperature processing is taken into account. As this applies to the present investigation the first term will be neglected further on. Boron and silicon can diffuse through the $TiSi_2$ material. The according continuity equations are

$$\tfrac{d}{dt}C_B^t = \nabla \cdot j_B^t - 2\tfrac{d}{dt}C_{TiB_2} \tag{5}$$
$$\tfrac{d}{dt}C_I^t = \nabla \cdot j_I^t + 2\tfrac{d}{dt}C_{TiB_2} - R_I^t. \tag{6}$$

The current densities are given by simple diffusion terms, i.e. $j_B^t = D_B^t \nabla C_B^t$, $j_I^t = D_I^t \nabla C_I^t$ where D_B^t, D_I^t are the diffusivities of boron and silicon in $TiSi_2$.

In analogy to point defect models in silicon substrate mobile silicon recombine to an equilibrium concentration $C_I^{eq,t}$. The relaxation rate is assumed to be $R_I^t =$

$(C_I^t - C_I^{eq,t})/\tau_I^t$ with relaxation time τ_I^t. Combining eq. (4)–(6) we obtain

$$\alpha_{11}\frac{d}{dt}C_B^t - \alpha_{12}\frac{d}{dt}C_I^t = \nabla \cdot j_B^t \quad (7)$$

$$\alpha_{21}\frac{d}{dt}C_I^t - \alpha_{22}\frac{d}{dt}C_B^t = \nabla \cdot j_I^t - R_I^t \quad (8)$$

with

$$\alpha_{11} = (1 + 4C_{TiB_2}/C_B^t) \qquad \alpha_{12} = 4C_{TiB_2}/C_I^t$$

$$\alpha_{21} = (1 + 4C_{TiB_2}/C_I^t) \qquad \alpha_{22} = 4C_{TiB_2}/C_B^t.$$

The boron transport in polysilicon is dominated by a two stream diffusion [3, 4]. Boron atoms diffuse in the grain interior C_B^{gi} and in grain boundaries C_B^{gb}. Diffusivity in grain boundaries is much faster than in grains. Dopants segregate between grain interior and grain boundaries with a rate G_s

$$\frac{d}{dt}C_B^{gi} = \nabla \cdot j_B^{gi} - G_s \quad (9)$$

$$\frac{d}{dt}C_B^{gb} = \nabla \cdot j_B^{gb} + G_s \quad (10)$$

$$G_s = t(C_{B.max}^{gb} - C_B^{gb})C_B^{gi} - e(C_{B.max}^{gi} - C_B^{gi})C_B^{gb}. \quad (11)$$

The trapping rate (kinetic coefficient t) of boron from grain interior to grain boundaries is proportional to free states on grain boundaries $(C_{B.max}^{gb} - C_B^{gb})$ and to filled states in grain interior C_B^{gi}. The emission rate (kinetic coefficient e) in the opposite direction is proportional to free states on grain interior $(C_{B.max}^{gi} - C_B^{gi})$ and to filled states in grain boundaries C_B^{gb}. $C_{B.max}^{gb}$ is the maximum boron concentration in grain boundaries and depends on the local grain size. $C_{B.max}^{gi}$ is the maximum boron concentration in grain interior and is given by the solid solubilty.

Usually polysilicon diffusion models do not include point defect kinetics. However the grain boundaries are known to absorbe interstitials very effectively [5]. From the reaction (1) it can be expected that much mobile silicon is generated during the formation of TiB_2. These silicon atoms enter the polysilicon as interstitials and recombine at grain boundaries. To account for this the polysilicon diffusion model was extended by a simple continuity equation for interstitials

$$\frac{d}{dt}C_I^p = \nabla \cdot j_I^p - R_I^p. \quad (12)$$

C_I^p is the concentration of intersitials in polysilicon. The recombination rate in polysilicon is $R_I^p = (C_I^p - C_I^{eq,p})/\tau_I^p$.

4 Results

The above equations have been implemented in PROMIS-NT [6]. Although some model parameters in $TiSi_2$ and polysilicon are not well known the proposed model provides insight into the underlying physical mechanisms:

At the beginning of the process B diffuses into $TiSi_2$ and starts reaction (1). The generated mobile silicon recombines in $TiSi_2$ (rate R_I^t in eq. (8)), outdiffuses into the polysilicon and recombines here too (rate R_I^p in eq. (12), Fig. 3). When the silicon concentration in $TiSi_2$ approaches its equilibrium value, Fig. 4, reaction (1) stops. Consequently the outdiffusion of B into $TiSi_2$ also stops. In effect the presented model for the $TiSi_2$/polysilicon bilayer system can explain the saturation of the B dose loss, Fig. 2.

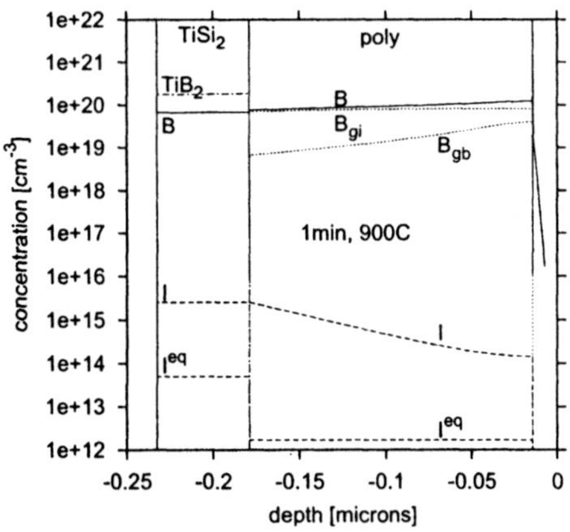

Figure 3: Distribution of solved model species boron (labeled as B in $TiSi_2$ and as B_{gb}, B_{gi}, B in poly), mobile silicon in $TiSi_2$ and silicon interstitials in poly (both labeled as I) after a 1 minute anneal at 900 °C. TiB_2 results from eq. (3).

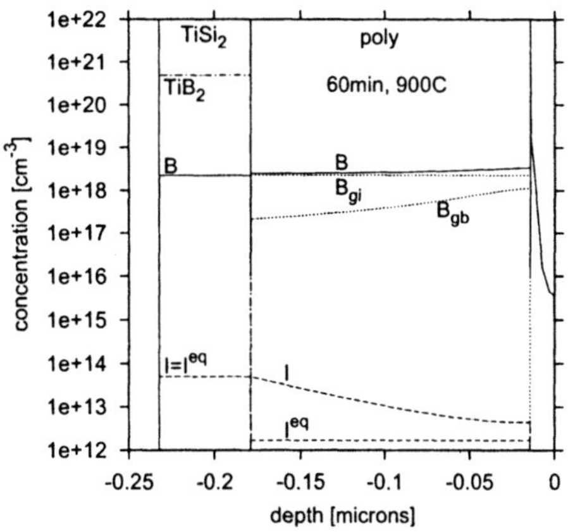

Figure 4: Distribution of model species after a 60 minute anneal at 900 °C.

References

[1] A. Kalnitsky, N. Brun, A. Brun, and J.-P. Gonchond. $TiSi_2$ integration in a submicron cmos process. ii. integration issues. *J. Electrochem. Soc.*, 142(6):1992–1996, 1995.

[2] V. Probst, H. Schaber, A. Mitwalsky, H. Kabza, B. Hoffmann, K. Maex, and L. van den Hove. Metal-dopant-compound formation in $TiSi_2$ and $TaSi_2$: Impact on dopant diffusion and contact resistance. *J. Appl. Phys.*, 70(2):693–707, July 1991.

[3] F. Lau. Modeling of polysilicon diffusion sources. In *International Electron Devices Meeting, Technical Digest*, pages 737–740, December 1990.

[4] F. Lau. Modeling of polysilicon diffusion sources during rapid optical annealing. *Applied Physics A (Solids and Surfaces)*, 54(2):139–146, 1992.

[5] S. Batra, K.H. Park, S.K. Banerjee, G.E. Lux, C.L. Kirschbaum, J.C. Norberg, T.C. Smith, J.K. Elliott, and B.J. Mulvaney. Effect of grain microstructure on p diffusion in polycrystalline-on-single crystal silicon systems. *Journal of Electronic Materials*, 21(2):227–231, 1992.

[6] R. Mlekus and S. Selberherr. Object-oriented algorithm and model management. In M.H. Hamza, editor, *Proceedings of the IASTED International Conference on Applied Modelling and Simulation*, pages 437–441, 1998.

Analysis of Statistical Fluctuations due to Line Edge Roughness in sub-0.1μm MOSFETs

S. Kaya[†], A. R. Brown, A. Asenov, D. Magot and T. Linton[‡]
Device Modelling Group, University of Glasgow, G12 8LT, Scotland, UK
[†] SEECS, Ohio University, Athens, OH 45701,USA, e-mail:kaya@ohio.edu
[‡] TCAD Division, Intel Corp., Santa Clara, CA 95052-8119, USA

Abstract

We present a full-3D statistical analysis of line edge roughness (LER) in sub-0.1 μm MOSFETs. The modelling approach for line edges and the parameters used in the analysis take into account the statistical nature of the roughness. The results indicate that intrinsic fluctuations in MOSFETs due to LER become comparable in size to random dopant effects and can seriously inhibit scaling below 50 nm.

1. Introduction

The intrinsic fluctuations in sub-0.1μm devices are normally associated with random dopant effects [1,2] and oxide thickness fluctuations [3]. However, as a random departure from the perfect geometry due to tolerances found in *every* lithography process, line edge roughness (LER) also introduces such intrinsic parameter fluctuations. Earlier works investigating the impact of LER in device simulations have been limited in terms of realism and sophistication. They all approximate LER in a 'square-wave' fashion in deterministic simulations [4,5] or employ faster 2D device simulations in a simplistic statistical approach [6]. In this work, we present a realistic 3D statistical approach to the simulation of LER in decanano MOSFETs. In the simulations LER is specified in terms of statistical parameters, *i.e.* rms amplitude (Δ) and correlation length (Λ). This allows both 3D and statistical aspects of LER to be naturally incorporated in a single simulation framework. We demonstrate the impact of LER in MOSFET scaling and show that it can further limit the integration of sub-0.1 μm MOSFETs, unless reduced significantly both in amplitude and correlation length.

2. LER Description and Model

Total LER amplitude is traditionally defined to be 3Δ, where the rms amplitude Δ of LER can be statistically obtained by inspection of lines generated by a given lithography process. Although not a new phenomenon, LER is rarely accounted for in device analysis, since the critical dimensions are at least an order of magnitude larger than the roughness in present day devices. However, as the aggressive scaling of Si-MOSFETs continues, LER of fixed amplitude may constitute an increasingly significant fraction of the gate length. Indeed, at the end of the Roadmap proposed by SIA [7], MOSFETs as small as 30 nm or less are anticipated. Data collected from different processes, summarised in Fig.1 attest to, at present, a minimum total LER limit of 5 to 6 nm (i.e. $\Delta \approx 2$ nm). This value, larger than the Roadmap requirement for devices below 100 nm, is alarming since metrology requirements are often more difficult to meet than actual line-width specifications for a given process.

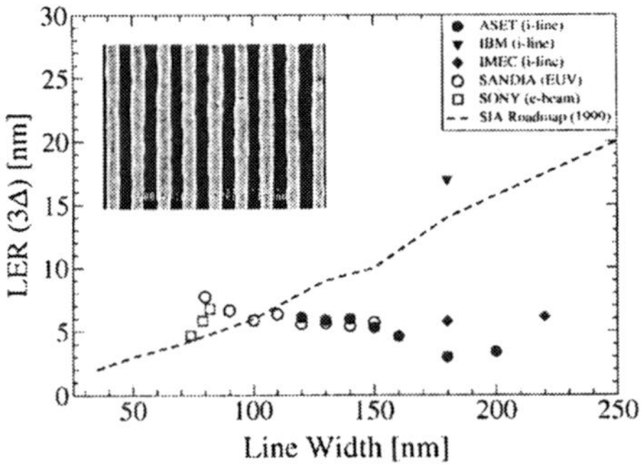

Fig.1. LER found in advanced lithography processes by various labs [6-12] and required by SIA roadmap (SIA) The inset shows LER found in sub-100 nm e-beam generated lines.

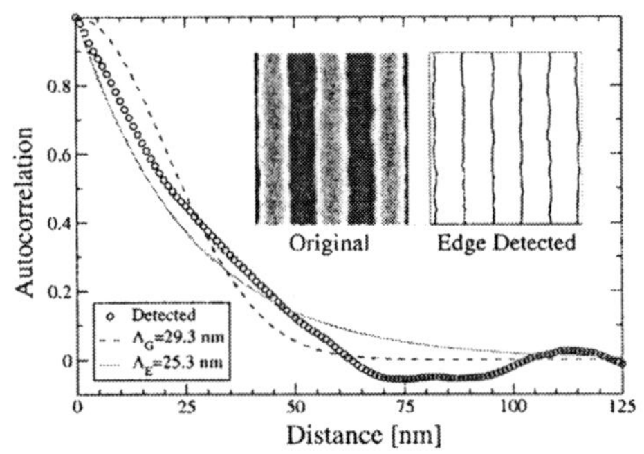

Fig.2. Autocorrelation of LER captured from 100 nm EUV lines. Correlation lengths obtained from Gaussian (Λ_G) and exponential (Λ_E) fits and detected edges are also given.

In contrast with rms values, significantly less is known regarding the correlation length of LER, which is reported to vary between 10 nm and 50 nm [7]. Our recent analysis on actual LER data obtained from EUV [8] and e-beam [9] lithography processes (see Fig. 2) indicate that typically Λ is approximately 30 nm. It is reasonable to assume that Λ may be reduced at higher resolution, which can utilise special resists and/or advanced exposure techniques for better LER performance. Gaussian and exponential models are found to perform equally well as least square fits to the captured autocorrelation data.

With reliable statistical data at hand, we reconstruct realistic source/drain junctions in 3D MOSFET simulations using a 1D Fourier synthesis approach. First a complex array with N elements is constructed, whose amplitude is determined by the power spectrum obtained from the adopted autocorrelation function. The phases of the elements are selected randomly, making each line unique. However, only (N/2 –2) elements of the array are independent. The rest are obtained by symmetry operations imposed so that, after inverse Fourier transform, the resulting height function $H(x)$ is real. Random line examples generated using this method are given in Fig.3 for typical values of Δ and Λ. Lines with a Gaussian autocorrelation are smoother due to a lack of high frequency components in this model, as also can be seen in Fig.3.

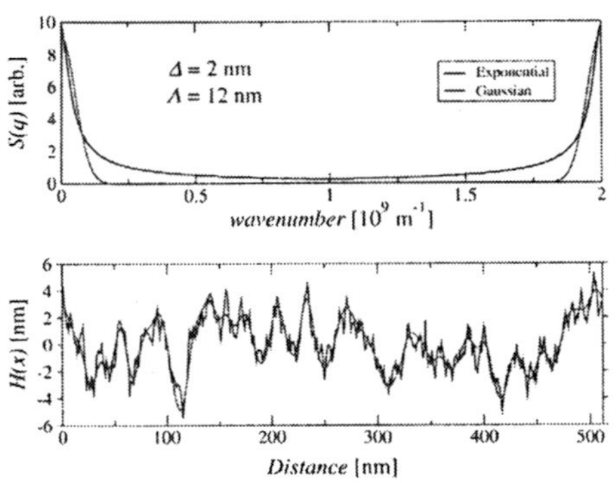

Fig.3. LER model used in the 3D device simulator. Both the power spectra (top) and actual random lines (bottom) are shown.

Fig.4. A typical 50×200 nm MOSFET used in 3D simulations. LER has parameters Λ=20 nm and Δ=3 nm.

3. Device Simulator

We employ the same 3D device simulator as in an earlier work investigating random dopant and oxide thickness fluctuation effects [2,3]. The LER modelling unit was incorporated within this to generate random junction patterns, without any numerical penalties owing to efficient FFT subroutines used in line generation. Throughout this work, ensembles of 200 MOSFETs with identical design parameters but unique junction patterns, such as the one in Fig.4, are simulated for statistical analysis. Unless indicated otherwise, Δ=2 nm and Λ=20 nm for all cases considered.

For the sake of simplicity and speed we use the drift-diffusion approximation, with constant mobility [2], and neither the quantum mechanical nor the atomistic doping options are enabled during simulations. A set of 200 devices takes 0.5–4 days to run on a single processor depending on the bias conditions, although in practice several processors are used in parallel. This is in marked comparison with 3 months reported for 70 MOSFETs by Oldiges *et al.* [6]. Using this efficient simulator, we focus on the variation of threshold voltage, V_T, leakage current, I_{off}, and drive current, I_{on}. Devices considered here comply with scaling requirements at each technology node and assume a Gaussian autocorrelation for LER used to construct the junctions.

4. Results and Discussion

Complementary to threshold fluctuations, LER causes fluctuations in drain on-state current, I_{on}, as can be seen in Fig.5. It is interesting to note that current distributions disclose a slight skew, which may be attributed to increased short channel effects in shorter elements of the ensemble. We verified that largest (smallest) I_{on} value in the distributions correspond to the device with the shortest (longest) effective gate length. For given device dimensions the standard deviation of threshold voltage fluctuations due to LER increase when Λ or Δ is increased. The latter dependence is given in Fig.6 for 30 and 50 nm MOSFETs. The inset in this figure plots the average threshold voltage lowering ($\langle V_T\rangle - V_{T0}$), where V_{T0} is calculated without LER. Thus, LER not only causes fluctuations but also lowers the threshold and its impact on threshold becomes larger as gate dimensions are reduced. Moreover, the fluctuations are comparable in magnitude to those resulting from random dopants in similar 30x30 nm devices studied by Asenov [2].

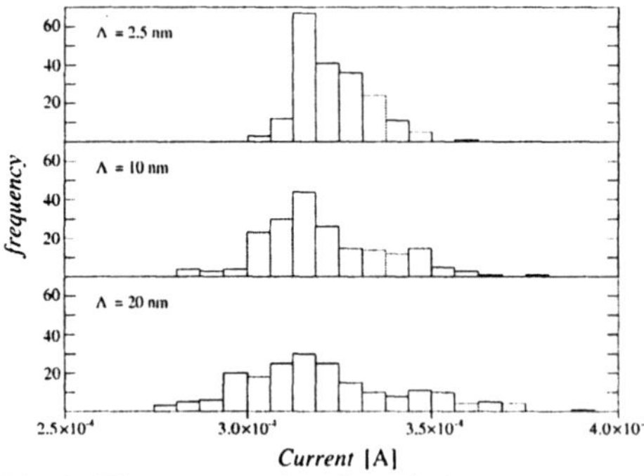

Fig.5. Histograms showing drain on-current distribution in 200 (30x30 nm) MOSFETs for three correlation lengths. The skew is due to short channel effects.

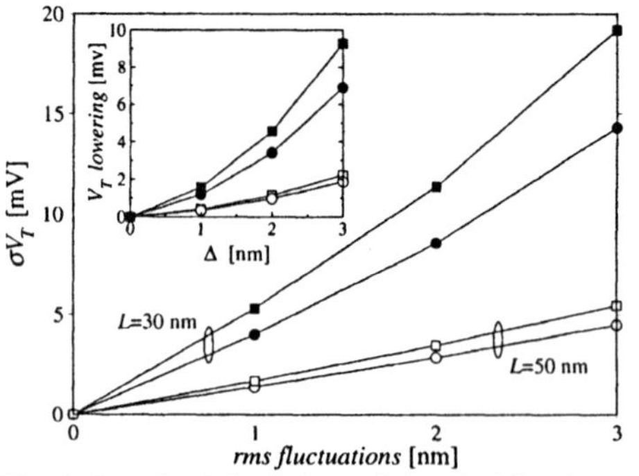

Fig.6. Standard deviation of threshold voltage for two decanano MOSFETs as a function of rms fluctuations at V_D=1.0V (squares) and V_D=0.1V (circles).

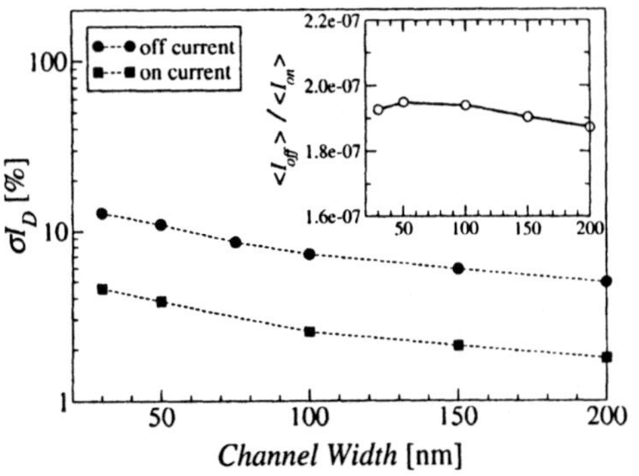

Fig.7. σI_D as a function of effective channel width in decanano MOSFETs. The inset shows ratio of average off- and on-currents. V_D=1.0 V, L_{eff}=50 nm, Δ=2 nm and Λ=20 nm.

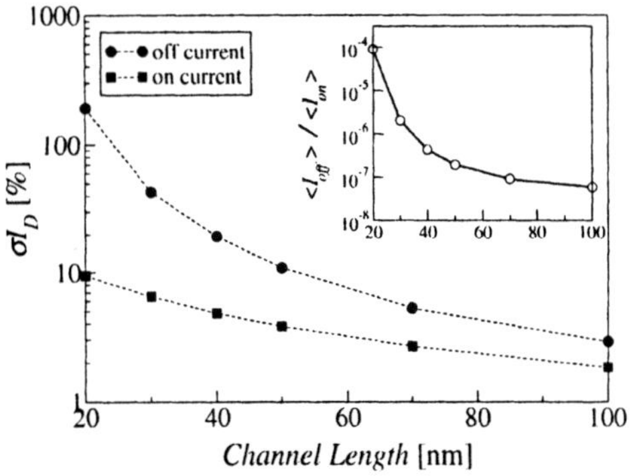

Fig.8. Dependence of drain current standard deviation on the effective gate length in decanano MOSFETs. The inset shows ratio of average off- and on-currents.

Fig.7 and Fig.8 display the dependence of the standard deviation of drain current on MOSFET gate width, W_{eff}, and length, L_{eff}, respectively for typical LER parameters. Both the on- and off-current dependencies are given for comparison. Clearly the W_{eff} dependence is an order of magnitude weaker than the L_{eff} dependence. Wider devices provide a progressively better average for the properties affected by the LER, thus reducing the uncertainty. The disproportionately large L_{eff} dependence is due to acute short channel effects in regions with drastic reduction in the local channel length introduced by LER. The simulated dependence becomes especially severe for devices with a gate length below 50 nm. Naturally in all cases off-current dependence is found to be more susceptible to the presence of LER.

The insets of Figs. 7 & 8 plot the ratio of average off-current to on-current. Since the average on-current remains relatively constant (see Fig.5), the ratio essentially provides a measure for the increased LER induced leakage. As in the case of standard deviation, the leakage current appears to be extremely sensitive to LER in MOSFETs with a gate length less than 50 nm. The width dependence of the same figure of merit is not significant, remaining below 5% overall. Therefore, both LER amplitude and correlation length must be reduced below the current projected minima to prevent excess leakage and fluctuations in MOSFETs with gate dimensions below 50 nm.

References

[1] Stolk, P A *et al.*, *IEEE Trans. Elec. Dev.*, **45**, 1960, (1998)
[2] Asenov, A, *IEEE Trans. Elec. Dev.*, **45**, 2505, (1998)
[3] Asenov, A and Kaya, S, *Proc. SISPAD*, Seattle, USA, 135, (2000)
[4] Linton, T, Giles, M and Packan, P, *Silicon Nanoelectr. Workshop*, 82, (1998)
[5] Linton, T, Yu, S and Shaheed, R, *VLSI Design*, in press.
[6] Oldiges, P *et al.*, *Proc. SISPAD*, Seattle, USA, 131, (2000)
[7] SIA, SIA Roadmap–Lithography, 147, (1999)
[8] Cardinale, G *et al.*, *J. Vac. Sci. Tech. B*, **17**, 2970, (1999)
[9] Thoms, S, Macintyre, D and McCarthy, M, *Microelec. Eng.*, **41/42**, 207, (1998)
[10] Mori, S *et al.*, *J. Vac. Sci. Technol. B*, **16**, 3739. (1998)
[11] Winkelmeier S *et al.*, *Proc. SPIE*, in press.
[12] Yoshizawa, M and Moriya, S, *Electr. Lett.*, **36**, 90, (2000)

Quantum Corrections in 3-D Drift Diffusion Simulations of Decanano MOSFETs Using an Effective Potential

J. R. Watling[1], A. R. Brown[1], A. Asenov[1] and D. K. Ferry[2]

[1] University of Glasgow, Glasgow, G12 8LT, Scotland

[2]Arizona State University, Tempe, AZ 85287-5706, USA

Abstract

As MOSFET devices are aggressively scaled into the deep submicron regime quantum mechanical effects become increasingly important. We compare the recently proposed effective potential formalism with the density gradient approach for first order quantum simulations of sub 0.1μm MOSFETs within a modified drift diffusion framework.

1. Introduction

MOSFETs scaled down to 30nm gate lengths have been successfully demonstrated [1]. Further evidence [2] indicates that double gate MOSFET architectures will remain operational down to gate lengths of 10nm and below. However, the combination of thin gate oxides and heavy doping in conventional devices, and the thin silicon body of the double-gate structures, will result in substantial quantum mechanical (QM) threshold voltage shift and transconductance degradation. Computationally efficient methods to include QM effects are required for the purpose of practical Computer Aided Design. First order quantum corrections based on density gradients (DG) have already been introduced in 2-D [3] and 3-D [4] drift-diffusion simulations. Recently a new *Effective Potential* (EP) approach for introducing quantum corrections in classical and semi-classical simulations has been proposed [5] and demonstrated in Monte Carlo MOSFET simulations [6]. This paper investigates the quantum influence on threshold voltage, carrier density profile and I_D-V_G current characteristics within a modified drift diffusion framework. The first section describes the new effective potential formalism. A general and efficient algorithm for including quantum corrections within the self-consistent drift diffusion framework is described in section 3. Results from the new effective potential algorithm are compared with those from the well established density gradient approach in section 4, while section 5 presents our conclusions and discussions.

2. The effective potential formalism

An alternative approach to density gradient for including first order quantum effects is the recently advanced effective potential approach [6] in which, unlike with DG, the electron equation state is not modified directly. Instead the *natural* non-zero size of an electronic wave-packet in the quantised system is used to construct an effective potential. The carriers are considered to be associated with a minimum dispersion Gaussian wavepacket, which is in turn convolved with the classical conduction band profile $V_{Classical}$ (obtained from the solution of Poisson's equation) to obtain an effective potential V_{eff} given by:

$$V_{eff} = \int V_{Classical}(\mathbf{x}+\mathbf{y})\, G(\mathbf{y},a_0)\, d\mathbf{y} \tag{1}$$

where G is a Gaussian with standard deviation a_0.

3. Implementation of quantum corrections within the drift diffusion framework

We use the Gummel algorithm for solving the semiconductor equations in the drift diffusion approximation, introducing quantum corrections to the electron density using either the EP or DG formalism. We have compared two techniques for including the quantum corrections within the standard Gummel algorithm: In the first scenario the EP or DG solution is calculated within each Newton iteration of the non-linear Poisson equation to calculate the electron concentration. The current continuity equation is then solved as normal. However, this approach is computationally expensive. The second scenario resembles the standard procedure used to self-consistently solve the Poisson and the Schrödinger equations, using a self-adjusting damping scheme [7], for the potential. This is motivated by the fact that the DG equation is a partial non-linear equation which closely resembles Schrödinger's equation. In this scheme the quantum correction to the electron concentration is calculated only once per Gummel iteration, either via the effective potential or via a solution of the density gradient equation of state. This is a far less computationally demanding approach, converging in less time over a wide range of bias conditions.

4. Results

We have carefully calibrated both the EP and DG approaches against the results of a 1-D Poisson-Schrödinger solver [8]. Although Poisson-Schrödinger simulations are more sophisticated they are not yet practical for 3-D device simulations. Fig. 1 shows the quantum mechanical threshold voltage shift for DG and EP as a function of substrate doping compared with the results of Jallepalli. Fig. 2 shows typical carrier concentration profiles obtained from the 1-D simulations. All show a peak in the concentration away from the Si/SiO_2 interface, although the effective potential produces a sharper drop-off at the Si/SiO_2 interface.

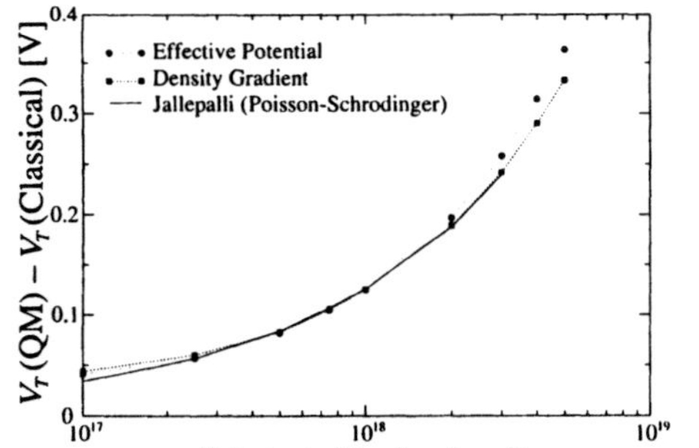

Fig. 1. Threshold voltage shift due to quantum effects versus substrate doping. Results for Density Gradient and Effective Potential are compared to those obtained from Poisson Schrödinger.

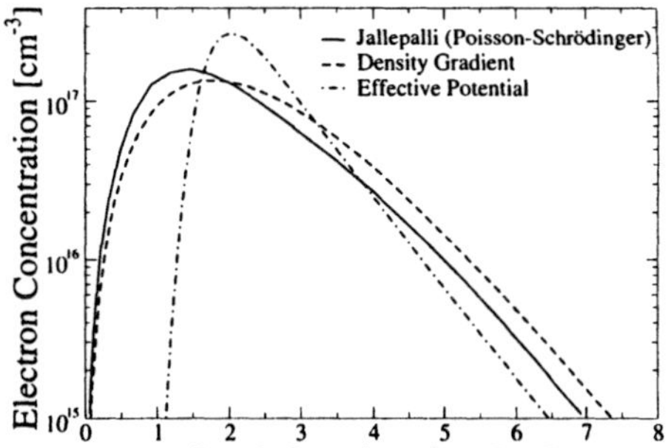

Fig. 2. Electron carrier concentration as a function of distance from the interface, for substrate doping of $5\times10^{17}cm^{-3}$. All have the same net sheet density.

Fig. 3, shows an I_D-V_G characteristic for a 30nm×30nm n-MOSFET obtained from our 3-D quantum simulator. This demonstrates the threshold voltage shift between the classical and the quantum simulations. We have also investigated the threshold voltage shift as a function of channel length, shown in Fig. 4. The variation with channel length is minimal at least down to 30nm, indicating the well scaled nature of our device structure.

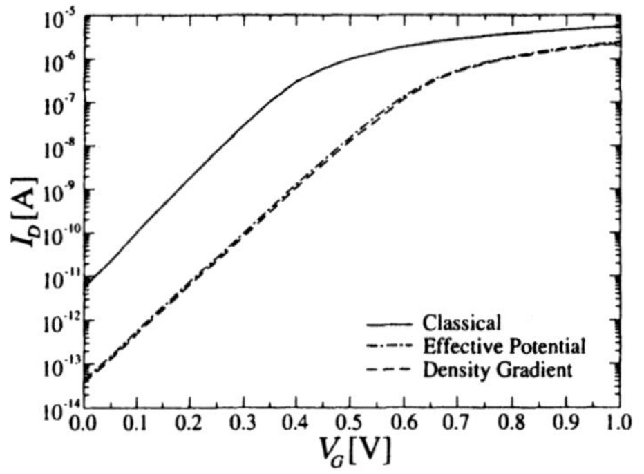

Fig. 3. I_D-V_G characteristic obtained from both classical and quantum simulators for a 30nm×30nm n-MOSFET, with V_D=0.01V and a substrate doping of 5×10^{18}cm^{-3}.

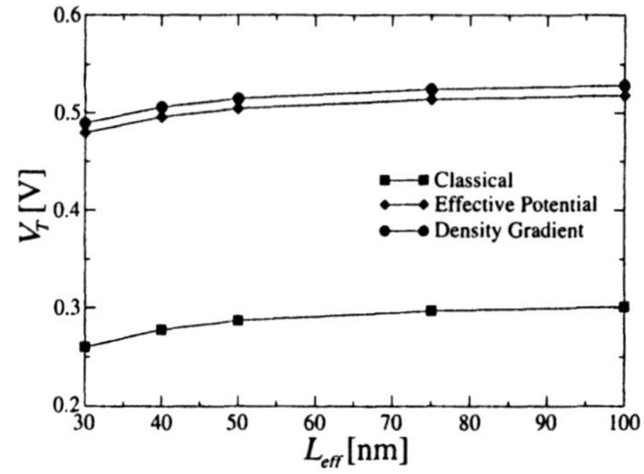

Fig. 4. Dependence of the threshold voltage on the channel length in MOSFETs with W_{eff}=50nm, $N_A=5\times10^{18}$cm^{-3} and t_{ox}=1.3nm. illustrating the quantum mechanical shift in threshold voltage.

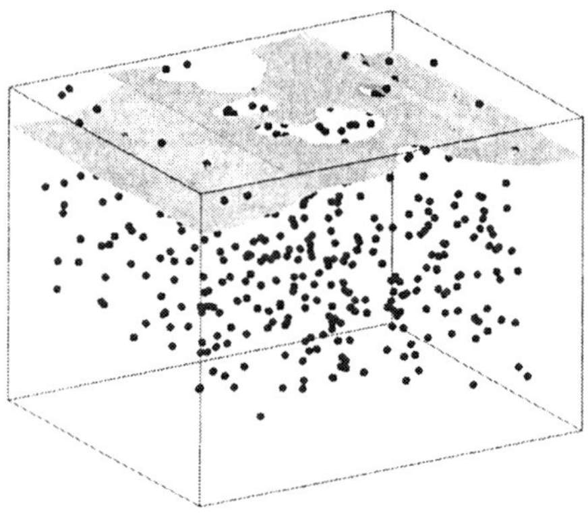

Fig. 5. An equi-concentration contour for a 30×50nm atomistic MOSFET, at threshold obtained using our effective potential simulator. Also shown are the individual acceptor positions throughout the structure.

Table 1. Results of atomistic simulations using classical, DG and EP methods

	Average threshold voltage $<V_T>$			Threshold voltage standard deviation σV_T		
Channel Length	Classical	Density Gradient	Effective Potential	Classical	Density Gradient	Effective Potential
30nm	0.179 V	0.376 V	0.370 V	28.2 mV	39.8 mV	40.6 mV
50nm	0.219 V	0.423 V	0.417 V	22.1 mV	30.8 mV	31.6 mV

We have also performed 3-D atomistic simulations, to compare the influence of EP on threshold voltage fluctuations as compared with DG [9]. Figure 5, shows a typical equi-concentration contour for a simulated atomistic MOSFET. Simulating 200 atomistic devices we have investigated threshold voltage and standard deviation at different channel lengths (Table 1). The EP simulations result in similar average threshold voltages and standard deviations compared to DG.

5. Conclusions

We have included both Density Gradient and Effective Potential quantum corrections within a self-consistent 3-D drift diffusion simulation, which includes a full solution of the current continuity equation. Both methods agree well with the available data from Poisson-Schrödinger simulations, although there is a better agreement between Density Gradient and Poisson-Schrödinger calculations in respect of the carrier densities. This, however, appears to have little discernable effect on threshold voltage and current characteristics. For the simulations presented here we have found that the Density Gradient approach is computationally more efficient due to the numerically intensive nature of the 3-D convolution involved in calculating the effective-potential (V_{eff}). However, it is difficult to make a definitive judgement on which is the most suitable method for including first order quantum effects within device simulations. An important next step is to investigate to what extent, if any, the two methods can cope with source-to-drain tunnelling in decanano MOSFETs.

References

[1] Chau, R., Kavalieros, J., Roberds, B., Schenker, R., Lionberger, D., Barlage, D., Doyle, B., Arghavani, R., Murthy, A. and Dewey, G., 30nm Physical Gate Length CMOS Transistors with 1.0 ps n-MOS and 1.7 ps p-MOS Gate Delays, IEDM Tech. Digest, 45-48, 2000

[2] Ren, Z., Venugopal, R., Datta, S., Lundstrom, M., Jovanovic, D. and Fossum J., The Ballistic Nanotransistor: A Simulation Study, IEDM Tech. Digest, 715-718, 2000

[3] Rafferty, C. S., Biegel, B., Yu, Z., Ancona, M. G., Bude, J. and Dutton, R. W., Multi-dimensional quantum effects simulation using a density gradient model and script-level programming technique, SISPAD'98, Eds De Meyer K, and Biesemans, 137-140, 1998

[4] Asenov, A., Slavcheva, G., Brown, A. R., Davies, J. H. and Saini, S., Quantum Mechanical Enhancement of the Random Dopant Induced Threshold Voltage Flucations and Lowering in Sub 0.1μm MOSFETs, IEDM Tech. Digest, 535-538, 1999

[5] Shifren, L., Akis, R. and Ferry, D. K., Correspondence between quantum and classical motion: comparing Bohmian mechanics with a smoothed effective potential approach, Physics Letters A, 274: 75-83, 2000

[6] Ferry, D. K., Akis, R. and Vasileska, D., Quantum Effects in MOSFETs: Use of an Effective Potential in 3D Monte Carlo Simulation of Ultra-Short Channel Devices, IEDM Tech. Digest, 287-290, 2000

[7] Stern, F., Iteration Methods for Calculating Self-Consistent Fields in Semiconductor Inversion Layers, Journal of Computational Physics, 6: 56-67, 1970

[8] Jallepalli, S., Bude, J., Shih, W. K., Pinto, M. R., Maziar, C. M. and Tasch Jr., A. F., Electron and hole quantization and their impact on deep submicron *p*- and *n*- MOSFET characteristics, IEEE Trans. Electron Devices, 44: 297-303, 1997

[9] Asenov, A., Slavcheva, G., Brown, A. R., Davies, J. H. and Saini, S., Increase in the Random Dopant Induced Threshold Fluctuations and Lowering in Sub-100nm MOSFETs Due to Quantum Effects: A 3-D Density Gradient Simulation Study, IEEE Trans. Electron Devices, 48: 722-729, 2001

Finite Element Simulation of 2D Quantum Effects in Ultra Short Channel MOSFETs with High-K Dielectric Gates

A. Poncet, B. Vergnet and M. Mouis
Laboratoire de Physique de la Matière, UMR-CNRS n° 5511, INSA-Lyon
20, Avenue Albert Einstein, F-69621 Villeurbanne, France
phone : (+33)4 72 43 87 30. e-mail poncet@lpm.insa-lyon.fr

Abstract - Ultra short channel MOS transistors with high permittivity gate dielectrics suffer from carrier quantum confinement in the channel and from electric field fringing in the dielectric layer. This paper deals with the 2D numerical simulation of the coupling between these two physical mechanisms. We shall demonstrate how they impact the threshold parameters in MOS transistors for metallurgical channel lengths ranging from 5 to 50 nanometers. A special attention has been paid to the control on numerical errors.

1 - Introduction

With the ultimate thinning of gate silicon dioxide layers, tunneling currents are becoming so large that going further in scaling-down requires new dielectric materials. HfO_2 for instance, increases the dielectric constant by one decade, and solutions are under investigation to get one decade more. These solutions will operate with channel lengths in the deca-nanometer range, leading to specific 2D effects, due to electric field fringing at gate edges which causes dramatic degradation of threshold voltages (V_t) and sub-threshold slopes (S) [3].

At the same time, quantum confinement in the channel may no longer be neglected. Carrier quantification has been extensively analyzed by means of numerical solutions of coupled Poisson and Schrödinger equations in 1D, in order to predict the impact of technological parameters on voltage-capacitance responses of *large* devices [1]. This has seldom been done for ultra-short gate MOSFETs [2], although device scaling rules would require a careful analysis of the couplings between quantum mechanical effects (QME) and short channel effects (SCE).

2- Numerical techniques

2-1 Finite elements

In this work, finite elements have been used to solve both Poisson and Schrödinger equations in 2D. This method is indeed especially well suited to analyze complex topographies. In addition, the Galerkin formulation of the equations makes the expression of most boundary conditions very straightforward. Vanishing probability current define boundary conditions along symmetry axes and ohmic contacts (unless carrier transport is not considered) which do not affect the hermitian nature of the Hamiltonian matrix. Transmission conditions, which are used to simulate wave function penetration into the gate dielectric, behave similarly .

Piecewise bilinear shape functions on quadrilateral elements were used in this work in order to highlight in a simple way the impact of mesh sizes near the silicon/dielectric interface. Meshes were graded on both sides perpendicularly to the interface, and constant in the lateral direction. Finite elements are worth being used with general purpose mesh generators, but we did not want to obscure the discussion by questions arising from local refinements or angle criteria.

2-2 Eigensystem

With finite elements, one has to face a 2-matrices eigenvalue problem, in which both matrices are equally sparse, unless the right-hand side matrix has been diagonalized by using a "mass-lumping" technique. Unfortunately, going further on, i.e. mapping this matrix to identity would break the symmetry of the Hamiltonian matrix. However, these are minor problems in comparison with non-sparse matrices arising from alternative methods based on Fourier series[5].

Simulating MOS devices in 2D brings another difficulty. As eigenvalues that are relevant to carrier quantification in the channel are fully mixed with those relative to source/drain areas [4], in contrast with the 1D case, it is necessary to sum up carrier densities over a large number of energy levels. This consumes both CPU time and storage space and it is mandatory to restrict the integration domain to its minimum.

2-3 Decoupling scheme

For solving Poisson and Schrödinger equations in a consistent way, fast decoupling schemes are required. Methods proposed in the literature are either slow (under-relaxation [5]) or difficult to implement (regula falsi [6]). Here, we implemented a new and very simple iterative scheme, which proved to be extremely fast : Once carrier densities have been deduced from eigen-energies and wave-functions, a "Fermi-like" potential is introduced and kept constant during Newton-Raphson iterations on Poisson's equation. This potential is calculated using a Boltzmann statistics, although it has no physical meaning in the given situation : this should only be seen as a simple mathematical trick to benefit from the fast convergence of Newton methods. This does not prevent the final solution to obey Fermi-Dirac statistics, with the density of states relevant to the given degree of confinement. As depicted in figure 1 for a 1D NMOS capacitor, convergence is extremely fast at any regime. Convergence remains good in 2D, so that simulations with up to 1500 mesh nodes was found affordable, even on a desktop computer .

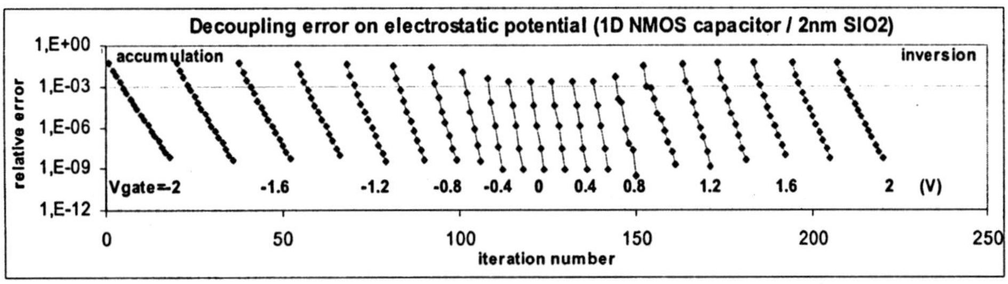

Figure 1 – Convergence of the decoupling scheme. 1D NMOS capacitor, with holes and electrons computation, from accumulation to inversion. Each symbol depicts one iteration.

2-4 Meshing constraints

The impact of the basic meshing parameters has been analyzed, with both classical and quantum models. Results can be summarized as follows :

- The impact of transverse mesh size in Si at the Si/dielectric interface is far higher when quantum effects are ignored (fig. 2). This is especially dramatic on capacitance values (fig. 2-b), and this can be related to the smoothing of carrier concentrations induced by their confinement in the vicinity of the interface.

- In short channel transistors, classical solutions are weakly sensitive to the lateral mesh size along the channel, but surprisingly, the quantum solutions are even less sensitive, especially with high-K dielectrics (fig. 3).

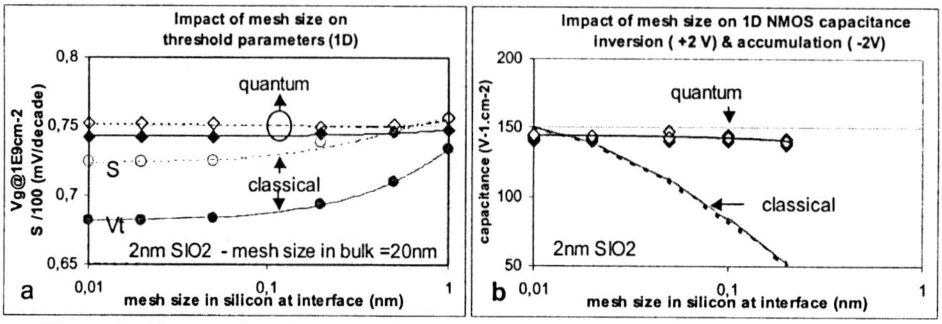

Figure 2 – Impact of *mesh size in silicon*, perpendicular to the channel at Si/SiO2 interface 1D structure - $10^{18}cm^{-3}$ channel doping - 2nm SiO2 **(a)** on S and V_t (estimated at 10^9cm^{-2} in the channel) **(b)** on capacitance in accumulation (-2V) and inversion (+2V)

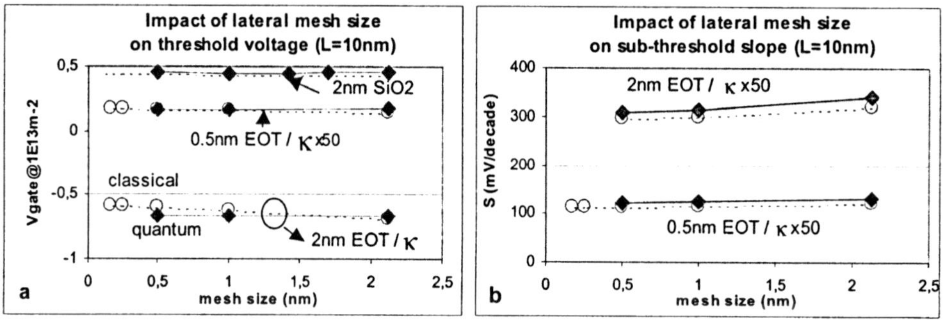

Figure 3 – Impact of the *lateral* mesh size - L_m=10nm , EOT=2nm.

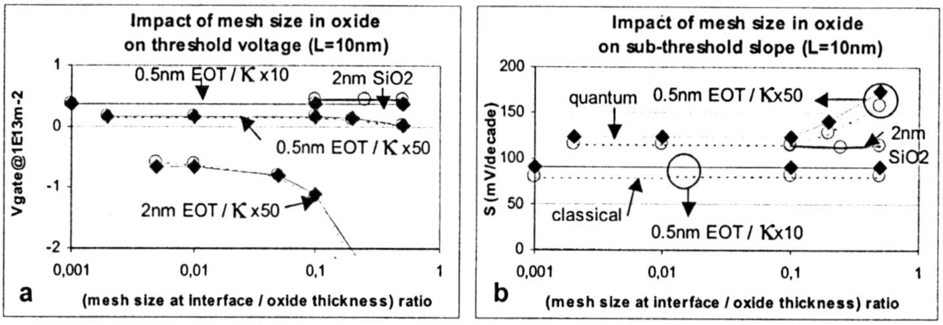

Figure 4 – Impact of the transverse *mesh size in the oxide* at Si/ SiO_2 interface, L_m= 10nm.

- The impact of the size of the simulated source/drain areas is weak as well (not shown here). It means that neglecting high frequency wave functions in the source or drain regions (because of a coarse lateral mesh) has a limited impact on carrier profiles in the channel and that the domain in which Schrödinger equation is to be solved can be shrunk without too much damage (keeping only 10nm apart from the ohmic contact edges has been proved to be sufficient).

- Using a too large transverse mesh size in the dielectric leads to over-estimate the degradation of V_t and S induced by a large thickness-to-length ratio (fig. 4).

3- Influence of channel length with high-K gates

The couplings between SCE and QME were analysed on NMOS transistors with metallurgical channel lengths (L_m) ranging from 5 to 50 nm and equivalent oxide thicknesses (EOT) from 0.5 to 2 nm. Dielectric constants were 1, 10 or 50 times larger than SiO_2 one. Doping levels were 10^{18}, 2.10^{19} and $10^{21}cm^{-2}$ in the channel, in the extensions and in the ohmic areas respectively. Simulations were performed with a fixed set of mesh size values : 0.5 nm lateral mesh size and graded transverse mesh with sizes ranging, in silicon, from 0.1nm at the interface to 50nm at bulk contact and, in the dielectric layer, from 1/100 to 1/4 of the layer thickness.

Our results agreed with simulations reported elsewhere [3], to predict large degradations of threshold parameters with increased permittivities. The major new information brought by our 2D simulations is that QME enhance SCE as follows :

Vt is underestimated by non-quantum models in long channel transistors, wheras it is overestimated for L_m<10 nm (fig. 5-a). On the other hand, quantum effects enhance the sub-threshold slope S (fig. 5-b).

With materials such as HfO_2 ($\kappa \simeq 30$), high permittivity and 2D carrier confinement have similar quantitative impacts on SCE; therefore, simulating the former while ignoring the later would be meaningless. Moreover, in contrast with the 1D case where QME reduce the carrier density at any regime, from accumulation to inversion (doted line in fig. 6), the charge in ultra-short channels is only reduced in a narrow range of gate biases, from V_t to the weak-to-strong inversion transition.

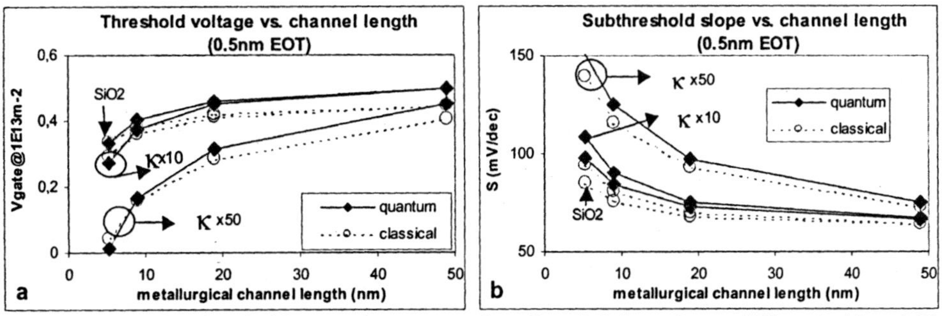

Figure 5 – NMOS threshold parameters versus L_m. Permittivity ratio vs. SiO_2 = 1, 10 and 50. EOT=0.5nm. **(a)** V_t (L_m), **(b)** S (L_m)

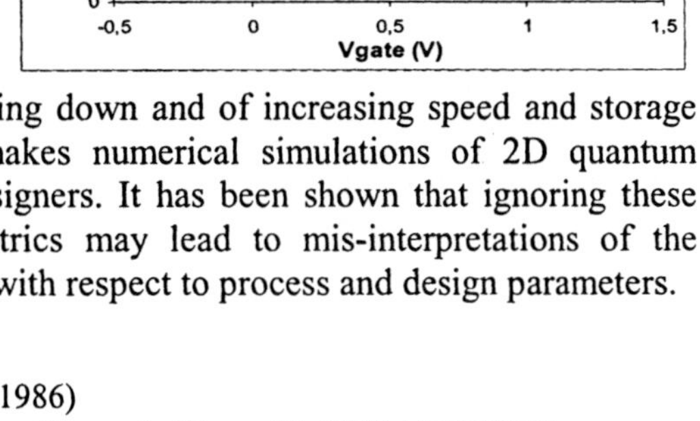

Figure 6 - Quantum / classical ratio on carrier density versus gate voltage in NMOS transistors with L_m=10nm and EOT=0.5nm. Permittivities are 1, 10 and 50 times SiO_2 one. The doted line depicts the 1D case.

4- Conclusion

The combination of devices scaling down and of increasing speed and storage capabilities of desk-top computers makes numerical simulations of 2D quantum effects affordable for MOS device designers. It has been shown that ignoring these effects in the case of high-K dielectrics may lead to mis-interpretations of the sensitivity of the device characteristics with respect to process and design parameters.

References

[1] M. Moglestue, J. Appl. Phys. 59, 3175 (1986)
[2] A.Abramo et. al. IEEE Trans. On Electron Dev. vol. 47, no. 10, 1858-1863(2000)
[3] B. Cheng et. al. IEEE/ED 46 (7) July 1999
[4] A. Poncet, C. Faugeras, M. Mouis. European J. of Physics , to be published (2001)
[5] A. Abou-Elnour, K. Schuenemann J. Appl. Phys. 74 (5) September 1993
[6] A. Trellakis, A. T. Galick, A. Pacelli, U. Ravaioli. J. Appl. Phys. 81 (12) 7880-7884 (1997)

Decananometer FDSOI Device Optimization including Random Variation

DANIEL CONNELLY (djconnel@ieee.org)
Acorn Technologies, Palo Alto, CA

Abstract

Here, for the first time, is described a method for including random variability in device parameters for the benchmarking and optimizing of a decananometer fully-depleted silicon-on-insulator (FDSOI) device. The effect of interdevice and intradevice variability on device performance and design optimization is discussed. These methods are readily extendible to other devices and technologies.

1 Introduction

Despite the fact that stochastics are an issue of increasing importance at state-of-the-art device dimensions [1-5], they continue to be neglected in device benchmarking and in assessment of technology performance. In addition to the traditional variation in gate length, variation in other device dimensions and doping density are critical factors in determining circuit performance and therefore must be considered when optimizing a device design. The design yielding the best circuit performance may not be the one with the best typical-device behavior.

2 Stochastic Effects

Two principal device responses are leakage and drive. Static leakage current depends on both interdevice and intradevice variation, as it is integrated across width [3]. Whether, for a given leakage current density, there is a single device of width $N \times W$ or N devices each of width W is unimportant.

Circuit performance will depend more on device-to-device variation than on variation across an individual device. In [1], a value of two standard deviations below the mean drive current was used to predict the performance of a simple parallel set of delay chains. However, the appropriate standard deviation is the interdevice, not intradevice, value. For example, applying the standard deviation of gate length as extracted from single cross-sectional SEM's from different devices could lead to an overestimation of interdevice variability, as it includes intradevice variability, as well.[1]

3 Method

3.1 Evaluation

To account for stochastic variation, devices are simulated for different representative values of the device parameters. The accumulated results, appropriately weighted, are used to generate, for each V_D and V_G, values of $\langle I_D \rangle$, $\langle \ln I_D \rangle$, and $\sigma_{\ln Id}$. The gate voltage yielding the target $\langle I_D \rangle$ is determined and a fixed $\Delta V_G = 1\,\text{V}$ overdrive is applied, at which point the $V_D = V_{Dlin} = 0.1\,\text{V}$ and $V_D = V_{DD} = 1.5\,\text{V}$ values of $\exp[\langle \ln I_D \rangle - 2\sigma_{\ln id}]$ are interpolated. A weighted

[1]This assumes cross-section positions are uncorrelated with intradevice gate length variation.

harmonic mean is calculated as the final benchmark, weighting the $V_D = V_{DD}$ to $V_D = V_{Dlin}$ values in an 8:1 ratio[2].

3.2 Parameter Space Sampling

To economize on simulations, a principle axis sampling strategy was selected as a discrete approximation to a continuous normal distribution. In one dimension, samples taken at $\Delta x \in (0, \pm\sqrt{3}\sigma)$, with weights of $(1,4,1)$ for $\Delta x \in (-1,0,1)\sqrt{3}\sigma$, yield matching of normal values of $\langle \Delta x^n \rangle$ through $n = 5$. For more than one dimension, points are sampled at $\Delta x_i \in (\pm\sqrt{3}\sigma)$ for i along each principle axis in turn, with $\Delta x_{j \neq i} = 0$, and at the center point. Samples are then given unity weight with the exception of the center point, which is weighted $6-2n$, where n is the number of dimensions. This results in the correct values of $\langle \Delta x_i^n \rangle$ through $n = 5$, for all i. It may, however, neglect the effect of deviations from normal behavior for large deviates, such as a catastrophic failure at a 3σ subnominal dimension.

4 Examples

4.1 L_G, t_{Si}, t_{box}

Simulations were performed using Medici version 2000.2.1 by Avanti with the *shiramob* mobility model and the *fldmob* longitudinal field dependence. No explicit quantum mechanical effects were considered; these will be addressed in a future work.

For all simulations the source and drain are length 50 nm, doped $10^{20}/\mathrm{cm}^3$ n-type, and aligned to the gate edge.

Results are shown versus gate length, silicon thickness, and buried oxide thickness in Figure 1. In each case, plots are shown for the case where variation is neglected and for the case where the x-axis parameter is assumed to be random with the appropriate σ and statistical dopant fluctuation, if the body is doped, is modeled. For t_{box} and t_{Si}, it is assumed all variation is interdevice, while for L_G, it is assumed the variance is partitioned equally between interdevice and intradevice components. Variation in doping, where applicable, is analytically estimated as described later. Note in each case the inclusion of stochastic effects into the analysis has a substantial effect on dependence of the benchmark on the associated parameter.

4.2 Gate Oxide Thickness

The same device, with $n_{Si} = 0$, $L_G = 50\,\mathrm{nm}$, $t_{Si} = 10\,\mathrm{nm}$, $t_{box} = 20\,\mathrm{nm}$, and $t_{ox} = 2\,\mathrm{nm}$, had its interdevice oxide variation varied from 0 to 0.3 nm. Since gate oxide thickness has such a strong influence on device performance, the variance in response associated with a variance in gate oxide thickness, both in the subthreshold and superthreshold regions, is substantial. To good approximation, the metric was degraded 6.5% for each 0.1 nm increase in σ_{tox}.

4.3 Silicon Doping

For the purposes of this work an analytic approximation to the effect of stochastic variation of atomic dopants on the effective doping density in

[2]This is intended to be a typical ratio; the optimal ratio is dependent on the circuit and on V_{DD} and V_{Dlin}.

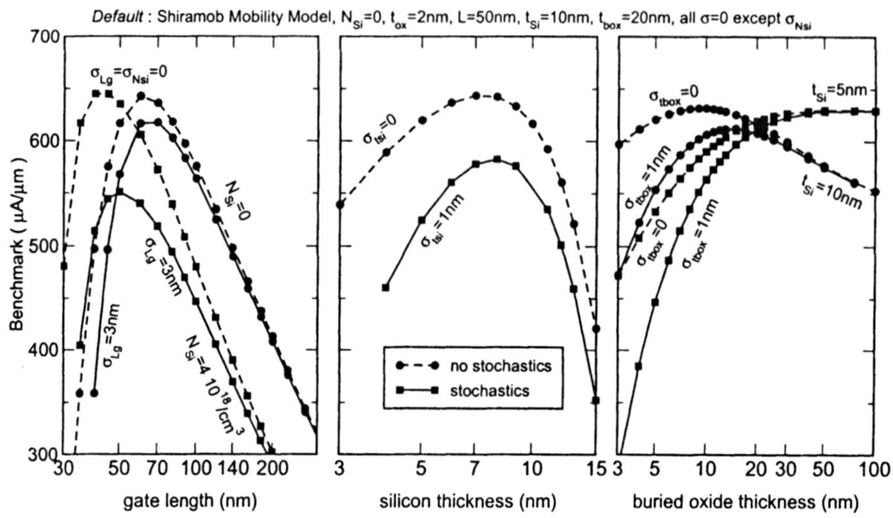

Fig. 1. Comparison of benchmark versus various parameters with and without modeling of variation of the x-axis parameter. In L_G plot, when gate length variation is suppressed, so is dopant stochastic modeling.

a slice of the channel, based on simple electrostatic analysis, was used[3], $\sigma_N^2 = N/\left[L_G t_{Si}\left(t_{Si} + 4t_{ox}\right)\right]$. Carriers traversing the channel along an interface segment from source to drain are considered to see an effective doping level proportional to the number of dopant atoms in a trapezoidal solid, roughly the set of channel points electrostatically closer to the segment than to the gate electrode, assuming $\epsilon_{Si}/\epsilon_{ox} = 3$. Thus the doping level is subject to local fluctuations in proportion to the Poisson-statistical variation in the atom count within the trapezoidal solid. The interdevice component of the resulting variance in the device response is in proportion to the ratio of the volume of this trapezoidal solid to that of the device body.

Figure 2 shows results of devices with either no stochastic analysis or stochastics applied only to doping. Note while higher doping is observed to help the performance of thicker-silicon devices when stochastics are neglected, the short-channel-effect benefits of the doping are offset when the modeled doping stochastic effects are included. Curves are shown both with and without mobility modeling to isolate the electrostatic/stochastic tradeoff.

4.4 Multivariate Optimization

Optimization of L_G, t_{Si}, t_{box} simultaneously, with $\sigma_L = 3\,\mathrm{nm}$ (50% interdevice variance), $\sigma_{tox} = \sigma_{tsi} = 1\,\mathrm{nm}$ (100% interdevice variance), and $n_{Si} = 0$, yielded an optimal design of $L_G = 70.56\,\mathrm{nm}$, $t_{Si} = 10.35\,\mathrm{nm}$, and $t_{box} = 27.64\,\mathrm{nm}$, with a benchmark current $I_{Dtest} = 600.4\,\mathrm{nA}/\mu\mathrm{m}$. When doping was allowed to vary as well, the optimal design had[4] $n_{Si} = 1.1 \times 10^{18}/\mathrm{cm}^3$, $L_G = 69.5\,\mathrm{nm}$, $t_{Si} =$

[3] This is intended to have the correct qualitative behavior. For better quantitative accuracy, validation against 3-d Monte Carlo simulations should be done. The formula assumes zero source/drain-to-gate overlap, consistent with the abrupt junctions used in this study.

[4] Reduced precision is provided due to the additional degree of freedom in the parameter space.

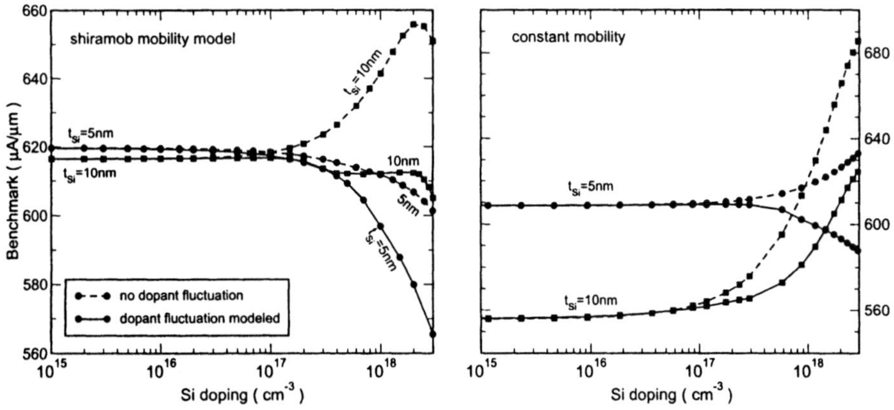

Fig. 2. Comparison of benchmark versus doping with and without modeling of statistical dopant fluctuations for $L_G = 50\,\text{nm}$ FDSOI nFETs with $t_{Si} = 5\,\text{nm}$ or $10\,\text{nm}$, $t_{box} = 20\,\text{nm}$. Plots are shown for *shiramob* mobility model and for a fixed electron mobilitie of $100\,\text{cm}^2/\text{V}\cdot\text{sec}$. The latter doesn't include the the effect of ionized impurity scattering with increaing doping density.

16.5 nm, and $t_{box} = 17.9\,\text{nm}$, yielding $I_{Dtest} = 603\,\text{nA}/\mu\text{m}$. A small performance improvement was gained by exploiting the improved short channel immunity of doped films, which allowed the silicon thickness to be increased, decreasing series resistance. At higher doping levels, the effect of doping stochastics and ionized impurity scattering cause a significant reduction in the benchmark from the optimal value.

5 Conclusion

Proper benchmarking is critical to device design optimization. The determination of a figure of merit to be optimized must include consideration of the variation of as well as the nominal value of device parameters. Variation includes both interdevice and intradevice components, each of which affect performance differently. Fully depleted silicon on insulator devices were examined, focusing on the effects on performance of gate length, buried oxide thickness, silicon thickness, and body doping. In each case modeling of the effects of variation in the parameters had a substantial influence both on the optimal choice of the parameter and on the performance expected at the optimal parameter value. Control of gate oxide thickness was also seen to be critical to performance.

Thanks to Dan Grupp (Acorn Technologies), Michael Kwong (Stanford University), Michael Duane (Applied Materials), and Dan Yergeau (Mixed Technology Associates) for their assistance with this project.

References

[1] D. Connelly, M Foisy; SISPAD 2000.

[2] A. Asenov, S. Kaya; SISPAD 2000.

[3] P. Oldiges, *et al.* ; SISPAD 2000.

[4] X. Tang *et al.*; 1999 IEEE Int. SOI Conference.

[5] Y. Taur *et al.*; IBM Journal of R&D, vol.39, no.1-2, pp 245-60.

Fully 2D quantum-mechanical simulation of nanoscale MOSFETs

A. Pirovano, A. L. Lacaita, and A. S. Spinelli[‡]
DEI, Politecnico di Milano, P.zza L. da Vinci 32, 20133 Milano, Italy
[‡] *Dip. di Scienze CCFFMM, Univ. dell'Insubria and INFM, Unità di Como via Valleggio 11, 22100 Como, Italy*

Abstract

We present results of fully 2D quantum-mechanical (QM) simulations of nanoscale MOSFET's. The validity of semiclassical transport models are first discussed. Then, QM effects on threshold voltage, subthreshold slope and short-channel performances are addressed. We show that QM effects significantly affect device performances in the nanoscale range.

1 Introduction

Careful evaluation of quantum mechanical (QM) effects has become a compelling issue in the design of aggressively scaled MOSFET's, where the confinement in the direction normal to the Si/SiO_2 interface affects important device parameters such as threshold voltage and gate capacitance. As the channel length is reduced below 50 nm, QM effects in the source-to-drain direction become also relevant, requiring a 2D treatment of the problem. To this aim, the effects of both the 2D QM charge distribution and QM charge transport should be assessed. However, recent studies suggested that semiclassical models for transport can be effectively used for channel lengths down to 10 nm or beyond [1,2]. Moreover, several important device parameters such as threshold voltage and subthreshold slope are only marginally affected by the actual transport model used. For these reasons, we have investigated QM effects on the performances of nanoscale MOSFET's relying on a fully 2D self-consistent Schrödinger-Poisson simulator coupled to a semiclassical transport model.

2 Numerical model

The inversion layer charge is computed solving the coupled 2D Schrödinger and Poisson equation system, while the transport is self-consistently included according to a drift-diffusion framework. Good convergence properties are obtained using the *effective intrinsic density* concept [5], where the local n_i is modified to account for the QM charge. This approach is also profitably used to resolve the inconsistency between zero QM charge at the boundaries (consequence of the closed-boundary conditions) and non-zero current flow: A classical charge distribution can be effectively recovered in the contact regions by a smoothing procedure on n_i. This approach results in a highly-effective fully 2D self-consistent QM model, achieving an exponential convergence rate. The accuracy of the solution against grid dimension was verified, as well as the influence of the boundary conditions, by changing the extension of the contact regions. Even if a detailed description of the charge transport was not our goal, a transport model based on the quantum transmitting boundary method [3] was also developed as a benchmark.

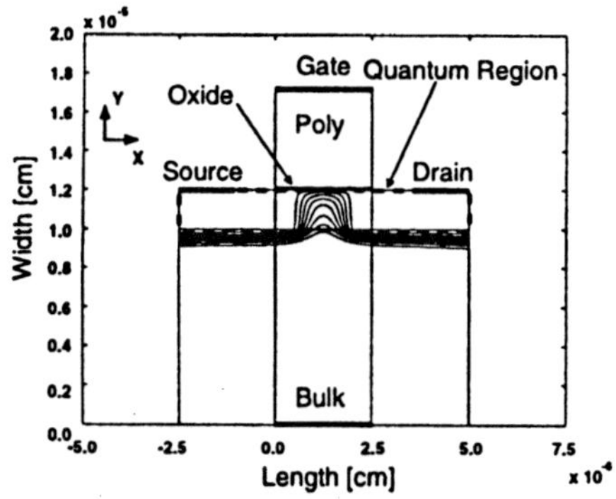

Figure 1: 25 nm MOSFET structure highlighting contacts, potential contours, and the box region where the QM charge is calculated.

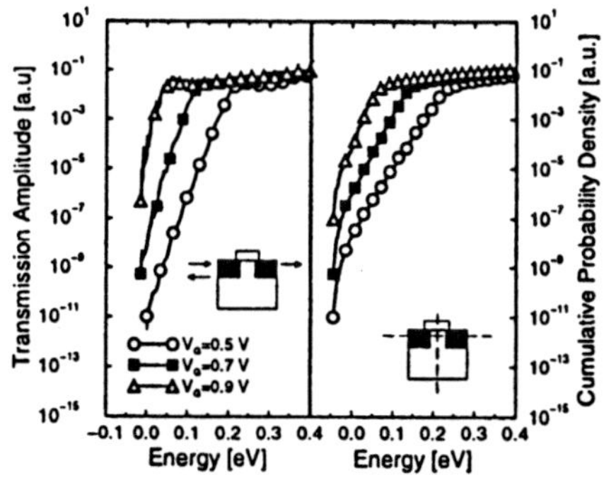

Figure 2: Comparison between TA and CPD at different gate biases for a 25 nm MOSFET (logarithmic scale).

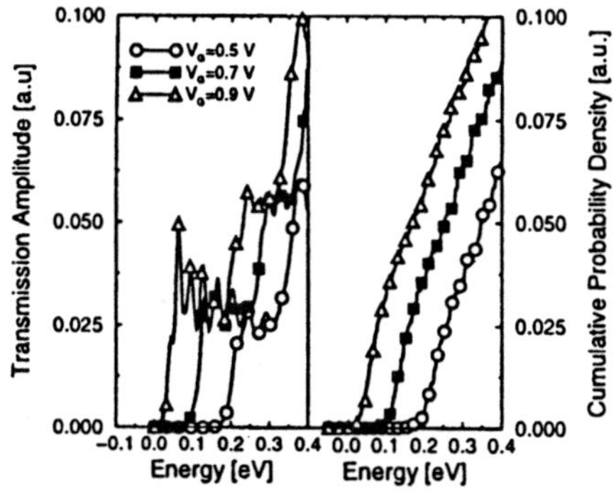

Figure 3: Same as Fig. 2 but in linear scale.

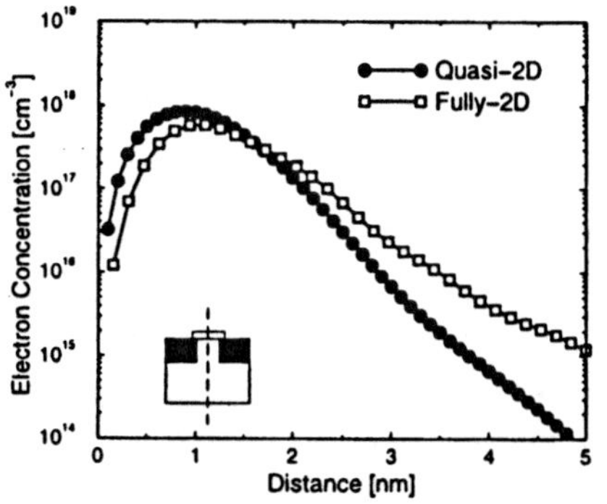

Figure 4: Electron concentration according to a fully-2D and to a quasi-2D QM model.

3 Impact of quantum transport

Fig. 1 shows the simulated structure, a 25 nm n^+-poly n-MOSFET with an oxide thickness of 1.5 nm and a uniform channel doping of 2×10^{19} cm^{-3}. The source/drain junction depth is 20 nm, with a lateral abruptness of 1.5 nm/decade, complying with the requirements for sub-100 nm node technology [4]. Fig. 1 also shows the box region where the QM charge is actually computed, the device contacts, and the potential contours. We have first investigated the impact of the transport model used on the device current: Figs. 2 and 3 show a comparison between the transmission amplitude (TA) for wavefunctions propagating along the channel direction (calculated with the quantum transmitting boundary method) and the cumulative probability density (CPD) in the middle of the channel (computed within the drift-diffusion framework), as a function of the energy and for different gate biases. Note that TA determines the current in a QM transport framework, while CPD is proportional to the current in the semiclassical model. In Fig. 2 TA and CPD are displayed on a logarithmic scale. A good agreement can be noticed, with an exponential growth due to tunneling of the wavefunctions from the contact regions into the channel. When looked on a linear scale (Fig. 3), TA shows a

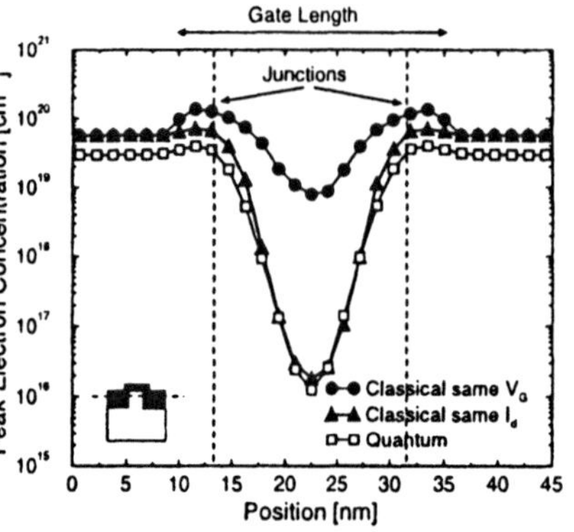

Figure 5: Peak electron concentration along the channel for QM and classical simulations.

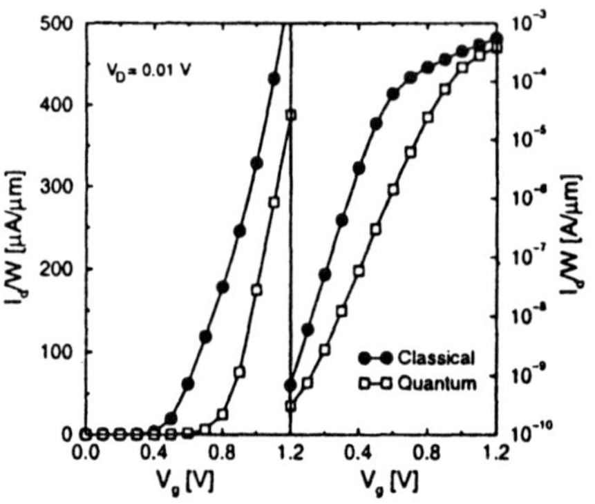

Figure 6: I-V characteristics at $V_D = 10$ mV for the device in Fig. 1.

step-like dependence in correspondence of the bound levels, while CPD is almost linear. However, these differences do not have any impact on the final result: To obtain the current, both TA and CPD must be weighted by the exponentially decreasing distribution function. Therefore, the main contribution to the current comes from the energy region lying within about k_BT from the first quantized level, which corresponds to the onset of TA and CPD. Since in both TA and CPD the onset is the same irrespective of the gate bias, we conclude that these differences will not affect strongly the charge distribution and device current. In the following, we will therefore refer to results obtained with the semiclassical transport model.

4 Device performances

Fig. 4 shows a comparison between the QM charge along a section in the middle of the channel computed according to our model and to a quasi-2D QM model, for the same device as in Fig. 1 biased at $V_G = 0.6$ V and $V_D = 0.01$ V. Non-negligible differences are obtained, demonstrating that a fully-2D treatment of QM effects is essential to reliably assess the charge distribution in nanoscale MOS devices. The peak charge concentration along the channel direction is instead shown in Fig. 5, according to 2D classical and QM calculations. At the same gate bias, QM effects result in less charge in the channel and in the spacer regions, where the confinement is still noticeable. Comparable channel charge is instead obtained at the same output current. However, even in this case, a small difference can be seen in the non-confined source/drain regions. This is due to the vanishing QM charge at the boundary, which determines a "dark space" where the carrier density is reduced compared to the equilibrium value. The classical distribution is in fact exactly recovered at a slightly larger distance from the boundary as well as at the device contacts. The drain current at $V_D = 10$ mV is shown in Fig. 6. A threshold voltage (V_T) shift is apparent, while a comparable transconductance above threshold is obtained. The V_T shift (300 mV) is significantly lower than what estimated by quasi-2D QM treatments (400 mV). This discrepancy can be ascribed to the modification in the potential shape induced by the short-channel effects (SCE), and can be captured only with a self-consistent 2D QM model. Moreover, the subthreshold slope

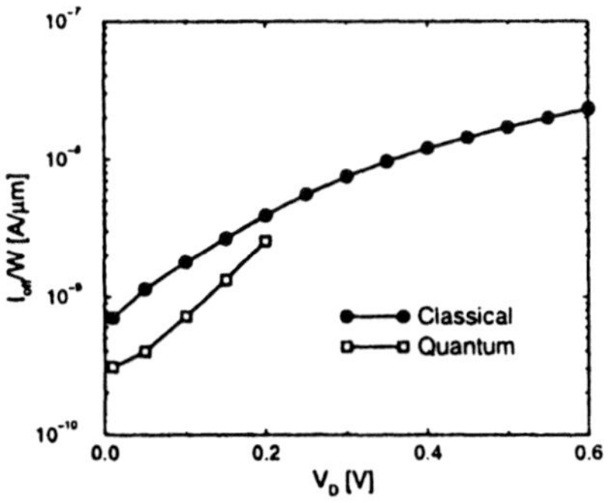

Figure 7: DIBL characteristics (off current) for the 25 nm device. Note the different dependences on V_D.

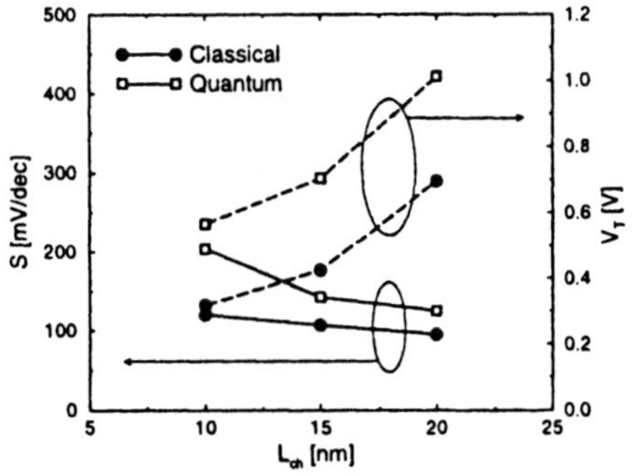

Figure 8: Short-channel performance of nanoscale MOSFET's as a function of the metallurgical channel length.

S changes from 107 mV/decade to 143 mV/decade for classical and QM simulations, respectively. This very high increase (about 33%) will significantly offset the traditional MOS scaling scenario and will accelerate the adoption of innovative structures.
QM effects on the short-channel performances have been also investigated. The off-state current of the device is shown in Fig. 7 as a function of V_D. As the computational time rapidly increases for high V_D, QM simulations were limited to $V_D = 0.2$ V. The off current is reduced in the QM model due to the increased V_T. However, DIBL performances are degraded due to the source-to-drain tunneling contribution to the current. Fig. 8 shows V_T and S as functions of the metallurgical channel length. Comparable V_T roll-offs are obtained, meaning that the QM charge distribution does not influence the barrier lowering in the channel. Significant degradation in the subthreshold slope is instead obtained in the QM simulations, in agreement with the results in Fig. 7.

5 Conclusions

We have developed a fully 2D QM model for simulation of nanoscale MOSFET's. A semiclassical transport model is adopted, supported by comparison against a QM model. Simulation results show that QM effects strongly influence important device performances such as threshold voltage, subthreshold slope, DIBL and SCE. We conclude that full QM simulations will become a mandatory issue for nanoscale MOSFET's modeling and design.

References

[1] Chen, W., Ouyang, Q., Register, L. F., Banerjee, S. K. (2000): Quantum effects *along* the channel of ultra-scaled Si-based MOSFETs?. In: IEDM 2000 Tech. Dig., pp. 291-294

[2] Fischetti, M. V. (1998): Theory of electron transport in small semiconductor devices using the Pauli master equation. J. Appl. Phys. 83: 270-291

[3] Lent, C. S., Kirkner, D. J. (1990): The quantum transmitting boundary method. J. Appl. Phys. 67: 6353-6359

[4] Semiconductor Industry Assoc. (1999): International technology roadmap for semiconductors

[5] Spinelli, A., Benvenuti, A., Pacelli, A. (1998): Self-consistent 2D model for quantum effects in n-MOS transistors. IEEE Trans. Electron Devices 45: 1342-1349

Ab-initio Electrodynamic Modeling of On-Chip Back-End Structures

Wim Schoenmaker
IMEC
Kapeldreef 75, B-3001 Leuven, Belgium
schoen@imec.be

Peter Meuris
IMEC
Kapeldreef 75, B-3001 Leuven, Belgium
meurisp@imec.be

Wim Magnus
IMEC
Kapeldreef 75, B-3001 Leuven, Belgium
magnus@imec.be

Abstract

The geometrical structure of electrodynamics is reviewed following the analogy with gravity. It is found that the vector potentials that represent magnetic fields can be identified as connections. As a consequence, these potentials should be assigned to the links of discretization grids. A ghost field is introduced to guarantee numerical stability in the solution scheme of solving electromagnetic field problems for interconnects and on-chip passives.

1 Introduction

Present-day integrated circuits are characterized by the down-scaling of device dimensions, the densification of device stacking as well as increased operation frequency. Furthermore, the integration of on-chip passive elements in the back-end process of the circuit design is a continuing activity. The high density device stacking combined with a higher clock frequency leads to unprecedented design difficulties because one needs to keep parasitic couplings below prescribed threshold levels (design rules). On the other hand, the integration of on-chip passives requires a detailed knowledge of the characteristic parameters such as the capacitance, inductance and resistance. These global parameters can be regarded as effective parameters representing objects that are obtained from integrating a local, i.e. space-time dependent quantity. As such, these parameters will be sensitive to the frequency under consideration.

In order to arrive at a full chip design, a hierarchical approach is advocated. The design-rule checking is done by partitioning the design into blocks and within each block one may zoom in to the critical paths and to study these sections in greater detail. The less critical parts can usually be handled by lumped-element descriptions or transmission line theory. The critical paths are sometimes referred to as 'tunnels' [1].

At the deepest level, the simulation of the back-end structures should be done by a field-solver approach. However, the circumstances at this level are rather unconventional from a field-solvers point of view.

We will summarize here the specific difficulties that a field solver should be able to deal with. *Geometry (1)* The structures that appear at this level are often essentially three dimensional. As a consequence, transmission-line theory is not applicable. *Geometry (2)* The patterns have a width and height of comparable size. Therefore, approaches that assume infinitely thin metallic layers are not justified from first principles. *Materials (1)* Metallic patterns will have finite resistivity and therefore the currents will flow with different densities in the metalic parts. These currents distributions are further modified by the skin effect and proximity effects if the frequency increases. *Materials (2)* Semiconducting materials are moderately conducting. This has a serious impact on the determination of effective parameters. The slow-wave mode is a consequence of putting a thin oxide layer between the metallic conductor and the semiconducting material.

Considering all the requirements listed above, we discuss in this paper a novel approach. Instead of building on existing approximation schemes derived from Maxwell's equations, we choosed to return to the Maxwell equations and to develop a solution scheme capable if dealing with the above problems. In addition it proves beneficial not only to return to the Maxwell equations, but also to carefully trace the physics that is hidden in the various symbols. In particular, the 'physical' identification of the vector potential is crucial for a successful numerical implementation. There exists an intimate connection between differential geometry and electrodynamics that we have to exploit in order to perform this task.

2 The geometry of electrodynamics

Electrodynamics was discovered as a phenomenological theory. Starting from early experiments one arrived after considerable effort at Gauss' law, Biot-Savart's law and Faraday's law of induction. Only Maxwell's law was obtained by theoretical reasoning and later experimentally confirmed by Herz. Maxwell's great achievement was later equalized by Einstein who proposed in the general theory of relativity that *gravity* $\equiv$ *curvature*. Ever since Einstein's achievement of describing gravity in terms of non-Euclidean geometry, theoretical physics has witnessed a stunning development based on geometrical reasoning resulting into the gauge theories having a geometrical interpretation very analogous to Einstein's theory of gravity. Electrodynamics is the low-energy limit of these gauge theories. Besides the aesthetic beauty that results from these insights, there is also pragmatic benefit. Solving electrodynamic problems on the computer, guided by the geometrical meaning of the variables is a decisive factor for the success of the calculation. This was already realized by Wilson [2] who initiated computer calculations on the quantum aspects of gauge theories. In order to perform computer calculations of the classical fields, geometry plays an equally important role as will be shown here.

The classical fields $\mathbf{E}$ and $\mathbf{B}$ as well as the sources ρ and $\mathbf{J}$ are invariant under gauge transformations and therefore their underlying geometrical meaning is hidden. We may identify the proper geometric character for these variables, such as scalars (zero-forms), force fields (one-forms), fluxes (two-forms) or volume densities (three-forms) as could be done for any other fluid dynamic system. This can be done without reference to the geometric nature of electrodynamics in the sense that $\mathbf{E}$ and $\mathbf{B}$ represent the *curvature* in the geometrical interpretation of electrodynamics. Therefore, in this paper we will consider the scalar potential and vector potential fields that *do* depend on gauge transformations and as such will give access to the geometry of electrodynamics.

2.1 Gravity as a gauge theory

The history of the concept of gauge invariance begins with the discovery of the principle of general covariance in general relativity. According to this principle the physical laws should maintain their form for all coordinate systems. Weyl [3] made an attempt to unify electrodynamics with gravity. According to the general theory of relativity, the gravitational field corresponds to curvature of space-time, and therefore, if a vector is parallel transported along a closed loop, the angle between the starting vector and the final vector will differ from zero. Furthermore, this angle is a measure for the curvature in the space. Weyl extended the Riemann geometry in such a way that not only the angle changes but also the *length* of the vector. The relative change in length is described by an anti-symmetric tensor and this tensor is invariant under changing the "unit of length". This invariance is closely related to charge conservation. Weyl called this "Maszstab Invarianz". The theory turned out to be contradictory and was abandoned, but the term "Maszstab Invarianz" survived (Maszstab = measure = gauge). Within the realm of quantum mechanics the principle of gauge invariance obtained its final interpretation: gauge invariance should refer to the phase transformations that may be applied on the wave functions. In particular, the phase transformation may be applied with different angle for different points in space and time. At first sight it looks as if we have lost the geometrical connection and that the link is only historical. However, a closer look at gravity shows that the link is still present.

Starting from the idea that all coordinate systems are equivalent, we may consider a general coordinate transformation $x^\mu \to x'^\mu = x'^\mu(x^\nu)$ and the transformation rule for coordinate differentials $dx'^\mu = \frac{\partial x'^\mu}{\partial x^\nu} dx^\nu$. A tuple transforming under a change of coordinates as the coordinate differentials is a *contravariant vector* and a *scalar* transforms in an invariant way. The *derivative* of a scalar $V_\mu = \frac{\partial \phi}{\partial x^\mu}$ transforms as $V'_\mu = \frac{\partial x^\nu}{\partial x'^\mu} V_\nu$. Any tuple transforming under a change of coordinates as the differentials of a scalar function is a *covariant vector* and in general, *tensors* transform according to a multiple set of pre-factors. The principle of general coordinate covariance can be implemented by claiming that all physical laws should be expressed as tensor equations. Since left and right hand side will transform with equal sets of pre-factors, their form invariance is guaranteed.

So far, we have only been concerned with the change from one arbitrary coordinate system to another. One might argue that this will hide well-known results in a thick shell of notational complexity. In order to peal off this shell and to reveal the physical implications one must refer to the *intrinsic* properties of the geometric structure. Occasionally, the intrinsic structure is simple, e.g. flat space time, and the familiar relations are recovered. It was Einstein's discovery that space-time is *not* flat in the presence of matter and therefore the physical laws are more involved.

Riemann geometry is a generalization of Euclidean geometry in the sense that locally one can still find coordinate systems, $\xi^\mu = (ict, \mathbf{x})$, such that the distance between two near-by point is given by Pythagoras' law, i.e. $ds^2 = \delta_{\mu\nu} d\xi^\mu d\xi^\nu$. In an arbitrary coordinate system the distance is given by

$$ds^2 = g_{\mu\nu}(x) dx^\mu dx^\nu \ , \quad g_{\mu\nu}(x) = \frac{\partial \xi^\alpha}{\partial x^\mu} \frac{\partial \xi^\beta}{\partial x^\nu} \delta_{\alpha\beta} \tag{1}$$

In the local coordinate system, ξ, the equation of motion of a freely falling particle is $\frac{d^2\xi^\mu}{ds^2} = 0$. In an arbitrary coordinate system, this equation becomes

$$\frac{d^2 x^\alpha}{ds^2} + \Gamma^\alpha_{\mu\nu} \frac{dx^\mu}{ds} \frac{dx^\nu}{ds} = 0 \tag{2}$$

where $\Gamma^\alpha_{\mu\nu}$ is the *affine connection* also referred to as the Christoffel symbols.

$$\Gamma^\alpha_{\mu\nu} = \frac{\partial x^\alpha}{\partial \xi^\beta} \frac{\partial^2 \xi^\beta}{\partial x^\mu \partial x^\nu} \tag{3}$$

The affine connection transforms under general coordinate transformations as

$$\Gamma'^\alpha_{\mu\nu} = \frac{\partial x'^\alpha}{\partial x^\rho} \frac{\partial x^\tau}{\partial x'^\mu} \frac{\partial x^\sigma}{\partial x'^\nu} \Gamma^\rho_{\tau\sigma} + \frac{\partial x'^\alpha}{\partial x^\rho} \frac{\partial^2 x^\rho}{\partial x'^\mu \partial x'^\nu} \tag{4}$$

The second term destroys the covariance of the affine connection, i.e. the affine connection is not a tensor. The metric tensor $g_{\mu\nu}(x)$ contains information on the local curvature of the Riemann geometry. Now consider a vector $V^\mu(\tau)$ along a curve $x^\mu(\tau)$. In the local-Euclidean coordinate system (ξ), the change of the vector along the curve is $dV^\mu/d\tau$. In another coordinate system (x'), we find from the transformation rules

$$\frac{dV'^\mu}{d\tau} = \frac{\partial x'^\mu}{\partial x^\nu} \frac{dV^\nu}{d\tau} + \frac{\partial^2 x'^\mu}{\partial x^\nu \partial x^\lambda} \frac{\partial x^\lambda}{\partial \tau} V^\nu(\tau) \tag{5}$$

The second derivative in the second term is an inhomogeneous term in the transformation rule that prevents $dV^\mu/d\tau$ from being a vector and contains the key to curvature. This term, is directly related to the affine connection. The combination

$$\frac{DV^\mu}{D\tau} = \frac{dV^\mu}{d\tau} + \Gamma^\mu_{\nu\lambda} \frac{dx^\lambda}{d\tau} V^\nu \tag{6}$$

does transform as a vector and is called the covariant derivative along the curve. In the restricted region where we can use the Euclidean coordinates, ξ, we may apply Euclidean geometrical methods, and in particular we can shift a vector over an infinitesimal distance from one point to another and keep the initial and final vector parallel. The components of the vector do not alter by the shift operation: $\delta V^\mu = 0$. Furthermore, in the local frame $x^\mu = \xi^\mu_{x(\tau)}$, the affine connection vanishes, i.e. $\Gamma^\alpha_{\mu\nu} = 0$. Therefore, the conventional operation of parallelly shifting a vector in the locally-Euclidean coordinate system can be expressed by the equation $DV^\mu/D\tau = 0$. Being a tensor equation, it holds for all coordinate systems. A vector, whose covariant derivative along a curve vanishes, is said to be *parallel* transported along the curve. The coordinates satisfy the following first-order differential equations.

$$\frac{dV^\mu}{d\tau} = -\Gamma^\mu_{\nu\lambda}\frac{dx^\lambda}{d\tau}V^\nu \tag{7}$$

The parallel transport of a vector V^μ over a small distance dx^ν changes the components of the vector by amounts $\delta V^\mu = -\Gamma^\mu_{\nu\lambda}V^\nu\delta x^\lambda$.

In general, if we want to perform the differentiation of a tensor field with respect to the coordinates, we must compare tensors in two nearby points. In fact, the comparison corresponds to subtraction, but a subtraction is only defined if the tensors are anchored to the same point. (In different points, we have different local coordinate systems.) Therefore we must first parallel transport the initial tensor to the nearby point before the subtraction can be performed. For example, the covariant differential of a vector field is

$$DV^\mu = dV^\mu - \delta V^\mu = \left(\frac{\partial V^\mu}{\partial x^\lambda} + \Gamma^\mu_{\lambda\kappa}V^\kappa\right)dx^\lambda = D_\lambda V^\mu dx^\lambda \tag{8}$$

So far, the general coordinate systems include both accelerations originating from non-uniform boosts of the coordinate systems as well as acceleration that may be caused by the gravitational field due to the presence of matter. In the first case, space-time is not really curved, in the second case it is curved. In order to find out whether gravitation is present one must extract information about the intrinsic properties of space-time. This can be done by the parallel transport of a vector field along a closed loop. If the initial and final vector differ, one can conclude that gravity is present. The difference that a closed loop transport generates is given by

$$\Delta V^\mu = R^\mu_{\rho\lambda\sigma}V^\rho\delta x^\lambda\delta x^\sigma \tag{9}$$

where $R^\mu_{\rho\lambda\sigma} = \Gamma^\rho_{\sigma\mu,\nu} - \Gamma^\rho_{\sigma\nu,\mu} + \Gamma^\rho_{\lambda\nu}\Gamma^\lambda_{\sigma\mu} - \Gamma^\rho_{\lambda\mu}\Gamma^\lambda_{\sigma\nu}$ is the *curvature* tensor or Riemann tensor. This tensor describes the intrinsic curvature in a point.

We are now prepared to consider the geometrical basis of electrodynamics and other gauge theories but we will first summarize a few important facts: *(1)* in each space-time point a local frame may be erected, *(2)* the affine connection is a path-dependent quantity, *(3)* the affine connection does not transform as a tensor, *(4)* the field strength (curvature) may be obtained by performing a parallel transport along a closed loop.

2.2 The geometrical interpretation of electrodynamics

Just as with the local Euclidean coordinate systems, we will consider the possibility to set up in each space-time point a local frame for fixing the phase of the complex wave function $\psi(\mathbf{x},t)$. Since the choice of such a local frame (gauge) is not unique we may rotate this frame without altering the physical content of the wave function. We can guarantee the latter by demanding appropriate transformation properties (c.f. the tensors in above section) of the variables. Changing the local frame for the phase of a wave function amounts to

$$\psi'(\mathbf{r},t) = \exp\left(\frac{ie}{\hbar}\chi(\mathbf{r},t)\right)\psi(\mathbf{r},t) \;\;,\;\; \psi'^{*}(\mathbf{r},t) = \exp\left(-\frac{ie}{\hbar}\chi(\mathbf{r},t)\right)\psi^{*}(\mathbf{r},t) \quad (10)$$

These transformation rules are similar to the contravariant and covariant transformation rules for vectors in the foregoing section. We can construct a 'scalar' by taking $\psi^{*}\psi$. The derivative of the wave function transforms as

$$\frac{\partial\psi'}{\partial x^{\mu}} = \exp\left(\frac{ie}{\hbar}\chi\right)\frac{\partial\psi}{\partial x^{\mu}} + \frac{ie}{\hbar}\frac{\partial\chi}{\partial x^{\mu}}\exp\left(\frac{ie}{\hbar}\chi\right)\psi \quad (11)$$

The second term prevents the derivative of ψ from transforming as a 'vector' under the change of gauge. However, now geometry will be of help to provide a solution for constructing gauge covariant variables from derivatives. We must postulate an 'affine connection', such that a covariant derivative can be made. For that purpose a connection A_{μ} is defined that transforms as $A_{\mu} = A_{\mu} + \frac{\partial\chi}{\partial x^{\mu}}$. The covariant derivative is

$$D_{\mu} = \frac{\partial}{\partial x^{\mu}} + \frac{ie}{\hbar}A_{\mu} \quad (12)$$

Just as with the gravitational affine connection, the field A_{μ} can be used to construct 'parallel' transport. Therefore, the field A_{μ} must be assigned to the *paths* along which the transport takes place. The curvature of the connection can also be revealed by integrating the affine connection around a closed loop. The result is $F_{\mu\nu}\delta x^{\mu}\delta x^{\nu} = \oint dx^{\mu}A_{\mu}$ and $F_{\mu\nu} = \frac{\partial A_{\mu}}{\partial x^{\nu}} - \frac{\partial A_{\nu}}{\partial x^{\mu}}$ is the electromagnetic field tensor.

3 Implications for numerical simulations

In order to perform numerical computations starting from the fields A_{μ} it is necessary to introduce a discretization grid. The simulation of a finite space and/or space-time domain requires that each grid point is separated a finite distance from its neighboring points. Then the differential operators that appear in the continuous field equations must be translated to the discretization grid by proper reference to the geometrical meaning of the variables. The connections A_{μ} should be assigned to the *links* in the grid.

The geometrical interpretation suggests that the link assignment is the only correct scheme for solving vector-potential field problems on the computer.

The numerical consequences of the above assignment will be considered in the following example, in which we will solve the steady-state equations $\nabla \times \mathbf{B} = \mu_0 \mathbf{J}$, $\mathbf{B} = \nabla \times \mathbf{A}$, $\mathbf{J} = \sigma \mathbf{E}$, $\mathbf{E} = -\nabla V$. The usual approach consists of substituting the second equation into the first one, adopting the Coulomb gauge $\nabla \cdot \mathbf{A} = 0$ and obtaining $-\nabla^2 \mathbf{A} = \mu_0 \, \mathbf{J}$. Analytic solution schemes address this equation as a three-fold Poisson equation. This approach is usually sustained in numerical solution schemes, assigning the three components A_x, A_y, A_z to the *nodes* of the discrete lattice. This approach is in conflict with the geometrical interpretation of electrodynamics, since three scalar fields and the three components of a vector field are geometrically different objects.

We will now discretize the set of equations on a regular Cartesian grid having N nodes in each direction following our new approach. The total number of nodes in D dimensions is $M_{nodes} = N^D$ and the number of links in the lattice is $M_{links} = DN^D(1 - \frac{1}{N})$.

As far as the description of the electromagnetic field is concerned, the counting of unknowns for the full lattice results into M_{links} variables (A_{ij}) for the links, and M_{nodes} variables (V_i) for the nodes. Since each link (node) gives rise to one equation, the naive counting is consistent. However, we have not yet implemented the gauge condition. The conventional Coulomb gauge $\nabla \cdot \mathbf{A} = 0$, constraints the link degrees of freedom and therefore not all link fields are independent. There are $3N^3(1 - \frac{1}{N})$ link variables and $3N^3(1 - \frac{1}{N}) + N^3$ equations, including the constraints. As a consequence, at first sight it seems that we are confronted with an overdetermined system of equations, since each node provides an extra equation for $\mathbf{A}$. However, the translation of the Maxwell-Ampère equation on the lattice leads to a singular matrix, i.e. not all rows are independent. The rank of the corresponding matrix is $3N^3(1 - \frac{1}{N})$, whereas there are $3N^3(1 - \frac{1}{N}) + N^3$ rows and $3N^3(1 - \frac{1}{N})$ columns. Such a situation is highly inconvenient for solving non-linear systems of equations, where the non-linearity stems from the source terms being explicitly dependent on the fields. Moreover, the application of the Newton-Raphson method requires that the matrices in the Newton equation be non-singular and square. In fact, the non-singular and square form of the Newton matrix can be recovered by introducing the more general gauge $\nabla \cdot \mathbf{A} + \nabla^2 \chi = 0$, where an additional scalar field χ, i.e. one unknown per node, is included. In this way the number of unknowns and the number of equations match again. In the continuum limit $(N \to \infty)$, the field χ and one component of $\mathbf{A}$ can be eliminated.

Though being irrelevant for theoretical understanding, the auxiliary field χ is essential for obtaining numerical stability on a discrete, finite lattice.

In other words, our specific gauge solely serves as a tool to obtain a discretization scheme that generates a regular Newton-Raphson matrix.

To summarize, instead of solving the problem

$$\nabla \times \nabla \times \mathbf{A} = \mu_0 \, \mathbf{J}(\mathbf{A}) \;\;, \;\; \nabla \cdot \mathbf{A} = 0 \tag{13}$$

we solve the equivalent system of equations

$$\nabla \times \nabla \times \mathbf{A} - \gamma \nabla \chi = \mu_0 \, \mathbf{J}(\mathbf{A}) \;\;, \;\; \nabla \cdot \mathbf{A} + \nabla^2 \chi = 0 \tag{14}$$

The equivalence of both sets of equations (13) and (14) was be demonstrated by [4]. Physical equivalence is guaranteed provided that $\nabla\chi$ does not lead to an additional current source. Therefore, it is required that $\nabla\chi = 0$. In fact, acting with the divergence operator on the first equation of (14) gives Laplace's equation for χ. The solution of the Laplace equation is identically zero if the solution vanishes at the boundary. We achieved to implement the gauge condition resulting into a unique solution and simultaneously to arrive at a system containing the same number of equations and variables. Hence a regular square Newton-Raphson matrix is obtained at each stage of the numerical solution scheme that solves the full set of non-linear equations.

3.1 Discretized operators

Integrated over a test volume ΔV_i surrounding a node i, the divergence operator, acting on vector potential A, can be discretized as a combination of 6 neighboring links

$$\int_{\Delta V_i} \nabla \cdot \mathbf{A} \, d\tau = \int_{\partial(\Delta V_i)} \mathbf{A} \cdot \mathbf{dS} \sim \sum_k^6 S_{ik} A_{ik} \tag{15}$$

The symbol $\sim$ represents the conversion to the grid formulation and $\partial\,(\Delta V_i)$ denotes the boundary of ΔV_i.
The *curl-curl* operator can be discretized for a link ij using a combination of 12 neighboring links and the link ij itself. As indicated in Fig. 1, the field $\mathbf{B}_i$ in the center of the 'wing' i, can be constructed by taking the circulation of the vector potential A around the wing i ($i = 1, 4$)

$$\mathbf{B}_i S_i = \sum_{j=1}^{3} \mathbf{A}_{ij} h_{ij} + \mathbf{A}_0 h_0 \tag{16}$$

where h_α is the length of the corresponding link α. Integration over a surface S_{ij} perpendicular to the link ij yields a linear combination of different $\mathbf{A}_{ij}$'s, the coefficients of which are denoted by Λ_{ij}.

$$\begin{aligned} \int_{\Delta S_{ij}} \nabla \times \nabla \times \mathbf{A} \cdot \mathbf{dS} &= \int_{\partial(\Delta S_{ij})} \nabla \times \mathbf{A} \cdot \mathbf{dr} \\ = \int_{\partial(\Delta S_{ij})} \mathbf{B} \cdot \mathbf{dr} &\sim \Lambda_{ij} \, A_{ij} + \sum_{kl}^{12} \Lambda_{ij}^{kl} \, A_{kl} \end{aligned}$$

For the nodal variables we apply the usual discretization methods.

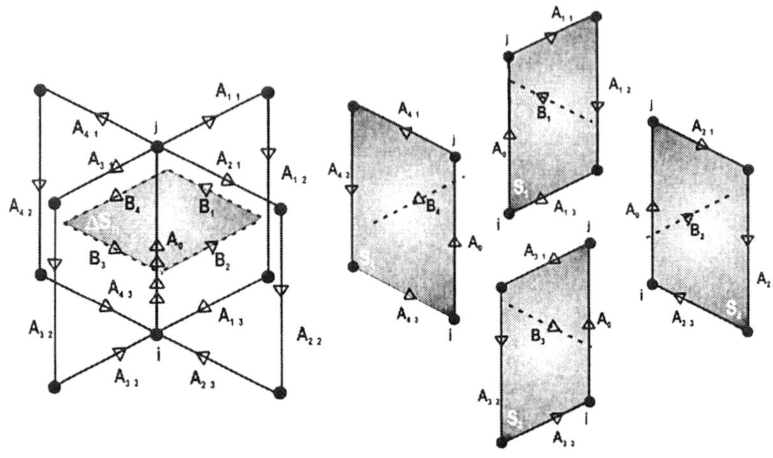

Figure 1: The assembly of the $\nabla \times \nabla \times$-operator using 12 contributions of neighboring links.

4 Application to the Spiral Inductor

We present an example demonstrating that the proposed potential formulation in terms of the Poisson field V, the vector field $\mathbf{A}$ and the ghost field χ, is a viable method to solve the Maxwell field problem. All subtleties related to that formulation, i.e. the positioning of the vector potential on links, and the introduction of the ghost field χ, are already encountered in constructing the solutions of the static equations.

A spiral inductor, as shown in Fig. 2 (left) was simulated. This structure also addresses the three dimensional aspects of the solver. The cross-section of the different lines is 1 μm $\times$ 1 μm. The overall size of the structure is 8 μm $\times$ 8 μm and the simulation domain is $23 \times 20 \times 9$ μm^3. The resistance is evaluated as $R = V/I$ and is 0.54 Ω. In Fig. 2 (right), the intensity of the magnetic field is shown at height 4.5 μm. From the results in Table II we obtain that the inductance of the spiral inductor is 4.23×10^{-11} Henry.

Electric energy (J)		*Magnetic energy (J)*	
$\frac{1}{2}\epsilon_0 \int d\tau\, E^2$	2.22E-18	$\frac{1}{2\mu_0} \int d\tau\, B^2$	3.80E-13
$\frac{1}{2} \int d\tau\, \rho\phi$	2.35E-18	$\frac{1}{2} \int d\tau\, \mathbf{J} \cdot \mathbf{A}$	3.90E-13

Table II: Some characteristic results for the spiral inductor.

5 Conclusions

We reviewed the geometrical aspects of electromagnetic field variables. Vector potentials are 1-forms. This has severe consequences for their discretization. The vector potentials need to be assigned to the links of the grid. Inclusion

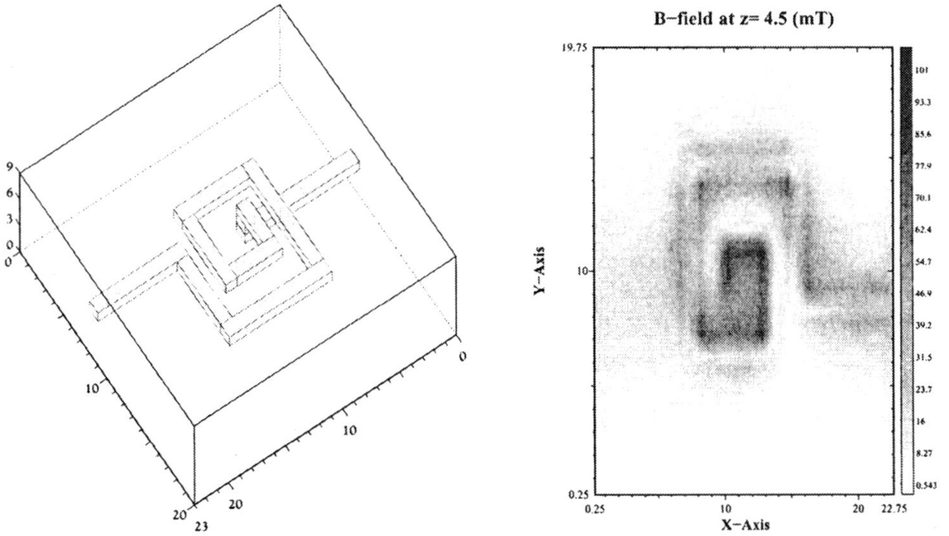

Figure 2: Layout of the spiral inductor structure (left) and the magnetic field strength in the plane of the spiral inductor (right).

of the gauge condition gives rise to non-symmetric Newton-Raphson matrices, when solving non-linear electromagnetic problems. This can be repaired by the inclusion of a 'ghost' field that has no physical effects but results into non-singular and square Newton-Raphson matrices.

Acknowledgement Part of this work is funded by the Flemish Institute for Science and Technology (IWT), Flanders (Belgium).

References

[1] Walker M.G. Modeling the wiring of deep submicron IC. IEEE Spectrum 2000; March edition.

[2] Wilson K. Confinement of Quarks. Phys. Rev. 1974; D10: 2445-2459

[3] Weyl H. Gravitation und Elektrizität, Sitzungberichte der Königlich-Preussischen Akademie der Wissenschaften. 1918; 26: 465-480

[4] Meuris P., Schoenmaker W., Magnus W. Strategy for Electromagnetic Interconnect Modeling. IEEE Trans. on CAD 2000; Vol. 20 No. 6

Analysis of Hot-Carrier-Induced Oxide Degradation in MOSFETs by Means of Full-Band Monte Carlo Simulation

Yoshinari Kamakura, Kazuaki Deguchi, and Kenji Taniguchi
Department of Electronics and Information Systems, Osaka University
2-1 Yamada-oka, Suita, Osaka 565-0871, Japan
E-mail: kamakura@eie.eng.osaka-u.ac.jp

Abstract

We demonstrate the investigations of oxide reliability by means of full-band Monte Carlo simulation. Firstly we discuss the accuracy of the scattering rates particularly for hot hole transport. It is shown that the quantum-yield experiment provides a means to verify the scattering models for hot carriers in MOS system. Secondly, the oxide breakdown is studied by using a substrate hot electron injection technique. Monte Carlo simulations are performed to examine the correlation between the oxide breakdown and the electron energy, and it is shown that the holes generated in the anode electrode play an important role in the oxide degradation. In addition, it is discussed that rigorous hot carrier simulations are necessary to study the reliability issues for ultra-thin oxide films used in advanced CMOS technology.

1 Introduction

The reliability of gate oxide is a major concern for further downscaling of MOS devices (Degraeve et al. 2000). Understanding the mechanism of oxide degradation and final breakdown is essential for accurate projection of oxide reliability. Many previous works (Chen et al. 1986, DiMaria et al. 2001) have pointed out that the hot carriers injected into the gate oxide originate the oxide degradation. Thus the rigorous modeling of the hot carrier behavior in MOS system is very important, e.g., to confirm the validity of the empirical projection manner using accelerated lifetime tests. In order to simulate the hot carrier transport in semiconductors using fewer assumptions and approximations, a full-band Monte Carlo (FBMC) technique has been extensively studied since the early 1980s (Shichijyo and Hess 1981, Fischetti and Laux 1988). Recently, the state-of-the-art FBMC simulators have been utilized by several groups for analyzing the oxide degradation (Bude et al. 1998, Ezaki et al. 2000). In the present work, we will firstly discuss the accuracy of the scattering models particularly for hot hole transport, which plays a key role in the oxide degradation. Then we demonstrate an example for the application of FBMC to studying oxide reliability, i.e., the investigation of breakdown induced by substrate hot electron (SHE) injection.

2 Verification of scattering models for FBMC simulation

In this work, we solve the Boltzmann transport equation for both electrons and holes in Si using a FBMC technique (Kunikiyo et al. 1994). The realistic band structure and the scattering rates essential for the hot carrier analysis, i.e., phonon scattering and impact ionization, are taken into account. All scattering rates have been derived on the basis of the band calculation (Cohen and Bergstresser 1966) and the resulting density of states. Figure 1 shows the scattering rates employed in our simulator. We used the electron scattering models reported by Kunikiyo et al. (1994), which are almost the same as those employed in the FBMC simulators of other groups (Abramo et al. 1994). They have been verified with various experimental data, e.g. the drift velocity, the ionization coefficient and the quantum yield, the soft x-ray photoemission spectroscopy, etc. (Cartier et al. 1993). On the other hand, although it has been pointed out that the hole holes significantly affects the oxide reliability, much less is known about the hole transport in Si (Kamakura et al. 2000). We assumed the theoretically computed ionization rate reported by Kunikiyo et al. (1996), and the phonon deformation potentials are determined by fitting the simulated drift velocity and the ionization coefficient to the experimental results (Ottaviani et al. 1975, Grant 1973). The states of secondary carriers after the impact ionization are obtained from the random-k approximation (Kane 1967).

In order to verify the scattering models, particularly for the hot hole transport, we simulated the quantum yield of the impact ionization. Quantum yield γ is defined as the number of electron-hole pairs generated by an injected hot electron (Chang et al. 1985). γ gives information about the ratio between the ionization and the phonon scattering rates, and hence it has played an important role in determining the scattering rates for hot electrons in Si (Cartier et al. 1993). γ as a function of electron energy can be measured by the charge separation technique as shown in Fig. 2. The number of pairs is derived from the ratio

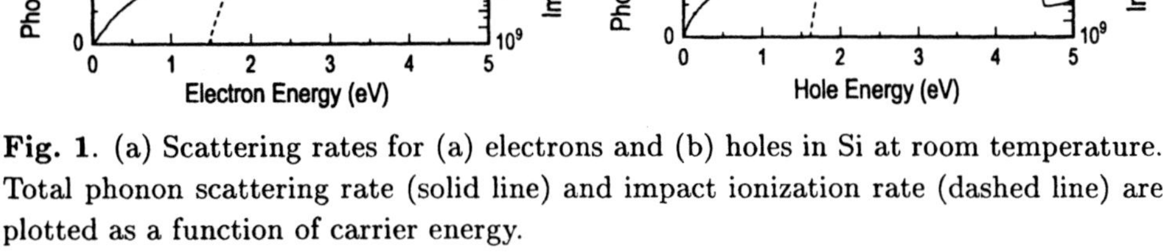

Fig. 1. (a) Scattering rates for (a) electrons and (b) holes in Si at room temperature. Total phonon scattering rate (solid line) and impact ionization rate (dashed line) are plotted as a function of carrier energy.

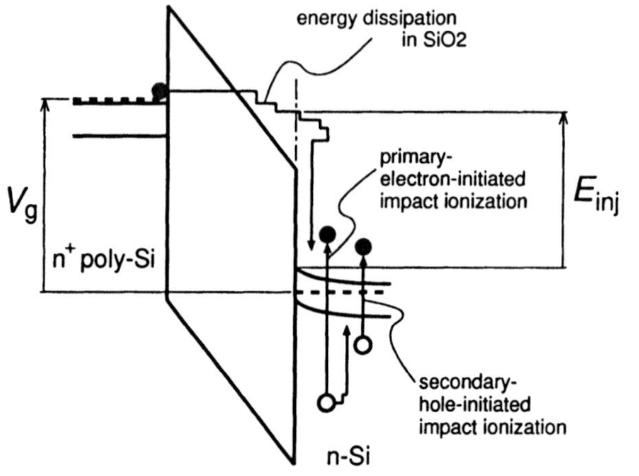

Fig. 2. Schematic energy band diagram illustrating quantum-yield experiment in FN tunneling regime.

of the channel (hole) current to the gate (electron) current of p-MOSFET. In this study our attention was focused on γ exceeding unity, which is obtained in the Fowler-Nordheim (FN) tunneling regime. Due to the high oxide voltage (> 3.2 V), the electrons are injected into the Si with high kinetic energies, and the secondary generated particles after impact ionizations can gain sufficient energy to produce additional pairs. The resulting γ contains the desired information about the impact ionization events induced by the secondary holes. We performed quantum-yield experiments by using MOSFETs with a heavily doped ($\simeq 8 \times 10^{17}$ cm^{-3}) Si substrate. Figure 3 (a) shows the experimental

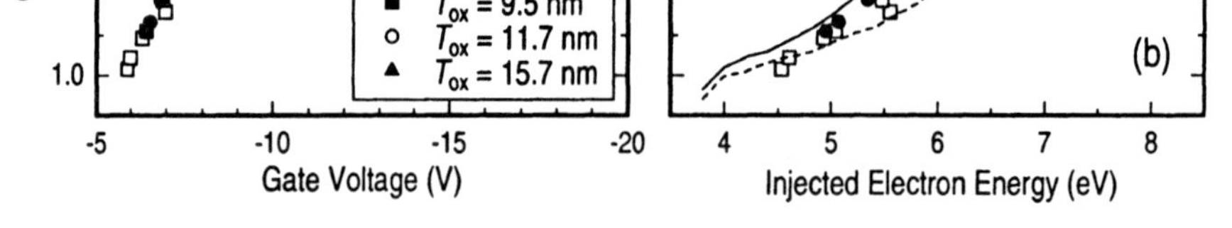

Fig. 3. (a) Measured γ as a function of gate voltage for the samples having various oxide thicknesses. (b) γ as a function of E_{inj}. Simulated results taking account of electron-initiated ionization process only (dashed line) and both electron- and hole-initiated ionization processes (solid line) are compared.

γ for various oxide thicknesses ranging from 6.1 to 15.7 nm. In thinner oxides ($t_{ox} \leq 7.7$ nm), a universal relationship between γ and V_g is found, irrespective of the oxide thickness indicating quasiballistic transport inside SiO_2. On the other hand, $\gamma - V_g$ characteristics for thicker oxides depend on thickness because of the strong energy dissipation in SiO_2. In the present study, we computed the initial electron energy immediately after the injection into Si, E_{inj}, by performing MC simulations in SiO_2 (Arnold et al. 1994). Note that a universal relationship irrespective of the oxide thickness is obtained, which ensures the validity of E_{inj} evaluation. The simulated results taking account of both electron- and hole-initiated impact ionization processes show good agreements with experiments. In this way, the quantum-yield experiment provides a means to verify the scattering models for not only hot electrons in Si but also hot holes in Si and hot electrons in SiO_2, whose accuracy significantly affects the modeling of oxide degradation.

3 Analysis of oxide reliability using FBMC simulations

3.1 Oxide breakdown induced by SHE injection

FN tunneling injection has been widely used to study the time-dependent dielectric breakdown of gate oxide films. During the FN stress as shown in Fig. 4 (a), hole current is observed at the substrate electrode. It has been suggested that this is caused by the hot hole injection from the anode interface. Experiments have shown that an oxide's lifetime t_{BD} can be projected from the time required for the hole fluence to reach some critical value Q_p, independent of the gate voltage (Chen et al. 1986):

$$\int_0^{t_{BD}} J_p(t)\, dt \equiv Q_p, \tag{1}$$

where $J_p(t)$ is the substrate hole current divided by gate area. This observation is the experimental basis for the anode-hole injection model, in which the oxide breakdown is explained by considering that holes are responsible for the damage

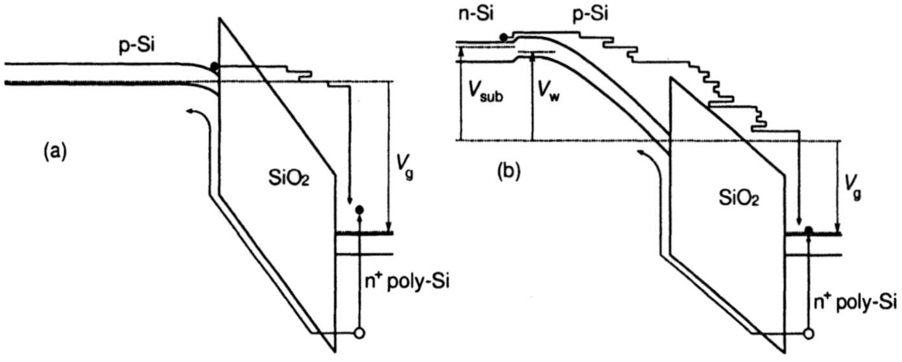

Fig. 4. Schematic energy band diagram illustrating (a) FN and (b) SHE injection.

leading to oxide degradation and final breakdown (Degraeve et al. 1998). However, because FN injection needs high oxide field (> 10 MV/cm), it is possible to consider that the electric field itself causes the damage creation (McPherson et al. 2000). In this study we investigated oxide breakdown using a different stress technique, i.e., SHE injection, with which we can independently control the oxide field and carrier energy. Figure 2 (b) shows the schematic energy band diagram illustrating SHE injection. Cold electrons, injected into the p-Si layer from the n-Si by forwardly biasing, are accelerated towards the Si/SiO_2 interface, and some of them gain sufficient energy to surmount the Si/SiO_2 barrier. Because the hole current through the oxide layer cannot be separated from the electron current, it is impossible to verify experimentally whether the holes also control the oxide breakdown induced by SHE stress. We calculated the hole current theoretically by means of MC simulation, and investigated the origin of the oxide degradation that eventually triggers the destructive breakdown.

The devices used are n-MOSFETs having conventional SiO_2 gate oxide, n^+ poly-Si gates, p-Si wells, and n-Si substrates. The oxide thickness is 5 nm, and the channel doping concentration is $5 \times 10^{17}\text{cm}^{-3}$. By using FN and SHE injection, we measured the gate current as a function of stress time until the occurrence of the oxide breakdown. In order to evaluate the hole current through the oxide layer, FBMC simulations were performed for the electron heating in the Si substrate and energy relaxation processes of electrons and holes in the poly-Si gate. Electrons in SiO_2 are computed with an analytical band MC model reported by Arnold et al. (1994). The hole injection into SiO_2/poly-Si barrier, which is lowered by an image force effect, is calculated with the tunneling probability based on the WKB approximation. Figure 5 shows the experimental charge-to-breakdown $Q_{\rm BD}$ as a function of V_g for FN and SHE injections. $Q_{\rm BD}$ is defined as the total charge fluence through the oxide layer until the occurrence

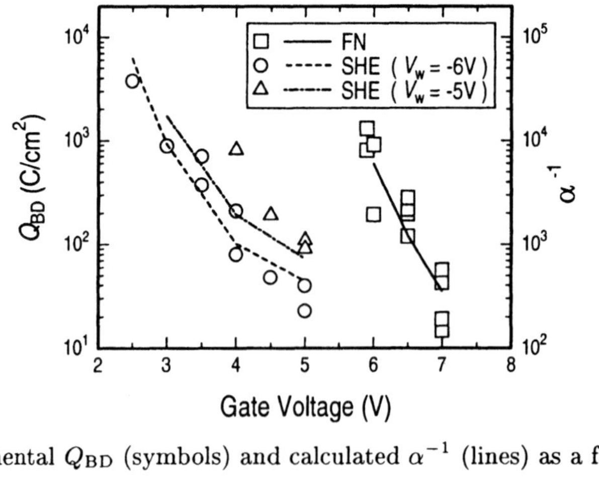

Fig. 5. Experimental $Q_{\rm BD}$ (symbols) and calculated α^{-1} (lines) as a function of gate voltage.

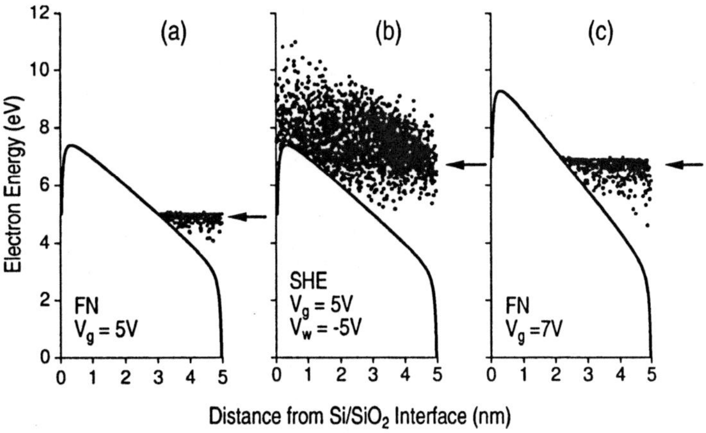

Fig. 6. Snapshots of the simulated electrons in the oxide layer for various stress conditions. The heavy lines indicate the conduction band edge of the gate oxide, and the energy is measured from the bottom of the conduction band in the gate poly-Si.

of the breakdown. The dependence of Q_{BD} on the bias conditions and the injection techniques can be understood in terms of the electron energy. Figure 6 shows the snapshots of the simulated electrons in the oxide layer. Under the same V_g (= 5 V) the electron energy at the anode interface for FN injection is much higher than that of SHE injection as shown in Figs. 6 (a) and 6 (b). In Fig. 5, Q_{BD} for FN injection is also much larger than that of SHE injection. We have to apply higher V_g for FN injection (see Fig. 6 (c)) to obtain the energy distribution similar to SHE injection. Note that under these conditions Q_{BD} for both injection techniques are comparable. These results indicate the strong correlation between the oxide breakdown and the electron energy at the anode. For further examination to the degradation mechanism, we computed the hole injection from the anode interface. In Fig. 5 we plot the simulated results of the inverse of the hole injection efficiency $J_p/J_g \equiv \alpha$, where J_g is the gate current density. On the basis of the anode-hole injection model, Q_{BD} is related to Q_p as $Q_{\mathrm{BD}} = Q_p/\alpha$. Good agreement between experiment and simulation is obtained with the same Q_p regardless of the injection technique. This indicates that the anode hole injection also controls the oxide's lifetime stressed by SHE injection.

3.2 Reliability projection of ultra-thin oxide

The aggressive scaling of oxide thickness into direct-tunneling regime has drastically reduced the margin for error in reliability projection (Degraeve et al. 2000). In order to understand how to extrapolate the results of accelerated lifetime tests to operation voltage, the mechanism of oxide degradation must be understood exactly. Even if we assume the anode-hole injection model, the situation is not so easy as for thicker oxides. Recent experiments have demonstrated that a strong correlation exists between the degradation of ultra-thin oxides and the hole energy injected by direct tunneling (Deguchi et al. 2000). That is, the lifetime of ultra-thin oxide (< 3 nm) is a function of not only the hole fluence as assumed in Sec. 3.1, but also the hole energy. Moreover, this means that the channel hot holes can create oxide defects via direct tunneling even though they have insufficient energy to surmount the energy barrier (DiMaria et al. 2001). The rigorous modeling of hot carrier behavior in Si is still important for discussing the reliability issues of advanced CMOS technology using reduced supply voltage.

4 Conclusion

We have demonstrated that the FBMC simulation is a very helpful tool for investigating the mechanism of oxide degradation. The MC simulator, whose accuracy have been verified by the quantum-yield experiment, has been utilized to analyze the oxide breakdown induced by FN and SHE injections. The strong correlation between the oxide breakdown and the electron energy was found, and it was shown that the holes generated in the anode electrode play an important role in the oxide degradation. In addition, it was indicated that the rigorous modeling of hot carrier behavior in Si is still important for discussing the reliability of aggressively scaled oxide films used in advanced CMOS technology.

Acknowledgement

This work was supported by Semiconductor Technology Academic Research Center (STARC).

References

[1] Abramo, A. , Baudry, L., Brunetti, R., Castagne, R., Charef, M., Dessenne, F., Dollfus, P., Dutton, R., Engl, W.L., Fauquembergue, R., Fiegna, C., Fischetti, M.V., Galdin, S., Goldsman, N., Hackel, M., Hamaguchi, C., Hess, K., Hennacy, K., Hesto, P., Higman, J.M., Iizuka, T., Jungemann, C., Kamakura, Y., Kosina, H., Kunikiyo, T., Laux, S.E., Lin, H., Maziar, C., Mizuno, H., Peifer, H.J., Ramaswamy, S., Sano, N., Scrobohaci, P.G., Selberherr, S., Takenaka, M., Tang, T.-W, Taniguchi, K., Thobel, J.L., Thoma, R., Tomizawa, K., Tomizawa, M., Vogelsang, T., Wang, S.-L., Wang, X., Yao, C.-S., Yoder, P.D., Yoshii, A. (1994): A comparison of numerical solutions of the Boltzmann transport equation for high-energy electron transport silicon. IEEE Trans. Electron Devices 41: 1646-1654

[2] Arnold, D., Cartier, E., DiMaria, D.J. (1994): Theory of high-field electron transport and impact ionization in silicon dioxide. Phys. Rev. B 49: 10278-10297

[3] Bude, J.D., Weir, B.E., Silverman, P.J. (1998): Explanation of stress-induced damage in thin oxides. In: Technical Digest of International Electron Devices Meeting, IEDM '98, San Francisco, CA. IEEE, Piscataway, pp. 179-182

[4] Cartier, E., Fischetti, M.V., Eklund, E.A., McFeely, F.R. (1993): Impact ionization in silicon. Appl. Phys. Lett. 62: 3339-3341

[5] Chang, C., Hu, C., Brodersen, R.W. (1985): Quantum yield of electron impact ionization in silicon. J. Appl. Phys. 57: 302-309

[6] Chen, I.C., Holland, S., Young, K.K., Chang, C., Hu, C. (1986): Substrate hole current and oxide breakdown. Appl. Phys. Lett. 49: 669-671

[7] Cohen, M.L., Bergstresser, T.K. (1966): Band structures and pseudopotential form factors for fourteen semiconductors of the diamond and zinc-blende structures. Phys. Rev. 141: 789-796

[8] Degraeve, R., Groeseneken, G., Bellens, R., Ogier, J.L., Depas, M., Roussel, P.J., Maes, H.E. (1998): New insights in the relation between electron trap generation and the statistical properties of oxide breakdown. IEEE Trans. Electron Devices 45: 904-911

[9] Degraeve, R. (ed.) (2000): Topical issue on oxide reliability. Semicond. Sci. Technol. 15: 425-490

[10] Deguchi, K., Ishida, A., Uno, S., Kamakura, Y., Taniguchi, K. (2000): Degradation of direct-tunneling gate oxide under hot-hole injection. Appl. Phys. Lett. 77: 1384-1386

[11] DiMaria, D.J., Stathis, J.H. (2001): Anode hole injection, defect generation, and breakdown in ultrathin silicon dioxide films. J. Appl. Phys. 89: 5015-5024

[12] Ezaki, T., Nakasato, H., Yamamoto, T., Hane, M. (2000): Simulation of hot hole currents in ultra-thin silicon dioxides: the relationship between time to breakdown and hot hole currents. In: Proceedings of the 2000 International Conference on Simulation of Semiconductor Processes and Devices, SISPAD 2000, Seattle, WA. IEEE, Piscataway, pp. 34-37

[13] Fischetti, M.V., Laux, S.E. (1988): Monte Carlo analysis of electron transport in small semiconductor devices including band-structure and space-charge effects. Phys. Rev. B 38: 9721-9745

[14] Grant, W.N. (1973): Electron and hole ionization rates in epitaxial silicon at high electric fields. Solid-State Electron. 16: 1189-1203

[15] Kamakura, Y., Kawashima, I., Deguchi, K., Taniguchi, K. (2000): Verification of hot hole scattering rates in silicon by quantum-yield experiment J. Appl. Phys. 88: 5802-5809

[16] Kane, E.O. (1967): Electron scattering by pair production in silicon. Phys. Rev. 159: 624-631

[17] Kunikiyo, T., Takenaka, M., Kamakura, Y., Yamaji, M., Mizuno, H., Morifuji, M., Taniguchi, K., Hamaguchi, C. (1994): A Monte Carlo simulation of anisotropic electron transport in silicon including full band structure and anisotropic impact-ionization model. J. Appl. Phys. 75: 297-312

[18] Kunikiyo, T., Takenaka, M., Morifuji, M., Taniguchi, K., Hamaguchi, C. (1996): A model of impact ionization due to the primary hole in silicon for a full band Monte Carlo simulation. J. Appl. Phys. 79: 7718-7725

[19] McPherson, J.W., Khamankar, R.B., A.Shanware, A. (2000): Complementary model for intrinsic time-dependent dielectric breakdown. J. Appl. Phys. 88: 5351-5359

[20] Ottaviani, G., Reggiani, L., Canali, C., Nava, F., Quaranta, A.A. (1975): Hole drift velocity in silicon Phys. Rev. B 12: 3318-3329

[21] Shichijo, H., Hess, K. (1981): Band-structure-dependent transport and impact ionization in GaAs, Phys. Rev. B 23: 4197-4207

Acceleration of Lattice Monte Carlo Simulations and Application to Diffusion/Clustering of As at High Concentrations

Scott T. Dunham and Zudian Qin
Department of Electrical Engineering, University of Washington
Seattle, WA 98195
dunham@ee.washington.edu

Abstract

Kinetic lattice Monte Carlo (KLMC) simulations enable practical atomic-scale modeling of device fabrication processes. In this paper, we discuss implementation of an acceleration algorithm which can provide orders of magnitude speed-up. We apply the result to diffusion and clustering of As at very high doping levels, which are critically important in the formation of ultrashallow junctions.

1. Introduction

Kinetic lattice Monte Carlo (KLMC) simulations are a promising approach for atomic-scale modeling of device fabrication processes [1–4]. By considering only the transitions (and not lattice vibrations) associated with the defects and impurities present in the system, KLMC is able to overcome the time-scale limits associated with molecular dynamics to consider macroscopic systems and time scales. In previous work, we have demonstrated the ability to model the fabrication of deep submicron device structures [2]. However, KLMC simulations are still much more CPU intensive than continuum simulations. In this paper, we discuss implementation of an acceleration algorithm, and we apply the result to a problem which has led to controversy in the literature [1, 5] and is critical for formation of low resistance shallow junctions: diffusion and clustering of As at very high doping levels.

The KLMC systems considers impurities and point defects located at silicon lattice sites. The energy of the system is defined based on the atomic scale arrangement, with parameters determined from *ab-initio* calculations and/or experimental observations. The system evolves through transitions from one atomic configuration to the next. The rates of these transitions can be determined based on the migration barriers and changes in system energy associated with the transition: $\nu = \nu_0 \exp(-E_m/kT)\exp[(E_i - E_f)/2kT]$, where E_m is the migration energy, E_i is the energy of the initial configuration and E_f is the configuration of the final state. At each time step, a transition is chosen from the full possible set based on the relative rates, and the time step is advanced by the inverse of the sum of the rates [1, 4].

2. Acceleration Algorithm

For heavily As-doped material, the simulation spends the bulk of its time transitioning among a set of coupled low energy states associated with As_nV

complexes, without resulting in long-range diffusion or advancing the system rapidly forward in time (Fig. 1). To overcome this bottleneck, we can generalize the concept of a state to include a group of states which approaches local equilibrium via extensive sampling within the expected time for a transition out of the set [4]. In this case, we can immediately jump to transitions out of this group of states, with a correspondingly large time acceleration. More formally, a set of configurations $\mathcal{K}$ can be replaced by a single effective state provided the transitions between members of the set are fast compared to transitions between a member of the set and configurations that are not members. In this case, the transition rate to a configuration j outside $\mathcal{K}$ is

$$\nu(\mathcal{K} \to j) = \sum_{k \in \mathcal{K}} p_k \nu(k \to j) = \sum_{k \in \mathcal{K}} \exp\left(-\left[E_k - E_{\mathcal{K}}\right]/kT\right) \nu(k \to j), \quad (1)$$

where p_k is the probability of configuration k within set $\mathcal{K}$, and $E_{\mathcal{K}} = -kT \ln\left[\sum_{k \in \mathcal{K}} \exp\left(-E_k/kT\right)\right]$. From a practical perspective, we check for existence of a group of tightly coupled states when the system revisits a given state with high frequency over a short time period. Starting from this state, the set of coupled states is then expanded automatically via the requirement that the rates out of the set are at least M times slower than rates within the set (typically, $M \sim 100$).

3. Simulation Results

We have applied this approach to simulation of diffusion of As at high concentrations. In our previous work based on As/V attraction at 1NN to 3NN separation [1], we found a rapid increase in diffusivity (beyond Fermi level effects) as the doping concentration increased above 10^{20}cm^{-3} which was in agreement with experimental results of Larsen *et al.* [6] (Fig. 2). However, List *et al.* [5] tried to reproduce these results and reported that the apparent As diffusivity actually dropped with doping concentration. They were only able to obtain enhancement as observed by Larsen *et al.* by eliminating 1NN binding, in contradiction to experimental [7] and theoretical [8] results. The reason for this discrepancy lies in the methods and time scales considered. In our previous work [1], we considered only very short time scales, before significant As/V cluster formation could occur. For longer times in a system with fixed V concentration (as analyzed by List *et al.*), the diffusivity drops sharply with doping due to formation of relatively immobile As_nV [3].

Because the period before clustering begins becomes extremely short as doping level rises, it is very difficult to get accurate statistics using this approach. We find that it is possible to consider more reasonable time scales and yet correctly extract the arsenic diffusivity versus concentration by recognizing that as clustering occurs, the equilibrium number of vacancies rises dramatically as well. We correct for this factor by scaling the time by the ratio of equilibrium to actual number of free vacancies in the system (Fig. 3). There is still a problem at high doping concentrations as vacancies get trapped in As_nV complexes. However, via use of the algorithm described above, we are able to accelerate

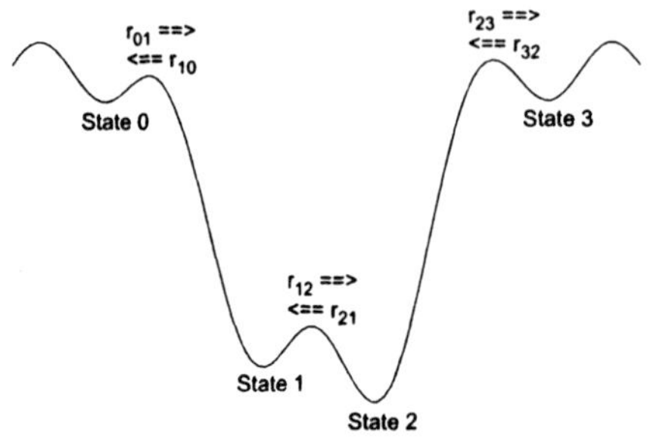

Fig. 1: Schematic of energy versus configuration for system with fast transitions between a small subset of states. If $r_{10}, r_{23} \ll r_{12}, r_{21}$, then a large number of steps will simply involve exchange betwen states 1 and 2. By detecting that a local equilibrium exists between these two states, we can ignore the r_{12}, r_{21} transitions and consider only the transitions out of the set (e.g., $r_{\mathcal{K}0} = p_1 r_{10} = r_{21} r_{10}/(r_{21} + r_{12})$).

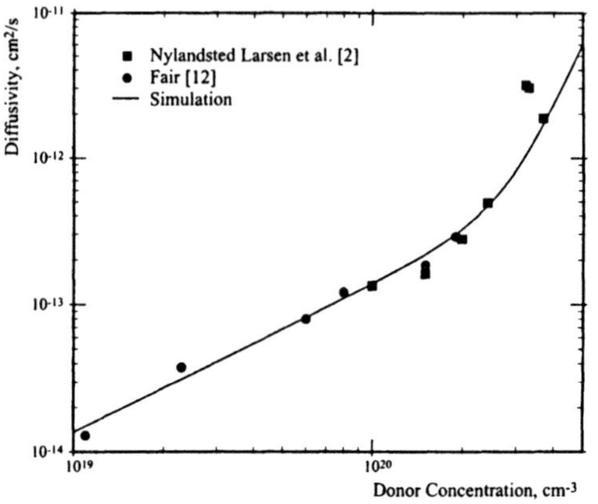

Fig. 2: Comparison of experimental measurement [6, 9] of arsenic diffusivity at 1050°C versus doping level to lattice Monte-Carlo simulations using short simulation times to avoid clustering [1]. The diffusivity for moderate doping levels is fit assuming diffusion via negatively-charged vacancies and then extrapolated to higher doping levels using the simulation results.

out of these states and thus simulate the long time periods needed to accurately extract equilibrium diffusivities.

For our simulations, we use arsenic/vacancy interactions out to 3NN from *ab-initio* calculations [8]. The results for normalized diffusivity versus arsenic concentration at 1000°C are shown in Fig. 4. We find a rapid increase in diffusivity above $2 \times 10^{20}\text{cm}^{-3}$, as in previous work [1] and experiments [6].

The acceleration algorithm enables practical simulation of large systems with strong binding energies. We use KLMC to analyze diffusion and clustering during annealing of laser or RTA activated junctions. In these systems, rapid arsenic deactivation is observed [10] due to As_nV cluster formation driven by the greatly enhanced diffusion at high concentrations described above. This process also requires vacancy generation as diffusion from the surface is much too slow to account for deactivation kinetics at short times and low temperatures. We also used KLMC simulations to explore the formation of As clusters which are favorable for enhanced Frenkel pair generation via kick-out of central silicon atom. We find that these structures arise naturally via interstitial-mediated diffusion of As which is enhanced by the same Si kick-out process. By including I kick-out, we are able to use KLMC for atomistic modeling of diffusion and clustering of As at high concentrations.

References

[1] S. T. Dunham and C. D. Wu, J. Appl. Phys. **78**, 2362 (1995).

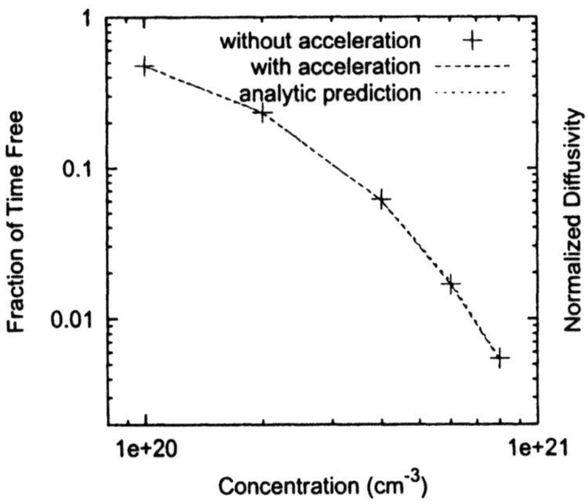

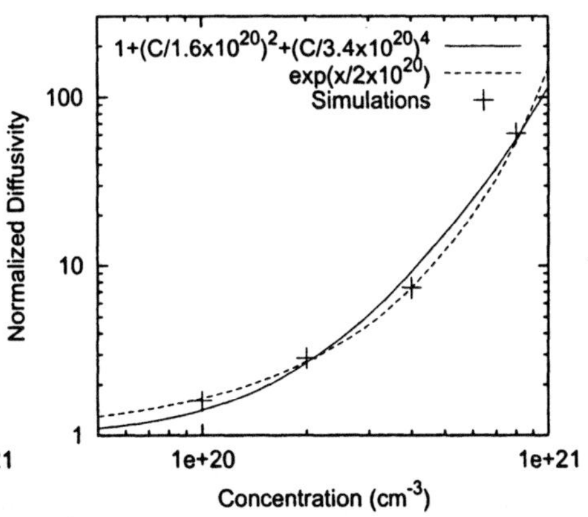

Fig. 3: Comparison of fraction of time that a vacancy is free from interactions with arsenic atoms as function of doping concentration calculated via three different methods: standard simulation without acceleration, simulation with acceleration, and analytic expression based on system probabilities. In all three cases, the results are indistinguishable, demonstrating the accuracy of both the acceleration algorithm and the analytic prediction.

Fig. 4: Simulated diffusivity of As versus doping concentration normalized by dividing by normal Fermi-level dependent diffusivity ($\propto n/n_i$). The results are based on *ab-initio* As/V binding energies at $T = 1000°C$. A rapid increase of diffusivity with polynomial or exponential doping dependence is seen.

[2] M. M. Bunea and S. T. Dunham, in **Semiconductor Process and Device Performance Modeling,** S. Dunham and J. Nelson, eds. (Mat. Res. Soc. Proc. **490**, Pittsburgh, PA, 1998) pp. 3–8.

[3] M. M. Bunea and S. T. Dunham, in **Defects and Diffusion in Silicon Processing,** S. Coffa, T. Diaz de la Rubia, C. Rafferty, and P. Stolk, eds. (Mat. Res. Soc. Proc. **469**, Pittsburgh, PA, 1997) pp. 353–358.

[4] A. Horsfield, S. T. Dunham and H. Fujitani, in **Multiscale Modeling of Materials,** V. Bulatov, T. de la Rubia, N. Ghoniem, E. Kaxiras and R. Phillips, eds. (Mat. Res. Soc. Proc. **538**, Pittsburgh, PA, 1999).

[5] S. List and H. Ryssel, J. Appl. Phys. **83**, 7585 (1998).

[6] A. Nylandsted Larsen, K. Kyllesbech Larsen, P. E. Andersen, B. G. Svensson, *J. Appl. Phys.* **73**, 691 (1993).

[7] M. Hirata, M. Hirata and H. Saito, *J. Phys. Soc. Jap.* **27**, 405 (1969).

[8] O. Pankratov, H. Huang, T. D. de la Rubia and C. Mailhiot, Phys. Rev. B, **56**, 13172 (1997); J. S. Nelson, P. A. Schultz and A. F. Wright, *Appl. Phys. Lett.* **73**, 247 (1998); M. Berding, A. Sher, *Phys. Rev. B* , **58**, 3853 (1998); J. Xie and S.P. Chen, *J. Appl. Phys.* **87**, 4160 (2000).

[9] R. B. Fair in **Impurity Doping Processes in Silicon**, ed. F. F. Y. Yang (North-Holland, New York, 1981) p. 315.

[10] Scott Luning, PhD thesis, Stanford University (1996).

Interstitial Cluster Evolution and Transient Phenomena in Si-crystal

Antonino La Magna, Salvo Coffa, Sebania Libertino, Matthias Strobel
CNR-IMETEM Stradale Primosole 50, I95121
Catania, Italy
lam@imetem.ct.cnr.it

Luciano Colombo
Dipartimento di Fisica, INFM and Università di Cagliari, Cittadella Universitaria,
Cagliari Italy

Abstract

The evolution of Interstitial (I) type defects in Si and its influence on out of equilibrium I super-saturation level is investigated. Two approaches complementary to Quantum Mechanics Calculations (QMC) are applied: the Kinetic Lattice Monte Carlo (KLMC) and the Non-Lattice Kinetic Monte Carlo (NKMC). Our simulations show that the behaviour of I-super-saturation during a far from equilibrium stage is strongly affected by the correspondent aggregate structural evolution. Therefore, even if KLMC and NKMC are based on the same energetics derived by QMC, they give a different prediction of the super-saturation behaviour.

1 Introduction

Transient phenomena occurring in Si, due to Interstitial (I) kinetic attracts a wide interest; and, nowadays, a large set of experimental data and theoretical results is available on this field [1-5]. In spite of this effort, a definitive interpretation of the experiments by means a reliable theory of the agglomeration kinetics is still lacking. Quantum Mechanics Calculations (QMC) evidenced a particular feature of stable I-clusters: i.e., in the small size regime, compact cluster formed by over coordinated atom are more stable that rod like defect [3-5]. One intriguing issue is the role of the structure-energetic relationship on the I kinetics evolution and, conversely, if a reliable description of such evolution can be obtained neglecting this relation.
In order to achieve such task, we compare the prediction of two kinetics approach based on the same energetics. In the first, the Non Lattice Kinetic Monte Carlo (NKMC) approach [6,7], one assumes that, in a capture (dissolution) event, a cluster of n defects instantaneously achieves the lowest energy configuration for a cluster with $n+1$ ($n-1$) members. In the second, the Kinetic Lattice Monte Carlo (KLMC) [7-8] approach, the relationship between energetic and cluster configuration is taken into account using an interaction model which matches the results of QMC. Thus, it can simulate structural transition of a given agglomerate between inherently different structures. These features allow to explore in detail the relationship between the structural evolution and the behavior free I density maintained by the clusters.

2 I cluster energetic and kinetic approaches

In the two stochastic approaches the kinetics is simulated by means a sequence of atomic transitions picked randomly according their own transition rate given by [6,7]

$$v(i \Rightarrow f) = v_\kappa \, x \, min\{1, exp[-(E(c_f) - E(c_i)/ k_B T]\} \qquad (1)$$

Here $E(c)$ is the total energy of the configuration c, T the reservoir temperature and v_κ is the attempt frequency of the κ-th transition allowing the transition from c_i to c_f. In the NKMC method, the change of energy is $E(c_f) - E(c_i) = E_b(n)$ for a dissolution event (where $E_b(n)$ is the binding energy, i.e. the energy cost for the detachment of one I from the more stable n-sized cluster); whilst $E(c_f) - E(c_i) \leq 0$ for other processes (capture, monomer random jump).

In the KLMC approach, the energy is evaluated using a defect-defect potential formulated on the basis of the static calculated by QMC [7-8]. The configuration are mapped in a reference (super)lattice formed by N_{sites} of a Si diamond lattice and N_{sites} associated tetrahedral (Td) sites [10]. The number of Monte Carlo particles is $N_{atoms} = N_{sites} + N_I$, where N_I is the number of Is. Kinetics is driven by two kinds of elementary transitions: a) the displacement of an atom from a Td site to a nearest neighbor (n.n.) Td site ($\kappa \equiv I_T \Leftrightarrow I_T$), b) the displacement of a regular atom to an empty n.n Td site or vice-versa ($\kappa \equiv N \Leftrightarrow I_T\text{-}V$). The latter process leads to the formation or to the dissolution of an instable $I_T\text{-}V$ complex (with an energy cost $\pm E_{IV}$), if the remaining n.n Td sites are empty. Otherwise, if defective Si atoms fill these Td sites, the formation of the $I_T\text{-}V$ complex activates the possibility of binding (with binding energy E_b,) the I, belonging to the complex, to those atoms. In order to match the energetic of compact clusters calculated by the QMC, the cost of over coordination must also be taken into account [4]. Therefore we introduce two energy levels: E_5 for a five fold coordinated atom and E_6 for a six fold coordinated one ($E_6 > E_5 > 0$) [8]. Chainlike structures (structural unit of {311} defects) are mapped as correlated adjacent I_Ts lying in a <110> chain of the Td sub-lattice. Their energetics can be modeled by means of two different energy contribution E_{ex} or E_{in} ($E_{ex} > E_{in}$) for atom at the extreme and in the inner part of the chain respectively.

The energy difference in (1) can be evaluated in term of the variations Δn_α of the number of the single contributions $E(c_i) - E(c_f) = \Sigma_\alpha \Delta n_\alpha E_\alpha + \Delta_\kappa$, where $\Delta_\kappa = 0, E_{IV}, -E_{IV}$ when $\kappa = I_T \Rightarrow I_T$, $N \Rightarrow I_T\text{-}V$, $I_T\text{-}V \Rightarrow N$, respectively. We can use the formations energy (calculated by QMC) of six I-defect configurations in order to evaluate the six KLMC parameters. Using the Tight Binding Molecular Dynamics TBMD [3,4] or the Local Density Approximation DFT-LDA [5] calculations, we obtain the following set of KLMC parameters

$$E_{IV} = 4.0;\ E_b = 4.6;\ E_5 = 2.0;\ E_6 = 4.3;\ \ E_{ex} = 0.4\ \ E_{in} = -2.7 \quad (TBMD) \qquad (2)$$

$$E_{IV} = 4.0,\ E_b = 4.5,\ \ E_5 = 2.2;\ E_6 = 4.2;\ \ E_{ex} = -0.9\ \ E_{in} = -2.83 \quad (DFT\text{-}LDA). \qquad (3)$$

The agreement between KMLC and QMC energetics is not restricted to the six configurations considered, in order to extract parameters, but it occurs (irrespective to the cluster size) for all I-type defect belonging to the two structure classes.

3 I system kinetics and super-saturation behaviour

Snapshots of the I kinetics evolution are shown in Fig.1 for a system at T=650°C temperature (in this case KLMC model is set in order to recover energetic calculated by means TBMD) [3,4]) . The initial state is a homogeneous distribution of N_I =32 Td Is and N_{sites}=262144 Si atoms sitting in regular lattice sites.

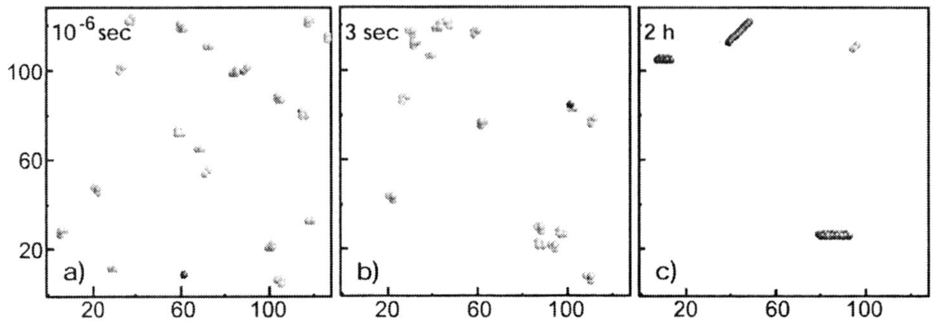

Fig.1 Snapshots of interstitial aggregates obtained by the simulated KLMC evolution for a closed system with N_{sites} = 262144, and N_I =32 at T=650°C at three different times. Bright spheres indicate atoms in compact cluster structures whilst dark spheres indicate atoms in chainlike structures.

Three stages characterize the kinetics: the nucleation of compact cluster initiated by Is diffusion along tetrahedral-dumbbell path (fig.1a); the formation of non correlated complexes formed by adjacent compact clusters (fig.1b), the structural transition from these complex to correlated chainlike structures, dominating the long time evolution (fig.1c). Structural transition to chainlike nucleus can occur only the system overcomes, by means thermal fluctuation, the kinetic barriers between the two configurations: this causes the metastable character of the stage dominated by compact cluster [8].

These kinetic features affect the behaviour of the super-saturation level S which the aggregates maintain in the space region where they are located. This is quantity correspond also to the diffusivity enhancement, during transient stages, for impurities which migrate by interstitials. By means of our simulation, we have calculated a parameter proportional to S: the rate $s(t) = \Delta\, t_{free}(t) / \Delta t_{sample}(t)$; here $\Delta\, t_{free}(t)$ is the time which the system spends to perform free I diffusive transitions during a suitable sampling time $\Delta\, t_{sample}(t)$. In Fig. 2 $s(t)$ is shown for a closed system (N_{sites}=262144, $N_I = 128$) with DFT-LDA energetics but with $E_{ex} = 0.1$ [5]. Compact clusters maintains the plateau of $s(t)$ between early stages (nucleation of compact clusters) and later stages (presence of chain structures). This behaviour $s(t)$ cannot be recovered by NKMC exploiting the same static (for the lowest energy states). Indeed it foresees a transition to the $s(t)$ level of rod-like defect just after $t > 10^{-5}$ sec. Anyway, the $s(t)$ dependences, calculated by KLMC, can be used to fit an $E^{Fit}{}_b(n)$ relation for a NKLC reproducing similar $s(t)$ behaviour (see fig. 2 dashed line and open circles). This fitting procedure gives the following value of the binding energy: $E^{Fit}{}_b(n) = 1.3, 1.0, 1.4, 1.0, 1.7, 1.1, 2.3, 1.3, 2.0 \ldots 2.2$ *for* $n=2,3,4,5,6,7,8,9,10,$

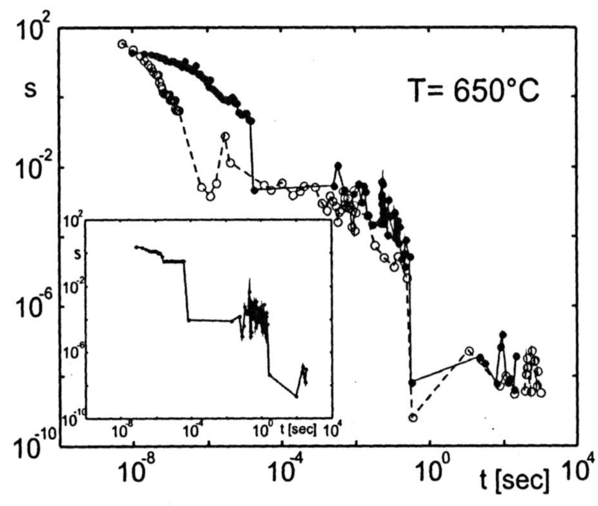

Fig 2 s(t) parameter for a closed system with N_{sites}=262144, N_I=128 and T = 650°C derived by a KLMC simulation (solid line and •) and by a NKMC simulation using with the binding energy dependence: $E_b^{Fit}(n)$ (dashed line and o). In the insert s(t) calculated by KLMC is for a system with N_{sites} = 262144, N_I=16.

...$n \geq 11$. Using the same $E^{Fit}{}_b(n)$ values a good agreement of the $s(t)$ dependences derived by KLMC and NKMC is recovered for the same system at different T. However, when a different density $\rho' \neq \rho$ of Is is considered, the NKMC model with the set of $E^{Fit}{}_b(n,\rho)$ values extracted gives an prediction of the $s(t)$ behaviour which is in total disagreement with that foreseen by KLMC. Thus, in order to recover again a similar behaviour, a different $E^{Fit}{}_b(n,\rho')$ dependence on n should be imposed. For the same system but having a eight fold lower density of Is (see insert in fig.2) we have $E^{Fit}{}_b(n,\rho')$ = *1.3, 1.0, 1.8, 1.0, 2.3, 1.1, 2.8, 1.3, 2.0 ...2.2 for* $2 \leq n \leq 10$ ...$n \geq 11$. These results make questionable the use of theories, which do not take into account the actual structural evolution of the cluster, to investigate non equilibrium I-kinetics: also the behaviour of average system properties (e.g. $s(t)$) is non predictable.

References

[1] S. Libertino, S. Coffa, J.L. Benton, K. Halliburton, D.J. Eaglesham, Nucl. Instr. Meth. B 1999, 148: 247-250.

[2] N.E.B Cowern et. al, Phys. Rev. Lett. 1999, 82: 4460-4463

[3] J.Kim, et. al., Phys. Rev. B 1997, 55: 16186-16195.

[4] A. Bongiorno, L. Colombo, F. Cargnoni, C. Gatti and M. Rosati, Europhys. Lett. 2000, 50: 608-611.

[5] J.Kim, F. Kirchhoff, J.W. Wilkins and F. S. Khan, Phys. Rev. Lett 2000, 84:503-506.

[6] L. Pelaz et al Appl. Phys. Lett 1997, 70: 2285-2287.

[7] A. La Magna, S. Coffa and L. Colombo Phys. Rev. Lett.1999, 82: 1720-1723.

[8] A. La Magna, S. Coffa and S. Libertino Mat. Res. Soc. Symp. series 2000, vol 610, B.11.

Monitoring Arsenic In-Situ Doping with Advanced Models for Poly-Silicon CVD

W. Pyka, C. Heitzinger, N. Tamaoki*, T. Takase*, T. Ohmine*, and S. Selberherr

Institute for Microelectronics, TU Vienna, Gusshausstrasse 27–29, A-1040 Vienna, Austria
Phone: +43-1-58801/36026, Fax: +43-1-58801/36099, Email: Pyka@iue.tuwien.ac.at
*TOSHIBA Corporate R&D Center, Kawasaki, Japan

Abstract

Experiments of As-doped poly-silicon deposition have shown that under certain process conditions step coverages > 1 can be achieved. We have developed a new model for the simulation of As-doped poly-silicon deposition, which takes into account surface coverage dependent sticking coefficients and surface coverage dependent As incorporation and desorption rates. The additional introduction of Langmuir type time-dependent surface coverage enabled the reproduction of the bottom-up filling of the trenches. In addition the rigorous treatment of the time-dependent surface coverage allows to trace the in-situ doping of the deposited film. Simulation results are shown for poly-Si deposition into $0.1\,\mu$m wide and $7\,\mu$m deep, high aspect ratio trenches.

1 Introduction

State-of-the-art memory cells are based on deep trench fabrication technologies. Thereby void-less filling of the deep trenches is one of the crucial process steps. Experiments of As-doped poly-silicon deposition have shown that under certain process conditions step coverages > 1 can be achieved. This means that the deposition rate down at the bottom of the trench is higher than at the feature opening which leads to an unexpected but highly desired bottom-up filling of the trench.

In order to trace the origin of this phenomenon and for a controlled application of this special behavior we have developed a new simulation model for As-doped poly-silicon deposition. The model takes into account surface coverage dependent sticking coefficients, surface coverage dependent As incorporation and desorption rates, and Langmuir type time-dependent surface coverage. In addition the in-situ doping of the deposited film is traced.

2 Model Setup

The model is based on the coupling between topography simulation and diffusion and reaction modeling carried out with the partial differential equation solver AMIGOS [1]. The most important expansion to the former works presented in [2] is the introduction of a time-dependent surface coverage. Due to the special surface chemistry of the process, assumptions for a fast achievement of a steady-state surface coverage did not hold any longer and a rigorous time-dependent calculation of the surface coverage was required. In contrast to the concentrations of the reactive gas species, the surface coverage is only defined and meaningful at this boundary of the simulation domain, which represents the feature surface. Therefore an overall formulation of the model was designed, where the surface coverage process is specified in a separate boundary model.

The relevant gas species for the poly-silicon deposition process are AsH_3, SiH_4, and SiH_2. Transport of the reactants in the high-pressure regime of some 10 Torr is characterized by the diffusion equation

$$\frac{\partial c_i}{\partial t} = \frac{\partial}{\partial x}(D_i \frac{\partial}{\partial x} c_i), \tag{1}$$

where i stands for the three types of reactants. The deposition velocity is governed by an adsorbed arsenic inhibition layer. The kinetics of the layer formation are given by

$$\frac{\partial C_s}{\partial t} = \eta_{\mathrm{AsH_3}} \Gamma_{\mathrm{AsH_3}} - k_s C_s - k_d C_s, \tag{2}$$

where C_s is the arsenic surface coverage, and $\eta_{\mathrm{AsH_3}}$ and $\Gamma_{\mathrm{AsH_3}}$ denote arsenic sticking coefficient and flux, respectively. k_d is the rate constant of arsenic desorption from the protective layer and k_s the incorporation rate of arsenic from the layer into the deposited poly-silicon film. The sticking coefficients for AsH_3 and SiH_4 are defined as

$$\eta_{\mathrm{AsH_3}} = \eta^0_{\mathrm{AsH_3}}(1 - C_s/C_0) \qquad \text{and} \qquad \eta_{\mathrm{SiH_4}} = \eta^0_{\mathrm{SiH_4}}(1 - C_s/C_0), \tag{3}$$

where $\eta^0_{\mathrm{AsH_3}}$ and $\eta^0_{\mathrm{SiH_4}}$ are the clean surface sticking probabilities and C_0 is the number of available surface sites for arsenic adsorption. The sticking coefficient of SiH_2 was assumed to be unity. The final deposition rate has three contributions, namely poly-Si deposition from surface reactions of SiH_2 and SiH_4 at the unprotected surface and the incorporation of As from the adsorption layer. Thus, the resulting deposition rate DR is given by

$$\mathrm{DR} = \eta_{\mathrm{SiH_4}} \Gamma_{\mathrm{SiH_4}} + \eta_{\mathrm{SiH_2}} \Gamma_{\mathrm{SiH_2}} + k_s C_s. \tag{4}$$

Again, η_i denote the sticking probabilities and Γ_i the corresponding fluxes. Coupling between gas concentrations c_i within the reactor space and the fluxes Γ_i towards the surface is given by the relations

$$\Gamma_i = \frac{1}{4}\nu_i c_i \qquad \text{with} \qquad \nu_i = \sqrt{\frac{8kT}{\pi m_i}}. \tag{5}$$

Thereby, ν_i is the average molecular velocity of gas species i with a molar mass m_i.

The modeling equations described above result in a set of partial differential equations for the concentrations of the reactants and the surface coverage which is iteratively solved for the profile evolving with time. Most importantly, by tracing the actual surface coverage the model allows to monitor the arsenic incorporation into the poly-Si film and thus to trace the in-situ doping of the deposited layer.

The boundary condition for the model are the concentrations of the reactants on top of the simulation domain. Since the presence of arsenic determines the protective film formation and thus the behavior of the deposition rate, the most important influence is given by the AsH_3 partial pressure.

3 Simulation Results

The simulation results for a 0.1 μm wide and 7 μm deep trench in Fig. 1 demonstrate the effect of increasing AsH_3 partial pressure (from left to right) which is controlled by the AsH_3 flow.

When no or little AsH_3 is present (left) no inhibition of the surface reaction is given and the usual step coverage < 1 and severe void formation are observed. With addition

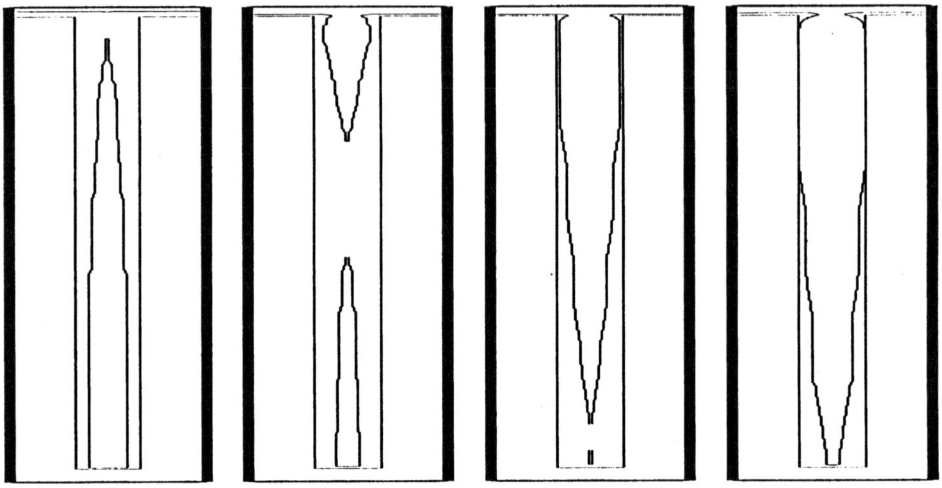

Fig. 1: Profile evolution of As-doped poly-Si deposition with AsH_3 partial pressure increasing from left to right. Note the distorted aspect ratio of the figures, originally it is 0.3 to 7.35 μm.

of AsH_3 an inhibiting layer is formed at the trench opening, lowering the deposition rate but shifting the effect of void formation to a lower region in the trench. With an even higher addition of AsH_3 the deposition is retarded in the complete trench but in the upper part with the higher As concentration more than in the lower part. Bottom-up filling is reached. Still increasing the AsH_3 partial pressure amplifies this effect but the possibility of the formation of a small overhang at the trench opening is given by SiH_2 contributions which under these process conditions become the predominant fraction of the deposition rate. Note that increasing the AsH_3 partial pressure results in stronger inhibition layer formation which reduces the deposition rate. Therefore longer deposition times are necessary to obtain a film thickness of 100 nm, to which the simulations are scaled. The deposition times range from 15 s to 180 s.

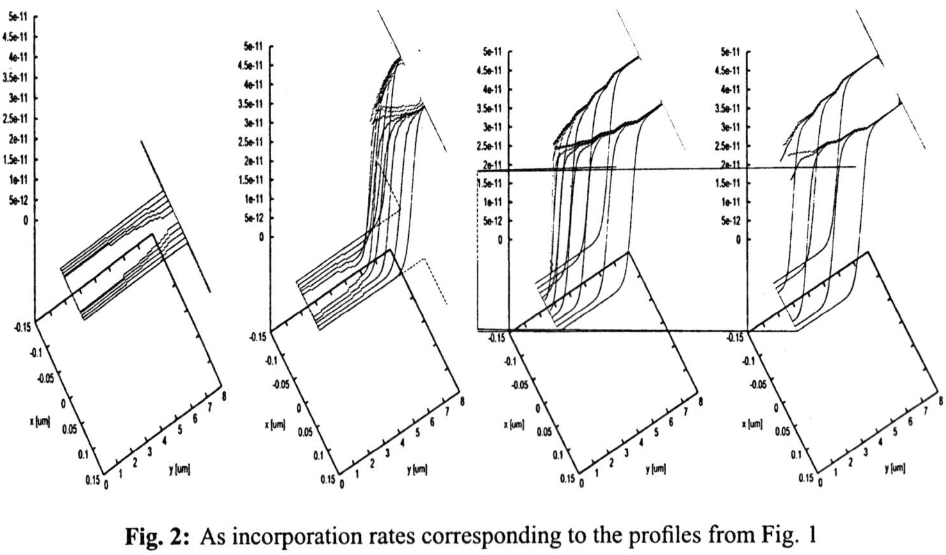

Fig. 2: As incorporation rates corresponding to the profiles from Fig. 1

In Fig. 2 the As incorporation rate corresponding to the profiles from Fig. 1 is shown. It is proportional to the surface coverage prevailing in the different trench regions and represents a measure for the in-situ doping of the deposited film. The figures also suggest that for low AsH_3 flows the surface coverage is concentration limited, whereas it is time limited for higher flows. Concentration limited (the two figures on the left hand side) means that the available As is completely consumed by the layer formation in the upper region of the trench. Even for longer times no significant surface coverage can be observed at the bottom of the trench. Time limitation for higher flows (two figures on the right hand side) means that enough As is present for covering the complete trench, just it takes some time until the As has diffused into the lower regions of the trench.

4 Discussion

The simulation results from above clearly show that the formation of a surface coverage layer has the strongest influence on the profile evolution. The surface layer formation is determined by the amount of As reaching the surface, which in the presented model is determined by three parameters, namely the concentration of AsH_3 on top of the simulation domain (Dirichlet boundary condition), the height of the simulation domain and the transport velocity, given by the diffusivity. Assessing these parameters is crucial for the simulation. In addition, the simulation domain has been split up into two regions for Knudsen diffusivity within the trench (the characteristic diffusion length is determined by the trench diameter) and regions with higher diffusivities outside the trench (boundary layer). This should allow for a more realistic approach of the transport phenomena and for an easier calibration of the parameters.

The simulations shown above demonstrate the fundamental effects of surface layer formation on profile evolution and therefore assume only a very thin boundary layer. Thus the ideal AsH_3 flow obtained from the simulations slightly differs from the experimentally observed one. It should be possible to further reduce this difference by an accurate calibration of the transport parameters from reactor scale (AsH_3 inlet gas flow) to the feature scale concentrations in the boundary layer and close to the trench surface.

5 Conclusion

We have shown that by introducing Langmuir type time-dependent surface coverage into CVD modeling it is possible to simulate the bottom-up filling behavior of As-doped poly-silicon trench filling. Furthermore, the rigorous treatment of the time-dependent surface coverage allows for a monitoring of the in-situ doping of the deposited film, which is meaningful for the characterization of the electrical properties of the deposited poly-Si film. Bottom-up filling can be achieved by increasing the AsH_3 fraction in the process gas. On the other hand this increase in the AsH_3 fraction also reduces the overall deposition velocity of the process. In order to obtain a satisfactory profile evolution the right balance between decreasing void formation probability and the reduction in deposition velocity with respect to manufacturing throughput has to be found.

Acknowledgment

Part of this work is supported by the "Christian Doppler Forschungsgesellschaft", Vienna, Austria.

[1] M. Radi, E. Leitner, and S. Selberherr, "AMIGOS: Analytical Model Interface & General Object-Oriented Solver," *IEEE J.Technology Computer Aided Design*, 1999. `http://www.ieee.org/products/online/journal/tcad/accepted/radi-feb99/`.

[2] W. Pyka, P. Fleischmann, B. Haindl, and S. Selberherr, "Three-Dimensional Simulation of HPCVD—Linking Continuum Transport and Reaction Kinetics with Topography Simulation," *IEEE Trans.Computer-Aided Design of Integrated Circuits and Systems*, vol. 18, no. 12, pp. 1741–1749, 1999.

Equipment and Process Simulation of Compound Semiconductor MOCVD in the Production Scale Multiwafer Planetary Reactor

M. Dauelsberg, M. Deufel, M. Reinhold, G. Strauch
AIXTRON AG,
Kackertstr. 15-17, 52072 Aachen, Germany
dau@aixtron.com

T. Bergunde
Ferdinand-Braun Institute,
Albert-Einstein-Str. 11, 12489 Berlin
bergunde@fbh-berlin.de

Abstract

The article addresses the use of computational modelling during the equipment design and process development of the Planetary Reactor®, an industrial production scale multiwafer reactor for the MOCVD (Metalorganic Chemical Vapour Deposition) of epitaxial compound semiconductor thin films. MOCVD equipment and process simulation is based on the coupled computation of gas flow field, heat transfer, including rf induction and thermal radiation, and chemical species mass transport and reaction kinetics on 2D axisymmetric and 3D computational domains.

1 Introduction

Compound semiconductor heterostructures are the basis for optoelectronic and electronic devices that can be found in numerous consumer products as well as in telecom and data transmission equipment, like lasers, modulators and detectors for fiber optics, or HBTs (Hetero Bipolar Transistors) and HEMTs (High Electron Mobility Transistor) as high frequency power amplifiers for wireless communication. In the near future Ultra High Brightness Light Emitting Diodes (UHB-LEDs) are likely to replace the conventional light bulb in practically any application from domestic and office lighting to traffic signals and outdoor advertising. Metal Organic Chemical Vapor Deposition (MOCVD) is the technique of choice to meet the requirements for high volume production in the compound semiconductor industry. During MOCVD processing a dilute mixture of reactive gases decompose and thin epitaxial layers of III-V materials, like AlGaAs, InGaAsP, InGaN and related compounds, are deposited in a low pressure reactor with substrate temperatures of 600°-800°C for conventional III-V materials and up to 1200°C for GaN based layers. [1] reviews the MOCVD growth of compound semiconductors.
The present paper introduces the use of equipment level computational modelling in the field of compound semiconductor MOCVD. Simulation has been applied to

further develop, improve and optimise equipment performance and process design of the Planetary Reactor® [2], an industrial production scale MOCVD reactor for multiwafer processing of up to 7x6" (or 49x2" equivalently) substrates. Model features and software requirements to describe the MOCVD process are discussed, model validity is demonstrated by comparison between predicted and measured growth rate distributions in the Planetary Reactor®, determined by X-ray diffractrometry. The impact of modelling and simulation on key design issues for the performance of MOCVD reactors is addressed, i.e. gas inlet design, control of reactor scale species depletion, and thermal management, in order to meet stringent production requirements, such as growth rate uniformity, compositional control of multinary III-V layers, and precursor species utilisation efficiency, while ensuring reproducibility and process stability in an industrial volume production environment.

2 Modelling approach

The computational model of the MOCVD growth process is based on first principle prediction of coupled flow, heat and mass transfer by solving the corresponding transport equations on 2D or 3D domains supplemented with appropriate boundary conditions. The MOCVD process model features multi-disciplinary physics and chemistry, as thermal radiation transfer, the electromagnetics for rf induction heating, gas phase and surface chemistry. See [3] for a comprehensive overview of CVD model approaches, and [4] for previous work on Planetary Reactor® modelling. Radiation propagation is computed by Monte-Carlo ray tracing; spectrally dependent optical properties of semitransparent walls and coated surfaces are taken into account. The transport equations are solved by a pressure based Finite-Volume method on unstructured grids [5].

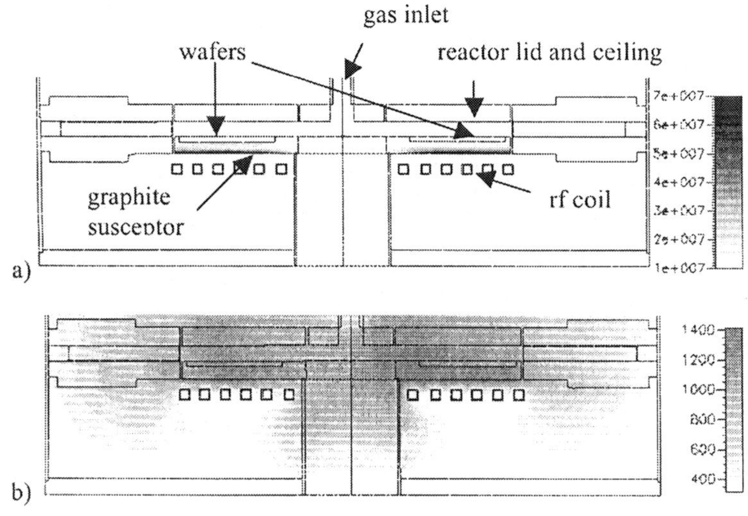

Fig. 1. Calculated inductive power density (W/m^3) distribution (a) and temperature (Kelvin) distribution (b) in the Planetary Reactor® at T = 1150°C (generic design of the reactor).

3 Results

The accurate prediction of detailed heat transfer in the process environment, including thermal radiation as the predominant mode of heat transfer and electromagnetic induction in a rf induction heated system, is a crucial prerequisite for the determination of species distribution, as well as layer thickness and composition. Thermal control and uniformity in the growth area up to a level of $\pm 1°C$ are needed to meet uniformity requirements during MOCVD growth of multinary layers, as e.g. InGaN or InGaAsP due the high temperature sensitivity of species incorporation, composition and resulting material properties. Fig. 1 shows the heat source distribution in the graphite susceptor of the Planetary Reactor® at typical process conditions for high temperature growth of GaN based layers, as resulting from the coupled computation of electromagnetic induction, heat transfer and flow.

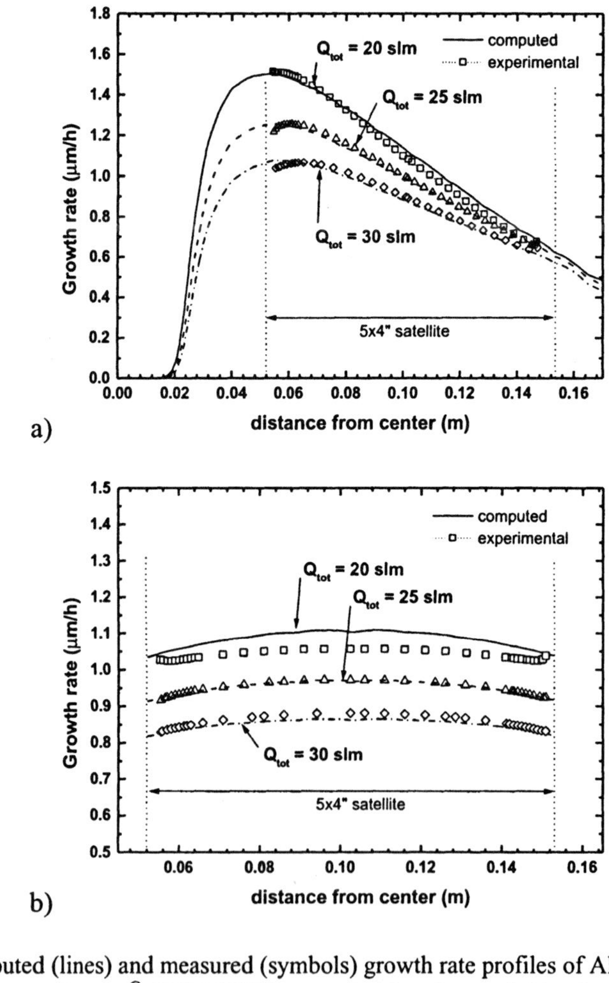

Fig. 2. Computed (lines) and measured (symbols) growth rate profiles of $Al_xGa_{1-x}As$ (x=0.28) in the Planetary Reactor® AIX 2400G3 on static (a) and rotated (b) substrates for various carrier gas flow rates. Growth temperature T = 600°C, process pressure p = 100 mbar.

Equipment and process modelling has been extensively used to reveal parametric dependencies of growth rate uniformity and precursor efficiency on a wide variety of process conditions and combinations thereof, as e.g. flow rates, flow rate split ratio between multiple gas inlets, and precursor partial pressures, as shown in Fig. 2 for carrier flow rates. The design of a multiple flow gas injector was also supported by the detailed 3D-simulation of coupled flow, heat transfer, and species mass transfer around the injection zone. A series of design criteria had to be satisfied, like the absence of local separation and return flow, separating different precursors, matching of flow rates, and thermal management to avoid pre-reaction and condensation of decomposition by-products. This allowed for the optimisation of a rather large number of geometry and process related parameters, which is prohibitive for a time consuming, purely empirical approach based on expensive cycles of re-designing, building of hardware components, and testing under process conditions.

4 Conclusion

Computational modelling of the MOCVD process is used for equipment and process optimisation in an industrial multiwafer Planetary Reactor®. The modelling approach is based on the solution of coupled incompressible gas flow, heat transfer, including rf induction and thermal radiation transfer, and chemical species mass transport and reaction kinetics. Key issues of modelling efforts include global heat transfer and thermal management in the rf induction heated system up to growth temperatures of 1150°C (e.g. during MOCVD of Nitride thin films), the analysis of parametric dependence of layer thickness uniformity, and the design of crucial components, like gas injectors. Equipment optimisation and process tuning resulted in a growth rate uniformity of GaAs of < 1% on 6" substrates and photoluminescence uniformity of < 1 nm standard deviation on the usable substrate area.

5 Acknowledgements

The authors would like to acknowledge the support by CFD Research Corporation.

References

[1] G. B. Stringfellow, Organometallic Vapor-Phase Epitaxy: Theory and Practice, 2nd edition, Academic Press, 1999

[2] P. M. Frijlink, J. Crystal Growth 93 (1988) 207 - 215

[3] C.R. Kleijn, Chemical Vapor Deposition Processes, in: Computational Modeling. In: Semiconductor Processing, Ed. M. Meyyappan (Artech House, 1995) ch. 4, pp. 97 – 230

[4] M. Dauelsberg et al., J. Crystal Growth 208 (2000) 85 - 92

[5] CFD-ACE+ User manual, CFD Research Corporation, 215 Wynn Drive, Huntsville, AL 35805, USA.

Numerical Simulation of Non-Equilibrium, Ultra-Rapid Heating of Si-thin films by Nanosecond-Pulse Excimer Lasers

A.T. Voutsas
Sharp Laboratories of America
Camas, Washington, USA
avoutsas@sharplabs.com

H. Kisdarjono and R. Solanki
Oregon Graduate Institute of Science and Technology
Beaverton, Oregon, USA
hidayat@ece.ogi.edu

A. Kumar
Fluent, Inc.
Evanston, Illinois, USA
ak@fluent.com

Abstract

In this paper we present our work on the numerical simulation of ultra-rapid heating (with phase-change) of silicon thin-films, which are irradiated with nanosecond-pulsed excimer laser. Our excimer-laser-annealing (ELA) modeling capability is based on a standard finite-element CFD software package, which, however, has been modified to accommodate the specific demands of very rapid heating of thin Si films. In that sense, we've abandoned the traditional equilibrium formulation (i.e. enthalpy method), for phase-change computations, and have adopted a new approach that allows superheated solid and undercooled liquid to exist during the various stages of the heating/cooling cycle. Our model has been successfully applied to predict the shape and temporal evolution of temperature profiles in the case of localized melting of silicon thin-films by excimer laser irradiation. Such scenario corresponds to conditions typically encountered in laser-induced lateral crystallization of a-Si films, a process that has recently attracted attention for the formation of high quality poly-Si films.

1 Introduction

A number of previous studies and analyses have established that excimer laser annealing (ELA) of thin Si-films is a highly non-equilibrium process in which spatially localized heating with large temperature gradients is produced [1, 2]. The nature of the physical system, i.e. a very thin Si film typically on top of a thick and insulating SiO_2 substrate, further complicates the heat transfer characteristics of the system and introduces two-dimensional thermal and solidification phenomena [2]. For this reason it is improper to simulate excimer laser annealing by applying the same type of one-dimensional models developed earlier for laser annealing of bulk Si wafers [3,4]. The complexity of ELA simulations is further compounded when considering the non-equilibrium nature of rapid heating and cooling, which favors superheating and undercooling phenomena, where solid or liquid exist above or below the bulk melting point, respectively [5]. The idiosyncrasies of excimer laser annealing manifest themselves particularly well in the case of highly localized complete melting of Si thin-films. These conditions favor lateral growth of Si crystals from the edges of the irradiated (molten) domain. During the melting and re-solidification stages a well-defined, sharp interface separates the molten (liquid) silicon from the solidified (solid) silicon. The interface moves laterally (i.e. parallel to the surface of the film) with a velocity that depends upon the superheating of the solid (during melting) and the undercooling of the molten silicon (during solidification).

The crystallization of the spatially localized molten silicon volume is dominated by two competing processes: (i) lateral growth from the edges of the irradiated domain, (ii) homogeneous/heterogeneous nucleation in the bulk. The probability of homogeneous nucleation increases exponentially with the degree of undercooling (i.e. higher undercooling results in higher probability for nucleation). Hence, depending upon the physical characteristics of the domain (i.e. domain width, bias temperature, etc.) and the parameters of the laser irradiation (i.e. energy density, duration of irradiation, etc.) part of the domain may cool sufficiently to trigger homogeneous (copious) nucleation. If this happens before the laterally grown crystal fronts have the chance to intercept each other, the domain will be crystallized partially by lateral growth and partially by copious nucleation. Such a situation is not desirable. Hence, understanding the mode of growth is an important aspect for the successful implementation of the LILC (laser-induced lateral crystallization) process. Furthermore, ability to predict the mode of growth as a function of controllable and/or "noise" parameters can provide a powerful tool to investigate the experimental space in an expeditious and cost-effective manner.

With this impetus we have developed a simulation tool that allows us to properly model LILC process and keep track of critical parameters such as the spatial and temporal variation of the interface, as well as the degree of undercooling in the molten silicon. In this work we report preliminary results obtained from this model showing its utility, strength and application in design of experiments.

2 Modeling Approach

In our approach, we have chosen a numerically robust, commercially available CFD code to accomplish this task (FIDAP 8.0 by Fluent, Inc.). This code, however, had to be appropriately modified and further refined to capture the specific features of ultra-fast thin-film heating with phase-change. The initial implementation of the code used the "enthalpy method" to determine phase changes and carry the associated computations for the release and absorption of the latent heat over the computational domain. The underlying assumption behind the enthalpy method is thermal equilibrium conditions in the modeled system. However, due to the short time scales, the strong localization of the heat source and the large, associated temperature gradients, this assumption is incorrect. In reality, during solidification, the thin film can experience strong undercooling away from the solid/liquid interface. These phenomena imply that the phase of the film is not uniquely determined by the melting point, but it is rather a function of the local degree of undercooling (defined as the difference between the melting point and the prevailing temperature at a given point within the domain). In order to capture these phenomena, we implemented a number of modifications, as suggested by Gupta et al. [6]. This was accomplished via a user-defined subroutine that controls the non-equilibrium absorption or release of latent heat, within the sub-elements of the computational domain. A significant function of this subroutine is the specification of the domains where the extra terms in the heat flow equation [eq. (1)] are valid. Additionally, the subroutine keeps tracks of the location of the solid/liquid interface and its evolution over time.

$$c_p\rho(T)\frac{\partial T}{\partial t} = \frac{\partial}{\partial x}\left(k(T)\frac{\partial T}{\partial x}\right)+\frac{\partial}{\partial y}\left(k(T)\frac{\partial T}{\partial y}\right)+S_{Laser}(T)+S_{\Delta H_m}(T) \quad (1)$$

where: c_p is the heat capacity, k the thermal conductivity, ρ the density. S_{Laser} is the source term representing the absorbed laser energy and $S_{\Delta Hm}$ is the source term that absorbs or releases latent heat. The subroutine computes $S_{\Delta Hm}$ by implementing a set of rules to determine exactly the amount of latent heat to be absorbed or released. It also uses a phase function to keep track of the fraction of material melted within a cell of the computational domain. The phase function can have value from 0 to 1, from complete solid to complete liquid, respectively.

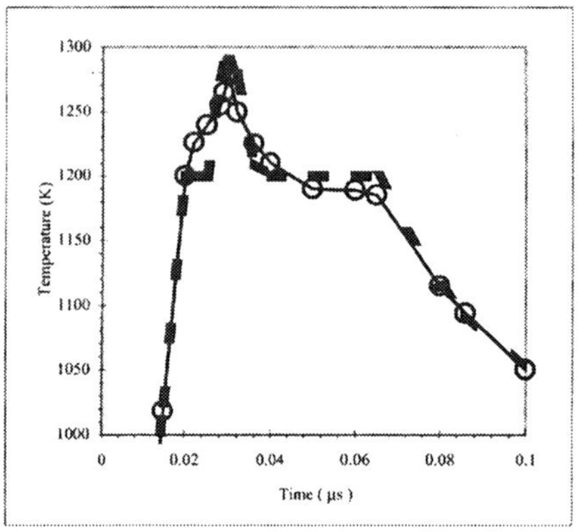

Fig. 1: Temperature history for the domain center from simulations using latent heat function (**o**) and enthalpy method (- -).

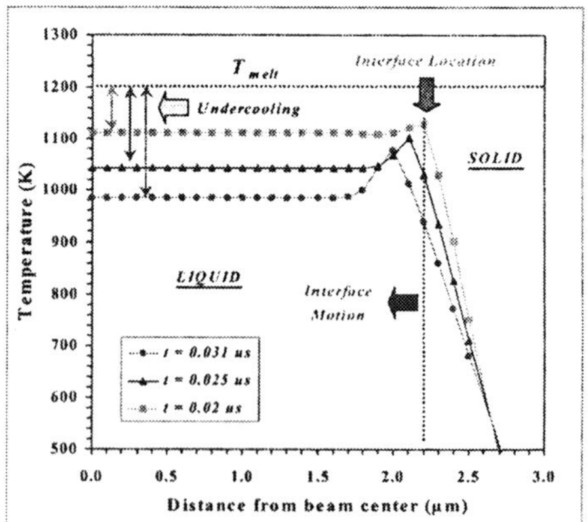

Fig. 2: Spatial temperature profiles across half-beam length at various time intervals during solidification.

As shown in figure 1, when the latent heat function method is used the temperature profiles demonstrate superheating and under-cooling. The enthalpy method does not allow this as is clearly shown by the plateau at the assumed melting point of 1200 K.

3 Key Results

Figure 2 shows an example of the simulated spatial temperature profiles along a 5μm-wide region irradiated by a 15nsec excimer laser (XeCl, 308nm) at 250mJ/cm². Only half-beam profiles are shown due to symmetry (i.e. "0" corresponds to the center of the beam). The plot indicates two important features: *(1) the precise location and motion of the solid/liquid interface as a function of time* (after the laser is switched off). *(2) The degree of undercooling and its spatial variation between the edge and center of the liquid region.*

Using the degree of undercooling as a measure of the nucleation probability, it is possible to estimate the extent of crystalline lateral growth (LGL) for a given set of irradiation conditions and film parameters. Results of such calculations are shown in figures 3 and 4. Precise modeling of lateral growth is essential for the implementation of LILC process, to obtain very high quality poly-Si microstructures. Our initial predictions agree reasonably well with experimental measurements. More work is currently in progress to improve and refine the predictive ability of our model.

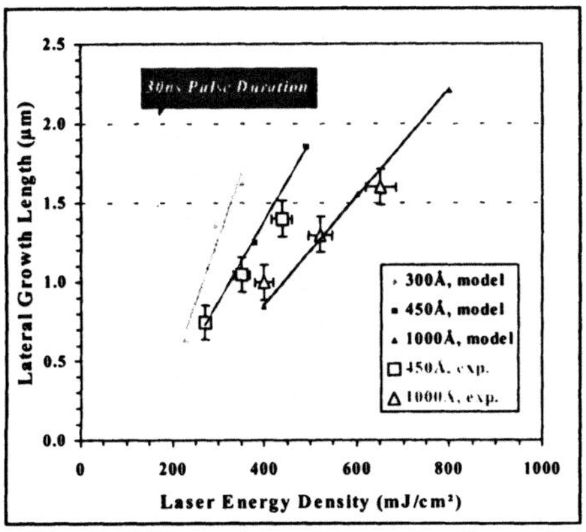

Fig. 3: LGL vs. laser energy density and film thickness.

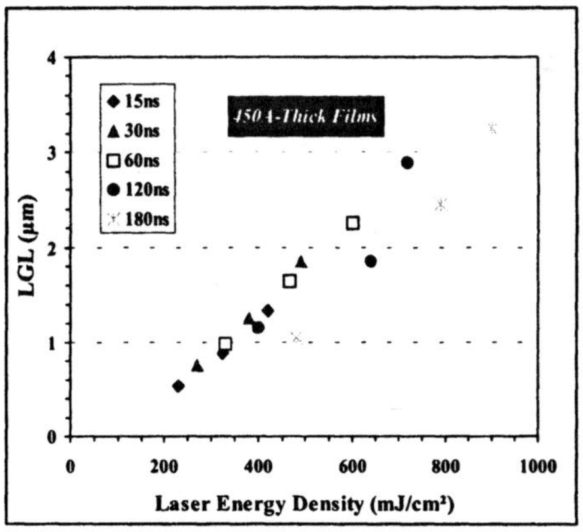

Fig. 4: LGL vs. laser energy density and laser pulse duration.

References

1. Im, J.S., Kim, H.J. and Thompson, M.O. (1993): Phase transformation mechanisms involved in excimer laser crystallization of a-Si films. Appl. Phys. Lett. 63: 1969-1971
2. Im, J.S. and Kim, H.J. (1994): On the super lateral growth phenomenon observed in excimer laser-induced crystallization of thin Si films: Appl. Phys. Lett., 64: 2303-2305
3. De Unamuno, S. and Fogarassy E. (1989): A thermal description of the melting of c- and a-silicon under pulsed excimer lasers. Appl. Surf. Sci., 36: 1-11
4. Baeri, P., Campisano, S.U., Foti, G. and Rimini, E. (1979): A melting model for pulsing-laser annealing of implanted semiconductors. J. Appl. Phys., 50: 788-797
5. Sposili, R., Im, J.S. (1996): Sequential lateral solidification of thin silicon films on SiO2. Appl. Phys. Lett., 69: 2864-2866
6. Gupta, V.V., Song H.J., Im J.S. (1997): Numerical analysis of excimer-laser-induced melting and solidification of thin Si films. Appl. Phys. Lett. 71: 99-101

2D Hierarchical Radio–Frequency Noise Modeling Based on a Langevin–Type Drift–Diffusion Model and Full–Band Monte–Carlo Generated Local Noise Sources

S. Decker[†], C. Jungemann, B. Neinhüs, and B. Meinerzhagen
University of Bremen, ITEM, Kufsteiner Straße, D-28334 Bremen
[†]Present address: Infineon Technologies AG, Otto-Hahn-Ring 6, 81739 Munich
E-mail: Stefan.Decker@infineon.com

Abstract

An accurate and efficient 2D drift-diffusion model for thermal noise simulation based on full–band Monte–Carlo (MC) generated local noise sources is presented. Good agreement of the new model and MC device simulations is found for NMOSFETs, whereas previously developed DD based noise models fail. Verification with experiment is shown for a SiGe HBT.

1 Introduction

Noise modeling of semiconductor devices in the framework of a 2D drift–diffusion (DD) model is based on an efficient numerical calculation of the admittance fields and full-band Monte-Carlo (MC) generated local noise sources. In this framework the modeling of the local noise sources for diffusion noise is investigated by comparing three different models consisting of the two approximations usually used in this framework and a new alternative employing directly the spectral density of the velocity fluctuations. For consistency all nonconstant coefficients of the local noise sources are derived from MC bulk simulations and MC device simulation is used to verify the DD based results for an ultra–short channel MOSFET. Finally, based on the new local noise model the noise behavior of a SiGe HBT is simulated and compared to measurement.

2 Theory

In this work, terminal current noise is modeled rigorously physics–based on a Langevin–type DD ansatz. The electron contribution — the hole contribution is analogous — has the following form:

$$S_{II}(\omega) = \int_{\Omega} (\nabla \underline{H}_I(\vec{r},\omega))^T q^2 n S_N \nabla \underline{H}_I^*(\vec{r},\omega) d^3\vec{r} = \int_{\Omega} \Delta S_{II}(\vec{r},\omega) d^3\vec{r}$$

S_{II} is the spectral density of the noise of the terminal current I, $\underline{H}_I$ the electron admittance field, n the electron density, Ω the device area, $\vec{r}$ a real space vector, ω the angular frequency, S_N the local diffusion noise source and ΔS_{II} the spatial distribution of S_{II}.

The most accurate formulation would be to replace S_N by the spectral density matrix of the velocity fluctuations at each point within the device. In all previous

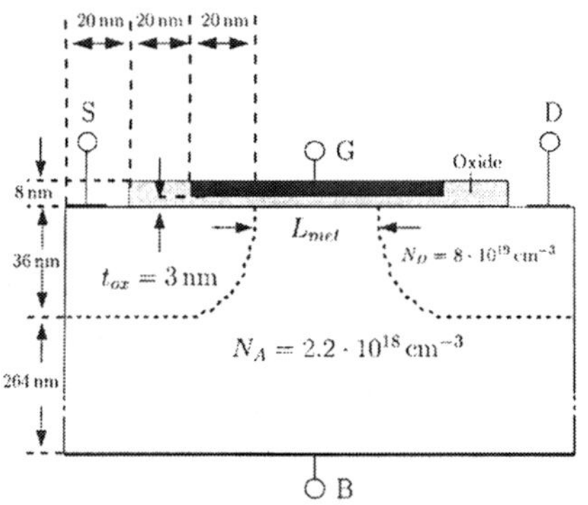

Fig. 1. n–MOSFET test structure

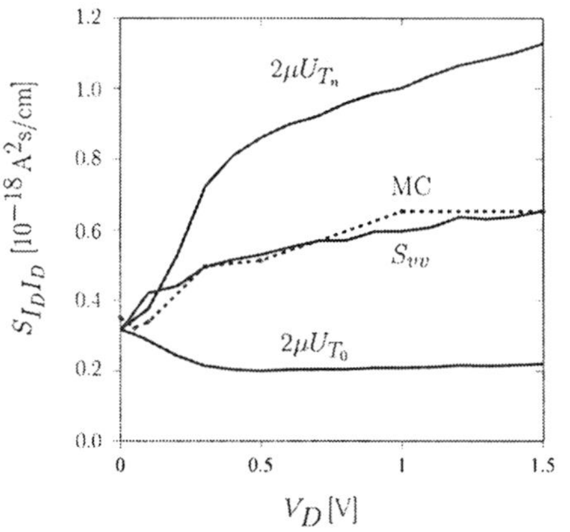

Fig. 2. MC and DD results for the spectral density of the drain current fluctuations as a function of V_D (V_G = $1.5V$)

studies simple isotropic approximations for this matrix were used [1, 2, 3, 4, 5]. S_N was replaced by either $2\mu_n U_{T_0}$ or $2\mu_n U_{T_n}$, where $U_T = k_B T/q$ is the thermal voltage, μ_n the electron (hole) mobility, T_0 the lattice, T_n the electron (hole) temperature, k_B Boltzmann's constant, and q the elementary charge. These approximations are only justified close to equilibrium. In this paper — for the first time — terminal current noise in 2D devices is modeled by a less stringent isotropic approximation. S_N is replaced by S_{vv}, which is derived from MC bulk simulations [6] and is the spectral density of the velocity fluctuations parallel to the driving field. S_{vv} is evaluated for a number of different driving fields, doping densities, germanium concentrations, and lattice temperatures from which a table model is generated that is subsequently used in the DD model. Since transport models of vastly different physical accuracy are used to establish this new approach, we call it hierarchcal numerical noise modeling. Moreover, this new method is validated by a direct comparison with the terminal current noise resulting from consistent MC device simulations. The overall 2D methodology described here is in some details similar to the 1D approach published in [7], but the local noise sources are evaluated in a different way. In order to simplify the comparison between the particle and the classical simulation all interface effects degrading transport are neglected. To allow a consistent comparison hierarchical numerical modeling is employed not only for S_{vv} but also μ_n and U_{T_n}.

3 Results

At first, an n–channel MOSFET with variable channel length L_{met} is investigated. L_{met} shown in fig. 1 varies between $100\,\mu$m and 40 nm. In fig. 2 the spectral density of the drain current fluctuations is shown for the different models, for the shortest device, and $V_G = 1.5$ V as a function of V_D ($V_S = V_B = 0$). The solid lines show the DD results using S_{vv} and the two approximations $2\mu_n U_{T_0}$ and $2\mu_n U_{T_n}$. The result of the MC device simulator shown by the dashed line is in close agreement with the DD result using S_{vv}. This good agreement validates the new efficient hierarchcal numerical noise modeling method based on S_{vv}.

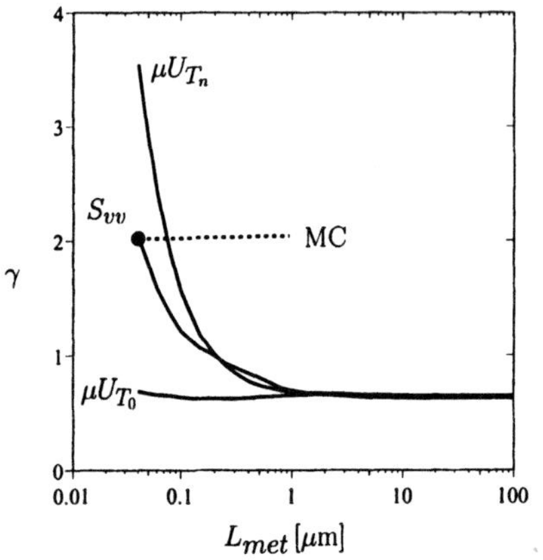

Fig. 3. Excess noise factor γ as a function of L_{met} ($V_D = 0$ and $V_D = 1.5\,\mathrm{V}$) The dot indicates the MC device simulation result

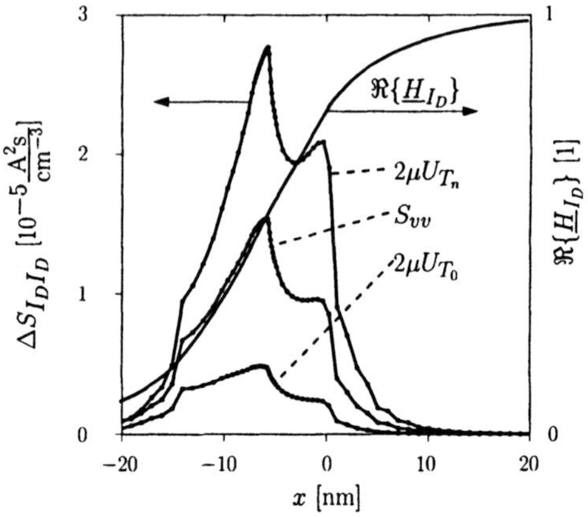

Fig. 4. Spatial distribution of the drain current noise and the real part of the transfer function field $\underline{H}_{I_D}$ (L_{met} = 40nm, $V_G = V_D = 1.5\,\mathrm{V}$). x is the position along the channel

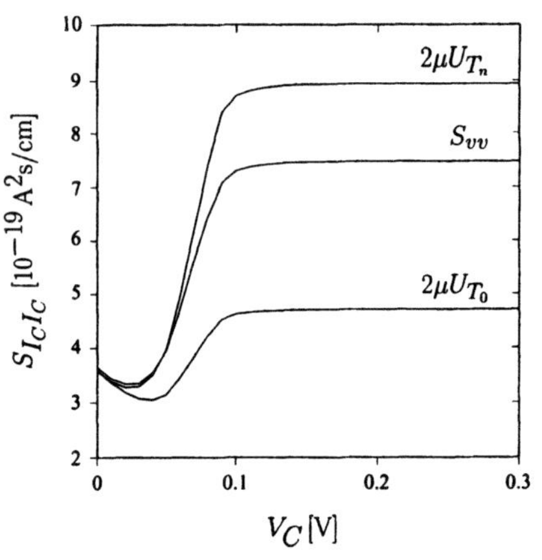

Fig. 5. Autocorrelation spectrum of the collector current as a function of V_C for $V_B = 0.85\,\mathrm{V}$ and $V_E = 0$

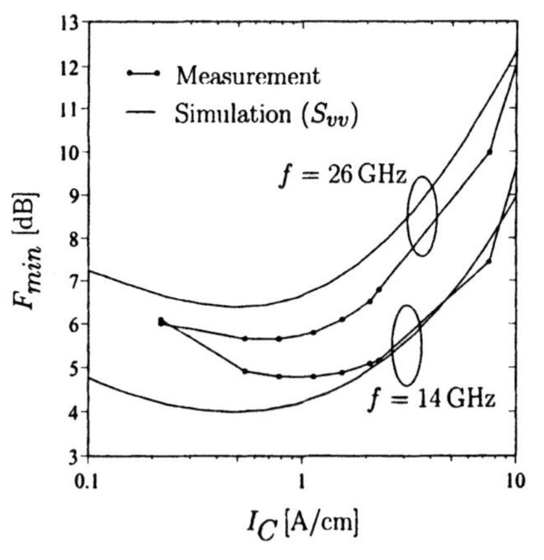

Fig. 6. Comparison of measured and simulated F_{min} for the SiGe HBT as a function of I_C ($V_C = 2\,\mathrm{V}$, $V_E = 0$)

Moreover, the large deviations of the other results shown in fig. 2 from the MC reference clearly indicate that the approximations for S_N typically used in previous studies (e.g. [1],[4]) lead to large errors. In fig. 3 the excess noise factor $\gamma = S_{I_D I_D}/(2k_B T_0 \Re\{Y_{DD}|_{V_D=0}\})$ [8] is shown for $V_G = 1.5\,\mathrm{V}$ as a function of L_{met} and V_D. All DD local noise source models investigated here yield $\gamma = 2/3$ for large L_{met} and $V_D = 1.5\,\mathrm{V}$ consistent with [9]. However, for small L_{met}

and $V_D = 1.5\,\mathrm{V}$ large deviations among the different local noise sources can be observed. In fig. 4 for $V_G = V_D = 1.5\,\mathrm{V}$ and $L_{met} = 40\,\mathrm{nm}$ the local contributions $\Delta S_{I_D I_D}$ in the channel region close to the oxide interface are plotted as a function of the channel position. The large deviations between the different local noise sources are again clearly visible. The dominating contributions to $S_{I_D I_D}$ originate from the source side of the channel consistent with [1], and not from the drain side as previously thought.

The SiGe HBT described in [10] is investigated as a second example. Fig. 5 shows the spectral density of the collector current fluctuations resulting from the three different local noise sources as a function of the collector voltage. The results are comparable to those in fig. 2. In Fig. 6 an experimental verification for the minimum noise figures F_{min} resulting from our new methodology is shown.

4 Conclusions

In conclusion, a new accurate and efficient methodology for terminal current noise modeling called hierarchical numerical noise modeling is presented. The accuracy of this approach is validated by MC device reference simulations as well as by experimental data. It is shown that the usual approximations for the local noise source — typically used in the framework of the DD model — lead to large errors.

Acknowledgment

Part of this work was funded by a research project sponsored by the semiconductor products sector of Motorola Inc.

References

[1] Goo JS *et al.* An accurate and efficient high frequency noise simulation technique for deep submicron MOSFETs. IEEE Trans. Elec. Dev. 2000; 47:2410–2419
[2] Scholten AJ *et al.* Accurate Thermal Noise Model for Deep-Submicron CMOS. IEDM Tech. Dig. 1999;155–158
[3] Klein P *et al.* An Analytical Thermal Noise Model of Deep Submicron MOSFET's. IEEE Elec. Dev. Let. 1999;20:399–401
[4] Donati S *et al.* Physics-Based RF Noise Modeling of Submicron MOSFETs. IEDM Tech. Dig. 1998;81–84
[5] Bonani F *et al.* An Efficient Approach to Noise Analysis Through Multidimensional Physics-Based Models. IEEE Trans. Elec. Dev. 1998;45:261–269
[6] Jungemann C *et al.* Efficient Full-Band Monte Carlo Simulation of Silicon Devices. IEICE Trans. Elec. 1999;870–879
[7] Starikov E *et al.* Transfer impedance calculations of electronic noise in two-terminal semiconductor structures. J. Appl. Phys. 1998;83:2052–2066
[8] Abidi AA. High-Frequency Noise Measurements on FET's with Small Dimensions. IEEE Trans. Elec. Dev. 1986;33:1801–1805
[9] Klaassen FM *et al.* Thermal Noise of MOS Transistors. Philips Res. Repts 1967;505–514
[10] Decker S *et al.* Investigation of High Frequency Noise in a SiGe Heterobipolar Transistor Based on Shockley's Impedance Field Method and the Hydrodynamic Model. MSM2000 2000;364-367

Variance and Covariance Estimation in Stationary Monte Carlo Device Simulation

H. Kosina, M. Nedjalkov, and S. Selberherr

Institute for Microelectronics, TU Vienna, Gusshausstrasse 27–29, A-1040 Vienna, Austria
Phone: +43-1-58801/36013, Fax: +43-1-58801/36099, Email: Kosina@iue.tuwien.ac.at

Abstract

This work deals with the Monte Carlo method for stationary device simulation, known as the Single-Particle Monte Carlo method. A thorough mathematical analysis of this method clearly identifies the independent, identically distributed random variables of the simulated process. Knowledge of these random variables allows usage of straight-forward estimates of the stochastic error. The presented method of error estimation is applicable to both distributed quantities and integrated quantities such as terminal currents.

1 Introduction

To assess the accuracy of Monte Carlo (MC) simulation it is necessary to estimate the probabilistic error bounds of the results. In the field of MC device simulation this problem has received renewed attention [1]. We present a new method of estimating the variance of the stationary MC method. The method is based on the identification of those random variables whose realizations are statistically independent. Knowing these random variables standard textbook formulae can be applied to estimate the variance.

2 Variance Estimation

In a thorough mathematical analysis of the single-particle MC method, outlined in [2][3], the Neumann series of the related integral equation is derived. Each term of that series describes the propagation of a carrier from the point of injection to the exit point at some contact. From this fact follows that only a complete numerical trajectory, that starts and terminates at the domain boundary, can be considered an independent realization of some random variable. In contrast, particle states generated on one trajectory are statistically dependent. The i-th realization of this random variable, say X, consists of all generated random variables for the i-th trajectory, such as the initial state at the domain boundary, $\mathbf{k}_0, \mathbf{r}_0$, the free flight times t_j, and the after-scattering states, $\mathbf{k}_j^a$.

$$x_i = \{\mathbf{k}_0, \mathbf{r}_0, t_0, \mathbf{k}_1^a, t_1, \dots \mathbf{k}_j^a, t_j, \dots\} \qquad j \le N_i \tag{1}$$

The considered trajectory consists of $N_i + 1$ free flight segments. Another random variable $Y(X)$ needed in the following is defined by its realizations

$$y_i = \{\mathbf{k}_1^b, \mathbf{r}_1, \dots \mathbf{k}_j^b, \mathbf{r}_j, \dots\} \qquad j \le N_i, \tag{2}$$

which contain all before-scattering states $\mathbf{k}_b^j$ and the particle locations at the times of scattering, $\mathbf{r}_j$.

With any quantity of interest, $A(\mathbf{k}, \mathbf{r})$, a random variable $\Psi_A(Y)$ is associated. Assuming the before-scattering method for average recording, the i-th realization of Ψ_A is of the form

$$\psi_{Ai} = \sum_{j=1}^{N_i} w_j \frac{A(\mathbf{k}_j^b, \mathbf{r}_j)}{\lambda(\mathbf{k}_j^b, \mathbf{r}_j)} \tag{3}$$

with λ being the total scattering rate. Summation is again over one complete trajectory starting and terminating at the domain boundary. If a statistical enhancement technique is used, the particle weighs w_j has to be accounted for, otherwise all weights equal one.

The result of a stationary MC device simulation can be expressed most generally as a ratio of statistical averages,

$$C = \frac{\langle\langle \mathrm{A} \rangle\rangle}{\langle\langle \mathrm{B} \rangle\rangle} \tag{4}$$

with the definition $\langle\langle\, . \,\rangle\rangle = \int_D \mathbf{dr} \int \mathbf{dk}\ .\ f(\mathbf{k},\mathbf{r})$. Here D denotes the simulation domain. The function A is typically a product of some $\mathbf{k}$-dependent function and an $\mathbf{r}$-dependent charge assignment function [4], whereas the denominator accounts for the normalization (cf. Table 1).

The random variable to be considered now is given by $\Psi_C = \Psi_A/\Psi_B$. In the MC simulation one has to generate the samples ψ_{Ai} and ψ_{Bi} using the rule (3). The so-called classical estimator of C is given by the ratio of the sample means.

$$\overline{C} = \frac{\overline{\psi}_A}{\overline{\psi}_B}, \qquad \overline{\psi}_\alpha = \frac{1}{N}\sum_{i=1}^{N} \psi_{\alpha i}, \qquad \alpha = A, B \tag{5}$$

Additionally, the sample variances, s_A^2 and s_B^2, and the sample covariance s_{AB}^2 have to be evaluated from the following definitions.

$$s_\alpha^2 = \frac{1}{N-1}\left[\sum_{i=1}^{N} \psi_{\alpha i}^2 - \frac{1}{N}\left(\sum_{i=1}^{N} \psi_{\alpha i}\right)^2\right], \qquad \alpha = A, B \tag{6}$$

$$s_{AB}^2 = \frac{1}{N-1}\left[\sum_{i=1}^{N} \psi_{Ai}\psi_{Bi} - \frac{1}{N}\sum_{i=1}^{N} \psi_{Ai}\sum_{i=1}^{N} \psi_{Bi}\right] \tag{7}$$

N is the number of trajectories constructed in the simulation. From these inputs the variance of the random variable Ψ_C can be estimated [5].

$$s_C^2 = s_A^2 - 2\overline{C}s_{AB}^2 + \overline{C}^2 s_B^2 \tag{8}$$

The error estimate for the result $\overline{C}$ is finally given by the standard deviation

$$\sigma\{\overline{C}\} = \frac{s_C}{\overline{\psi}_B\sqrt{N}}. \tag{9}$$

3 Application and Discussion

The following example demonstrates how the presented method of error estimation can be applied to both distributed quantities and integrated quantities such as terminal currents. As an example an n+nn+ silicon diode has been simulated, processing $5 \cdot 10^8$ scattering events, which resulted in the simulation of $N = 5.48 \cdot 10^7$ trajectories. The variances of the physical quantities collected in Table 1 have been calculated.

In Table 1 the number of physical particles in the simulation domain is denoted by N_D, $W_p(\mathbf{r})$ is the charge assignment function for grid point $\mathbf{r}_p$ and $V_p = \int W_p(\mathbf{r})\mathbf{dr}$. The

quantity	C	A	B
carrier concentration	n	$(N_D/V_p)W_p(\mathbf{r})$	1
current density	$\mathbf{j}$	$(qN_D/V_p)\mathbf{v}(\mathbf{k})W_p(\mathbf{r})$	1
terminal current	I_l	$(qN_D/V_p)\mathbf{v}(\mathbf{k})\cdot\nabla h_l(\mathbf{r})$	1
mean velocity	$\langle\mathbf{v}\rangle$	$\mathbf{v}(\mathbf{k})W_p(\mathbf{r})$	$W_p(\mathbf{r})$
mean energy	$\langle\epsilon\rangle$	$\epsilon(\mathbf{k})W_p(\mathbf{r})$	$W_p(\mathbf{r})$

Table 1: Examples of physical quantities used in (4).

current calculation employs the weight function method with $h_l(\mathbf{r})$ being the weight function for contact l.

Fig. 1 shows the electron concentration and the distribution of the standard deviation. In Fig. 2 the current density is depicted, which exhibits large fluctuations in the left and right contact region. Consistent with the observed fluctuation is the estimated standard deviation which is considerably higher in the contact regions than in the n-region.

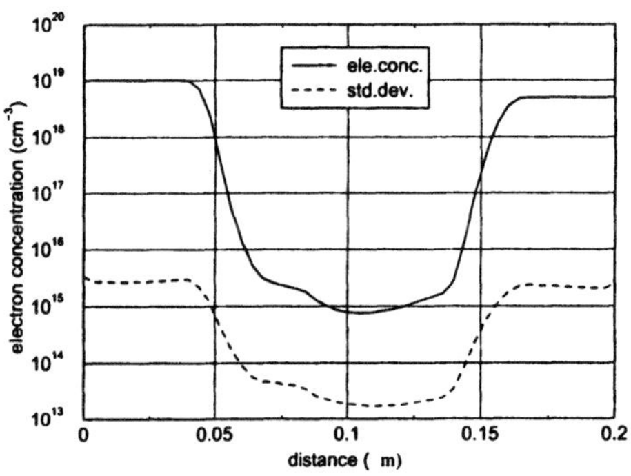

Fig. 1: Electron concentration and standard deviation in an n+nn+ diode with a 0.2 eV built-in barrier.

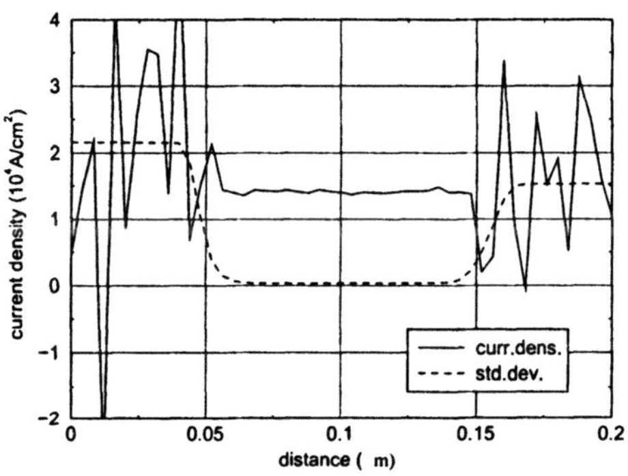

Fig. 2: Current density and standard deviation.

When computing the mean velocity using (4), ⟨⟨A⟩⟩ represents the particle current density and ⟨⟨B⟩⟩ the particle density. The correlation factor $r_{AB} = s^2_{AB}/(s_A s_B)$ of the MC estimate of the two densities is plotted in Fig. 3 (dashed line). An interesting result is the high positive correlation of the energy density and the particle density (solid line), which gives a significant reduction in standard deviation according to (8). A consequence of positive correlation is that a mean value per carrier has less variance than a mean value per unit volume, This is demonstrated in Fig. 4. Accounting for the fluctuations of both the energy density and the particle density gives the lower standard deviation (curve std.dev), while by neglecting the fluctuation of the particle density the standard deviation is clearly increased (curve std.dev.A).

The presented method has been used in conjunction with two statistical enhancement methods, the trajectory repetition scheme due to Philips and Price [6], and the recently investigated method of event biasing [3]. The presented method is parameter free in the sense that neither an unknown parameter of a stochastic process such as the correlation time needs to be estimated, nor the particle's history needs to be divided into sub-histories of some artificially predefined length.Instead, the total history is divided naturally into independent sub-histories at the times when the particle reenters the sim-

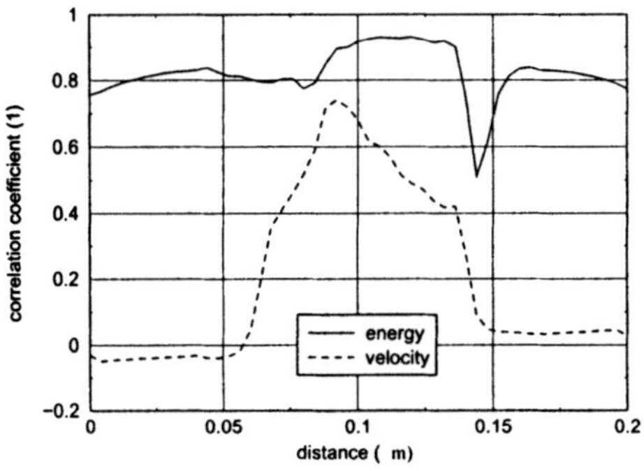

Fig. 3: Correlation coefficients of the the energy density $n\langle\epsilon\rangle$ and n, and of the particle current $n\langle v_x\rangle$ and n.

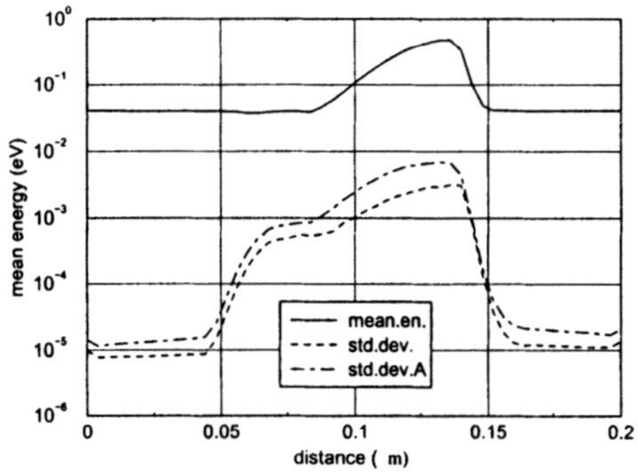

Fig. 4: Mean energy and its standard deviation, including the fluctuations of energy density and particle density (std.dev), and neglecting the fluctuation of particle density (std.dev.A).

ulation domain. A process with such property is referred to as regenerative stochastic process [5].

4 Conclusion

For the first time the independent, identically distributed random variables underlying the Single-Particle Monte Carlo method for device simulation have been identified. Knowing these random variables allows this Monte Carlo method to be supplemented with the natural stochastic error estimate. Without variance estimation, the statistics can be collected after each free flight of the test particle, whereas variance estimation requires that over one particle trajectory a sub-statistics is collected, which is added to the total statistics when the trajectory terminates at the domain boundary. Variance estimation of both distributed and integrated quantities is demonstrated. If mean values per particle on a mesh are to be computed, correlation of the mean value per volume and the particle number per volume has to be taken into account, which leads to partial cancellation of statistical fluctuations.

Acknowledgment

This work has been partly supported by the IST program, project NANOTCAD, IST-1999-10828.

[1] C. Jungemann and B. Meinerzhagen, "Efficiency and Stochastic Error of Monte Carlo Device Simulations," in *Int.Electron Devices Meeting*, pp. 109–112, 2000.

[2] H. Kosina, M. Nedjalkov, and S. Selberherr, "Theory of the Monte Carlo Method for Semiconductor Device Simulation," *IEEE Trans.Electron Devices*, vol. 47, no. 10, pp. 1898–1908, 2000.

[3] H. Kosina, M. Nedjalkov, and S. Selberherr, "Variance Reduction in Monte Carlo Device Simulation by Means of Event Biasing," in *Proc. Modeling and Simulation of Microsystems, MSM 2001* (M. Laudon and B. Romanowicz, eds.), pp. 11–14, Computational Publications, Mar. 2001.

[4] R. Hockney and J. Eastwood, *Computer Simulation Using Particles*. Bristol and Philadelphia: Adam Hilger, 1988.

[5] R. Rubinstein, *Simulation and the Monte Carlo Method*. John Wiley and Sons, 1981.

[6] A. Phillips and P. Price, "Monte Carlo Calculations on Hot Electron Tails," *Applied Physics Letters*, vol. 30, pp. 528–530, May 1977.

Analysis of Gate Tunneling Current in MOS Structures using Quantum Mechanical Simulation

Matsuto Ogawa and Tanroku Miyoshi

Department of Electrical and Electronics Engineering,
Kobe University
1 Rokkodai, Nada, Kobe, 657-8501, JAPAN
phone & facsimile: +81-78-803-6074
E-mail: ogawa@eedept.kobe-u.ac.jp

Abstract

We report studies of quantum transport in n^+Si-SiO_2-p Si Metal-Oxide-Semiconductor (MOS) structures based upon a non-equilibrium tight-binding Green's function method. As a result, the quasi-bound states at the SiO_2-p Si interface are found to be lower than those calculated by the conventional Shrödinger-Poisson analysis, since the wavefunctions in this region are coupled with the Bloch functions in the electrodes. It is also found that the leakage current through the oxide consists of not only the intraband tunneling but also the interband tunneling current.

1 Introduction

The recent MOS scaling less than sub-100 nm gate length has changed operating voltage less than 1V and an oxide thickness less than 3 nm, which corresponds to several layers of atom. The progress in the miniaturization has revealed such crucial problems as the gate tunneling, deterioration in device performance i.e. loss of inversion charge and transconductance due to the quantization of carriers in the channel, and polysilicon gate depletion. So far, semi-analytical models [1] and quantum correction models [2] have been applied to analyze such problems. In order to understand these aspects more physically, we have to simulate the transport of carriers quantum-mechanically taking into account the possible tunneling paths as well as quantization in the MOS structure.

In this paper, we present simulations of quantum electron transport in a MOS structure based on a non-equilibrium tight-binding Green's function (NEGF) method.

2 Modeling and Simulation Results

We confine ourselves within an n^+ poly Si-SiO_2-p Si MOS diode as schematically illustrated in Fig. 1. We have employed a tight-binding 2-band model which describes the conduction band (CB) and valence band (VB), and the pure imaginary band. Figure 2 (a) and (b) show the complex band structures of Si and SiO_2, respectively. The pure imaginary wave number $\mathbf{Im}k_z$ in the band gap implies the decay constant of evanescent electron waves. The evanescent waves should be appropriately matched between Si-SiO_2 interfaces.

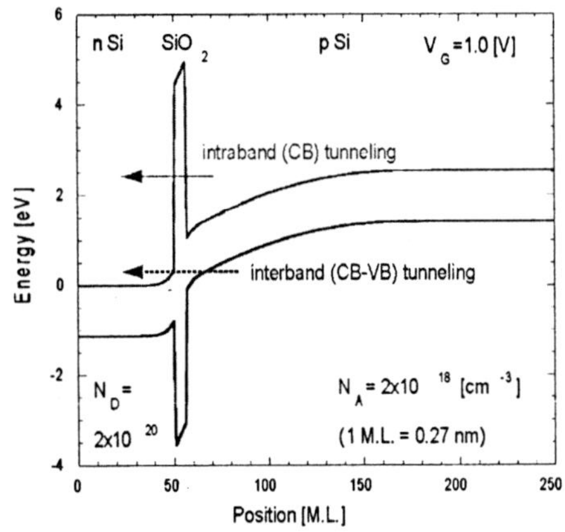

Figure 1: Schematic structure of an n^+ poly Si-SiO_2-p Si MOS diode.

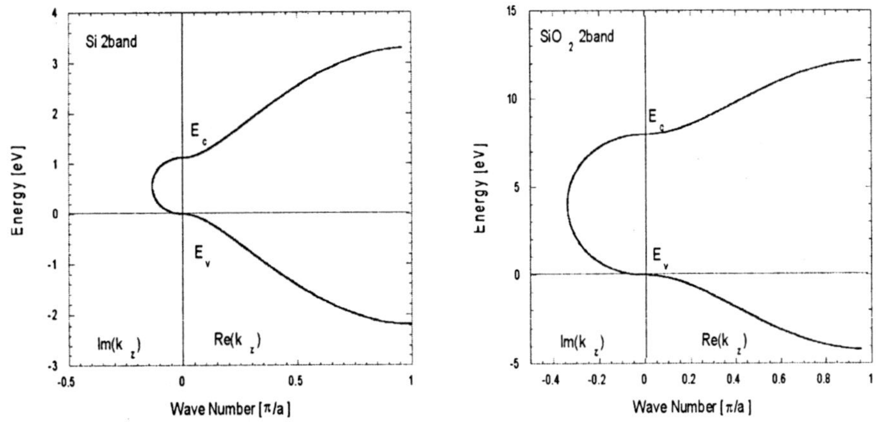

Figure 2: Complex band structures of (a) Si and (b) SiO_2 calculated by the 2B model. The real bands (**Re** k_z) are drawn on the right portion, whereas the imaginary bands (**Im** k_z) are on the left.

Using the band structures, carrier concentrations n and p, current density J, and a spectral function or local density of states (LDOS) A at lattice position L are straightforwardly expressed in terms of the NEGF[3,4]:

$$n_L = -\frac{2i}{S\Delta}\sum_{\boldsymbol{k}_\parallel}\int\frac{dE_z}{2\pi}\mathbf{Tr}\left[G^<_{L,L}(\boldsymbol{k}_\parallel,E_z)\right], \quad (\text{for } p_L,\ G^< \longrightarrow G^>), \quad (1)$$

$$J_L = \frac{2e}{\hbar S}\sum_{\boldsymbol{k}_\parallel}\int\frac{dE}{2\pi}2\mathbf{Re}\left\{\mathbf{Tr}\left[-t_{L,L+1}\cdot G^<_{L+1,L}(\boldsymbol{k}_\parallel,E_z)\right]\right\}, \quad (2)$$

$$A_{L,L} = 2\mathbf{Im}[G^R_{L,L}(E)] \quad (3)$$

where S is the cross sectional area, Δ monolayer spacing (=0.27 nm), e the electronic charge, $t_{L,L+1}$ is the **hopping matrix** which is related to the matrix elements of the Hamiltonian, and Tr denotes trace of the matrix. Poisson's equation is simultaneously solved to include the space charge effect for selfconsistent calculation.

Figure 3 shows the LDOS around the oxide region. Darker portions have higher density of states. Several quasi-bound states are observed without assuming infinite wall boundaries at the electrodes. For comparison, Fig. 4 shows the result calculated by the Schrödinger-Poisson analysis (SPA) with infinite wall boundaries. Note that in the present NEGF analysis, each quasi-bound state lies below corresponding one calculated by the SPA, since the NEGF correctly takes into account the open boundary conditions at electrodes through self-energies which include couplings between the electrodes and the device region. The shift of the quasi-bound states may change the threshold voltage of the MOS predicted by the conventional SPA.

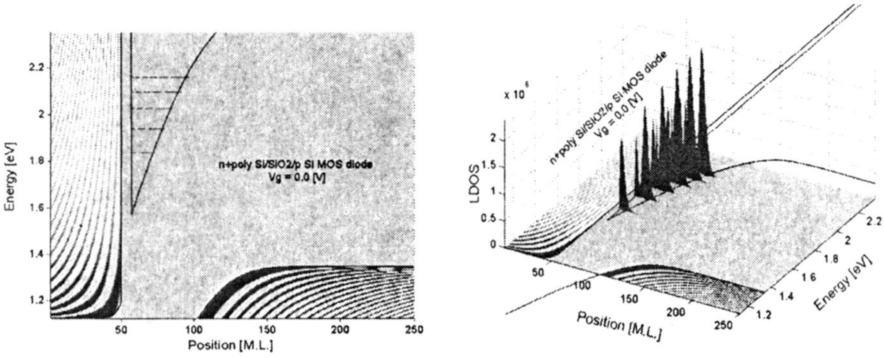

Figure 3: Local density of states (LDOS) in the MOS diode at gate bias $V_g = 0$ V. Six quasi-bound states are drawn in the triangular quantum well from which electrons are injected into the gate electrode when intraband tunneling occurs. The contribution of states from the transverse energies have been ignored for clarity.

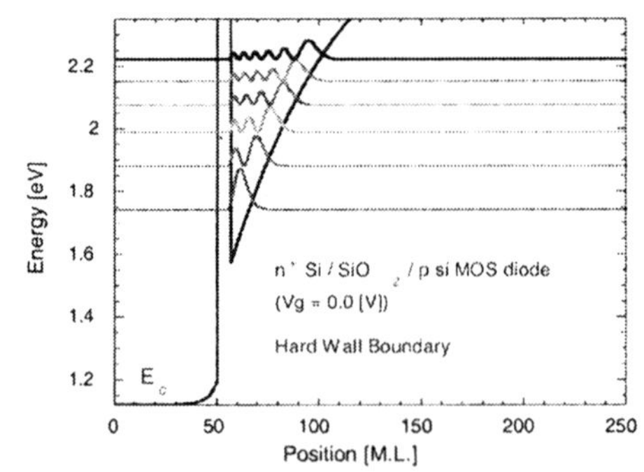

Figure 4: Bound-states calculated by the Schrödinger-Poisson analysis with infinite-wall boundaries at the ends of the electrodes. Probability densities are also shown.

Figure 5 (a) shows the I-V characteristics of the MOS diode with $t_{ox} = 1.6$ and 2.5 nm, which is qualitatively in agreement with the experimental data[5]. The current density under the inversion condition is fairly large due to the tunneling current. The tunneling current is found to be made up of both the injection from the quasi-bound states and the VB-CB interband tunneling as shown in the LDOS in Fig 5(b). The interband tunneling can be understood from the overlap of the local density of state between the CB and VB.

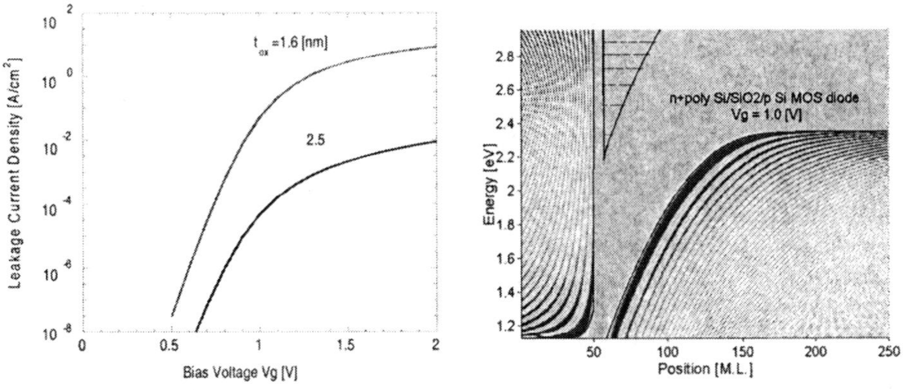

Figure 5: (a) Oxide thickness dependence of the tunneling current ($t_{ox} = 1.6$ and 2.5 nm). (b) Local density of states at $V_g = 1.0$ V. Overlapping of the states can be observed arount $E = 1.2$ eV.

3 Conclusion

We have applied the non-equilibrium tight-binding Green's function (NEGF) formalism to analyze tunneling in the MOS structure. As a result, the NEGF method is found to include correctly the open boundary conditions. The quasi-bound states in the MOS triangular well are found to become lower due to the boundary condition, which may change the threshold voltage predicted by the conventional method. The amount of the tunneling current under the inversion biased condition is so large that we have to be careful in comparing the simulated C-V characteristics with experimental data.

References

[1] C-H Choi, Y.W. Wu, J-S. Goo, Z. Yu, and R.W. Dutton, IEEE Trans. on Electron Devices, 2000; **47** : 1843-1850.

[2] M.G. Ancona, Z. Yu, R.W. Dutton, P.J.V. Voorde, M. Cao, and D. Vook, "Density-Gradient Analysis of MOS Tunneling, **ibid** : 2310-2319.

[3] R. Lake, G. Klimeck, R.C. Bowen, and D. Jovanovic, "Single and multiband modeling of quantum electron transport through layered semiconductor devices," 1997; **81** : 7845-7869.

[4] M. Ogawa, T. Sugano, R. Tominaga, and T. Miyoshi, "Simulation of Multi-Band Quantum Transport Reflecting Realistic Band Structure," IEICE Trans. on Electron Devices, 2000; **E83-a**: 1235-1241.

[5] H.S. Momose, M. Ono, T. Yoshitomi, T. Ohguro, S. Nakamura, M. Saito, H. Iwai, "1.5 nm direct-tunneling gate oxide Si MOSFET's," IEEE Trans. on Electron Devices 1996; **43** : 1233-1242.

2-D Self-Consistent Solution of Schrödinger Equation, Boltzmann Transport Equation, Poisson and Current-continuity Equation for MOSFET

Chung-Kuang Huang and Neil Goldsman
Department of Electrical Engineering, University of Maryland, College Park, MD 20742
Email: ckhuang@glue.umd.edu, neil@glue.umd.edu

Abstract

We present a method of modeling quantum confinement effects in MOSFET's by solving the Schrödinger, Boltzmann, Poisson and current-continuity equations self-consistently.

1 Introduction

As MOSFET dimensions shrink into the nanometer regime, and oxide thicknesses reduce to the angstrom level, significant quantization effects occur in the motion of carriers perpendicular to the interface. We need to calculate the electronic properties of the quasi two-dimensional electron gas(2-DEG) and their effects on the MOSFET characteristics. Spinelli et al.(1998) presented a self-consistent two-dimensional drift-diffusion(DD) model for carrier quantization effects in the channel of highly-doped nMOSFET's. A quantum mechanical treatment of electron inversion layers of nMOSFET's is also incorporated in the hydrodynamic(HD) transport model by Wang et al.(1998). While these methods represent important contributions, they assume an equilibrium form at lattice or electron temperatures for the distribution function. However, it is well known that the channel distribution function is often far from equilibrium, which can lead to difficulties when populating the inversion layer subbands. Here we incorporate carrier quantization effects in the channel of highly-doped n-MOSFET's by solving the Schrödinger, Poisson, Boltzmann and Current-continuity equations self-consistently. The new method naturally accounts for highly nonequilibrium effects including velocity overshoot, and we populate the subbands according to the self-consistent nonequilibrium distribution function we obtain by solving the Boltzmann transport equation(BTE) (Liang et al. 1997).

2 Theory and realization

First we solve the Poisson equation, electron BTE and the hole current-continuity equation with the Spherical Harmonic Boltzmann(SHBTE) method. After getting self-consistent simulation results, we can solve the Schrödinger equation based on the potential energy $V(y)$ from the Poisson equation result. Using the new quantum-mechanical electron density, and a quantum mechanically updated form for the potential, we solve the Poisson equation, electron BTE and holes current-continuity equation again until these four equations converge. The flowchart of the QM-SHBTE is shown in Fig. 1. The complete model is given by:

$$\nabla^2_{\vec{r}}\phi(\vec{r}) = \frac{q}{\epsilon_s}\left[n_q(\vec{r}) - p(\vec{r}) + N_A(\vec{r}) - N_D(\vec{r})\right] \quad (1)$$

$$\frac{\partial f(\vec{r},\vec{k},t)}{\partial t} = -\frac{1}{\hbar}\nabla_{\vec{k}}\varepsilon \cdot \nabla_{\vec{r}} f(\vec{k},\vec{r},t) - \frac{q}{\hbar}\nabla_{\vec{r}}\phi(\vec{r},t)\cdot\nabla_{\vec{k}} f(\vec{k},\vec{r},t) - \left.\frac{\partial f(\vec{r},\vec{k},t)}{\partial t}\right|_c \quad (2)$$

$$\frac{\partial p(\vec{r},t)}{\partial t} = \nabla_{\vec{r}}\cdot[\mu_p p(\vec{r},t)\nabla_{\vec{r}}\phi(\vec{r},t) + \mu_p V_t \nabla_{\vec{r}} p(\vec{r},t)] - R(n,p) \quad (3)$$

$$-\frac{\hbar^2}{2m^*}\frac{d^2}{dy^2}\psi_i(y) + V(y)\psi_i(y) = E_i\psi_i(y) \quad (4)$$

where $\phi(\vec{r},t)$ is potential, n_q is the quantum electron density, $f(\vec{r},\vec{k},t)$ is electron distribution function, $\left.\frac{\partial f(\vec{r},\vec{k},t)}{\partial t}\right|_c$ is the collision term, $p(\vec{r},t)$ is hole concentration, m^* is effective mass along the direction perpendicular to the interface, $V(y)$ is the potential energy of the electrons, and E_i and ψ_i are the eigenenergy and eigenfunction of the ith subband respectively.

To solve the Schrödinger equation, we use a QL decomposition algorithm and a constant effective-mass approximation. After getting the energy levels and envelope functions, we populate the subbands and compute the electron density with the following

$$n_q(y) = \sum_i |\psi_i(y)|^2 \int_{E_i}^{\infty} f(E)D_i(E)dE \quad (5)$$

where $f(E)$ is the distribution function calculated directly by the Spherical Harmonic BTE solver, and $D_i(E)$ is the quasi 2-D density of states for electrons. Fig. 2 is the energy dispersion of a quantum well with subband energies. The integration of eqn.(5) corresponds to the area under distribution function curve between E_i and ∞ as shown in Fig. 3, which gives the electron concentration of each subband.

3 Results

Fig. 4 shows the classical and quantum electron concentration of the nMOSFET. The bias condition is $V_d = 0.05V$ and $V_g = 2.5V$. The subthreshold characteristics($\log(I_d)$ vs. V_g) are shown in Fig. 5. We can see in this figure that the threshold voltage is increased by the quantization effect. Also the larger V_g is, the larger the difference between classical and QM results. We show I_d vs. V_g for high drain bias(V_d=1.0 V) in Fig. 6. As can be seen in this figure the difference between classical and QM result is reduced by increasing drain bias since DIBL will decrease the depth of the quantum potential well and pinchoff widens it. The I-V characteristics are shown in Fig. 7 and Fig. 8 for $L_g = 0.1\mu m$ and $L_g = 0.25\mu m$ respectively. They show that for these particular devices, the shorter the device length, the larger the difference between classical and QM results. Current density is shown in Fig. 9 and 10. We can see clearly the QM effect that the peak value of current density is several $\mathring{A}$ away from the SiO_2/Si interface, and the smooth current density usually implies a well-converged result.

Acknowledgment We are grateful to Surinder Singh for helpful discussion and Intel for supporting the project.

References

[1] Liang, W., Goldsman, N., Mayergoyz, I., Oldiges, P.(1997): 2-D MOSFET modeling including surface effects and impact ionization by self-consistent solution of the Boltzmann, Poisson and Hole-continuity equations. IEEE Trans. Electron Dev. 44: 257-276.

[2] Spinelli, A. S., Benvenuti, A. and Pacelli, A.(1998): Self-consistent 2-D model for quantum effects in n-MOS transistors. IEEE Trans. Electron Dev. 45:1342-1349.

[3] Wang, H., Shih, W.-K., Green, S., Hareland, S., Maziar, C. M. and Tasch Jr., A. F.(1998): Hydrodynamic (HD) simulation of N-channel MOSFET's with a computationally efficient inversion layer quantization model. In: Proceedings of the (IWCE-5), Notre Dame, Indiana. pp. 423-428.

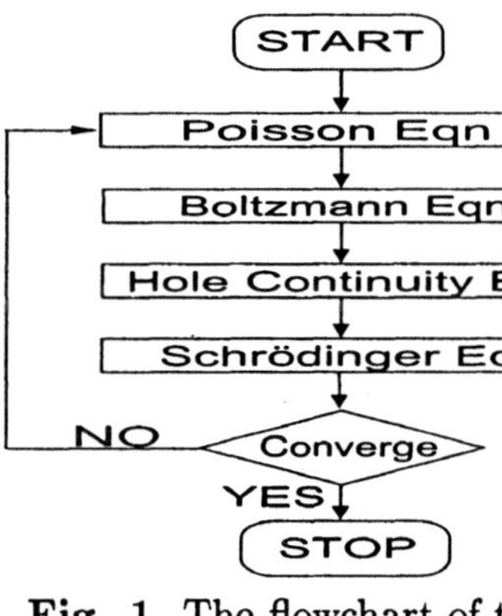

Fig. 1. The flowchart of the quantum BTE simulator

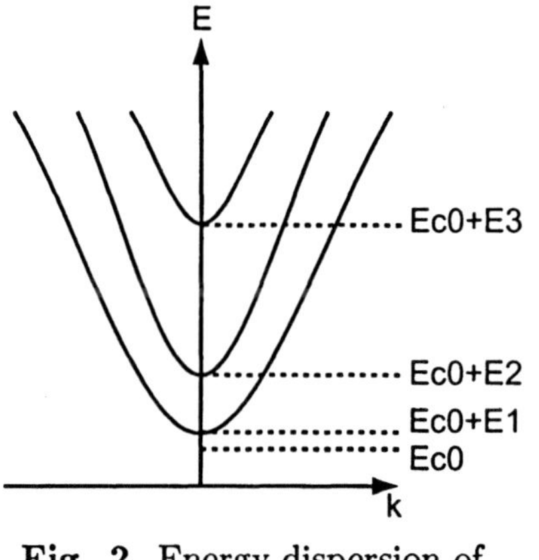

Fig. 2. Energy dispersion of quantum well.

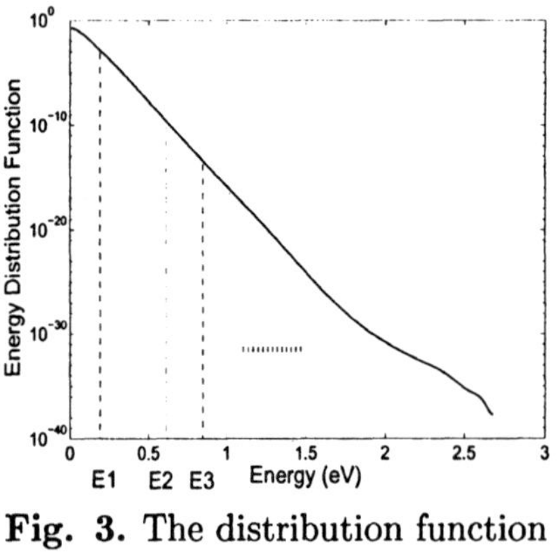

Fig. 3. The distribution function at surface for $V_g = 2.5V, V_d = 0.05V$.

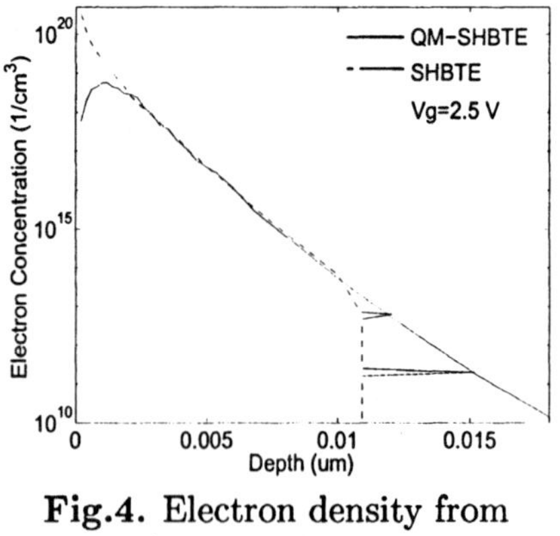

Fig.4. Electron density from surface to substrate at $V_g = 2.5V, V_d = 0.05V$.

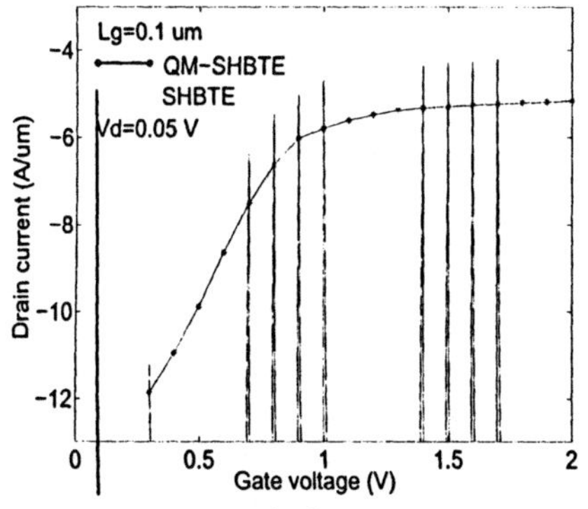

Fig.5. $\log(I_d)$ vs. V_g for $V_d = 0.05V, L_g = 0.1\mu m$.

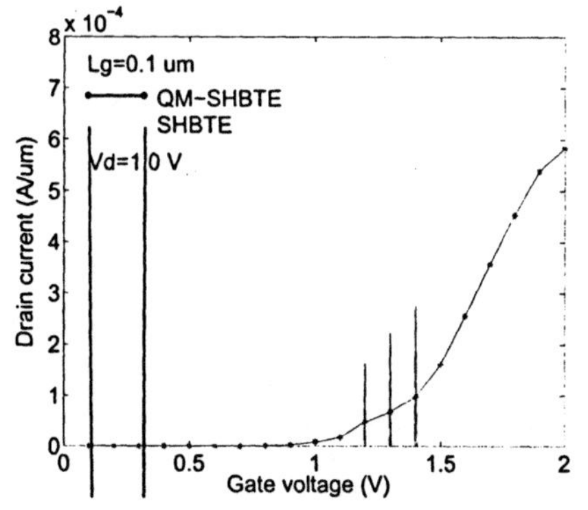

Fig.6. I_d vs. V_g for $V_d = 1.0V, L_g = 0.1\mu m$.

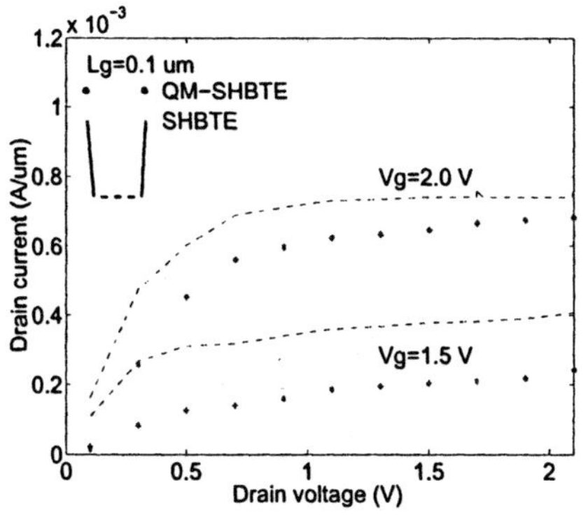

Fig.7. I-V characterictics for $L_g = 0.1\mu m$.

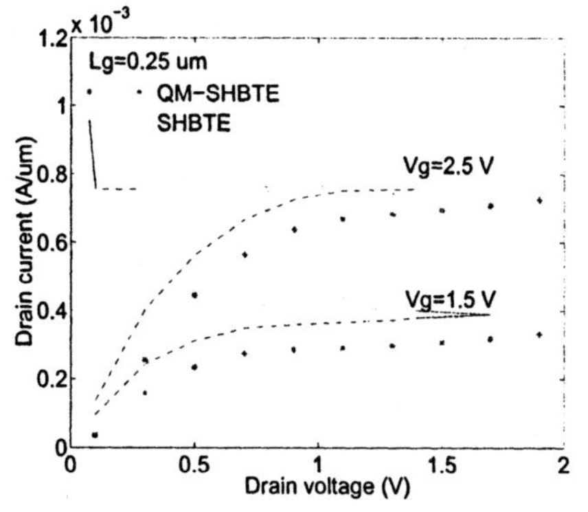

Fig.8. I-V characterictics for $L_g = 0.25\mu m$.

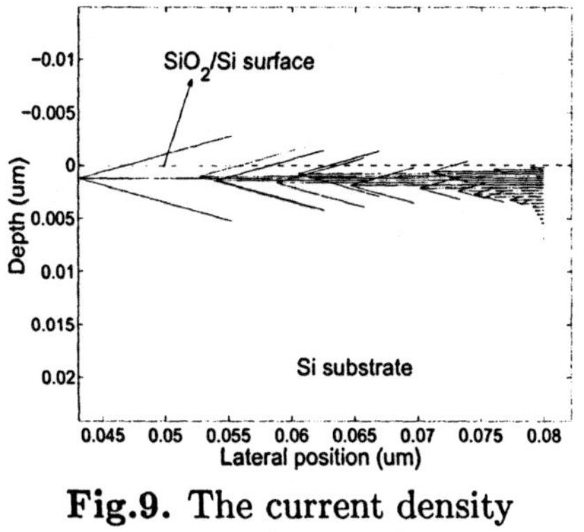

Fig.9. The current density maxium below surface of nMOSFET(QM-SHBTE).

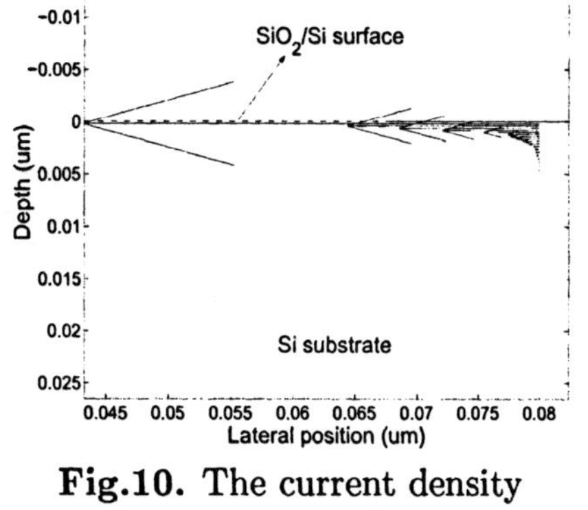

Fig.10. The current density maxium at surface of nMOSFET(SHBTE).

Boundary Condition Models for Terminal Current Fluctuations

M. Nedjalkov, T. Grasser, H. Kosina, and S. Selberherr
Institute for Microelectronics, TU Vienna, Gusshausstrasse 27–29, A-1040 Vienna, Austria
Phone: +43-1-58801/36014, Fax: +43-1-58801/36099, Email: Nedjalkov@iue.tuwien.ac.at

Abstract

A stochastic approach to a recently proposed model of terminal current fluctuations is presented. Two kinds of boundary conditions suitable for noise simulations in semiconductor devices are proposed. The properties and the domain of application of the two models are investigated and the conclusions are drawn from numerical experiments.

1 Introduction

The importance of current fluctuations in semiconductor devices and the physical and numerical complexity of their characterization stimulated a mutual development of the basic models and appearance of novel models for the current noise. Fundamental is the Ensemble Monte Carlo (EMC) method which provides both a model and numerical approach to the phenomena. The method is based on the notion that a direct emulation of the stochastic processes underlying the transport phenomena provides along with the physical mean values also their fluctuations. If the transport process is emulated and the current $i(t)$ is recorded in the time interval $(0,T)$ [1], the autocovariance function which characterizes the fluctuations is retrieved from its basic definition:

$$C_i(t,\tau) = \langle i(t)i(t+\tau)\rangle - \langle i\rangle(t)\langle i\rangle(t+\tau) \Rightarrow C_i(\tau) = \frac{1}{T}\int_0^T i(t)i(t+\tau)dt - \langle i\rangle^2 \quad (1)$$

where the brackets denote ensemble averages. A stationary process is considered ($\langle i\rangle$ = const) so that C_i becomes independent on the reference time t and the ensemble average is replaced by the time average in (1). The advantages and the drawbacks of this stationary EMC model are well known. The latter are particularly due to the fact that boundary conditions (BC) determine the device behavior. The EMC which is appropriate for evolution problems requires an initial transient period until the ensemble becomes stationary inside the device. The leaving particles must be re-injected to maintain the stationary process. Different models for re-injection from the boundaries are studied [2], [3] and it is shown that they affect the fluctuation characteristics.

Recently an alternative model for the current noise has been proposed [4], [5]. The autocovariance function has been obtained as a statistical average:

$$C_i(\tau) = \langle i\rangle \int d\mathbf{k} \int_L dx\, i(\mathbf{k},x)g(\mathbf{k},x,\tau) - \langle i\rangle^2; \quad \langle i\rangle = \int d\mathbf{k} \int_L dx\, i(\mathbf{k},x)f_s(\mathbf{k},x) \quad (2)$$

The space coordinate x is for a one-dimensional device with a length L and $\mathbf{k}$ denotes the wave vector. Here $i(\mathbf{k},x)$ is the current contribution from a particle in the particular phase space point according to the Ramo-Shockley theorem and g is an effective distribution function. The latter is the solution of the time dependent Boltzmann equation for an initial condition (IC) $g_0(\mathbf{k},x) = i(\mathbf{k},x)f_s(\mathbf{k},x)/\langle i\rangle$, f_s being the stationary solution in the device. In [3] the model has been applied to a bulk semiconductor using a deterministic method.

2 Boundary conditions

As applied to devices, the model requires proper BC. They are formulated in the framework of the stochastic method developed for evaluation of (2). A Monte Carlo approach to the latter simulates an evolution of the effective distribution function in contrast to the direct emulation of the current fluctuations.

The first BC model can be deduced from the limit $C_i \to 0$ when $\tau \to \infty$. The effective distribution g must evolve from g_0 to f_s in this limit. Practically C_i becomes zero after some time t_c typical for the concrete device. The stationary distribution f_s is ensured by the physical BC and it is thus concluded that they are the proper BC for (2). The particles that leave the device are re-injected according the BC in the way also used by the stationary EMC. The following effect can be encountered in this picture: particles exiting the device contribute to C_i after being reinjected with the BC, commonly taken to be the equilibrium distribution. If the exiting particles are not thermalized in the contact, the sudden cooling by the BC affects the autocovariance function. This problem is typical also for the stationary EMC which relies on the re-injection to maintain the steady state current. Special algorithms have been developed to avoid this effect [3].

The second BC model is obtained from (2) by noting that C_i contains as an integrand the difference $\phi(\mathbf{k}, x, \tau) = g(\mathbf{k}, x, \tau) - f_s(\mathbf{k}, x)$. Both, g and f_s satisfy the Boltzmann equation which has the following integral form:

$$
\begin{aligned}
f(\mathbf{k}, x, \tau) &= \int_0^\tau dt' \int d\mathbf{k}' f(\mathbf{k}', x(t'), t') S(\mathbf{k}', \mathbf{k}(t')) e^{-\int_{t'}^{\tau} dy \lambda(\mathbf{k}(y))} \\
&+ f_0(\mathbf{k}(0), x(0)) e^{-\int_0^\tau dy \lambda(\mathbf{k}(y))} + f_b(\mathbf{k}(t_b), x(t_b)) e^{-\int_{t_b}^{\tau} dy \lambda(\mathbf{k}(y))}
\end{aligned} \tag{3}
$$

with a trajectory determined by the electric force F and velocity $\mathbf{v}$:

$$
\mathbf{k}(t') = \mathbf{k} - \int_{t'}^{\tau} F(x(y)) dy \qquad x(t') = x - \int_{t'}^{\tau} v_x(\mathbf{k}(y)) dy \tag{4}
$$

The initial condition f_0 and the boundary conditions f_b participate explicitly in the integral form (4). f_b is zero inside the device and is specified only on the boundaries, while f_0 is zero if $x(0)$ is placed outside the device. The time t_b is determined from the position where the Newton trajectory (4) crosses the device boundary $x(t_b) = x_b$. f_s is a solution of (4) for IC given by $f_0 = f_s$, while for the function g the IC are given by $f_0 = g_0$. Both, f_s and g utilize the same BC given by the term f_b. It follows that the equation for ϕ does not contain f_b:

$$
\begin{aligned}
\phi(\mathbf{k}, x, \tau) &= \int_0^\tau dt' \int d\mathbf{k}' \phi(\mathbf{k}', x(t'), t') S(\mathbf{k}', \mathbf{k}(t')) e^{-\int_{t'}^{\tau} dy \lambda(\mathbf{k}(y))} \\
&+ g_0(\mathbf{k}(0), x(0)) e^{-\int_0^\tau dy \lambda(\mathbf{k}(y))} - f_s(\mathbf{k}(0), x(0)) e^{-\int_0^\tau dy \lambda(\mathbf{k}(y))}
\end{aligned}
$$

This equation describes a purely transient problem, where two ensembles with initial conditions g_0 and f_s evolve in time as $\phi_0(\mathbf{k}, x, \tau)$ and $\phi_s(\mathbf{k}, x, \tau)$ and give the solution as the difference $\phi = \phi_0 - \phi_s$. The boundaries absorb all particles that leave the device without re-injection.

The autocovariance (2) becomes:

$$C_i(\tau) = \langle i\rangle \int d\mathbf{k} \int_L dx\, i(\mathbf{k},x)\,(\phi_0(\mathbf{k},x,\tau) - \phi_s(\mathbf{k},x,\tau))$$
$$= \langle i\rangle \int d\mathbf{k} \int_L dx\, i(\mathbf{k},x)\phi_0(\mathbf{k},x,\tau) - \langle i\rangle(\tau)\langle i\rangle(0) \qquad (5)$$

where

$$\langle i\rangle(\tau) = \int d\mathbf{k} \int_L dx\, i(\mathbf{k},x)\phi_s(\mathbf{k},x,\tau); \qquad \langle i\rangle(0) = \langle i\rangle$$

By using the definition of g_0 it is seen that (5) is the autocovariance function of the process of evolution of an ensemble of particles initially distributed according the stationary distribution f_s in the device. With the absorbing BC the particles leak trough the device boundaries which corresponds to a transient process. (5) resembles the basic definition of the autocovariance function given by the first equation in (1).

The stochastic method consists of alternative MC algorithms which obtain the IC, simulate the evolution process, and obey the BC. The main feature is that a One Particle MC (OPMC) simulation is used to obtain the stationary distribution f_s inside the device. The latter gives rise to initial points of an ensemble of particles, whose evolution provides the current autocovariance function. In this way the initial transient simulation required by the stationary EMC can be saved since OPMC is much more efficient in obtaining f_s. Indeed, while the EMC method collects the information at the end of each trajectory, the OPMC collects the information from each trajectory segment. In the present formulation of the task the ensemble is followed until the time t_c which is commonly few orders less than the averaging time T of the stationary EMC. The evolution process imposed by the first kind of BC resembles the physical transport in the device and thus can be simulated self-consistently by a coupling with the Poisson equation. Here we compare the two kinds of BC using non-selfconsistent simulations. In this case the electric field is stationary, so that the ensemble trajectories can be mapped onto the OPMC trajectory similar to the way discussed in [6].

The numerical experiments for the two BC models are summarized in Fig. 1 and Fig. 2 respectively. A silicon n^+nn^+ structure with a length of each segment of 0.2μm and doping concentration $10^{17} : 10^{16}$ cm^{-3} is considered at different applied voltages. The autocovariance functions are normalized as $C_i(\tau)/C_i(0)$. The curves for the absorbing and the injecting BC overlap each-other and relax to zero as expected. However this is not true for a resistor with the same total length $0.6\mu m$, considered on the right of Fig. 1. At equilibrium the results of the two models coincide. At $0.25V$ there is already a difference in the corresponding curves which increases with the increase of the applied voltage. At $1V$ the autocovariance for the absorbing BC relaxes to zero, while for the injecting BC remains well above zero as seen on the left picture of Fig. 2. We assign this to the influence of the ohmic contact on the autocorrelation during the process of re-injection: the carriers are hot when leaving the device and are injected back thermalized. This conclusion is supported by additional experiments showing a decrease of the effect when the length of the structure increases under a constant electric field inside. In this case the portion of the leaving carriers decreases with respect to the total number of particles in the resistor. Another experiment is shown on the right of Fig. 2. The simualted structure is formed by two tiny highly doped $0.05\mu m$

regions attached on both sides of the resistor which cool the exiting carries. The two autocorelation functions coincide again.

We conclude that the injecting BC can be applied if the contact regions thermalize the carriers before they leave the device. The absorbing BC model is universal and physically more transparent since it refers to the dwelling time of the particles in the device.

Acknowledgment

This work has been supported by the IST program, project NANOTCAD, IST-1999-10828.

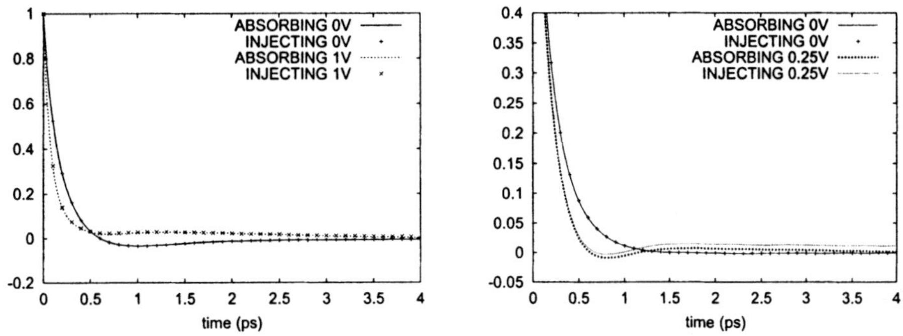

Fig. 1: Current Autocovariance for the n^+nn^+ diode (left) and the resistor (right)

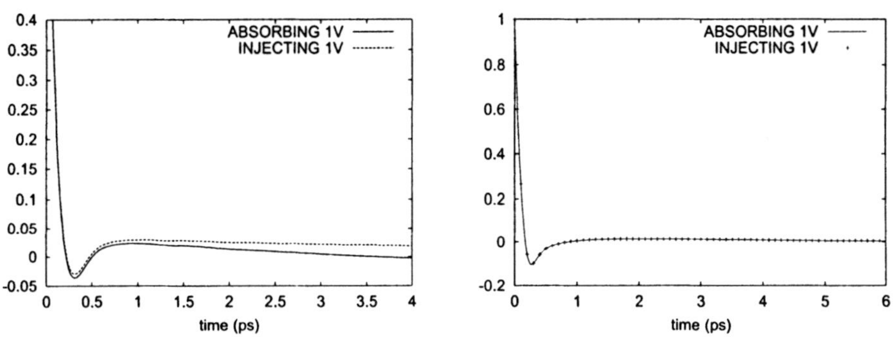

Fig. 2: Current Autocovariance at 1 Volt for the resistor (left) and the n^+nn^+ structure (right)

[1] L. Varani, T. Kuhn, L. Reggiani, and Y. Perles, "Current and Number Fluctuations in Submicron n+nn+ Structures," *Solid-State Electron.*, vol. 36, no. 2, pp. 251–261, 1993.

[2] T. Gonzalez and D. Pardo, "Ensemble Monte Carlo with Poisson Solver for the Study of Current Fluctuations in Homogeneous GaAs Structures," *J.Appl.Phys.*, vol. 73, no. 11, pp. 7453–7464, 1993.

[3] J. Adams, T.Tang, and L. Kay, "Monte Carlo Simulation of Noise in GaAs Semiconductor Devices," *IEEE Trans.Electron Devices*, vol. 41, no. 4, pp. 575–581, 1994.

[4] C. Korman and I. Mayergoyz, "Semiconductor Noise in the Framework of Semiclassical Transport," *Physical Review B*, vol. 54, no. 24, pp. 17620–17627, 1996.

[5] A. Piazza, C. Korman, and A. Jardeh, "A Physics-Based Semiconductor Noise Model Suitable for Efficient Numerical Implementation," *IEEE Trans.Computer-Aided Design*, vol. 18, no. 12, pp. 1730–1740, 1999.

[6] H. Kosina, M. Nedjalkov, and S. Selberherr, "A Monte Carlo Method for Small Signal Analysis of the Boltzmann Equation," *J.Appl.Phys.*, vol. 87, no. 9, pp. 4308–4314, 2000.

Electron Velocity in sub-50-nm Channel MOSFETs

Dimitri A. Antoniadis, Ihsan J. Djomehri, and Anthony Lochtefeld
Microsystems Technology Laboratories
Massachusetts Institute of Technology
Cambridge, MA 02139, USA

Abstract

Inverse modeling of state-of-the-art NMOSFETs is used to investigate electron transport models and in particular to extract the effective velocity of electron injection from source to channel. It is found that this velocity is less than 50% of the maximum possible velocity, i.e. the thermal velocity of electrons in the source. Based on the Landauer formulation, as adapted by Lundstrom to silicon MOSFETs this indicates that modern NMOSFETs are quite far from their ballistic transport limit and therefore their current is still limited by momentum scattering as manifested in the electron mobility. Investigation of mobility in those transistors reveals that it is reduced with channel length, most likely due to Coulomb scattering by the ionized dopant atoms in the source and drain halos that are necessary for well-tempered ultra-short-channel MOSFETs, and possibly remote scattering by the source dopants.

1 Introduction

Investigation of electron transport models by comparison of simulated to experimental devices requires accurate knowledge of the experimental device structure and doping configuration. To-date there are no direct doping characterization techniques for deeply scaled MOSFETs, and therefore one has to resort to indirect characterization methods. Inverse modeling (IM) based on 2-D device simulation has proven to be a very powerful such method. While at first sight this may appear as a vicious circle, in fact using electrical measurements in a regime where device operation is very sensitive to electrostatic potential and not very sensitive to transport models resolves the dilemma. Weak-inversion is such a regime for I-V characteristics [1], and C-V also has no dependence on carrier transport.

In this work, combination of weak-inversion I-V and C-V measurements of devices from two different technologies, that here will be referred to by their physical oxide thickness, t_{ox}, 3.3 nm and 1.5 nm. Accurate doping S/D extension and channel profiles are obtained by this method and are then used to investigate electron transport in these devices.

IM-extracted values of the source series resistance, R_S, are used in a new method to extract electron velocity nearest to the source-channel injection point than any other method, and this velocity is compared with the theoretical values of thermal injection velocity in order to assess how close to the ballistic limit modern NMOSFETs operate. Further, IM-extracted values of effective channel length, L_{eff}, are used along with values of R_S to extract low-longitudinal-field mobility in extremely short channel devices.

Given that all standard methods for L_{eff} and R_{SD} characterization fail in the case of modern sub-100-nm devices, IM provides new insight into mobility at short channels.

2 Inverse Modeling for MOSFET Chararterization

Figure 1 shows the inverse modeling flow diagram. Because the subsequent log(I)-V data is sensitive to the gate stack, the first step is to accurately account for it by extracting the oxide thickness and poly doping from the accumulation and inversion regimes of the C_{gg}-V curve, respectively. Quantum Mechanical effects are captured via an approximate band gap shift [2]. Now, 1-D IM is performed to extract the channel doping on a long-channel device. No process simulation or SIMS were availabe here, but could be used to help construct a good initial guess to the 2-D profiles of the short-channel devices. Next, detailed TEMs were necessary to completely characterize the gate stack and spacer dielectric structures to capture the effect of the fringing capacitances. Finally, the combined log(I)-V and C-V IM approach alternated between a few iterations of each data set using a standard non-linear least-squares optimization algorithm. The effect of small variations in the 2-D doping parameterization on the RMS error are used to obtain a converging solution to the parameterization (which consists of two 2-D gaussians for S/D extensions and halos, plus the long-channel profile).

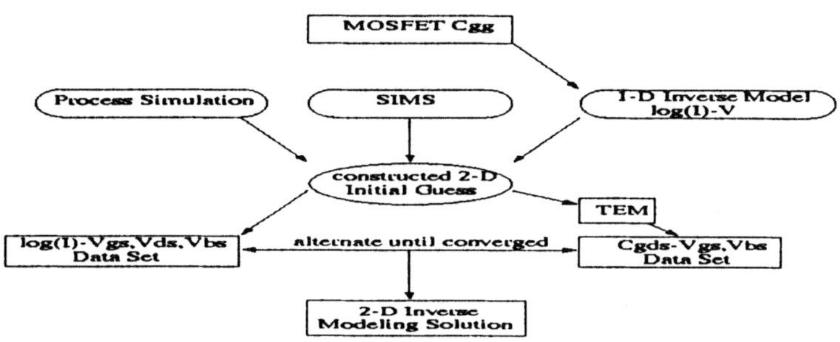

Figure 1. Flowchart delineating ideal comprehensive log(I)-V & C-V IM methodology

Once reliable doping profiles are obtained for several devices of different gate length in each technology, it is possible to proceed with the investigation of electron transport. First, electron mobility vs. transverse electric field, the so-called universal mobility is calibrated for each technology using long channel devices and the split-C-V method for extraction of μ_{eff} vs. E_{eff}. Note that since different nitridation processes were used in the fabrication of the two technologies, the universal effective mobility curve is only universal within each technology. The generalized low-field mobility model [3] from MEDICI is used to give the minimum of the bulk mobility, assumed to be for minority carriers in bulk, and the universal mobility. The universal vertical field dependence is assumed to have an inverse cube root term and an inverse to a higher power which is calibrated along with the magnitudes of these terms.

With the μ_{eff} vs. E_{eff} model in hand, next the source/drain series resistance R_{SD} is extracted by fitting simulated to measured I_D-V_{GS} at low V_{DS} (typically 50 mV). Note that there are several components of R_{SD}. The V_{GS}-dependent component of the source/drain extension is included in the simulation via the the IM-extracted S/D doping profile, but the deep-S/D and silicide contact resistance are lumped into a single resistance value. For the two technologies under investigation we find the R_{SD} is 220 and 245 $\Omega-\mu$ m, for t_{ox} 1.5 and 3.3 nm at V_{GS} 1.0 and 1.8 V, respectively.

Then, the high longitudinal field transport model is investigated. First, for moderately short-channel devices (L_{eff} > 100 nm), the fitting parameters beta and v_{sat} in the Caughey-Thomas expression for the EB mobilitly model [4] are obtained by matching strong inversion I-V at high V_{DS}. Next, a single value for the carrier relaxation time is found that matches the strong inversion data for the entire family of short devices.

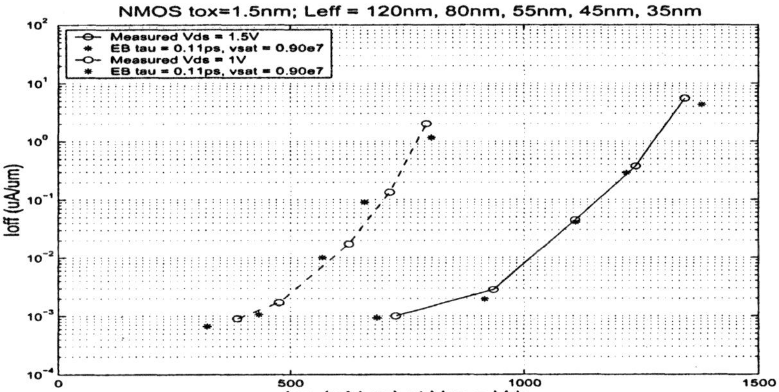

Figure 2. Energy balance fit to experimental I_{on} vs. I_{off} curves at two V_{DD}s

Figure 2 shows the resulting I_{Dsat} vs. I_{off} (V_{DS}=V_{GS}=V_{DD}) for the t_{ox} = 1.5 nm technology. For all devices the high longitudinal field transport model parameters are the same, while the μ_{eff} vs. E_{eff} is unique to each technology. While the Energy Balance model is a simplification it is gratifying to observe that the values for v_{sat} and low-T_e relaxation time constant are both unique and consistent with experimental and theoretical values, respectively [5, 6].

3 Source-Channel Injection Electron Velocity

It has been shown that electrons in a nMOSFET channel can exceed significantly v_{sat} for isotropic field regions (approx. 10^7 cm/s) [7]. The ultimate limit of drain current is, rather, thought to be determined by the thermal injection velocity, v_{inj}, from the source accumulation region into the channel [8]. This can be stated as

$$I_{on}/W = v_{inj}Q_i(x_o)(T/(2-T)) \qquad \text{Eq. (1)}$$

Where $Q_I(x_o)$ is the normalized inversion charge at the conduction-band peak at $x=x_o$ at the source side of the channel. Eq. 1 then defines the *effective source-channel injection velocity* at x_o as $v_{eff}=v_{inj}T/(2\text{-}T)$, where T is the transmission coefficient at x_o which depends on momentum scattering rate, and barrier width. T=1 and therefore $v_{eff}=v_{inj}$, represents the fully ballistic limit where there is no backscattering of electrons into the source. Clearly, v_{eff} is purely diffusive because the longitudinal electric field is 0 at x_o, and is of great significance in understanding the ultimate limit of MOSFET performance but is also very difficult to measure experimentally.

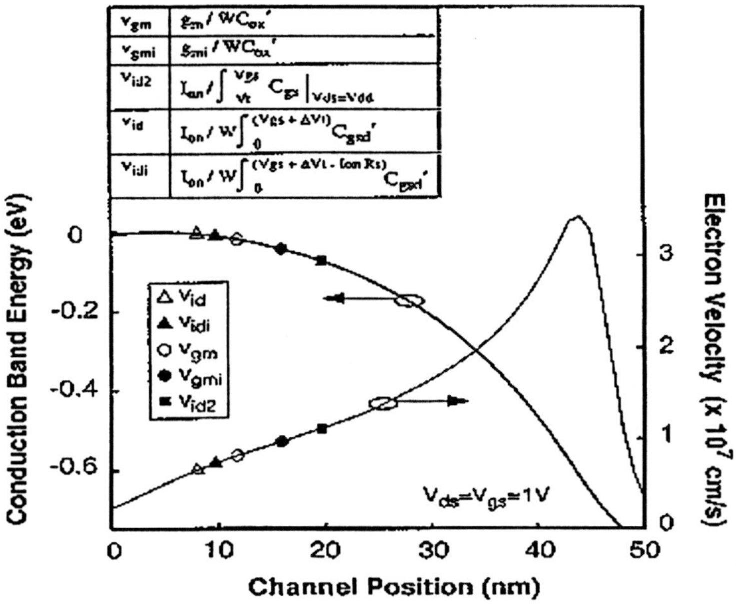

Figure 3. Conduction band and carrier velocity versus position in bulk L_{eff} = 50 nm NMOS.

Several methods have been reported for measuring near-source carrier channel velocity experimentally [7, 9, 10]. In order to evaluate these experimental velocities relative to the well-defined v_{eff} we have simulated measurement techniques and plotted the extracted electron velocity for each on the simulated velocity vs. channel position curve, for a modern nMOSFET similar to the shortest t_{ox}=1.5 nm inverse-modeled ones. Figure 3 shows the results, and the inset summarizes the experimental velocity extraction methods. Two observations can be made: 1) None of the techniques yields a velocity at x_o but the most recent method [10], with or without correction for R_S yields velocities closest to x_o. 2) Both v_{gm} and v_{gmi}, the latter from *intrinsic* g_m corrected for R_S [11], while very easy to obtain represent velocities rather far downstream from x_o.

It has been pointed out [12] that velocity vs. DIBL is a characteristic merit curve unique to each technology, similar to I_{Dsat} vs. I_{off} ($V_{DS}=V_{GS}=V_{DD}$); in both cases, channel length is an implicit variable. Independent of technology, the highest performance well tempered MOSFETs generally exhibit DIBL less than 100-150 mV/V, and therefore is best to compare devices from different technologies at constant DIBL. We arbitrarily define v_{idi} at DIBL=100 mV/V to represent the upper limit of v_{eff} for a given technology. Table 1 shows the values of this v_{eff} for the two experimental technologies as well as for a Monte Carlo simulated 25 nm nMOSFET [13]. As can be seen the value of v_{eff} and therefore the transmission coefficient is well below the ballistic limit, and appears to be decreasing with scaling. Since it is desirable to increase v_{eff} for commensurate device performance with scaling we examine next why it decreases instead.

Table 1. v_{eff} *for: Monte Carlo; Tech. A =* t_{ox} *1.5 nm,* V_{DD} *1 V; Tech. B =* t_{ox} *3.3 nm* V_{DD} *1.8 V*

	25 nm M.C.	Tech. A	Tech. B
v_{eff} (cm/s)	$6.1\text{-}7.0\times10^6$	6.7×10^6	7.9×10^6
v_{eff} / v_{inj}	0.35-0.40	0.39	0.49
T	0.52-0.57	0.56	0.66

← Decreasing L_{eff}, T_{ox}
← Increasing v_{inj}

4 Electron Mobility in Ultra-Short-Channel MOSFETs

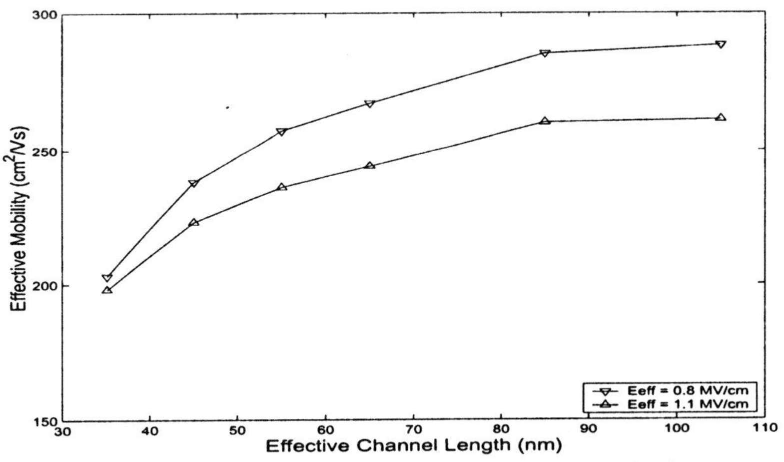

Figure 4. Measured μ_{eff} *vs.* L_{eff} *for different* E_{eff} *on the* t_{ox} *= 1.5 nm technology.*

As discussed earlier, the transmission coefficient, T, depends on the rate of momentum scattering and the width of the conduction band barrier at the source-end of the channel. It should therefore be closely related to the effective low longitudinal field mobility in that region. There are two reasons why the effective mobility would be decreasing with scaling: 1) In modern MOSFETs strong doping halos are used to improve electrostatic integrity and therefore coulomb scattering can increase to where it dominates the transverse-field-controlled surface scattering. 2) Remote scattering by source dopants as been pointed out as possible mechanism [14]. However, it is well known that all the standard methods for mobility characterization fail for modern short-channel MOSFETs, so these effects have not been verified to date. In this work we take advantage of inverse modeling, and therefore knowledge of parasitic resistance and channel length to do such a characterization as a function of channel length in the t_{ox}=1.5 nm technology. Figure 4 shows the results of these measurements for two different effective transverse fields. The device details are given in the caption. It is quite clear that the effective channel mobility indeed decreases with channel length and in fact becomes independent of transverse field for the shortest devices. This suggests, that the surface scattering mechanism is no longer dominant for short lengths and indeed local and/or remote coulomb scattering dominates. It can be concluded then that conventionally extracted long-channel mobility is not sufficient for understanding transport in modern MOSFETs. The corollary to this observation is that in order to approach the ballistic limit it will be necessary to find methods to increase mobility in ultra-short-channel MOSFETs, either through the introduction of silicon-compatible high-mobility materials such as strained Si on relaxed SiGe [e.g. 15], or by the removal of dopants [e.g. undoped-channel double-gate structures], or a combination of both.

5 Conclusion

Inverse modeling of modern MOSFETs can be used not only as an engineering tool but also in order to shed light on transport mechanisms via simulation of devices with high fidelity to the real experimental ones. It is found that the energy balance model approximation is adequate for modeling a very broad range of devices with a unique set of model parameters. It is also found that source-channel injection velocity is still far below the thermal velocity limit even in sub-50-nm nMOSFETs, and the reason appears to be increased scattering from dopants near the source.

Acknowledgement
This work has been supported by DARPA

References

[1] Z. Lee et al., IEDM proceedings, p. 683, 1997.
[2] M. J. Van Dort et al., Solid-State Electronics, Vol. 37 No. 3, 1994, p 411.
[3] S. A. Mujtaba et al., NUPAD V, Honolulu, Hawaii, June 5-6, 1994
[4] A. Forghieri et al., IEEE Trans. CAD, Vol. 7 No. 2, Feb. 1988.
[5] S. M. Sze, Physics of Semiconductor Devices, 2nd ed.(New York:John Wiley & Sons, 1981).
[6] M. V. Fischetti, IEEE TED, Vol. 38, 1991, p 634.
[7] G. Shahidi, D. Antoniadis, H. Smith, IEEE EDL, Feb. 1988, p 94-96.
[8] F. Assad, Z. Ren, S. Datta, M. Lundstrom, P. Bendix, 1999 IEDM Digest, p 547-550.
[9] T. Mizuno, R. Ohba, Journal of Applied Physics Vol. 82 No. 10, Nov. 1997, p 5235-5240.
[10] A. Lochtefeld and D. A. Antoniadis, IEEE EDL, Vol. 22 No. 2, Feb. 2001, p 95.
[11] S. Y. Chou and D. A. Antoniadis, IEEE TED-34, 1987, p 448.
[12] H. Hu, J. Jacobs, L. Su, D. Antoniadis, IEEE TED, April 1995, p 669-677.
[13] Y. Taur, C. Wann, D. Frank, 1998 IEDM Technical Digest, p 789-792.
[14] M. V. Fischetti and S. E. Laux, J. of Applied Physics, Vol. 89 No. 2, Jan. 2001, p 1205.
[15] K. Rim, J. Hoyt, J. Gibbons, IEEE TED, July 2000, p 1406-1415.

3D Statistical Simulation of Intrinsic Fluctuations in Decanano MOSFETs Introduced by Discrete Dopants, Oxide Thickness Fluctuations and LER

Asen Asenov
Department of Electronics and Electrical Engineering, University of Glasgow
Glasgow G12 8LT, Scotland, UK
A.Asenov@elec.gla.ac.uk

Abstract

The need for statistical 3D simulations to study intrinsic parameter fluctuations in aggressively scaled MOSFETs introduced by discreteness of charge and atomicity of matter is discussed. We describe a hierarchical implementation of such a 3D 'atomistic' simulation approach, which includes quantum mechanical corrections based on the Density Gradient algorithm. Simulation examples of intrinsic parameter fluctuations associated with random discrete dopants in the active region of the device and in the polysilicon gate, oxide thickness fluctuation within the gate area, and line edge roughness (LER) of the gate are presented. We speculate about the challenges ahead in understanding and accurately simulating the atomistic effects in the next generation of MOSFETs.

1 Introduction

The International Technology Roadmap for Semiconductors [1] forecasts that semiconductor devices will reach decanano (i.e. sub-100 nm) dimensions in the first decade of the new millennium. At the same time, doubts have been expressed by leading IC manufacturers, such as Intel [2], that conventional device architectures can be scaled to deep decanano dimensions and that Moore's Law can be maintained. It becomes extremely important to examine, by means of predictive device modelling and simulation, possible scaling scenarios, pitfalls, and promising device architectures in order to give direction to the urgently needed, and very costly, technology developments of the next decade, and thus save unnecessary effort and expense. Predictive modelling and simulation face numerous challenges associated with the complex geometry, high fields, non-equilibrium transport and quantum effects exhibited by any decanano device architecture.

A fact not yet widely realised is that the scaling of devices in integrated circuits will reach a stage where the granularity of the electric charge and the atomicity of matter begin to introduce substantial 'intrinsic' variation in individual device characteristics, and must be included in the simulations. The variation in number and position of dopant atoms in the active region of MOSFETs will make each transistor microscopically different and introduce significant intrinsic parameter variation from device to device on a single chip [3]. In addition, the gate oxide thickness becomes equivalent to several atomic layers with a typical interface roughness of the order of 1-2 atomic layers [4]. This will introduce a substantial variation in the oxide thickness in the gate region resulting in a microscopically different oxide thickness pattern in each transistor. The unique pattern of the oxide roughness in each decanano MOSFET will affect both device electrostatics and the surface roughness limited mobility from device to device. The granularity of the gate material and the photo-resist, together with other factors, will introduce unavoidable roughness in the gate pattern definition and statistical variations in geometry between devices. When combined together the variations in dopant statistics, oxide thickness pattern, gate material and geometry will have a crucial impact on the functionality, yield and reliability of the corresponding circuits and systems.

From discreteness of charge to granularity of material, in this paper we survey some of the significant aspects of the atomistic picture of MOSFETs in the sub-100 nm regime. We will first highlight important, and practical, issues pertaining to the 3D atomistic simulator development in a statistical and quantum mechanical framework. This will be followed by a series of simulation results highlighting the effects of discrete dopants, rough interfaces and imperfect definition of edges on the intrinsic parameter fluctuations in decanano MOSFETs. Finally, we will point out issues with regard to 'atomicity' and simulator development in MOSFETs that require further attention.

2 Simulation approach

The statistical, atomic scale variations in decanano devices shifts the paradigm of numerical device simulations [5]. It is no longer sufficient to simulate a single device with continuous doping distribution, uniform oxide thickness and unified dimensions to represent one macroscopic design. Each MOSFET is microscopically different at the level of dopant distribution, oxide thickness and gate pattern, so an ensemble of macroscopically identical but microscopically different devices must be characterised. The aim of the numerical simulation shifts from predicting the characteristics of a single device towards estimating the mean values and the variance of basic device parameters, such as threshold voltage, subthreshold slope, transconductance, drive current, etc. for a whole ensemble of microscopically different devices in the system. It must be emphasised that even the mean values obtained from statistical atomistic simulations are not identical to the values corresponding to continuous charge simulation [6, 7]. The simulation of a single device with random dopants, oxide thickness and gate pattern variation requires a 3D solution with fine grain discretisation. The requirement for statistical simulations transforms the problem effectively into a four dimensional one where the fourth dimension is the size of the statistical sample.

We use drift diffusion (DD) simulations to study intrinsic parameters fluctuations in decanano MOSFETS. Quantum mechanical confinement effects in the inversion layer are taken into account using the density gradient (DG) formalism [8, 9]. The drift-diffusion approach does not properly represent the non-equilibrium carrier dynamics and ballistic transport effects and hence underestimates the drain current. However, it can be used with confidence to estimate the threshold voltage based on a current criterion in the subthreshold region, where the current is exponentially controlled by the gate and its underestimation produces a minute error in the calculated value for the threshold voltage. Above threshold the quantum corrected DD simulations provide a sufficiently accurate estimate for the variation in the device parameters resulting from the electrostatics of random discrete dopants, local oxide thickness variations and LER.

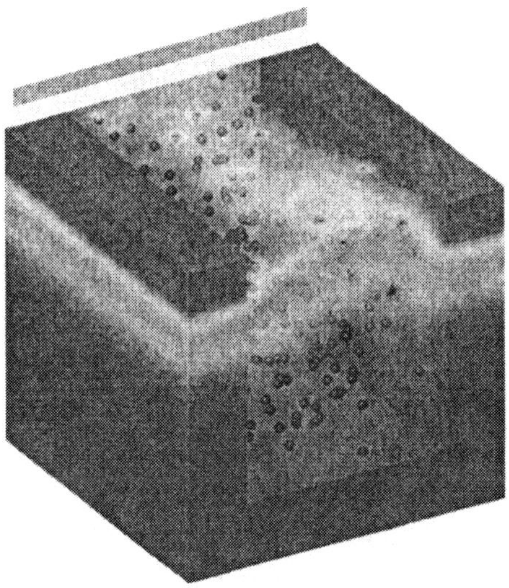

Fig. 1. Potential distribution at threshold voltage obtained from atomistic DG simulations of a 30×50 nm MOSFET.

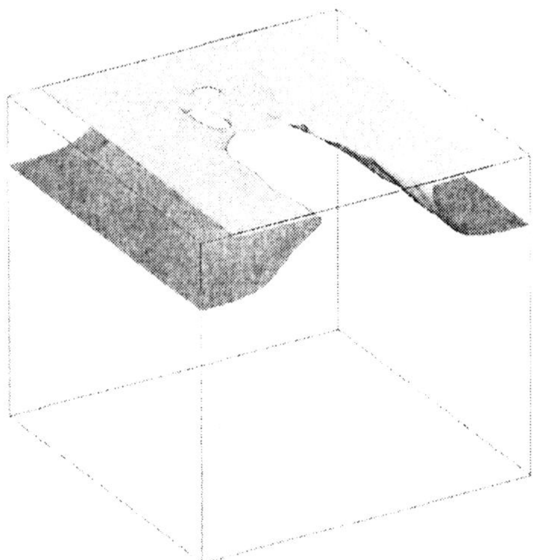

Fig. 2. One equi-concentration contour corresponding to the potential distribution in Fig. 1.

A typical atomistic solution domain is shown in Fig 1. The discrete dopants are placed in the outlined channel region between the source and drain. In the rest of the simulation domain the doping charge has a continuous distribution. Although the best way to introduce the doping distributions in the atomistic simulations would be to use the output from an atomic scale process simulator [10], we use a simpler approach. The expected number of dopants in the atomistic region is estimated by integrating the continuous doping distribution, obtained for example from a standard process simulator. The actual number of dopants in each MOSFET from the simulated ensemble is chosen randomly from a Poisson distribution with the above mean. Then, using a rejection technique, the dopants are placed randomly according to the initial continuous doping distribution. The potential distribution at gate voltage equal to the threshold voltage is superimposed in Fig. 1. Strong potential fluctuations at the Si/SiO_2 interface associated with the discrete dopants can be observed. The current in the device percolates through the valleys of the potential landscape. One electron equi-concentration contour which corresponds to this solution is presented in Fig. 2. The equi-concentration contour highlights the basic features of the quantum charge distribution. The quantum confinement in the channel results in a smoothing of the carrier density profile with a maximum in the electron concentration, located approximately 1.5 nm below the interface.

3 Random discrete dopants

The dependence of the threshold voltage on oxide thickness, obtained from classical and quantum DG simulations, is presented in Fig. 3 for a 50×50 nm MOSFET. With channel doping concentration 5×10^{18} cm^{-3}. Such devices have an average of 170 atoms in the channel depletion region. Samples of 200 microscopically different devices are simulate to extract the average threshold voltage and its standard deviation for each macroscopic combination of device design parameters. Results for the average threshold voltage, obtained from atomistic simulations, and for the threshold voltage, obtained from continuous charge simulations are compared for metal and poly-silicon gate devices. The quantum mechanical threshold voltage shift decreases with the reduction in the oxide thickness t_{ox}. The inclusion of the poly-Si gate in the simulations results in an additional increase in the threshold voltage. Most importantly the random dopant induced threshold voltage lowering, inherent to the atomistic simulations, and associated with percolation of the channel current through valleys in the potential fluctuations, increases in the quantum case.

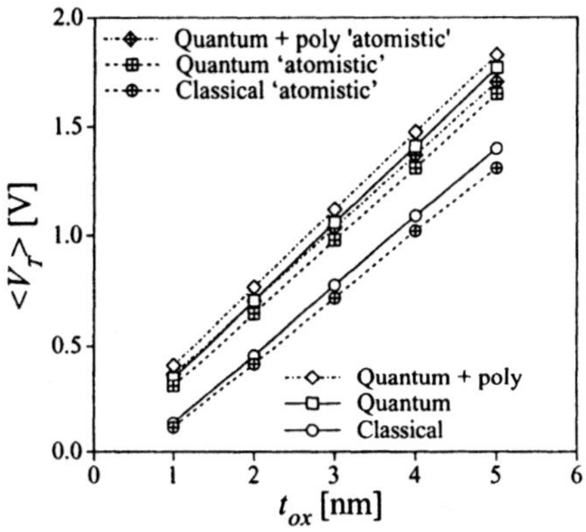

Fig. 3. Dependence of average threshold voltage on the gate oxide thickness in a 50×50 nm MOSFET.

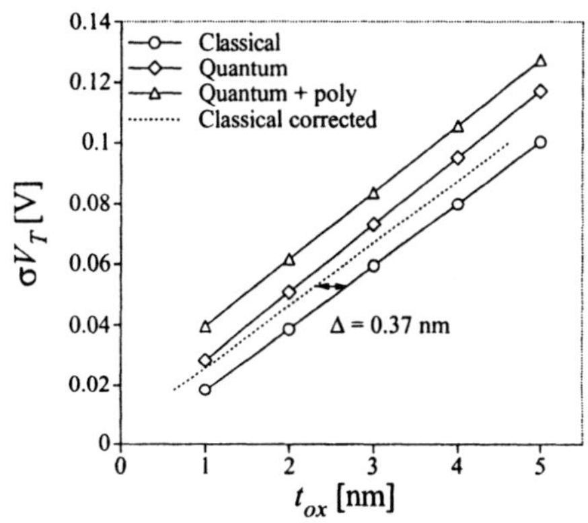

Fig. 4. Threshold voltage standard deviation as a function of the oxide thickness in a 50×50 nm MOSFET.

The dependencies of the threshold voltage standard deviation, σV_T, as a function of the oxide thickness, extracted from classical and from quantum atomistic simulations, are compared in Fig. 4. In the classical simulations of metal gate devices σV_T scales linearly to zero with the corresponding scaling of t_{ox} as a result of the screening from the gate. The values of σV_T corresponding to the quantum simulations are shifted up with respect to the classical simulations, and the shift increases slightly with the increase in the oxide thickness. The inclusion of the poly-silicon gate in the simulations results in an additional increase in σV_T, which, in combination with the increase associated with the quantum mechanical effects, almost doubles the fluctuations for oxide thicknesses below 2 nm. Thus the inclusion of quantum effects paints a more pessimistic picture than anticipated until now from purely DD simulations.

4 Oxide thickness fluctuations

In decanano MOSFETs the gate oxide thickness will be equivalent to a few silicon atomic layers with a typical interface roughness of one to two atomic layers [11]. The roughness can introduce more than 50% variation in the oxide thickness in devices with a 1 nm gate oxide. The unique random pattern of the gate oxide in each MOSFET, and the related surface landscape and potential fluctuations, will have a substantial contribution to the intrinsic parameter variation in such devices. A typical random Si/SiO_2 interface used in the simulation of a 30×30 nm MOSFET, is shown at the top of Fig. 5. The interface has been re-constructed using a 2D Fourier synthesis approach and a power spectrum corresponding to a Gaussian correlation function. Continuous charge was used, and only roughness of the Si/SiO_2 interface was introduced in the simulations. The potential distribution at threshold voltage is shown at the bottom of the same figure. One equiconcentration surface corresponding to electron charge density 1×10^{17} cm^{-3} is plotted in the middle illustrating the quantum confinement effects in both vertical and lateral directions.

The dependence of the threshold voltage standard deviation σV_T on the correlation length Λ obtained from classical and DG simulations is compared in Fig. 6. The

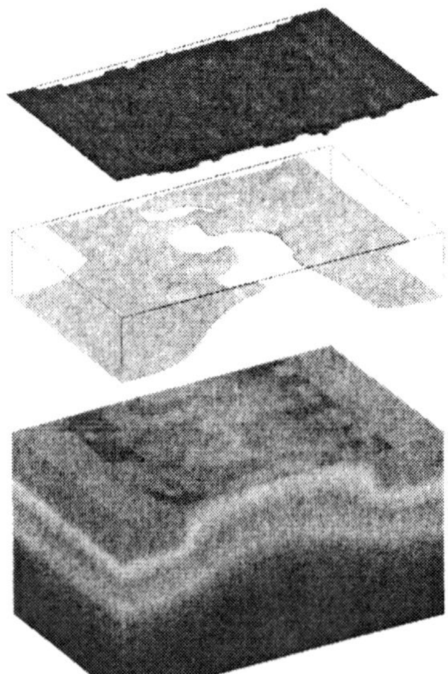

Fig. 5. The profile of the random Si/SiO_2 interface in a 30×30 nm MOSFET with t_{ox}= 1 nm (top), an equiconcentration contour obtained from DG simulation and the potential distribution (bottom).

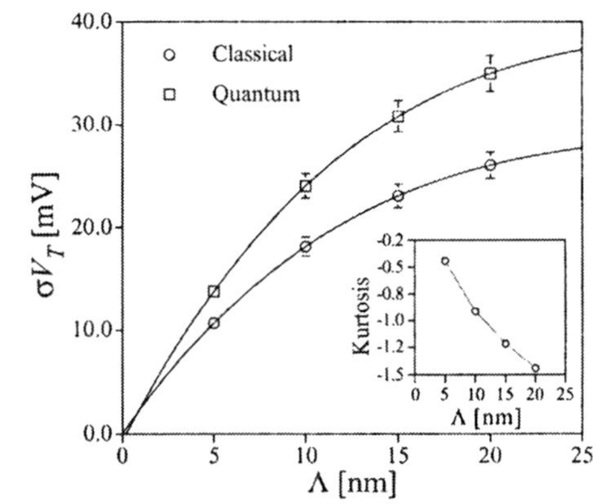

Fig. 6. Dependence of the threshold voltage standard deviation σV_T on the correlation length Λ for the 30×30 nm MOSFETs as in Fig. 5. Classical and DG simulation results are compared.

introduction of quantum corrections results in an increase in the threshold voltage variation. We believe that this is related to the lateral confinement effects, which narrow the current percolation paths. In both cases the dependence of σV_T on Λ is linear for correlation lengths much smaller than the characteristic MOSFET dimensions and saturates for large Λ. The dependence of the kurtosis of the threshold voltage distribution as a function of Λ in the DG case is presented as an inset in the same figure. The increasingly negative values of the kurtosis are an indication for the flattening of the σV_T distribution with the increase in correlation length. For devices with characteristic dimension below 30 nm the oxide thickness induced threshold voltage fluctuations become comparable to the fluctuations induced by random discrete dopants, particularly when the contribution of the both interfaces is taken into account and the larger correlation length suggested by AFM measurements are adopted.

5 Line edge roughness

Line edge roughness (LER), resulting from granularity of resists and gate material and from factors limiting the tool performance used in lithography processes, is difficult to reduce below 5 nm. As the aggressive scaling of Si MOSFETs continues to the sub-100 nm regime LER will constitute a large fraction of the gate length and will introduce significant intrinsic variations in the device parameters in its own right. As a natural extension to the statistical 3D simulations described in this paper, we approach the simulation of LER in decanano MOSFETs in a statistical fashion. LER in the simulations is specified by rms amplitude, Δ, and correlation length, Λ. This allows both 3D and statistical aspects of LER to be naturally incorporated in a single simulation framework. The reconstruction of realistic gate edges is based on a 1D Fourier synthesis approach, similar to that used in the generation of the Si/SiO_2 interface. A typical 30×30 nm MOSFET with continuous doping and LER is shown in Fig.7 for Λ = 10 nm and LER = 3Δ = 6 nm.

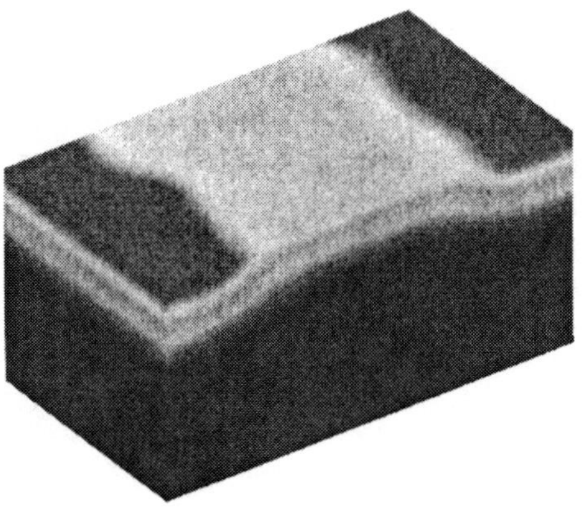

Fig. 7. Potential distribution at threshold in a well scaled 30×30 nm MOSFET with line edge roughness of the gate of 6 nm.

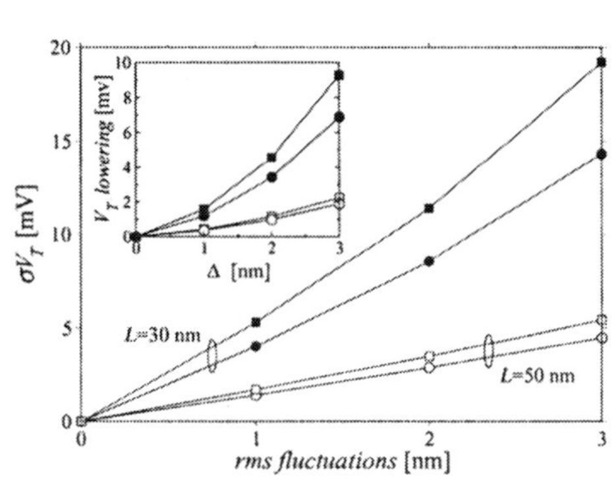

Fig. 8. Threshold voltage fluctuations associated with LER as a function of the rms amplitude Δ.

LER causes threshold fluctuations similar to those resulting from random dopants or oxide thickness fluctuation, as can be seen in Fig. 8. Fluctuations increase when Λ or Δ is increased. The average threshold voltage is also affected, with a pronounced threshold voltage lowering taking place. The fluctuations are comparable in magnitude to those resulting from random dopants in similar devices.

6 The challenges ahead

In any simulation, the accuracy is only as good as the description of the simulation domain and the complexity of various physical models in use. Hence, predictive atomistic process simulators should be natural companions to further attempts in atomistic device modelling, providing atomic scale information for the device structures. A main challenge in studying decanano devices therefore lies in interfacing the atomistic device modelling tools with predictive atomistic process simulations, resulting in a more reliable description of the problem at hand.

Based on the same philosophy, improvements to the physical models employed in the transport problem are also required. For instance higher-moments of the Boltzmann transport equation such as hydrodynamic models may be incorporated in the atomistic simulations to correctly account for the non-equilibrium conditions in decanano devices. Particle simulations may be especially useful to uncover effects related to the discrete nature of carriers and noise, which would affect the device performance at an atomistic level, and to resolve *ab-initio* long range electrostatic effects, carrier-carrier and carrier-impurity interactions. More elaborate or exact treatment of quantum effects is still missing in our simulations. At present, the trade off is between the efficiency of numerical simulations and the range of applicability in quantum mechanical simulation strategies.

The increased understanding of different components of fluctuations in device characteristics points to an increasing need for developing new device architectures that can suppress them. Several novel device concepts such as double gate MOSFETs or SiGe heterojunction MOSFETs must be simulated to assess their immunity to atomistic processes. Ultimately, the device performance is affected by the interplay of many different sources of fluctuations discussed within this paper. Thus, the correlation between the contributions of different components of fluctuations should be investigated in atomistic simulations by concurrently incorporating different processes in a single device. In doing so, any other processes which are relevant to the decanano regime, but have not been treated fully here, must also be included.

Acknowledgement

This work is supported by NASA-Ames Grant No. NAG 2-1241 and SHEFC Research Development Grant VIDEOS. The author is grateful to A. R. Brown, S. Kaya, J. H. Davies and G. Slavcheva for their numerous contributions to this work.

References

[1] The International Technology Roadmap for Semiconductors, SIA, 1999 Edition

[2] P. A. Packan, Pushing the limits, Science 1999; 285:2079

[3] T. Mizuno, J. Okamura and A. Toriumi, Experimental study of threshold voltage fluctuation due to statistical variation of channel dopant number in MOSFET's, IEEE Trans. Electron Devices 1994; 41:2216

[4] H. S. Momose, M. Ono, T. Yoshitomi, T. Ohguro, S. Nakamura, M. Sato, and H. Ivai, 1.5~nm direct-tunnelling gate oxide Si MOSFETs, IEEE Trans. Electron Devices 1996; 43:1233

[5] A. Asenov, A. R. Brown, J. H. Davies and S. Saini, Hierarchical approach to atomistic 3D MOSFET simulation, IEEE Trans. CAD Integ. Circuits Sys. 1999; 18:1558

[6] A. Asenov, Random dopant induced threshold voltage lowering and fluctuations in sub 0.1 micron MOSFETs: A 3D 'atomistic' simulation study, IEEE Trans. Electron Dev. 1998; 45:2505

[7] A. Asenov, S. Kaya, Effect of oxide roughness on the threshold voltage fluctuations in decanano MOSFETs with ultrathin gate oxide, Proc. SISPAD 2000, p 135

[8] C. S Rafferty, B. Biegel, Z. Yu, M. G. Ancona, J. Bude and R. W. Dutton, Multi-dimensional quantum effects simulation using a density-gradient model and script-level programming technique, in Proc SISPAD 1998, Eds. K. De Meyer and S. Biesemans, p 137

[9] A. Asenov, G. Slavcheva, A. R. Brown, J. H. Davies and S. Saini, Increase in the random dopant induced threshold fluctuations and lowering in sub 100 nm MOSFETs due to quantum effects: A 3-D density-gradient simulation study, IEEE Trans. Electron Dev. 2001; 48:722

[10] G. H. Gilmer, L. Palaz, C. Rafferty and M. Jaraiz, Multiscale modeling of the implantation and annealing of silicon devices, in Proc SISPAD 1998, Eds. K. De Meyer and S. Biesemans, p 46

[11] S. M. Goodnick, D. Ferry, C. W. Wilmsen, Z. Liliental, D. Fathy, and O. L. Krivanek, Surface roughness at the Si(100) - SiO_2 interface, Phys. Rev. 1985; B32: 8171

Modeling of Reactive Ion Etching for Si/SiO_2 Systems

S. Hamaguchi and H. Ohta
Department of Fundamental Energy Science, Kyoto University,
Uji, Kyoto 611-0011.

Abstract

Molecular dynamics (MD) simulations have been used to study surface reaction dynamics of Si and SiO_2 etching by halogens. To perform classical MD simulations for SiO_2 etching reaction by halogens (Cl or F), we have constructed new sets of two– and three–body interatomic potential functions based on potential energy data obtained from *ab initio* quantum mechanical calculations of the electronic states. Etching yields for Si and SiO_2 targets are obtained from MD simulations for both beam etching and reactive ion etching (RIE). The obtained yields are in good agreement with experimental observations.

1 Introduction

Patterned etching of Si is one of the most important fabrication processes employed in the semiconductor chip manufacturing. To transfer desired submicron patterns to the Si surface efficiently, one needs to use highly selective etching processes. As the minimum dimensions of required etched patterns are now close to 0.1 μ under the current state–of–the–art manufacturing technologies and are diminishing rapidly, development of more reliable, high–throughput processes with good critical–dimension control is pursued aggressively in the semiconductor industry.

One of the most important issues for the control of critical dimensions in such processes is to understand the surface reaction mechanisms during actual processing conditions. In the present paper, we focus on Si and SiO_2 etching processes by plasmas containing halogen (Cl or F) atoms. Using molecular dynamics (MD) simulations, we attempt to elucidate details of etching processes under realistic etching conditions.

Classical MD simulations with empirical potentials have been used by various authors to analyze details of surface reaction dynamics [1]–[9]. Data obtained from such simulations can compensate for scarcity of some experimental data. As to beam etching and/or reactive ion etching (RIE) of Si by halogens, Stillinger-Weber (SW) type interatomic potentials [1, 2] are known to represent many aspects of surface reactions during etching processes with reasonable accuracy.

To extend such MD simulations to etching of SiO_2 surfaces by halogens, we have developed new potential sets for Si-O-F and Si-O-Cl systems [10], which are consistent with previously known interatomic potentials of SW type for Si

[1], Si-Cl [2], Si-F [3], and Si-O [11] systems. As the SW-type potential functions include three-body interactions, we have determined ternary interactions by fitting a functional form to potential energy data obtained from *ab initio* calculations of electronic states of such systems.

2 Molecular Dynamics Simulation

The simulation method is now presented briefly. Atoms are normally injected to a target which is placed in a simulation cell with periodic boundary conditions imposed in the horizontal directions. The atoms in the bottom layer are rigidly fixed to prevent the drift of the entire simulation cell. Since energetic ions impinging on the surface are expected to be neutralized near the target surface due to an Auger emission process, we only consider charge–neutral atoms as impinging species. It is also assumed that the target surface is kept charge–neutral during the process.

After injection of each atom, we let the system evolve for 0.7 ps under the constant total–energy conditions. Most transient processes such as rapid release of the kinetic energy of the injected atom to the target are almost completed during this period. We then artificially cool the entire system (by decreasing the velocities of all atoms) for 0.3 ps to reduce the target temperature to the initial temperature (i.e., 300° K) in order to prevent the system from being unrealistically heated up.

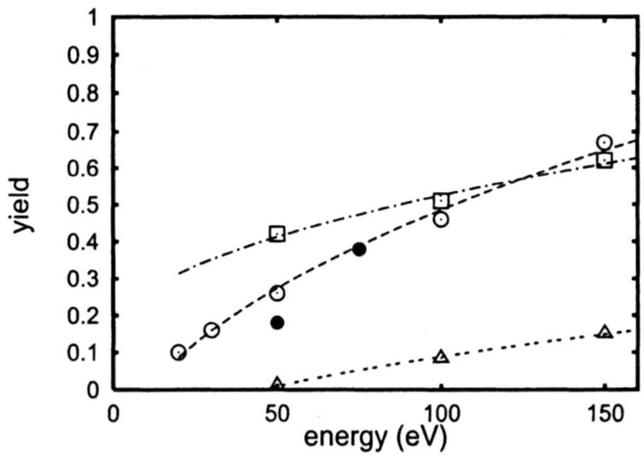

Fig. 1. Etching yields of the silicon target for Cl (○), F (□), and Ar (△) beam etching. The filled circles represent etching yields for a Cl RIE process with Ar bombardment.

3 Conclusions

We have performed MD simulations of Si and SiO_2 etching by Cl and F monomer beams as well as Ar beams with various impinging energies. Etching yields obtained from our MD simulations are shown in Figs. 1 and 2. Here Si etching yields (i.e., the number of Si atoms emitted from the surface per impact) that we have obtained are in good agreement with experimental observations and previous MD simulation results [9]. There are not many experimentally obtained SiO_2 etching yields by Cl or F monoenergetic beams, with which we could compare our simulation results. However our observation of the threshold energy for SiO_2 etching by Ar impact (which is about 50 eV) is in good agreement with earlier experimental observations [12, 13].

We have also simulated RIE processes. The RIE simulation results are indicated by filled circles in Figs. 1 and 2. Here we have injected about 100 Cl atoms with low kinetic energy (1eV) into the target surface in each interval between two successive Ar injections. As shown in both Figs. 1 and 2, the etching yields by Ar impact increase up to those by Cl beam etching since the low energy Cl atoms cover the target surface and terminate Si and O bonds.

In summary, we have performed MD simulations to study surface reactions during Si and SiO_2 etching by halogens. The obtained results are consistent with various experimental observations that have been obtained to date.

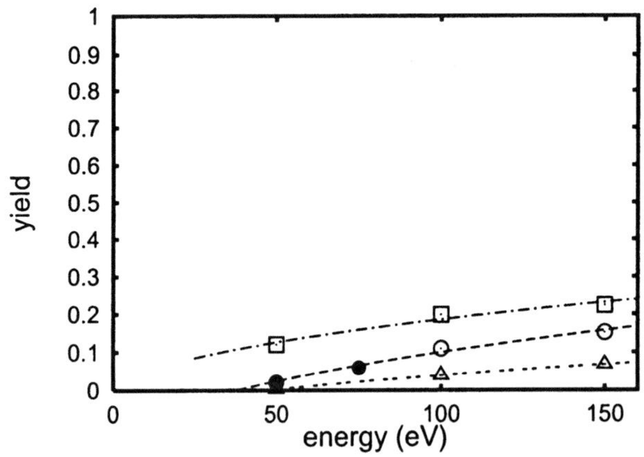

Fig. 2. Etching yields of the SiO_2 target for Cl (○), F (□), and Ar (△) beam etching. The filled circles represent etching yields for a Cl RIE process with Ar bombardment.

References

[1] F. H. Stillinger and T. A. Weber, Phys. Rev. B **31**, 5262 (1985); J. Chem. Phys. **88**, 5123 (1988); Phys. Rev. Lett. **62**, 2144 (1989).

[2] T. A. Weber and F. H. Stillinger, J. Chem. Phys. **92** 6239 (1990).

[3] H. Feil, J. Dieleman, and B. J. Garrison, J. Appl. Phys. **74**, 1303 (1993).

[4] R. Smith, D. E. Harrison, Jr., and B. J. Garrison, Phys. Rev. B **40**, 93 (1989).

[5] T. A. Schoolcraft and B. J. Garrison, J. Vac. Sci. Technol. A**8**, 3496 (1990); J. Am. Chem. Soc. **113**, 8221 (1991).

[6] M. E. Barone and D. B. Graves, J. Appl. Phys. **77**,1263 (1995); J. Appl. Phys. **78**,6604 (1995).

[7] D. E. Hanson, A. F. Voter, and J. D. Kress, J. Appl. Phys. **82**, 3552, 1997.

[8] N. A. Kubota and D. J. Economou, S. J. Plimpton, J. Appl. Phys. **83**, 4055, 1998.

[9] H. Ohta and S. Hamaguchi, submitted to J. Vac. Sci. Technol. A (2001).

[10] H. Ohta and S. Hamaguchi, submitted to J. Chem. Phys. (2001).

[11] T. Watanabe, H. Fujiwara, H. Noguchi, T. Hosono, and I. Ohdomari, Jpn. J. App. Phys. **L38**, 366 (1999).

[12] W. M. Holber and J. Forster, in *Proceeding of Symposium on Dry Process*, (ed. by Y. Horiike *et al.* The Institute of Electrical Engineers of Japan, Technical Committee on Electron Devices, Tokyo) p. 9 (1989).

[13] D. J. Oostra, R. P. van Ingen, A. Haring, A. E. de Vries, and G. N. A van Veen, Appl. Phys. Lett. **50** 1506 (1987)

Simulation and Prediction of Aspect Ratio Dependent Phenomena during SiO_2 and Si Feature Etching in Fluorocarbon Plasmas

George Kokkoris,[1,2] Evangelos Gogolides,[1] Andreas G. Boudouvis.[2]
[1]Institute of Microelectronics, National Center for Scientific Research "Demokritos", Athens, Greece
[2]Department of Chemical Engineering, National Technical University of Athens, Athens, Greece
gkok@imel.demokritos.gr, evgog@imel.demokritos.gr, boudouvi@chemeng.ntua.gr

Abstract

A simulator to calculate etching rates in SiO_2 and Si features in fluorocarbon plasmas has been developed. This simulator links the gas (plasma) phase composition with the etching rate inside features and it is used for predicting aspect ratio dependent phenomena during SiO_2 and Si feature etching.

1 Introduction

Etching of SiO_2 and silicon containing low k dielectrics is a very critical process step in microelectronics fabrication. SiO_2 and Si etching is also extremely important for optoelectronics and microsystem fabrication and is performed in fluorocarbon plasmas. It is useful to understand and predict the etching phenomena in order to confront the problems during an etching process

Phenomena such as Reactive Ion Etching lag [1] (i.e. etching rate reduction in narrower trenches and holes, *RIE lag*), *etch stop* [1], and *inverse RIE lag* [2], are usually observed during SiO_2 and Si feature etching and can cause several problems; both RIE lag and inverse RIE lag lead to different etched depths, and overetching is needed, which could result in the damage of the underlying layer. The above-mentioned phenomena have been observed to depend on the Aspect Ratio (AR) of the feature (depth/width of the feature) rather than the absolute feature size and are included in the general term Aspect Ratio Dependent Etching (ARDE). The desirable elimination of RIE lag and inverse RIE lag is denoted by the term Aspect Ratio Independent Etching (ARIE); i.e. etching rate constancy as etching time increases and consequently as AR increases.

A model to calculate etching rates in SiO_2 and Si features in fluorocarbon plasmas has been developed. The model includes a) a surface model [3] for open area etching of SiO_2 and Si, b) a flux calculator [4], which calculates local fluxes on each elementary surface of the feature being etched, and c) a coupling of the two models (a) and (b) to calculate the local etching rate inside features. Based on this model the effect of gas phase composition on ARDE and ARIE is simulated and investigation of processes' window can be done.

2 The surface model

An adsorption-type model [3] accounting for etching and deposition in fluorocarbon plasmas during Si and SiO_2 open area etching is used to describe the surface physical and chemical processes. In this phenomenological model, fluorine atoms (F), fluorocarbon radicals (CF_x), and polymer (produced on the surface) are considered bonded to surface atoms. The basic equations of the model are F, CF_x, and polymer site balances and the model coefficients have been calculated by fits [3] to beam experiments' results. The independent variables are a) the ratios of neutral species (F, CF_x) fluxes to ion flux, b) ion energy, and c) ion composition. If absolute value of etching rate is to be calculated, ion flux value is needed. Besides etching yield (Si atoms/ion) and rate, effective sticking coefficients, S_E, of the neutral species are also outputs of the surface model. S_E of a species represents [5] the net loss ($S_E>0$) or creation ($S_E<0$) of this species during surface reactions.

3 The flux calculator

The flux calculator [4] is used for the calculation of ion and neutral flux on each elementary surface of a structure. It takes into account shadowing of ion and neutral flux and re-emission of neutral flux in features. Charging effects are not considered, but are simulated by an increased ion angular spread. The total flux of a species i (ion or neutral) at an elementary surface s of a feature (trench with infinite length or hole with cylindrical symmetry) is given by the equation

$$j_i(s)=j_{direct,i}(s)+\int_{\text{feature profile}} g_i(s,s')[1-S_{E,i}(j_1,j_2,\ldots,j_N)]j_i(s')ds', \qquad (1)$$

where $g_i(s, s')$ is a function which depends on the re-emission mechanism of the species i and the geometry of the structure being etched. $j_{direct,i}$ is the flux of the neutral species i coming directly from the plasma and expresses the effect of shadowing on the flux. The integral of Eq. 1 stands for the flux coming at s from all other elementary surfaces s′ by re-emission.

The calculation of local fluxes and local etching rates *requires a coupling* [4] *of the surface model with the flux calculator.* The key point of the coupling algorithm is the simultaneous calculation of a) $S_{E,i}$ by the surface model [5] and b) total flux for every species by a system of N integral equations such as Eq. 1 (N is the number of species).

4 Results

The proposed model essentially links the gas phase composition with ARDE in SiO_2 and Si trenches. The gas phase composition is defined by the neutral (F atoms, CF_x radicals) and ion fluxes coming from the gas phase of a fluorocarbon plasma and reaching an open area ($j_{F,0}$, $j_{CFx,0}$, and $j_{ION,0}$) and is expressed by the ratios: $R_{F,0}=j_{F,0}/j_{ION,0}$, $R_{CFx,0}=j_{CFx,0}/j_{ION,0}$.

In Fig. 1 *maps of ARDE and ARIE* in SiO_2 and Si trenches are presented. This new approach divides the gas phase composition into regions leading to a specific AR dependent phenomenon. For example, in Fig. 1a, which refers to SiO_2 trench etching, deposition happens in region A, intense RIE lag and etch stop in C, RIE lag with no etch stop in B, inverse RIE lag in D, and ARIE up to a specific AR in E. A similar division of gas phase composition has been done in Fig. 1b for Si trench etching.

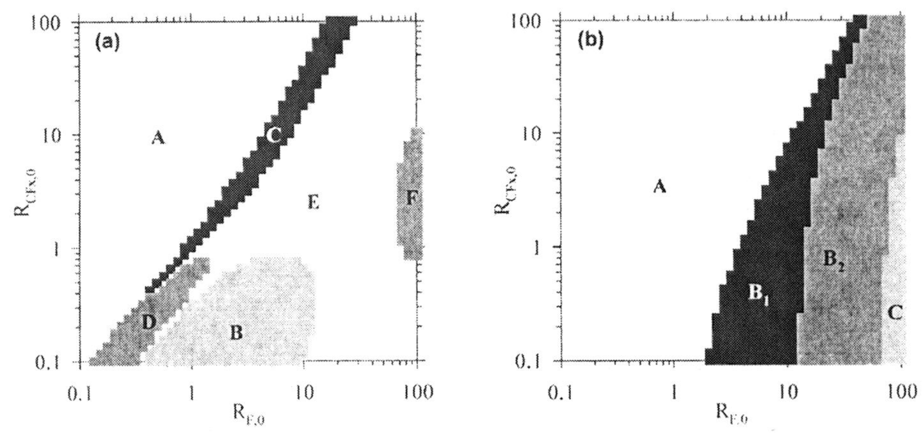

Fig. 1. a) Effect of gas phase composition (i.e. ratios $R_{F,0}$, $R_{CFx,0}$) on SiO_2 trench etching: Region *A*: deposition; *B*: regular RIE lag with no etch stop; *C*: intense RIE lag and etch stop; *D*: inverse RIE lag; *E*: ARIE (less than 5% decrease in etching rate) up to AR=3 to 7 depending on the distance from deposition region. The further from deposition region the higher the AR for which ARIE is observed. *F*: ARIE up to AR=7. A slight inverse RIE lag can be observed for some compositions. **b)** The same as (a) for Si trench etching. Region *A*: deposition. B_1: intense RIE lag and etch stop. B_2: RIE lag. Larger trench AR is needed to observe etch stop. *C*: Mild RIE lag, and ARIE up to AR=3 to 4 for high $R_{F,0}$ and low $R_{CFx,0}$. Ion energy is 100eV, ion composition is 10% CF_3^+, 85% CF_2^+, 5% CF^+.

In Fig. 2a etching or deposition is predicted in SiO_2 trenches: the contours of zero etching rate at SiO_2 trench bottom as a function of gas phase composition are shown for different AR values. Each curve represents a boundary between deposition (left side) and etching (right side). *Knowing the gas phase composition and the trench AR, one can predict what will happen at the bottom of the trench: etching or deposition.*

The proposed model can be used for a preliminary investigation of processes' window satisfying demands on ARIE, etching rate magnitude, and etching rate selectivity. Fig. 2b refers to a process window during etching of SiO_2 trenches over a Si layer have to be etched. In the black region of Fig. 2b etching rate is greater than 240 nm/min, etching rate selectivity (SiO_2/Si) greater than 10, and etching rate deviation from the open area value less than 5%.

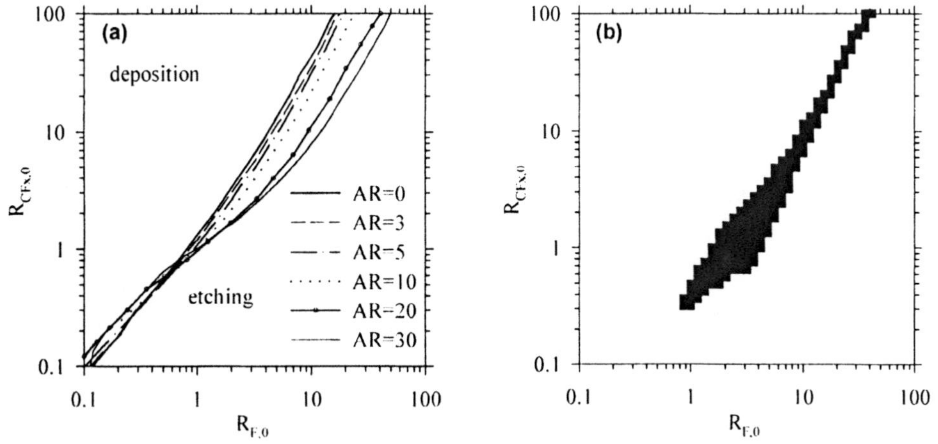

Fig. 2. a) Contours of zero etching rate at a SiO_2 trench bottom as a function of gas phase composition for different trench AR. Each curve represents a boundary between deposition (left side) and etching (right side). **b)** Range of gas phase compositions of an etching process of SiO_2 trenches on a Si layer. The trenches AR lies in the range of AR=1 to 3 during etching. In black region etching rate is greater 240 nm/min (ion flux is 17 mA/cm^2), etching rate selectivity (SiO_2/Si) is greater than 10, and etching rate deviation from the respective open area value is less than 5%. Ion energy and composition are the same as in Fig. 1.

5 Conclusions

Aspect ratio dependent phenomena during SiO_2 and Si feature etching in fluorocarbon plasmas are predicted and preliminary investigation of processes' window can be achieved by a simulator for feature etching that couples a surface model for open area etching with a calculator of fluxes inside features.

References

[1] Joubert O., Oehrlein G.S., Zhang H., J. Vac. Sci. Technol. A 1994; 12:658-664.

[2] Doemling M.F., Rueger N.R., Oehrlein G.S., Appl. Phys. Lett. 1996; 68:10-12.

[3] Gogolides E., Vauvert P., Kokkoris G., Turban G., Boudouvis A.G., J. Appl. Phys. 2000; 88:5570-5584.

[4] Kokkoris G., Gogolides E., Boudouvis A.G., 'Etching of SiO_2 and Si features in fluorocarbon plasmas. I: Effect of gas phase composition on aspect ratio dependent phenomena', submitted for publication to J. Appl. Phys.

[5] Kokkoris G., Gogolides E., Boudouvis A.G., 'Etching of SiO_2 and Si features in fluorocarbon plasmas. II: Explanation of aspect ratio dependent phenomena', submitted for publication to J. Appl. Phys.

System Level Modeling of an Electrostatic Torsional Actuator

Robert Sattler, Gerhard Wachutka
Institute for Physics of Electrotechnology, Munich University of Technology
80290 Munich, Germnay
sattler@tep.ei.tum.de, wachutka@tep.ei.tum.de

Florian Plötz
Infineon Technologies AG
81730 Munich, Germany
florian.ploetz@infineon.com

Sebastian Hoffmann
Optical Communications and High Frequency Engineering
33095 Paderborn, Germany
hoffmann@ont.upb.de

Abstract

We present an efficient methodology for setting up MEMS macromodels which are based on a physical device description and lead to tractable mathematical relations for the device operation. Since design and technology parameters are input parameters of the resulting model, our approach is in particular suited for design studies. In addition to the reduction in degrees of freedom and, hence, the reduced simulation time, macromodels can easily be coupled with the electronic circuitry and the entire device can be simulated on system level. The methodology is demonstrated with reference to an electrostatic torsional actuator.

1. Introduction

A systematic methodology of formulating transparent compact models is, among others, provided by a thermodynamic system description in terms of driving forces and resulting flows of the relevant physical quantities [1]. This leads to a full system description as a "Generalized Kirchhoffian Network". This, in turn, is equivalent to a system of ordinary differential–algebraic equations for the node variables which can be solved using a standard analog network simulator. We demonstrate the practicality of the latter method by an exemplifying problem,

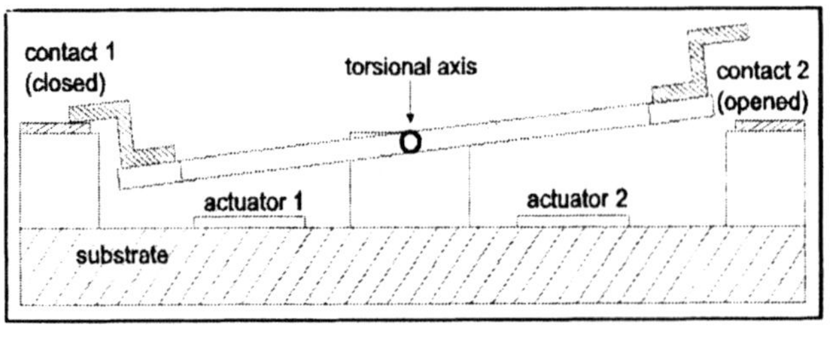

Fig. 1: Schematic cross section of a torsional actuator [2].

namely an electrostatic torsional actuator (fig. 1), which is a widely used basic structure employed in tiltable micromirrors [2] or microswitches [3]. Electrostatic actuation is especially favoured for switching applications in telecommunication because of its low power consumption.

2. Torsional Actuator

The torsional actuator is considered to be a tiltable, rigid body. Hence, the only degree of freedom is the angle of torsion φ. Neglecting fringing fields at the edges of the actuator, the total torque M acting on the upper electrode results in [5]:

$$M=\frac{1}{2}\epsilon\frac{U^2L^2}{D^2}w\frac{1}{\gamma^2}\left(\frac{\gamma}{1-\gamma}+\log(1-\gamma)\right) \qquad (1)$$

Here ε denotes the electric permittivity, U the applied voltage, L and w the length and the width of the actuation electrode, and D the gap between the electrodes. γ is the normalized angle of torsion.
The total torque is composed of the electrostatic torque (eq. 1), the mechanical torque exerted by the tethers by which the tiltable electrode is suspended, and the torque induced by the inertia and the damping of the actuator. In the notation of Kirchhoffian Network variables, the quantity conjugate to the generalized force ("across") variable dφ/dt (angular velocity of torsion) is the torque M which plays the role of the generalized flow ("through") quantity.

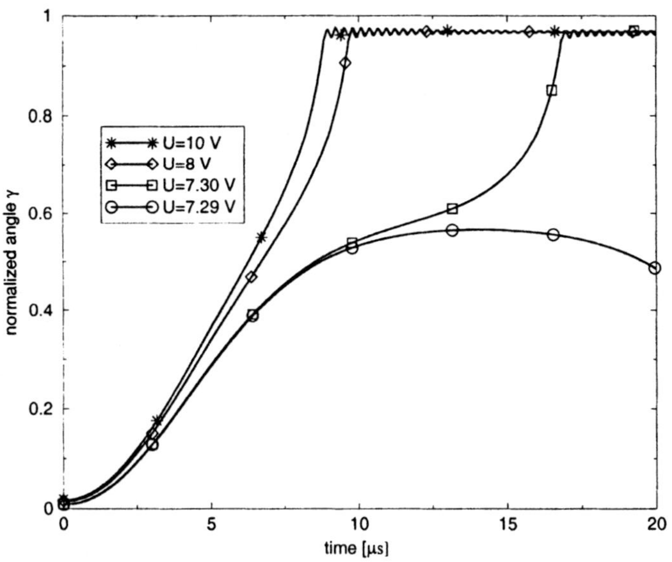

Fig. 2: Pull–in dynamics for various actuation voltages. The static pull–in voltage U_{pi} = 8 V.

2.1. Results

Performing a DC analysis allows to compute the pull–in hysteresis of the actuator. Below the static pull–in voltage V_{pi} exists an equilibrium between the nonlinear electrostatic torque and the restoring mechanical torque of the tethers. A compari–

son with experimental data is shown in [5] and shows good agreement.

Performing a transient analysis the snap down occurs at a dynamic pull–in voltage $U_{dpi} < U_{pi}$ due to the influence of the inertia (fig. 2).

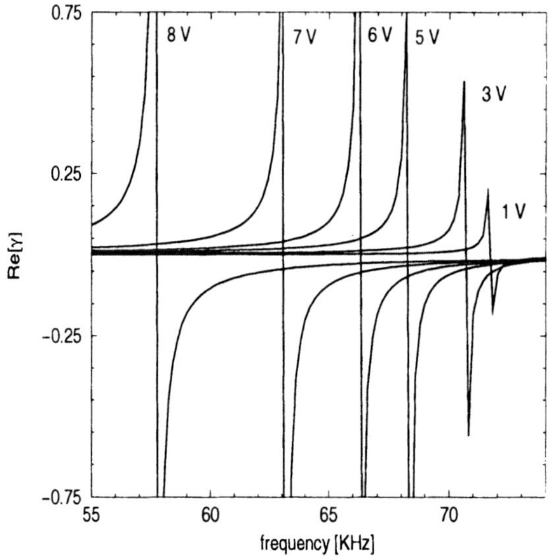

Fig. 3: Small signal analysis showing the shift of the resonance frequency caused by spring softening for various voltages.

For an electromechanical modal analysis and resonance measurements of electrostatically actuated devices, the effect of electrostatic spring softening has to be taken into account. Using our compact model the shift of the resonance frequency effected by the bias voltage can easily be analysed (fig. 3) and conforms well with the measured frequency of 72 kHz [3].

3. Charge pump

Due to technical constraints the available supply voltage is in many cases too low to operate the switch. This problem can be solved by means of a charge pump. A cascade of several capacitors and diodes is driven by an alternating voltage generated from the DC supply voltage. At each stage, the voltage is increased, but the available output current is limited by the corresponding demand of charge per cycle ($I_{max}=U_0C_1f$). Therefore, a charge pump can be described as a voltage source with a virtual source impedance, leading to a voltage drop but not associated with resistive loss. Also the ripple increases, if the load current is high compared to the available capacitors.

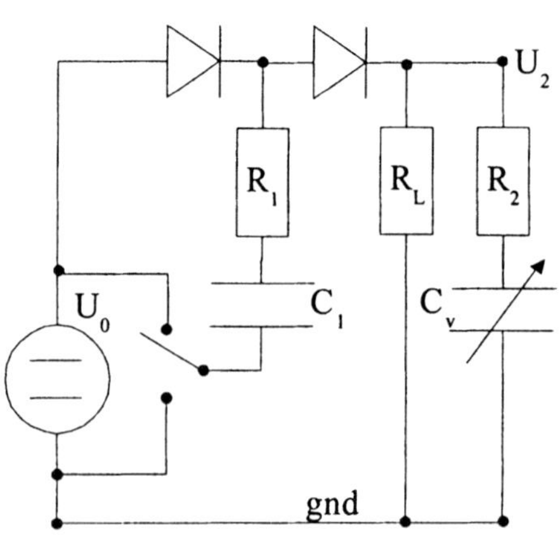

Fig. 4: Schematic of a simple charge pump.

4. System level simulation

Because charge pumps are sensitive to the behaviour of the load they drive, it is important for the circuit designer to have a reliable simulation model for the interaction of these two components. Hence a combination of a simple charge pump and the torsional actuator mentioned above was chosen as an example (fig. 4). The charge pump has only one stage, resistors (R1, R2, R_L) are included to represent the losses in a non–ideal circuit. The torsional actuator is displayed as a variable capacitance. As the actuator is coded in SpectreHDL[4] (analog hardware description language similar to the standard VHDL–AMS), it can easily be included in the circuitry of the charge pump. The influence of the pump frequency on the behaviour of the system is depicted in fig. 5.

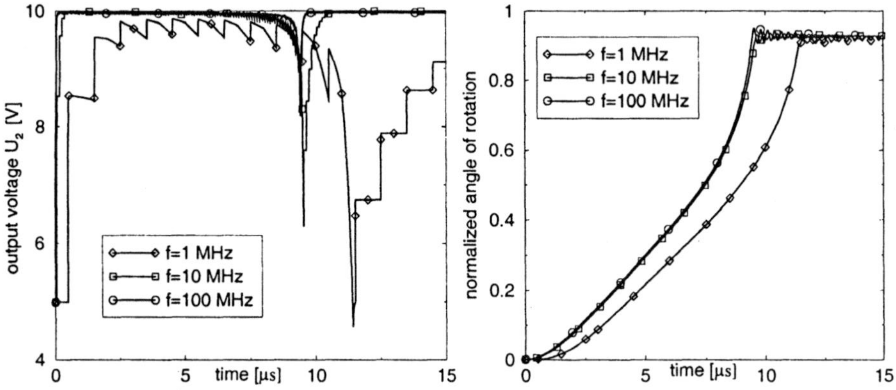

Fig. 5: Evolution of the output voltage U_2 of the charge pump and the pull–in characteristics for various pump frequencies.

5. Conclusion

System level modeling based on physical device description provides valuable insight and understanding of the dynamic behaviour of the device and allows one to perform parameter studies (geometry of the actuator and specifications of the components of the charge pump) in order to optimize the system with respect to performance, technological constraints and chip area.

References

[1] G. Wachutka, "Tailored modeling: a way to the 'virtual transducer fab'", Sensors and Actuators, vol. A46–47, pp. 603–612, 1995.

[2] J. Bühler, "Deformable Micromirror Arrays by CMOS Technology", PhD Thesis No. 12139, ETH Zürich, ISBN: 3–907574–16–8, 1997.

[3] F. Plötz, S. Michaelis, R. Aigner, H.–J. Timme, J. Binder, R. Noé, "A Low–Voltage Torsional Actuator for Application in RF–Microswitches", Proc. Eurosensors XIV, August 2000, Copenhagen, Denmark, pp. 297–300.

[4] http://www.cadence.com

[5] R. Sattler, F. Plötz, G. Wachutka, "Macromodeling of an Electrostatic Torsional Actuator", Proc. Transducers '01, June 2001, Munich, Germany, in press.

Impact of Substrate Resistance on Drain Current Noise in MOSFETs

Jung-Suk Goo, Simona Donati[†], Chang-Hoon Choi, Zhiping Yu,
Thomas H. Lee, and Robert W. Dutton
Center for Integrated Systems, Stanford University, Stanford, CA, USA
goojs@gloworm.stanford.edu
[†]Dipartimento di Elettronica, Politecnico Di Torino, Torino, Italy

Abstract

This paper identifies the physical origin and contribution mechanism of substrate induced channel thermal noise in MOSFETs. Resistance of the substrate generates potential fluctuations that in turn produce additive channel noise via the channel depletion capacitor. The additive noise may result in a frequency dependence of the drain current noise due to a pole associated with the R_{sub}-C_{depl} network. Its bias and length dependencies conforms to those of reported excess noise, it thus may exaggerate the amount of the channel thermal noise factor.

1 Introduction

It has been known that short channel MOSFETs yield larger drain current noise (S_{i_d}) [1] than the value predicted by the van der Ziel model, which agrees with long channel MOSFETs. However, the amount of the increase is still controversial: some studies have demonstrated a substantial increase while other studies have claimed that less than a factor of two increase has been observed for MOSFETs with 0.17μm [2].

While circuit design commonly assumes that the drain current noise is frequency independent, experimental data has shown a frequency dependence of noise resistance ($R_n \triangleq \frac{S_{i_d}}{4kT|Y_{21}|^2}$) that directly represents the drain current noise in MOSFETs, in the GHz frequency range [3], [4]. However, it has been totally ignored or attributed to measurement errors [4]. This paper investigates the physical origin of the frequency dependence for drain current noise, using physically-based two-dimensional noise simulations. Implications on noise modeling are also discussed.

2 Noise Simulation

In this work, the IFM (Impedance Field Method) is implemented in a general-purpose multi-dimensional device simulator called PADRE [5]. Since the primary interest of the present work is the substrate induced thermal noise, the drift-diffusion model is used and other noise sources, such as $1/f$ noise and generation-recombination noise, are ignored. The MOSFET structure used for noise simulation is illustrated in Fig. 1 (a). This structure, while quite simple, still provides useful analysis results and physical insights. To minimize secondary effects, the junction length (L_{sd}) is reduced.

3 Drain Current Noise Model

It has been a common assumption that the drain current noise only exhibits the $1/f$ noise ($S_{i_d,1/f}$) and white channel thermal noise ($S_{i_d,channel}$) contributions; they dominate regions I and III in Fig. 1 (b), respectively. Although local noise sources generate white noise up to the quantum limit, as frequency becomes much higher than the cutoff

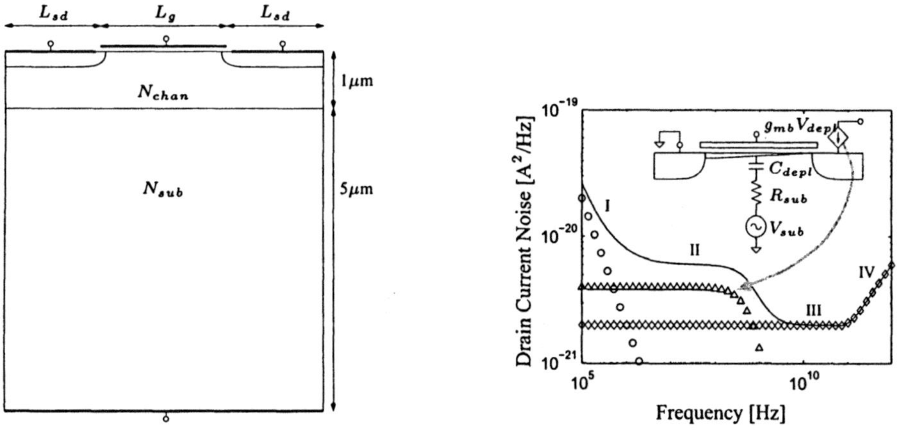

(a) (b)

Fig. 1. (a) Cross-section of the MOSFET structure used for simulation. (b) Anticipated general drain current noise power spectrum for bulk MOSFETs. The solid line is total noise power, the circle symbol is the $1/f$ noise component ($S_{i_d,1/f}$), the triangle symbol is the additive component induced by the substrate ($S_{i_d,sub}$), and the diamond symbol is the channel thermal noise component ($S_{i_d,channel}$). The inset circuit diagram explains the contribution mechanism of the substrate thermal noise to the drain current noise.

frequency, the capacitive coupling between the channel and gate makes propagation of local fluctuations stronger, consequently channel thermal noise increases as region IV in Fig. 1 (b) [6].

In addition, spreading resistance (R_{sub}) of the substrate generates thermal fluctuations given by:

$$S_{v_{sub}} = 4kTR_{sub} \tag{1}$$

where k is Boltzmann's constant and T is the lattice temperature [7]. These fluctuations in the substrate potential modulate the channel charge of the depletion capacitor (C_{depl}) and subsequently produce additive channel noise ($S_{i_d,sub}$), amplified by substrate transconductance (g_{mb}). Since the $R_{sub}C_{depl}$ network in Fig. 1 (b) is a low-pass filter, the amount of additive channel noise is given by

$$S_{i_d,sub} = \frac{4kTR_{sub}g_{mb}^2}{1+(\omega R_{sub}C_{depl})^2} \tag{2}$$

This equation suggests that $S_{i_d,sub}$ has a low frequency noise plateau ($4kTR_{sub}g_{mb}^2$) dominating region II in Fig. 1 (b) and a pole at $f = (2\pi R_{sub}C_{depl})^{-1}$. Thus $S_{i_d,sub}$ effectively increases the drain current noise factor ($\gamma \triangleq \frac{S_{i_d}}{4kTg_{d0}}$). This phenomenon was first discussed by Jindal [7] but only the low frequency noise plateau was considered. For SOI MOSFETs, Faccio *et al.* [8] has introduced a parallel shunting capacitor, which is the sum of the capacitances of the front gate oxide (C_{ox}) and buried oxide (C_{BOx}).

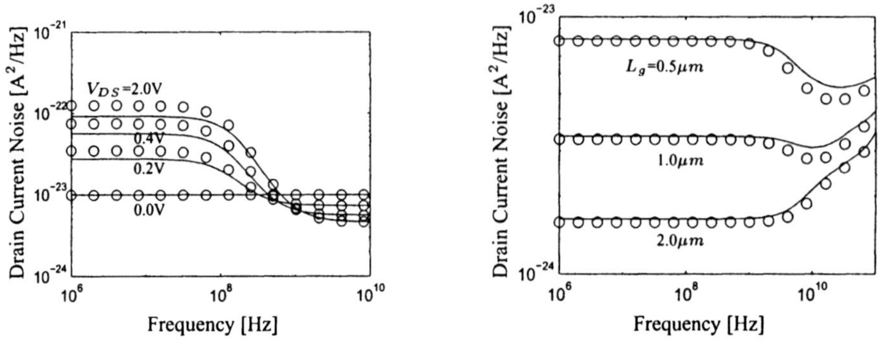

(a) (b)

Fig. 2. Simulation results of drain current noise power spectrum. The gate oxide thickness is 60nm, N_{chan} is $5.0\times10^{17}\text{cm}^{-3}$, L_{sd} is 0.01μm, and V_{GS} is 1.5V. The symbols are PADRE simulation results and lines are calculated using (2). (a) A 0.5μm MOSFET for different drain bias conditions. N_{sub} is $1.0\times10^{15}\text{cm}^{-3}$. (b) MOSFETs with different gate lengths. N_{sub} is $3.2\times10^{16}\text{cm}^{-3}$, and V_{DS} is 2.5V.

For bulk MOSFETs, using similar physical reasoning, Kishore *et al.* [3] have explained the phenomenon by adding an RC network between the gate and substrate. However, the channel charge shields the substrate network from the gate electrode under strong inversion conditions. Therefore C_{ox} must be replaced with C_{depl}.

4 Results and Discussions

In Fig. 2 (a) and (b), noise spectra are reproduced by adding the additive component, calculated by using (2), to the simulation results for MOSFETs with a highly doped substrate ($N_{sub} = N_{chan}$). The calculated spectra fit the low frequency noise plateau and pole location, introducing less than 30% error, with C_{depl}-R_{sub}-g_{mb} values extracted from ac and dc simulation results of PADRE. R_{sub} is extracted from a structure without junctions and gate oxide; C_{depl} is approximately equal to $C_{db} + C_{sb}$ if the junction area becomes negligibly small. It is important to note that $S_{v_{sub}}$ indirectly contributes to S_{i_d} via the channel depletion capacitor. Due to the bias dependence of g_{mb}, $S_{i_d,sub}$ is zero under zero drain bias, increases as drain bias increases, and finally saturates for higher drain bias, as shown in Fig. 2 (a). It also largely increases the drain noise factor at low gate bias conditions. Moreover $S_{i_d,sub}$ becomes more significant as the channel length decreases in Fig. 2 (b) since $S_{i_d,sub}$ is proportional to g_{mb}^2 while $S_{i_d,channel}$ is proportional to g_m. Interestingly, these tendencies conform with the reported description of excess drain current noise in short channel MOSFETs [1].

It is worth mentioning that those studies reporting a substantial increase of the drain noise factor have collected data at frequencies below 100MHz due to the bandwidth limits of the transimpedance amplifier. Moreover the samples used relatively primitive technologies, that employ only shallow threshold adjustment doping on a highly resistive substrate. Therefore, it is now suggested that the amount of channel thermal noise has been exaggerated due to the additive thermal noise contribution from the substrate. Thus to characterize the pure channel thermal noise, data needs to be obtained either in region III or from MOSFETs with a highly doped substrate.

As the transition between regions II and III approaches the GHz range in modern bulk CMOS technologies, modeling using a single RC lump may not be successful. The actual $S_{i_d,sub}$ spectrum is the sum of multiple components with different poles due to the three-dimensional distributed effect of the substrate network; the slope of the transition is much smoother than -20dB/decade, typically below -5dB/decade. Thus prediction of the $S_{i_d,sub}$ characteristic poses an ongoing challenge. Nonetheless, the use of three-dimensional TCAD simulations can be expected to provide additional data in support of new compact modeling.

5 Conclusion

This paper investigates the physical origin, controlling mechanism, and modeling implications of substrate induced channel thermal noise in MOSFETs. Resistance of the substrate generates potential fluctuations that in turn produce additive channel noise by modulating the inversion charge via the channel depletion capacitor. For low-doped substrates, the additive noise may result in a frequency dependence of the drain current noise due to a pole associated with the R_{sub}-C_{depl} network. Since it conforms to the tendency of the reported excess noise, it may exaggerate the value being reported for the channel thermal noise factor. Hence, careful characterization is required to achieve accurate modeling.

Acknowledgments

The authors would like to thank Dr. Mark R. Pinto, Dr. Conor Rafferty, and Dr. Muhammad A. Alam of Agere Systems for providing PADRE. Special thanks go to Texas Instruments for mentoring and promoting this work as a customized SRC project.

References

[1] R. P. Jindal, "Hot-Electron Effects on Channel Thermal Noise in Fine-Line NMOS Field-Effect Transistor," *IEEE Transactions on Electron Devices*, vol. 33, no. 9, pp. 1395–1397, Sept. 1986.

[2] A. J. Scholten, H. J. Tromp, L. F. Tiemeijer, R. van Langevelde, R. J. Havens, P. W. H. de Vreede, R. F. M. Roes, P. H. Woerlee, A. H. Montree, and D. B. M. Klaassen, "Accurate Thermal Noise Model for Deep-Submicron CMOS," in *Technical Digest of IEDM (International Electron Devices Meeting)*, Washington, DC, Dec. 1999, pp. 155–158.

[3] S. V. Kishore, G. Chang, G. Asmanis, C. Hull, and F. Stubbe, "Substrate-Induced High-Frequency Noise in Deep Sub-Micron MOSFETs for RF Applications," in *Proceedings of IEEE CICC (Custom Integrated Circuits Conference)*, San Diego, CA, May 1999, pp. 365–368.

[4] J. J. Ou, X. Jin, C. Hu, and P. R. Gray, "Submicron CMOS thermal noise modeling from an RF perspective," in *Proceedings of the Symposium on VLSI Technology*, Kyoto, Japan, June 1999, pp. 151–152.

[5] S. Donati, M. A. Alam, K. S. Krisch, S. Martin, M. R. Pinto, H. H. Vuong, F. Bonani, and G. Ghione, "Physics-Based RF Noise Modeling of Submicron MOSFETs," in *Technical Digest of IEDM (International Electron Devices Meeting)*, San Francisco, CA, Dec. 1998, pp. 81–84.

[6] J.-S. Goo, C.-H. Choi, F. Danneville, E. Morifuji, H. S. Momose, Z. Yu, H. Iwai, T. H. Lee, and R. W. Dutton, "An Accurate and Efficient High Frequency Noise Simulation Technique for Deep Submicron MOSFETs," *IEEE Transactions on Electron Devices*, vol. 47, no. 12, pp. 2410–2419, Dec. 2000.

[7] R. P. Jindal, "Distributed Substrate Resistance Noise in Fine-Line NMOS Field-Effect Transistors," *IEEE Transactions on Electron Devices*, vol. 32, no. 11, pp. 2450–2453, Nov. 1985.

[8] F. Faccio, F. Anghinolfi, E. H. M. Heijne, P. Jarron, and S. Cristoloveanu, "Noise Contribution of the Body Resistance in Partially-Depleted SOI MOSFET's," *IEEE Transactions on Electron Devices*, vol. 45, no. 5, pp. 1033–1038, May 1998.

An Efficient Frequency-Domain Analysis Technique of MOSFET Operation

Kyu-Il Lee, Jinsoo Kim, *Hyungsoon Shin, **Chanho Lee, Young June Park, and Hong Shick Min
School of EE and CS, Seoul National Univ., Seoul, Korea, E-mail : macjr@bawi.org
*Dept. of IEE, Ewha Womans Univ., Seoul, Korea
**School Of EE, Soongsil Univ., Seoul, Korea

Abstract
We propose a harmonic balance technique for the frequency-domain analysis of MOSFET operation. Our approach is based on the charge-sheet and the non-quasistatic(NQS) MOSFET models in the channel region with the harmonic balance(HB) technique applied to the channel charges. Lateral field effect is considered in the formulation to analyze the short channel MOSFET devices. It is shown that the proposed method renders a computationally efficient tool to analyze the harmonic distortion occurrence in the MOSFET devices due to the nonlinear response of the channel charges.

1. Introduction

The modeling of the distortion in the MOSFET under the multi-tone input signal becomes an important issue as the CMOS RF circuit technology is merged as a main stream in recent RF technology. The conventional time-domain device simulation is not adequate for this purpose since too many integration steps are required to trace the multi-tone signal response[1]. Hence, the frequency-domain approaches including harmonic balance technique have been used in the two-dimensional numerical simulators to cope with the difficulties. But, as can be known from PISCES-HB[2], the numerical burden in HB analysis becomes extremely heavy, which may cause efficiency problem in the practical applications.
Aiming at an efficient RF device simulation, we have proposed a new HB method[3] based on the charge-sheet[4] and the non-quasistatic[5] MOSFET models with long-channel approximation, which greatly reduces numerical burden but gives most of the essential features of the harmonic behavior of MOSFET. In this paper, the method to extend the scheme to the short channel device will be proposed.

2. Frequency-Domain Formulation of MOSFET Model

We employ a one-dimensional nMOSFET model as the basic framework of our method, which describes the normalized charge density at a location x in the channel and a time t as[5]

$$\begin{aligned} q_e(y,t) &= V_{GB}(t) - V_{FB} - \phi_S(y,t) - \gamma\sqrt{\phi_S(y,t) - V_t} \\ &\cong V_{GB}(t) - V_{FB} - \phi_{S0}(t) - \gamma\sqrt{\phi_{S0}(t) - V_t} - F_B(t)\cdot\{\phi_S(y,t) - \phi_{S0}(t)\} \end{aligned} \quad (1)$$

For numerical approach, boundary conditions $\phi_{s0}(t)$ and $\phi_{sL}(t)$ (surface potentials at both ends of the channel) are calculated from the MOSFET charge equation. Consequently, the current continuity equation can be written as

$$\frac{\partial q_e(y,t)}{\partial t} = \frac{\mu_e}{2F_B(t)} \cdot \frac{\partial^2}{\partial y^2}\left[\{q_e(y,t) + F_B(t)\cdot V_t\}^2\right], \quad (2)$$

where the electron mobility μ_e is assumed to be a constant value of *500* $[cm^2/V\cdot sec]$. Our goal is to find a numerical approximant q_e satisfying Eq. (2) for given boundary conditions. Instead of time-domain treatment, we pose the problem in frequency

domain dual to time or space variable and consider the corresponding numerical scheme together with the effect of the NQS component.
Pursuing *Park*'s work[6], we can decompose the normalized charge density into two parts, the quasistatic and the non-quasistatic components:

$$q_e(y,t)=q_e^{QS}(y,t)+q_e^{NQS}(y,t). \tag{3}$$

The quasistatic component q_e^{QS} is obtained analytically by setting the left-hand-sided time-derivative term in Eq. (2) to be zero. Let us assume that the inversion charges at both source and drain ends of the channel respond instantaneously according to the changes of biases, the non-quasistatic component q_e^{NQS} can then be approximated by finite Fourier sine series as

$$q_e^{NQS}(y,t)=\sum_{m=1}^{M}Q_m(t)\cdot\sin\left(\frac{m\pi y}{L}\right) \text{ and } Q_m(t)=Q_{m,0}+\sum_{h=1}^{H}\left\{Q_{m,h}^{R}\cos(\omega_h t)-Q_{m,h}^{I}\sin(\omega_h t)\right\}. \tag{4}$$

The Fourier coefficients of $Q_m(t)$ compose our solution vector. With the above equations, the least-squared solution of the nonlinear system from Eq. (2) is obtained by Newton-Raphson method via $(2H+1)$ time and M space samples. For the comparison purpose in terms of the accuracy and computational efficiency, we also follow the finite-difference method (FDM) adopted in the usual harmonic balance simulation tools and *Park*'s mathematical approximation for SPICE.
For the modeling of the short channel effect, the lateral field effect has to be considered in the formulation[7]. Thus the model equations are changed to include lateral field term which contains the first and second derivatives of the surface potential. For the numerical approach, we assume the linear relationship between the potential and the charge density and the averaged value of the first and second derivatives of the surface potential are used in the numerical scheme. The resulting form of the master equation for iteration is the same as Eq. (2) except for the modification in the correct factor F_B. Thus the numerical complexity and burden are unchanged compared to the long channel formulation. Further consideration of the short channel effect is under development.

3. Results and Discussion

Device and bias conditions for simulation are shown in Table 1.
Figures 1 and 2 compare the source and drain currents obtained by our method, FDM, *Park*'s approach, and MEDICI for rising and falling transitions of ramp input signal on 10[μm] MOSFET, where a general agreement is observed except small deviation caused by the slight differences in physical parameters. The peak values of the NQS charge density at each node are depicted in Fig. 3 to show the charge-transit time effect. In the turn-on stage, the absolute value of the NQS component becomes larger toward the drain end due to the charge-transit time. In the turn-on stage, as the inversion charges are supplied from the source, the charges near drain end cannot be induced quickly according to the bias change. In the turn-off stage, however, the NQS component is not maximal in the vicinity of the channel ends because the inversion charges remote from both ends of the channel suffer the propagation delay. Since boundary potentials at both ends are not identical, this phenomenon is not symmetric around the center of the channel. However, FDM using 10 space nodes in the present example requires longer simulation time than the NQS methods. The simulation of the short channel devices using FDM requires fine discretization and more nodes in the channel(especially near junction) at the expense of simulation efficiency as the charge density changes drastically in the channel.

Table 1. Device parameters and bias conditions for simulation

	N_A	N_D	T_{OX}	V_{DS}
L=10 [μm]	2.1×10^{16} [cm^{-3}]		180 [Å]	1[V]
L=5 [μm]	5.0×10^{16} [cm^{-3}]	5.0×10^{20} [cm^{-3}]	150 [Å]	
L=0.5 [μm]	2.0×10^{17} [cm^{-3}]		80 [Å	5[V]
Ramp	From 0 [V] to 5 [V], rising & falling time = 2 [nsec], 20[psec]			
Single-tone	$V_{gb} = 3 + 1 \cdot \sin(2\pi f_1 t)$ [V]			
Two-tone	$V_{gb} = 3 + 1 \cdot \sin(2\pi f_1 t) + 1 \cdot \sin(2\pi f_2 t)$ [V]			

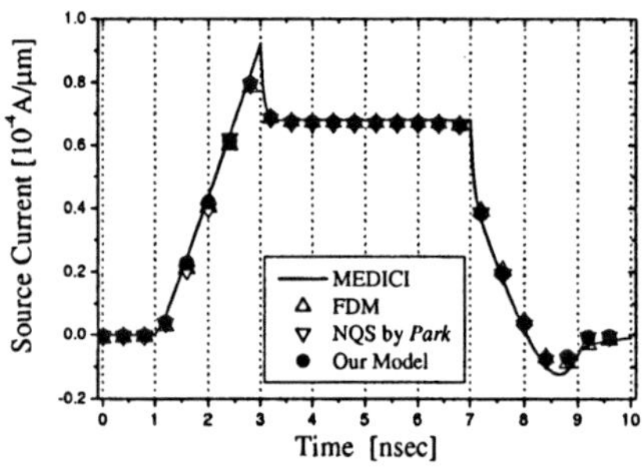

Fig. 1. Source currents of the 10[μm] MOSFET for ramp input

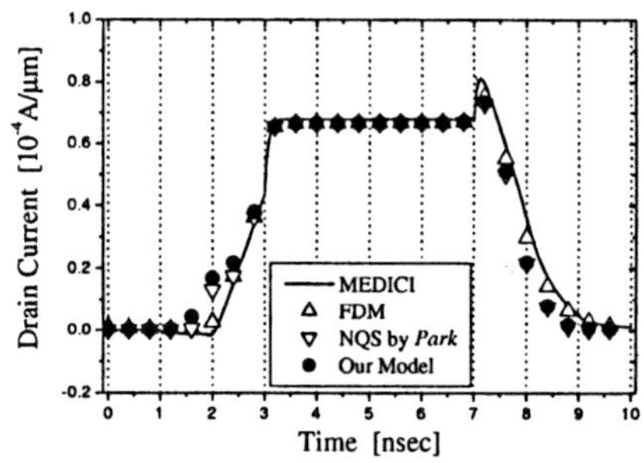

Fig. 2 Drain currents of the 10[μm] MOSFET for ramp input

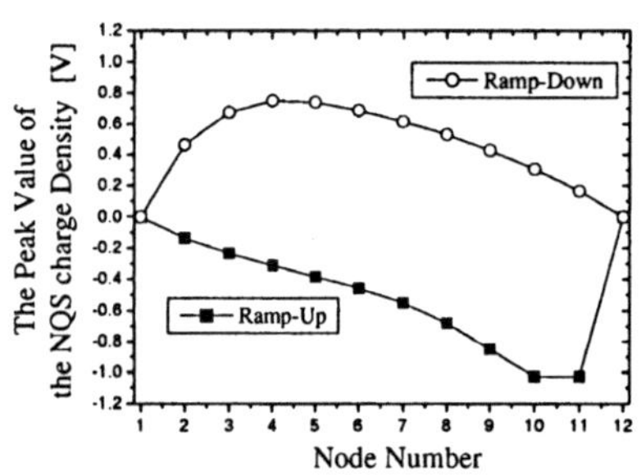

Fig. 3 Peak value of the NQS charges in the channel (1: Source, 12 : Drain)

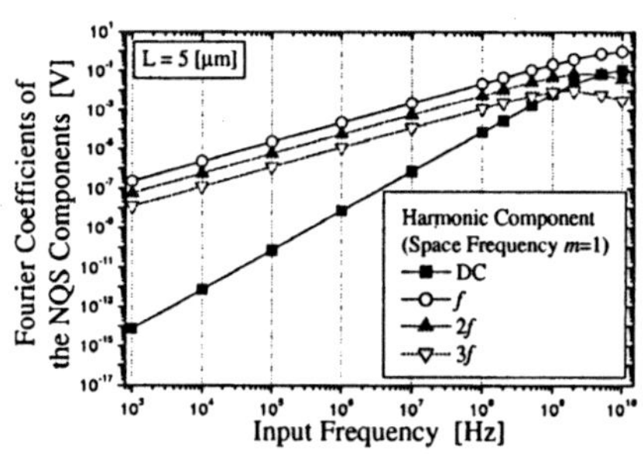

Fig. 4 Fourier Coefficients of the NQS component for single-tone input

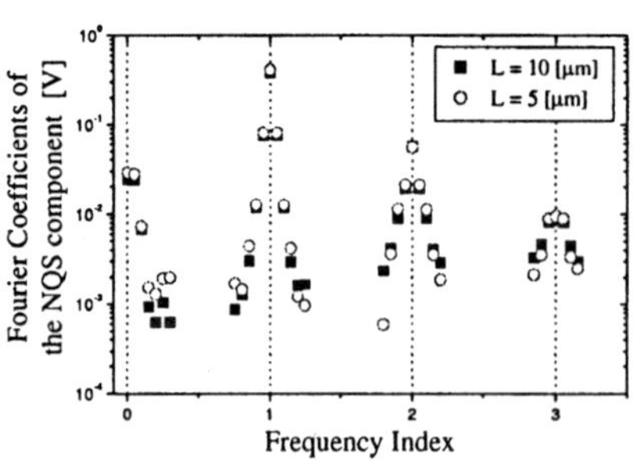

Fig. 5 Fourier Coefficients of the NQS component for two-tone input

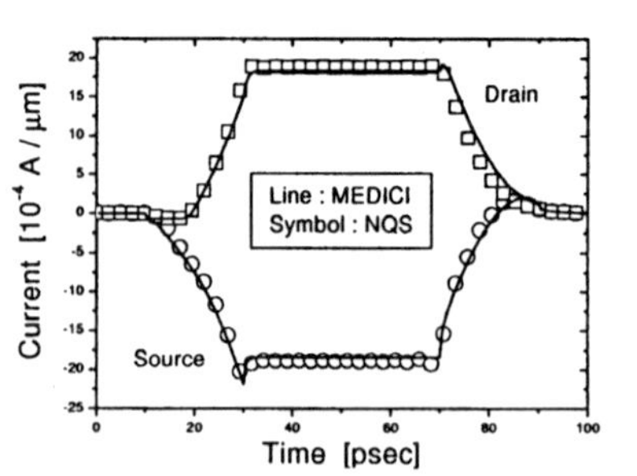

Fig. 6 Currents of the 0.5[μm] MOSFET for ramp input

In Fig. 4(L=5[μm]), the DC and the first three terms of Eq. (5) are shown for the case of space frequency $m=1$, which are the dominant terms of q_e. When input frequencies are higher than the intrinsic cut-off frequency, the NQS components are responsible for harmonic distortion. It has been reported that the third-order harmonic distortion component shows the notch behavior[8]. The value of the third harmonic Fourier cosine coefficient changes from positive to negative and again to positive as the input frequency increases. The change in sign causes the notch behavior in the harmonic distortion.

In Fig. 5, the Fourier coefficients of the NQS charges in case of two-tone input on gate are shown for four dominant harmonic components of f_2(the higher frequency, 100[MHz]), around each of which has several harmonic components of f_1(the lower frequency, 10[kHz]). These harmonic and intermodulated components will give rise to the nonlinear distortion to the output currents. However, two-tone large-signal input signal in this example is not tractable in most device simulation tools.

Figure 6 compares the current of 0.5[μm] MOSFET in thé saturation region (V_D=5[V]) with the result of MEDICI. The current values from our method are consistent with those from MEDICI when the MOSFET is operated in the saturation condition(high lateral field). This example confirms the applicability of the modification in the correction factor F_B for the short channel MOSFET.

4. Conclusion

It is shown that our simple but novel and physically reasonable method has many merits over the conventional approaches in the analysis of AC signals in view of the numerical burden; computation time and memory usage. It is also shown that the simple modification on the model is acceptable in analyzing the short channel effect and further consideration of the short channel effect is under development. The error that may be caused by the simplification of the model formulation is compensated by the non-quasistatic treatment. Also the proposed model can be extended to the direct combination of the device simulation with the circuit simulation.

Acknowledgment

This work has been supported by the National Research Laboratory Project of the Ministry of Science and Technology and by the Brain Korea 21 Project.

References

[1] Kevenaar, T. A., and Maten, E. J. W. ter (1999), RF IC Simulation : state-of-the-art and future tends, SISPAD, pp.7-10

[2] Troyanovsky, B., Yu, Z., Yu, and Dutton, R. W. (1995), Large Signal Frequency Domain Device Analysis via the Harmonic Balance Technique, SISPAD, pp.114-117

[3] Lee, K. -I., Kim, J., Park, Y. J., and Min, H. S., A Simple Frequency-Domain Analysis of MOSFET – Including Non-Quasistatic Effect, IEEE TCAD, To-be-published in July 2001

[4] Brews, J. R. (1978) : A Charge-sheet Model of the MOSFET, SSE vol.21, pp.345-355

[5] Park, H. J. (1989) : Charge-sheet and Non-quasistatic MOSFET Models for SPICE, Ph. D. Dissertation, U. C. Berkeley

[6] Park, H. J., Ko, P. K., and Hu, C. (1986) : A Measurement–based Charge-sheet capacitance Model of Short-channel MOSFET's for SPICE, IEDM, pp.40- 43

[7] Miura-Mattausch, M., Feldmann, U., Rahm, A., Bollu, M., and Savignac, D. (1996) : Unified Complete MOSFET Model for Analysis of Digital and Analog Circuits, IEEE TCAD vol. 15, pp. 1-7

[8] Pu, L. –J., and Tsividis, Y. P. (1990) : Harmonic Distortion of the Four-terminal MOSFET in Non-quasistatic Operation, IEE Proceedings vol. 137, pp.325-332

Modeling of Bias Dependent Fluctuations of Flicker Noise of MOSFETs

Ken'ichiro Sonoda, Motoaki Tanizawa, Katsumi Eikyu, Kiyoshi Ishikawa, Toshio Kumamoto*, Hiroyuki Kouno*, and Masahide Inuishi
*ULSI Development Center, System LSI Div.**,
Mitsubishi Electric Corp., Hyogo 664-8641, Japan
sonoda@lsi.melco.co.jp

Abstract
Drain and gate bias dependent fluctuations of flicker noise of MOSFETs are explained in terms of carrier concentration distributions in a MOSFET channel. A proposed model well describes the increase of the fluctuation in the saturation region of operation. In addition, the gate bias dependence of the fluctuation in the saturation region can also be calculated using the model. Our model predicts that for any gate voltage change, the fluctuation in the saturation region will be 2–2.5 times that in the linear region.

1 Introduction

A typical noise spectrum of MOSFETs exhibits $1/f$ dependence below 1MHz, so that noise characteristics of CMOS circuits in the audio frequency range is determined by the flicker noise. Moreover, RF characteristics such as noise figure and oscillator phase noise are affected by the flicker noise through upconversion of low frequency noise (Hajimiri and Lee 1998). For RF application, long channel or large gate area options, which are usually adopted for low noise application in low frequency range, are no longer acceptable because of low cut-off frequency and large parasitic capacitance. As the gate area shrinks, the number of interface traps in the area decreases and the relative fluctuation in that number becomes remarkable, which leads to an increase in the variation of flicker noise from sample to sample. In a previous work (Brederlow et al. 1999), the fluctuation is expressed as a function of $V_{ds}/(V_{gs} - V_{th})$ in the linear region and is independent of bias in saturation region. However, our experimental results show that the fluctuation depend on gate bias even in the saturation condition. In this paper, a model is proposed to explain the fluctuation in both linear and saturation regions, and the gate voltage dependence of the fluctuation.

2 Measurement

A setup for flicker noise measurement is schematically shown in Fig. 1. A DUT is biased to form a source-follower amplifier. Measured noise voltage, v_o, is divided by the voltage gain of the amplifier to give input referred noise voltage. The gain can be measured with the same setup, except for a small-signal AC voltage source which is attached to the gate of the DUT. The output voltage of this configuration is insensitive to power supply noise because the gate electrode, which is the most sensitive to the power supply noise, is tied directly to the ground. In common-source configuration, on the contrary, the noise voltage applied to the gate electrode is amplified by g_{m}-times and appears at the output, which may mask the flicker noise signal to be measured.

An example of measured input referred noise voltage spectrum $S_V \equiv S_I/g_{\mathrm{m}}^2$ is shown in Fig. 2. The fluctuation of the noise spectrum clearly depends on bias conditions, which will be discussed in the next section. From our measurement results, the fluctuation of g_{m} is much smaller than that of S_V for all bias conditions, so that the relative fluctuations of S_V can be regarded as the same as those of S_I.

3 Fluctuation Modeling

The capture-emission process induces a fluctuation in the number of carriers and a correlated mobility variation, which is called the correlated number-mobility fluctuation (ΔN-$\Delta\mu$) model, on which the BSIM3 flicker noise model is based. According to the ΔN-$\Delta\mu$ model, the flicker noise current power spectrum, S_I, at frequency f is given by (Hung et al. 1990)

$$S_I = \frac{I^2}{fL_{\mathrm{eff}}} \int_0^{L_{\mathrm{eff}}} \eta dx, \tag{1}$$

where $\eta \equiv (1/L_{\mathrm{eff}}W_{\mathrm{eff}})(1/n + \alpha\mu)^2 N_{\mathrm{t}}kT/\gamma$, I is the drain current, n is the carrier density per unit area, N_{t} is the trap density per unit volume and energy, and γ is the attenuation coefficient of the electron wave function in the oxide. Measured input referred noise is shown in Fig. 3 with simulation results of ΔN-$\Delta\mu$ (Eq. (1)), SPICE2 ($S_I = \mathtt{KF} I_{\mathrm{d}}^{\mathtt{AF}}/C_{\mathrm{ox}}L_{\mathrm{eff}}^2 f^{\mathrm{EF}}$), and Gray-Meyer (Gray and Meyer 1993) ($S_I = \mathtt{KF} g_{\mathrm{m}}^2/C_{\mathrm{ox}}W_{\mathrm{eff}}L_{\mathrm{eff}} f^{\mathrm{EF}}$) models. The ΔN-$\Delta\mu$ model with the mobility fluctuation parameter $\alpha = 0$ well describes the gate voltage dependence of S_V, which means the mobility fluctuation is negligible for our samples. The other two models cannot accurately predict the bias dependence. Note that the increase of S_V with V_{gs} comes from the gate voltage dependence of g_{m}, or μ (Ziel and Park 1979).

The gate area dependence of S_V and its standard deviation is shown in Fig. 4. The noise spectrum S_V is inversely proportional to the gate area, $W_{\mathrm{eff}}L_{\mathrm{eff}}$. The normalized standard deviation is proportional to $1/(W_{\mathrm{eff}}L_{\mathrm{eff}})^{1/2}$ as expected from Pelgrom's rule (Pelgrom et al. 1989).

The drain bias dependence of the noise fluctuation is shown in Fig. 5. As the drain voltage increases, the fluctuation increases and saturates. The saturation level depends on the gate voltage, as shown in Fig. 6. The bias dependence of the noise is modeled as

$$\frac{\sigma(S_V)}{S_V} = \left(\frac{\sqrt{\frac{1}{L_{\mathrm{eff}}} \int_0^{L_{\mathrm{eff}}} \eta^2 dx}}{\frac{1}{L_{\mathrm{eff}}} \int_0^{L_{\mathrm{eff}}} \eta dx} \right) \left(\frac{1}{\sqrt{N_{\mathrm{t}}kTL_{\mathrm{eff}}W_{\mathrm{eff}}/\gamma}} \right). \tag{2}$$

In the linear region limit of $V_{\mathrm{ds}} \to 0$, n and η can be regarded as constant along the channel and $\sigma(S_V)/S_V \to 1/\sqrt{N_{\mathrm{t}}kTL_{\mathrm{eff}}W_{\mathrm{eff}}/\gamma} \equiv (\sigma(S_V)/S_V)_0$ as expected.

As V_{ds} increases, n near the drain decreases and η increases from the source to the drain, which leads to the increase of the first term of Eq. (2). The fluctuation calculated with Eq. (2) is plotted in Figs. 5, 6. The carrier concentration n in the variable η is calculated from $n(x) = C_{ox}(V_{gs} - V_{th} - V(x))/q$, where the channel potential $V(x)$ is obtained from the drain current formula based on the gradual channel model with velocity saturation effect taken into account (Ko 1989). The simulated data explain the drain and gate bias dependence of the noise fluctuation. As V_{gs} increases, the η distribution approaches the uniform one shown in Fig. 7, and the noise fluctuation given by Eq. (2) is reduced.

4 Conclusion

Drain and gate bias dependent fluctuations of flicker noise of MOSFETs have been explained in terms of carrier concentration distributions in a MOSFET channel. The fluctuation in the saturation region is 2–2.5 times that in the linear region.

References

Brederlow, R., Weber, W., Landsiedel, D. S., Thewes, R. (1999): Fluctuations of the Low Frequency Noise of MOS Transistors and their Modeling in Analog and RF-Circuits. International Electron Devices Meeting Tech. Dig. pp. 159–162

Gray, P. R., Meyer, R. G. (1993): Analysis and Design of Analog Integrated Circuits. John Wily & Sons, New York

Hajimiri, A., Lee, T. H. (1998): A General Theory of Phase Noise in Electrical Oscillators. IEEE J. Solid-State Circ. 33: pp. 179–194

Hung, K. K., Ko, P. K., Hu, C., Cheng, Y. C. (1990): A Unified Model for the Flicker Noise in Metal-Oxide-Semiconductor Field-Effect Transistors. IEEE Trans. Electron Devices 37: pp. 654-665

Ko, P. K. (1989): Approaches to Scaling. In: Einspruch, N. G., Gildenblat, G. S. (eds.): VLSI Electronics Microstructure Science. Academic Press, San Diego, pp. 1–37 (Advanced MOS Device Physics, vol. 18)

Pelgrom, M. J. M., Duinmaijer, A. C. J., Welbers, A. P. G. (1989): Matching Properties of MOS Transistors. IEEE J. Solid-State Circ. 24: pp. 1433–1440

Ziel, A. v. d., Park, H. S. (1979): A discrepancy in the elementary theory of MOSFET modeling. Appl. Phys. Lett. 35: pp. 942–944

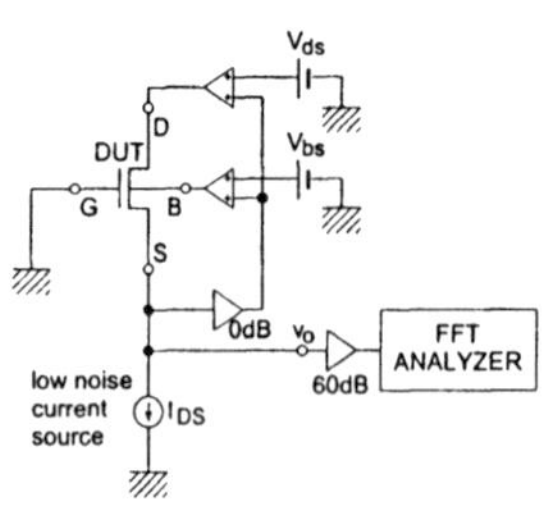

Fig. 1. Schematic of noise measurement setup.

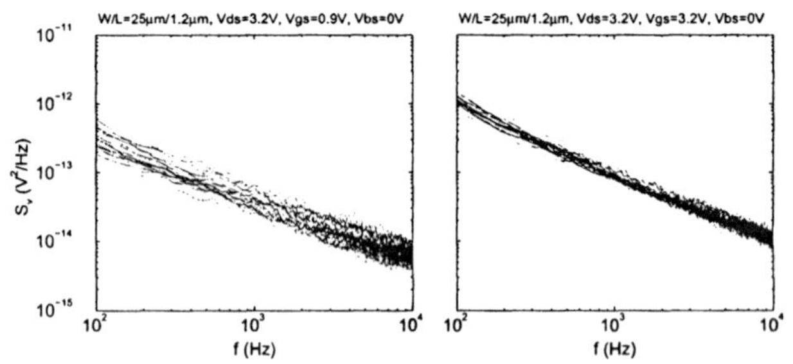

Fig. 2. Examples of measured input referred noise voltage spectrum of 35 samples for each bias condition. $V_{ds} = 3.2V$, $V_{gs} = 0.9V$(left), 3.2V(right).

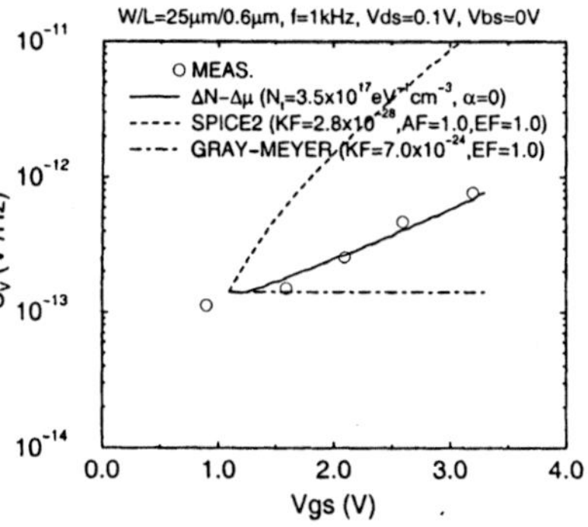

Fig. 3. The gate bias dependence of the input referred noise voltage.

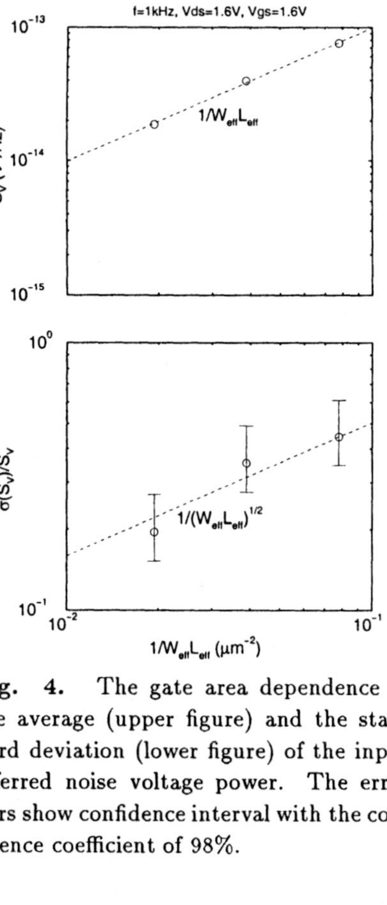

Fig. 4. The gate area dependence of the average (upper figure) and the standard deviation (lower figure) of the input referred noise voltage power. The error bars show confidence interval with the confidence coefficient of 98%.

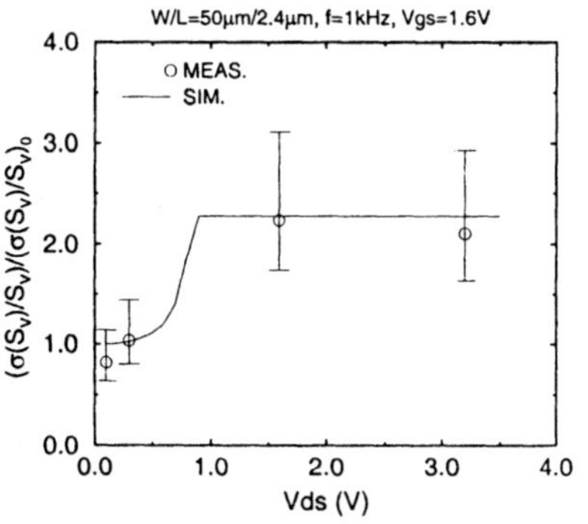

Fig. 5. The drain voltage dependence of the normalized standard deviation of the input referred noise voltage power. The measured data have been divided by $1/\sqrt{N_t kT L_{\rm eff} W_{\rm eff}/\gamma}$.

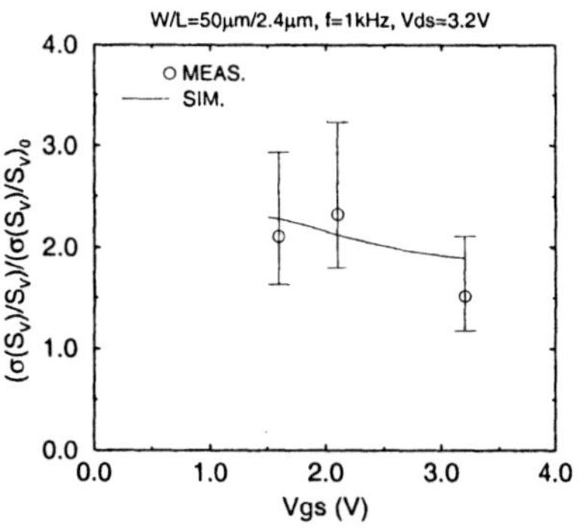

Fig. 6. The gate voltage dependence of the normalized standard deviation of the input referred noise voltage power.

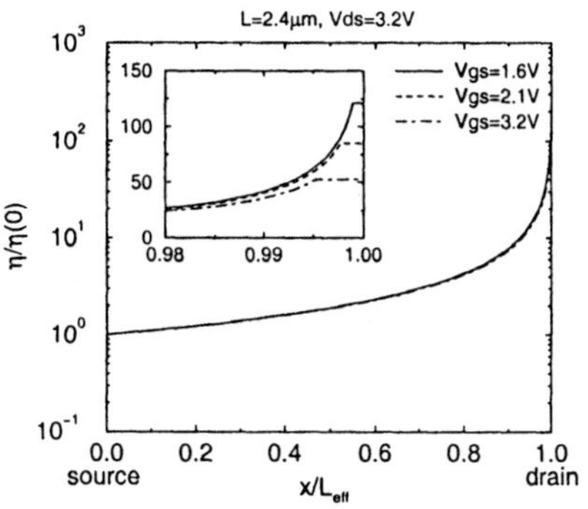

Fig. 7. The parameter η as a function of position in the channel.

Compact MOS Modelling for RF CMOS Circuit Simulation

A.J. Scholten, R. van Langevelde, L.F. Tiemeijer,
R.J. Havens, and D.B.M. Klaassen
Philips Research Laboratories
Prof. Holstlaan 4, 5656 AA Eindhoven, The Netherlands
E-mail: andries.scholten@philips.com

Abstract

Modern CMOS technologies are becoming increasingly attractive for RF applications. This imposes stringent requirements on compact models used in circuit simulation: not only currents and charges, but also noise, power gain, impedances, and harmonic distortion must be modelled accurately. In this paper several of these issues will be addressed with the help of Philips' new public-domain compact MOS model, MOS Model 11.

1 Introduction

As modern CMOS is rapidly progressing towards sub-100 nm dimensions, RF performance of MOS devices is improving strongly [1, 2, 3]. As a consequence, CMOS has become a viable option for RF applications. In order to exploit the RF capabilities of CMOS, compact models for circuit simulation are required that describe all the relevant quantities the RF designer is interested in: not only currents and charges, but also noise figure, power gain, impedance levels, and harmonic distortion [4, 5, 6].

In this paper, several of these issues will be addressed with the help of Philips' new compact MOS model, MOS Model 11. This model gives a good description of the currents and all its derivatives up to the 3rd order, enabling accurate distortion modelling (see Section 2). Using this model, we show that thermal noise can be described very well without the need of any additional parameters (see Section 3). Next, we show how non-quasi-static effects can be modelled using the concept of channel segmentation (see Section 4).

2 RF Distortion

The compact MOS Model named 'MOS Model 11' (MM11) [7], the successor of MOS Model 9, has been developed in order to accurately describe the MOSFET currents themselves as well as their higher-order derivatives (up to 3rd order) with respect to all terminal voltages. The model is based on a continuous description of the surface potential throughout all operating regimes, including the increasingly important moderate inversion regime. Compared to contemporary models such as BSIM4 [8] and MOS Model 9 [7], the new model contains improved expressions for mobility reduction and velocity saturation. All relevant physical effects in today's MOSFETs, such as poly depletion, bias-dependent overlap capacitances, quantum-mechanical effects, and gate leakage current are included. In this section we will show that the model gives an excellent description of harmonic distortion, both at low and RF frequencies.

2.1 Experimental method

As illustrated in Fig. 1, a sinusoidal gate voltage leads to a distorted drain current, containing the ground harmonic (HD1) as well as unwanted higher-order harmonics (HD2, HD3...). This has been measured from 16 MHz to 1 GHz using an RF network analyzer (HP8753E) with an extension to measure the 2nd and 3rd harmonic power levels [9]. Measurements are de-embedded up to the test-structure terminals [9]. Compact model parameters for MM11 are extracted from the DC current and its 1st-order derivatives, and from low-frequency CV measurements. The compact model has been extended with parasitic resistances to describe polysilicon [10, 11] and bulk resistance [12] effects.

2.2 Results

Measured and modelled results for a 0.35 μm n-MOS transistor as a function of DC gate bias are shown in Fig. 2 for an intermediate frequency (f=16 MHz), and in Fig. 3 for an RF frequency (f=1 GHz). RF results for an 0.18 μm n-MOS transistor in 0.18 μm technology are shown in Fig. 4. In all cases it is observed that MM11 (solid lines) gives an accurate description of HD1, HD2, and HD3. Simulations of HD3 using BSIM3v3 and MOS Model 9 are also shown for reference. In contrast to what has been observed for bipolars [14], it was found that at f=1 GHz the description of distortion is still mainly determined by the drain current model of the MOSFET, while the parasitic and intrinsic charges only contribute about -57 dBm to HD3. Accurate modelling results have also been obtained for different values of drain bias, different technologies, and for PMOS, see for instance Ref. [13].

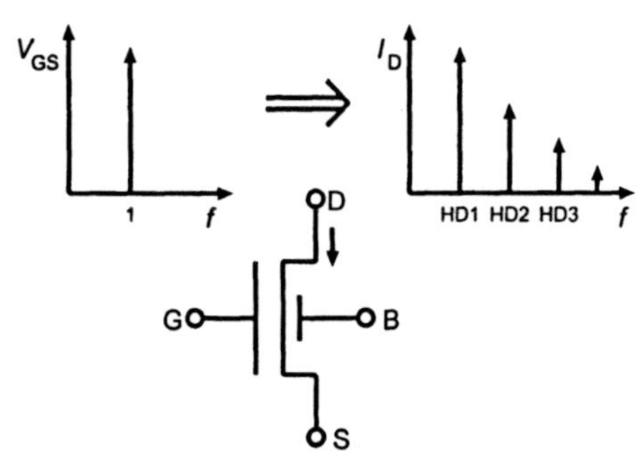

Fig. 1: A sinusoidal V_{GS} results in a distorted drain current I_D containing not only the ground harmonic (HD1) but also unwanted higher-order harmonics (HD2, HD3...), due to the non-linear dependence of I_D on V_{GS}.

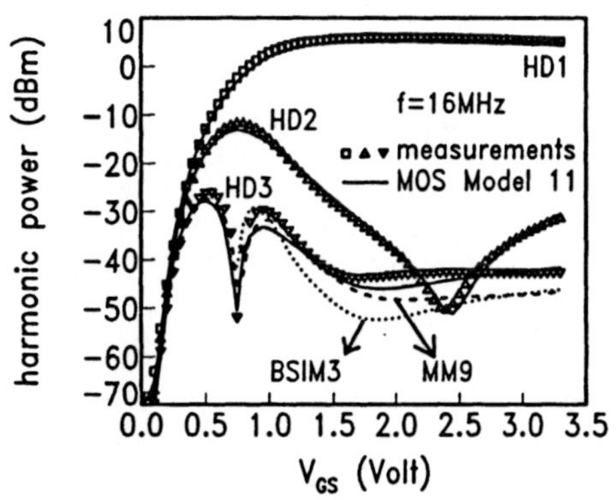

Fig. 2: LF distortion (f = 16 MHz) as a function of V_{GS} of a 16 × 10/0.35 μm n-channel device in 0.35 μm technology. V_{DS} = 3.3 V; P_{in} = −5 dBm.

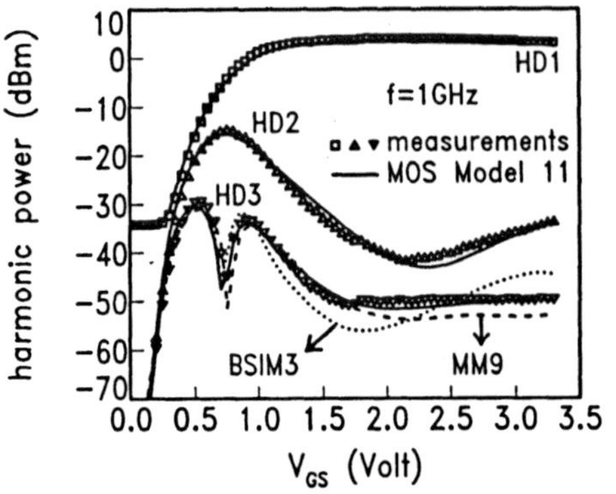

Fig. 3: RF distortion ($f = 1$ GHz) as a function of V_{GS} on the same device as in Fig. 2, i.e. a $16 \times 10/0.35$ μm n-channel device in 0.35 μm technology. $V_{DS} = 3.3$ V; $P_{in} = -5$ dBm.

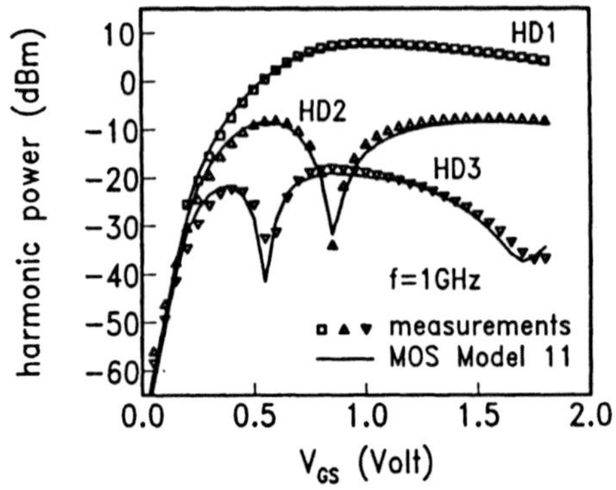

Fig. 4: RF distortion ($f = 1$ GHz) for a $64 \times 3/0.18$ μm n-channel device in 0.18 μm technology. $V_{DS} = 1.8$ V; $P_{in} = -5$ dBm.

3 Thermal noise

Accurate modelling of thermal noise is a prerequisite for the application of modern CMOS technologies to low-noise RF circuit design. In some publications [15, 16], it has been reported that the thermal noise in submicron MOSFETs is increased considerably w.r.t. the predictions based on long-channel thermal noise theory. The reported enhancements amount to more than a factor of 10 for n-channel devices as long as 0.7 μm, and are often attributed to hot-carrier effects. Contrary to this, we will show that the drain current thermal noise of deep-submicron MOSFETs can be described accurately *without* invoking carrier heating or introducing any parameters (see also Ref. [17]). This is achieved by using the Klaassen-Prins formula [18, 19] combined with the appropriate equations for velocity saturation [17]. For the resulting expressions we refer to [7].

3.1 Experimental method

Drain current thermal noise is investigated at an intermediate frequency (248 MHz) where both low-frequency 1/f noise, and high-frequency induced gate noise can be neglected. We study the 50-Ohm noise figure F_{50} (i.e. the noise figure at 50 Ω generator impedance), which is dominated by the MOSFET drain current thermal noise. The on-wafer noise measurements are performed using a conventional noise figure test set. A 50-Ω resistor is used to match the output impedance to the noise figure meter input. Besides the noise measurements, standard DC-characterization has been performed in order to determine the compact model parameters for each device. Besides the noise contribution of the MOSFET itself, all the device and interconnect parasitics, RF-probes, cables and bias-tees and the 50-Ω matching resistor have been characterized experimentally and included in the simulations.

3.2 Results

In Figs. 5 and 6 F_{50} is plotted as a function of V_{GS} for a number of channel lengths in 0.35 μm technology. It is seen that the thermal noise model, as described above, gives an excellent fit to the data. The same conclusion is drawn from Figs. 7 and 8, where F_{50} is plotted as a function of V_{GS} and V_{DS}, respectively, for a 0.18 μm device in a 0.18 μm technology. Note that all parameters used in the noise model are obtained from DC and CV measurements; all our experimental results are described accurately *without* introducing additional noise parameters, or invoking carrier heating or new physical phenomena. Of course we do not deny the existence of the physical effect of carrier heating; we simply find that the effect is not large enough to enhance the thermal noise to a measurable extent. The interesting question remains of course why Refs. [15, 16] *do* find such a large enhancement. Possibilities are (i) noise of the weak avalanche current [20] (ii) noise due to bulk resistance (included in our model) [21] or (iii) a generation-recombination noise contribution.

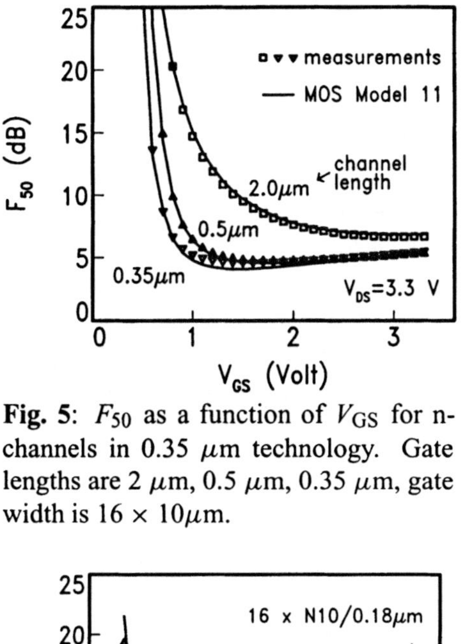

Fig. 5: F_{50} as a function of V_{GS} for n-channels in 0.35 μm technology. Gate lengths are 2 μm, 0.5 μm, 0.35 μm, gate width is 16 × 10μm.

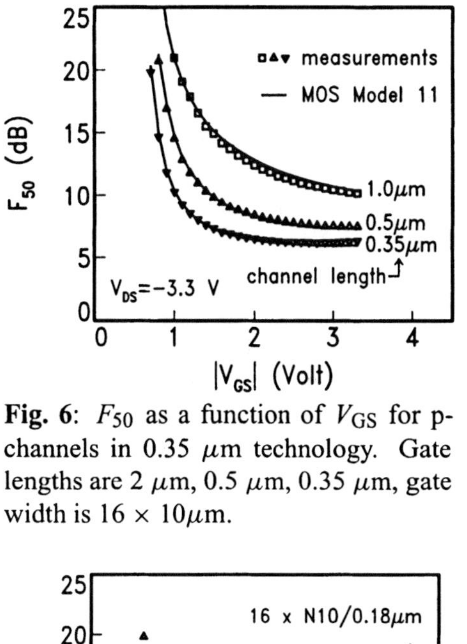

Fig. 6: F_{50} as a function of V_{GS} for p-channels in 0.35 μm technology. Gate lengths are 2 μm, 0.5 μm, 0.35 μm, gate width is 16 × 10μm.

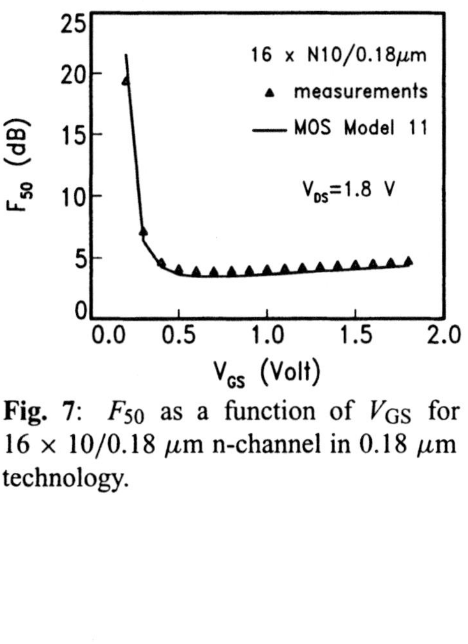

Fig. 7: F_{50} as a function of V_{GS} for 16 × 10/0.18 μm n-channel in 0.18 μm technology.

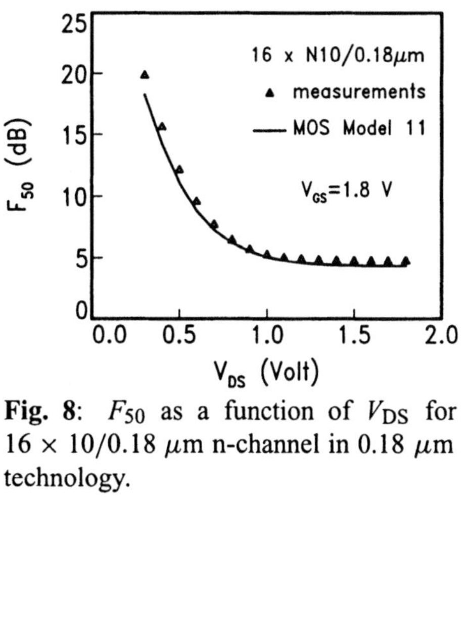

Fig. 8: F_{50} as a function of V_{DS} for 16 × 10/0.18 μm n-channel in 0.18 μm technology.

4 Non-quasi-static effects

To facilitate RF CMOS circuit design, RF modelling tools should give an accurate description of quantities like power gain, input impedance and the phase delay between drain current and gate voltage. To achieve this, not only parasitic resistances [10, 12], but also effects of finite channel transit times have to be taken into account. These so-called non-quasi-static (NQS) effects are included in various small-signal models (see e.g. [19, 22, 23]), which are, however, not suited for large-signal, transient and harmonic-balance simulations. An alternative model is the NQS extension of BSIM3 [24], which is computationally efficient, but is restricted to large-signal analysis. We follow the approach illustrated in Fig. 9, which is based on the idea of channel segmentation [19, 25, 26]. Advantages of this approach are that (i) it is physically sound, (ii) it can be used in all kinds of circuit analysis (DC, small-signal, transient, and noise analysis), and that (iii) all parameters are determined from DC- and low-frequency CV-measurements.

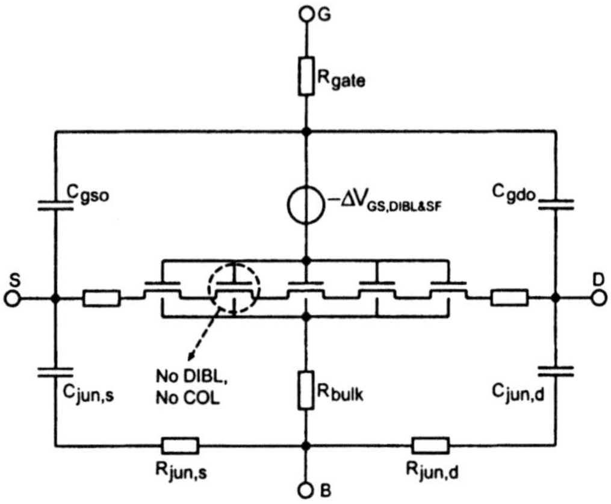

Fig. 9: NQS model consisting of five MM11 segments.

4.1 Experimental method

Using an HP8510 network analyzer, small-signal S-parameters [27] have been measured on a number of devices in ground-signal-ground configuration, fabricated in an 0.18 μm CMOS technology. The measured S-parameters are converted into Y-parameters [27] and de-embedded up to the test-structure terminals. From the Y-parameters, relevant quantities such as input impedance and maximum stable/available power gain are calculated.

4.2 Input resistance

One of the key parameters for the RF CMOS designer is the input impedance. Here we consider the input impedance in case the output is ac-shortcircuited, which is given by $R_{in} = \text{Re}(1/Y_{11})$. This quantity is of special interest since it is equal to zero for a QS model without parasitic resistances. In reality, however, it has a finite value (see Fig. 10), which is partly due to parasitic resistances, but mainly due to the NQS effect. This is shown in Fig. 10 by varying the number of channel segments in our NQS model: by increasing the number of segments, the modelled curve rapidly approaches the measurements. In the linear regime of operation, it can be shown analytically that for frequencies well below f_T:

$$R_{in}(N_{seg}) = R_{in}^{NQS} \cdot \left(1 - \frac{1}{N_{seg}^2}\right) \quad (1)$$

where $R_{\rm in}^{\rm NQS}$ is the full NQS input impedance ($N_{\rm seg} = \infty$). We have found empirically that Eq. (1) holds also for the saturation mode. From Eq. (1) we see that, using 5 segments, the modelled $R_{\rm in}$ only deviates 4% from $R_{\rm in}^{\rm NQS}$. Using 5 segments an accurate description is also obtained for $R_{\rm in}$ as a function of $V_{\rm GS}$ for a number of channel lengths (see Fig. 11).

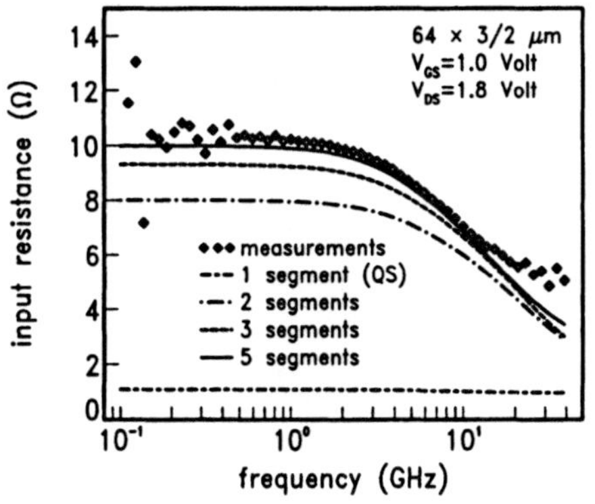

Fig. 10: Symbols: input resistance versus frequency for a $64 \times 3/2$ μm n-channel in 0.18 μm technology. Lines are calculated with different number of channel segments.

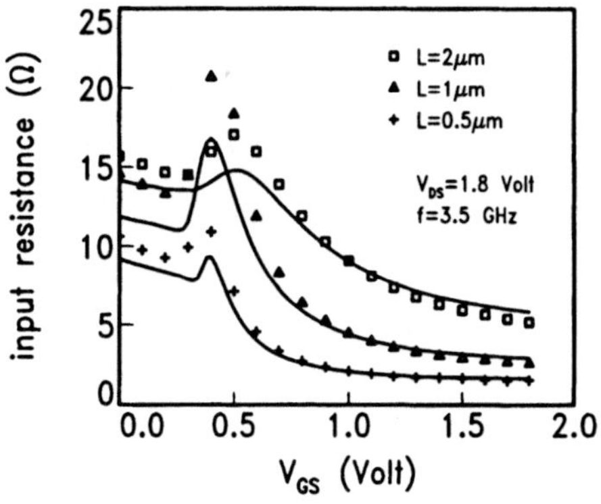

Fig. 11: Symbols: input resistance at 3.5 GHz versus $V_{\rm GS}$ for a number of channel lengths. Gate width is 64×3 μm. Lines are calculated with a 5-segment NQS model.

4.3 Maximum power gain

Another quantity of interest is the maximum stable/available power gain [27], which we define as:

$$G_{\rm msg/max} = \begin{cases} |\frac{Y_{21}}{Y_{12}}| & \text{if } K < 1 \\ |\frac{Y_{21}}{Y_{12}}| \cdot \left(K - \sqrt{K^2 - 1}\right) & \text{otherwise} \end{cases} \tag{2}$$

where K is the Rollett stability factor, given by

$$K = \frac{2{\rm Re}(Y_{11}){\rm Re}(Y_{22}) - {\rm Re}(Y_{21}Y_{12}))}{|Y_{21}Y_{12}|} \tag{3}$$

The frequency at which $G_{\rm msg/max} = 1$ defines the maximum oscillation frequency $f_{\rm max}$. In Fig. 12, it is seen that a quasi-static model is not able to predict $f_{\rm max}$ for an $L = 2$ μm device. Using the NQS model, however, it is seen that the fit between model and data improves considerably when the number of segments is raised to five. In general, the number of segments needed depends on the channel length and the operation frequency. Using 5 segments only, we see in Fig. 13 that $G_{\rm max}$ is modelled very well for a range of channel lengths $L < 2$ μm over the full gate bias range in saturation.

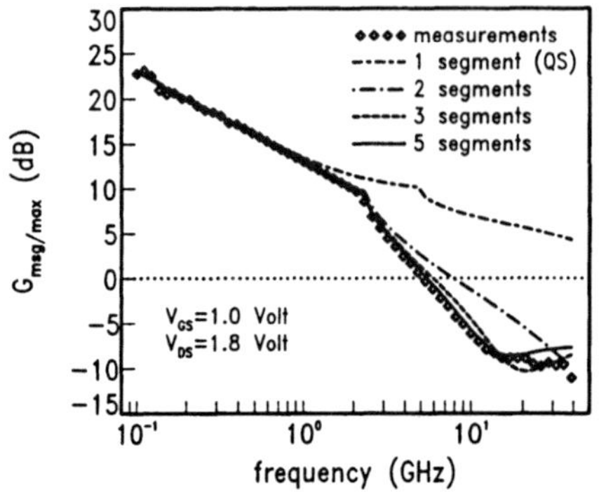

Fig. 12: Symbols: maximum stable/available power gain versus frequency for a 64 × 3/2 μm n-channel in 0.18 μm technology. Lines are calculated with different number of channel segments.

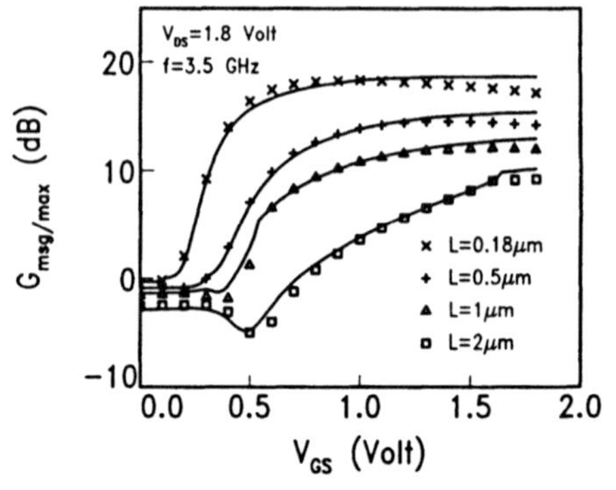

Fig. 13: Symbols: maximum stable/available power gain at 3.5 GHz versus V_{GS} for a number of channel lengths. Gate width is 64 × 3 μm. Lines are calculated with a 5-segment NQS model based on MM11.

4.4 High-frequency distortion and non-quasi-static effects

Finally we discuss the impact of NQS effects on RF distortion. In Fig. 14, 1 GHz distortion data for a 2μm n-channel transistor are shown. At low values of gate bias the QS model no longer gives good results. In this bias region the harmonic frequency exceeds the cut-off frequency f_T and as a result the quasi-static approach is no longer valid. Fig. 15 shows that the NQS-model with N=10 segments gives accurate results over almost the whole bias region.

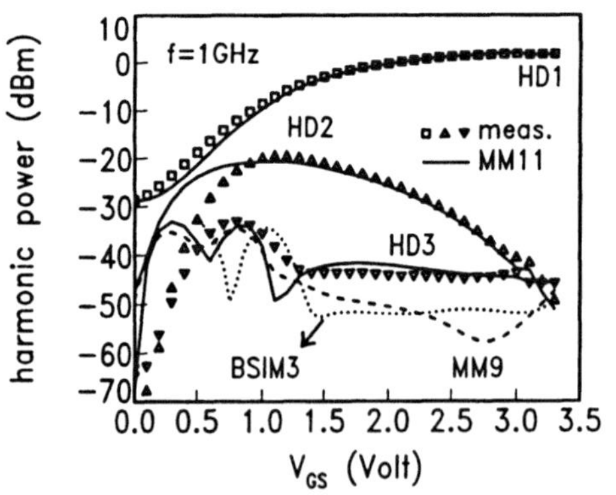

Fig. 14: First, second, and third harmonic distortion power levels of a W/L = 160/2 μm n-channel device in 0.35 μm technology. V_{DS} = 3.3 V; P_{in} = −5 dBm.

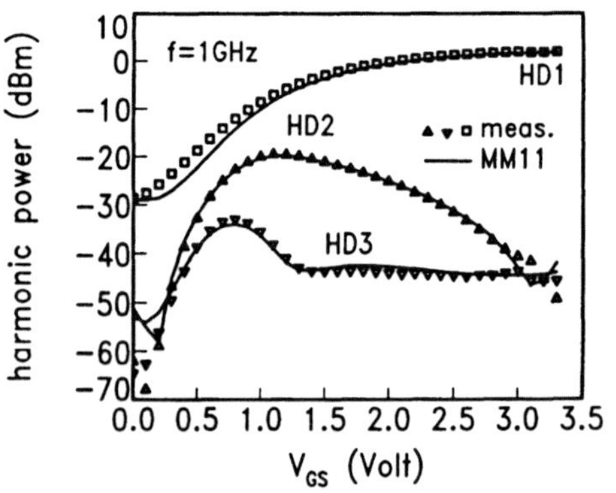

Fig. 15: Same measurements as in Fig. 14. Solid line is calculated with a 10-segment NQS model based on MM11.

5 Conclusion

In this paper, various aspects of compact modelling for RF CMOS have been addressed. First, we have shown that MOS Model 11 forms an excellent basis for RF CMOS circuit design, because, in contrast to other contemporary compact models, it is able to describe accurately the higher-order derivatives of the drain current, which results in an excellent description of third-order harmonic distortion. Moreover, we have shown that drain current thermal noise can be *predicted* with MOS Model 11: carrier heating effects or other additional physical effects do not need to be included to get a good description of the noise data. Finally, we have presented a simple yet effective method to turn a QS model into an NQS model by applying channel segmentation.

References

[1] H.S. Momose, E. Morifuji, T. Yoshitomi, T. Ohguro, M. Saito, T. Morimoto, Y. Katsumata, and H. Iwai, IEDM'96, p. 105.

[2] E. Morifuji, H.S. Momose, T. Ohguro, T. Yoshitomi, H. Kimijima, F. Matsuoka, M. Kinugawa, Y. Katsumata, and H. Iwai, 1999 VLSI Symp., p.163.

[3] P.H. Woerlee, R. van Langevelde, A.H. Montree, D.B.M. Klaassen, L.F. Tiemeijer, and P.W.H. de Vreede, ESSDERC 2000, p. 576.

[4] D.B.M. Klaassen, R. van Langevelde, A.J. Scholten and L.F. Tiemeijer, ESSDERC'99, p. 95.

[5] L.F. Tiemeijer, L.M.F. de Maaijer, R. van Langevelde, A.J. Scholten, and D.B.M. Klaassen, Proc. AACD'99, p.129.

[6] Y. Cheng , M. Schröter, C. Enz, M. Matloubian, and D. Pehlke, ESSCIRC'98, p.416.

[7] http://www.semiconductors.philips.com/Philips_models

[8] http://www-device.eecs.berkeley.edu

[9] L.F. Tiemeijer, R. van Langevelde, O. Gaillard, R.J. Havens, P.G.M. Baltus, P.H. Woerlee, and D.B.M. Klaassen, ESSDERC 2000, p. 464.

[10] B. Razavi, R.H. Yan, and K.F. Lee, IEEE Trans. on Circuits and Systems-I: Fundamental Theory and Applications **41**, No. 11, p. 750 (1994).

[11] R. Vanoppen, J.A.M. Geelen, and D.B.M. Klaassen, IEDM'94, p. 173.

[12] L.F. Tiemeijer, and D.B.M. Klaassen, ESSDERC'98, p. 480.

[13] R. van Langevelde, L.F. Tiemeijer, R.J. Havens, M.J. Knitel, R.F.M. Roes, P.H. Woerlee, and D.B.M. Klaassen, IEDM 2000, p. 807.

[14] M. Schröter, D.R. Pehlke, and T-Y. Lee, ESSDERC'99, p. 476.

[15] A.A. Abidi, IEEE Trans. El. Dev. **33**, p. 1801 (1986).

[16] P. Klein, IEEE El. Dev. Lett. **20**, p. 399-401 (1999).

[17] A.J. Scholten, H.J.Tromp, L.F. Tiemeijer, R. van Langevelde, R.J. Havens, P.W.H. de Vreede, R.F.M. Roes, P.H.Woerlee, A.H. Montree, and D.B.M.Klaassen, IEDM'99, p. 155.

[18] F.M. Klaassen and J. Prins, Philips Res. Repts. **22**, p. 504, (1967).

[19] Y.P. Tsividis, *Operation and modeling of the MOS transistor*, McGraw-Hill Inc. (1987).

[20] A. van der Ziel and E.R. Chenette, Adv. in El. and El. Phys. **46**, p. 313 (1978).

[21] J.S. Goo, S. Donati, C.H. Choi, Z. Yu, T.H. Lee, and R.W. Dutton, SISPAD 2001.

[22] T. Smedes and F.M. Klaassen, Solid-State-Electronics **38**, p. 121 (1995).

[23] L.F. Tiemeijer, P.W.H. de Vreede, A.J. Scholten, and D.B.M. Klaassen, ESSDERC'99, p. 652.

[24] M. Chan, K.Y. Hui, C. Hu, and P.K. Ko, IEEE Trans. Electron Dev. **45**, p. 834 (1998).

[25] A.J. Scholten, L.F. Tiemeijer, P.W.H. de Vreede, and D.B.M. Klaassen, IEDM'99, p. 163.

[26] R. Gillon, D. Vanhoenacker, J.-P. Colinge, Proc. AACD'99, p.227.

[27] G. Gonzales, *Microwave transistor amplifiers*, Prentice Hall, 1997.

Statistical Analysis of VLSI Using TCAD

Naoyuki Shigyo

System LSI Design Division, Toshiba Corporation Semiconductor Company
STE Bld., 2-5-1, Kasama, Sakae-ku, Yokohama 247-8585, Japan
E-mail: naoyuki.shigyo@toshiba.co.jp

Abstract

Statistical process fluctuations influence upon device and circuit performance of VLSI with device miniaturization. Thus, robust process, device and circuit designs are needed. This article describes the statistical analysis of MOSFET and interconnect using Technology CAD (TCAD).

1 Introduction

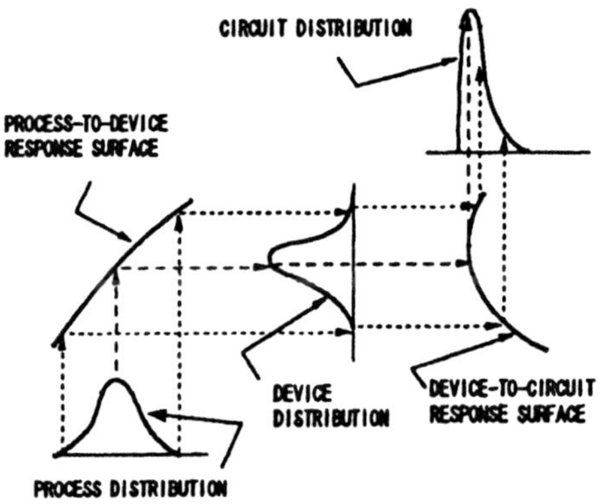

Fig. 1 Process fluctuation and circuit performance [2].

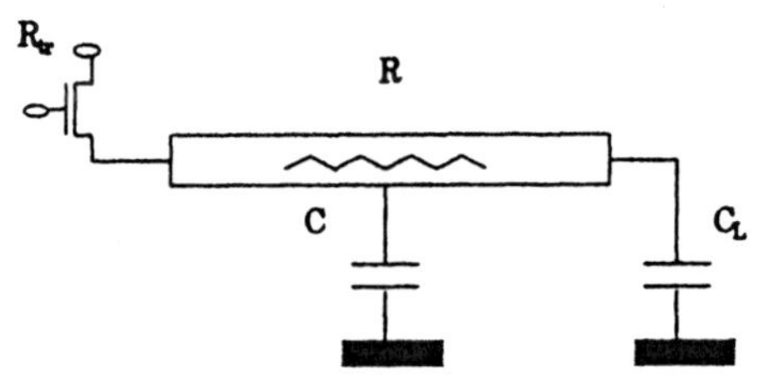

Fig. 2 Variability of device and interconnect.

The dependence of the parametric yield of VLSI on statistical process fluctuations such as a critical dimension (CD) and an ion implantation, has been increased. The statistical simulation is needed for robust designs of process, device and circuit [1].

Fig. 1 shows a schematic explanation of the relation between the process fluctuation and circuit performance [2]. The process fluctuation is related to device performance through a process-to-device *response surface*. The device performance is related to circuit performance through a device-to-circuit response surface. Process/device simulators can be used for representing the process-to-device response surface. Also, a circuit simulator is used for the device-to-circuit response surface.

This article presents TCAD based statistical analysis for the variability of device and interconnect, as shown in Fig. 2. The circuit delay t_{delay} is approximated as 2.3 $R_{tr}C$ + RC where R_{tr} is the MOSFET resistance [3]. In the case of short interconnect length

L, $R_{tr}C$ dominates t_{delay}. On the other hand, the interconnect RC delay dominates t_{delay} for long L. In Section 2, the worst case circuit model for MOSFET and problems for the statistical simulation are described. The interconnect variability is treated in Section 3.

2 Device Variability: MOSFET

2.1 Worst case circuit model

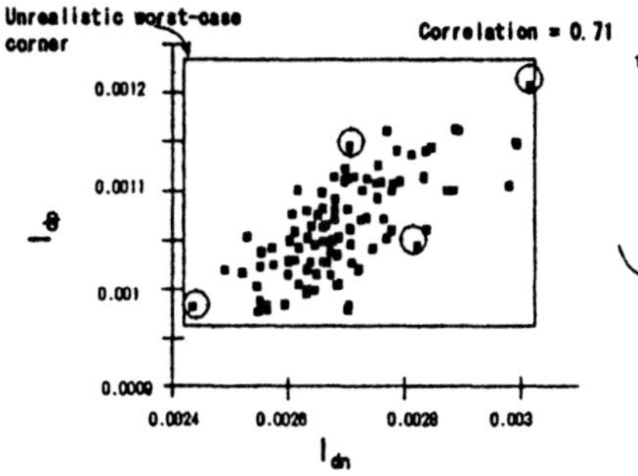

Fig. 3 Unrealistic worst-case corners [4].

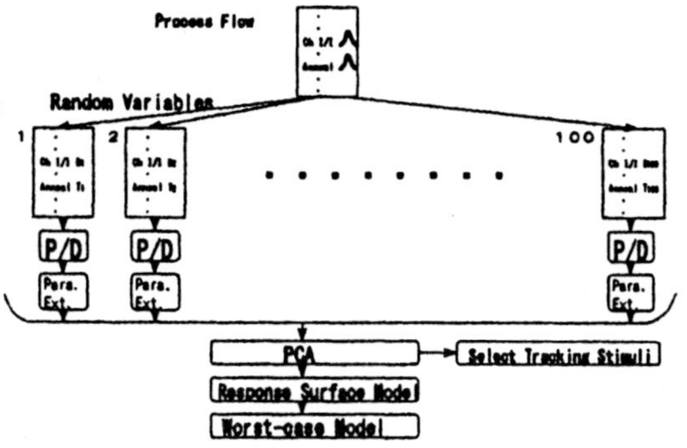

Fig. 4 Schematic explanation for statistical analysis and generation of worst-case models [5].

Fig. 3 shows a scatter plot of the n-ch and p-ch saturation drain current, I_{dn} and I_{dp} [4]. A circuit designer has used min-max worst-case corner models. In this example, however, there was a strong correlation between I_{dn} and I_{dp}, whose correlation coefficient was 0.71. Thus, the upper-left and lower-right corners were *unrealistic* worst-case corners.

Fig. 4 shows a schematic explanation for the statistical analysis and generation of worst-case BSIM3v3 models [5]. Before Monte Carlo (MC) runs, process conditions for a simulation were assigned separately to each process step using random numbers with assuming process fluctuations of the normal distribution. The number of MC runs is related to the accuracy [6].

MC runs resulted in 97 sets of BSIM3v3 parameters. At each run, a set of 144 (= 72 × 2) model parameters for n-ch and p-ch MOSFET's was extracted. Hence, 97 points which corresponded to the extracted parameter sets, were placed in a 144-dimensional space; a BSIM3v3 space. Independent *factors* were selected using the principal component analysis (PCA) [7]. Then, tracking stimuli were selected from the process variables, which had strong correlation to the factors. Finally, the response surface model (RSM), hence the worst-case models were generated as combinations of the tracking stimuli using the values of σ for the process variables. It is noted that the conventional method gives the worst-case models in the sense of the corners of the *process* fluctuations, so that they do *not always* correspond to those for the circuit performance, as shown in Fig. 1.

Table 1 shows 11 factors, which were needed to cover the 80 % of the BSIM3v3 space. An increase in the number of factors *m* enlarges the coverage of the BSIM3v3 space. However, this results in an increase in the worst-case corners, hence, the number of required circuit simulations. For example, 128 (= 2^{11-4}) circuit simulations are required even using the design-of-experiments (DOE) of resolution V. For a large circuit, a reduction of *m* is important to decrease the circuit simulation time.

The total coverage of BSIM3v3 space using factors 1, 2 and 3 was only 43 %. In order to supplement the small coverage, the factors were selected not by its coverage, but by its sensitivity to the performance of a small but critical circuit. The *sensitivity* analysis was carried out for reducing *m*. A ring-oscillator was used as an example.

Fig. 5 shows the sensitivities of the factors to the delay time *tpd* with ± 3σ process fluctuations. The characters of the lateral axis indicate the sign of each factor. For example, *f01+* means that factor 1 was changed +3σ from the nominal value. The short-channel effect was pronounced in the case of *f01+*. Factors 2, 1, and 8 had higher sensitivities to *tpd*, where the tracking stimulus for factor 8 was the gate oxidation temperature. On the other hand, factor 3 (LDD spacer) had a small contribution to *tpd*. This is because LDD spacer is related to

Table 1 Tracking stimuli and coverage of BSIM3v3 space for the factors [5].

Name	Tracking Stimulus	Coverrage of BSIM3 Space [%]
Factor 1	Gate CD	19
Factor 2	P-ch As I/I Dose	12
Factor 3	LDD SpacerThickness	12
Factor 4	N-ch B I/I Dose	8
Factor 5	Gate Oxidation Time	7
Factor 6	N-ch LDD I/I Dose	6
Factor 7	P-ch B I/I Dose	4
Factor 8	Gate Oxidation Temp.	4
Factor 9	SiN Depo. Thickness	4
Factor 10	None	3
Factor 11	Dummy Oxidation Time	2

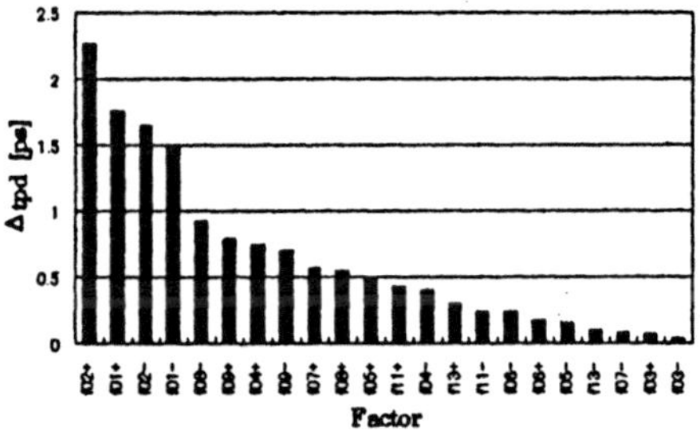

Fig. 5 Sensitivities of factors to delay time *tpd* [5]. $|V_{DS}|$ = 3.3 V, W / L = 5 / 0.3 for n-ch, and W / L = 10 / 0.3 for p-ch.

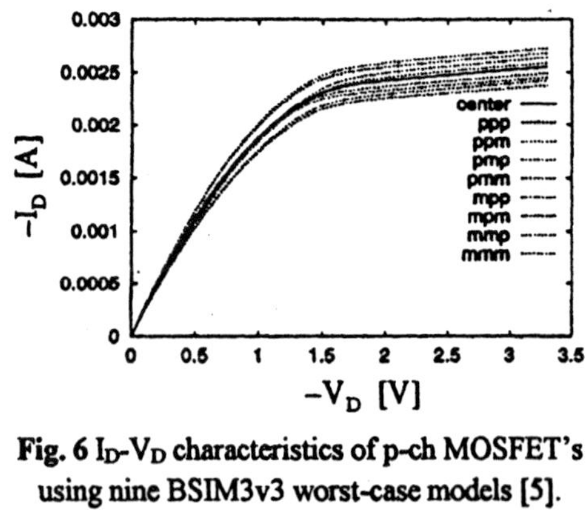

Fig. 6 I_D-V_D characteristics of p-ch MOSFET's using nine BSIM3v3 worst-case models [5]. V_{GS} = −3.3 V, V_{BS} = 0 V and *W*/*L* = 10/ 0.3.

both the source/drain sheet resistance and the overlap capacitance. There was a tradeoff between the two, so that factor 3 had small influence on *tpd*. Factors 1, 2, and 8 were selected for reducing *m*. In general, the factor with high coverage of the BSIM3v3 space does not always correspond to the high sensitivity to the circuit performance.

Fig. 6 shows the p-ch I_D-V_D characteristics of the nominal and 2^3 worst-case models using factors 1, 2, and 8, based on $\pm 3\sigma$ process fluctuations. Factor 1 (CD) had large influence on both I_{dn} and I_{dp}. Factor 2 (p-ch As I/I dose) had largest influence on I_{dp} due to the buried-channel structure of p-ch MOSFET's ($L = 0.3$ μm).

2.2 Tasks for statistical simulation

The following list describes problems for realizing the statistical simulation.

(1) Definition of the worst-case model in the sense of circuit performance [8].
(2) Estimation of the standard deviation σ of the process fluctuation.
(3) TCAD predictability.
(4) Numerical noise such as the meshing noise [9] .

The problem (4) is treated in this article. The meshing noise is very crucial problem for the sensitivity and statistical analysis.

Fig. 7 shows the threshold voltage V_{th} as a function of the channel length L [9]. There were discontinuities in V_{th}. This is so-called meshing noise. The meshing noise is due to the change in the potential barrier between the source and channel, which is induced by the abrupt change in mesh points along the lateral direction. The default mesh was automatically generated. Hence, the abrupt change in the number of mesh points at some L caused these V_{th} discontinuities. There were only five mesh points in the channel (solid line). When fixing the number of mesh points in the channel as 20 (broken line), a real potential barrier was obtained. The difference in a maximum potential barrier between default and fixed meshes was 18 mV, which corresponded to the discontinuity in V_{th}. This is a typical example of the meshing noise.

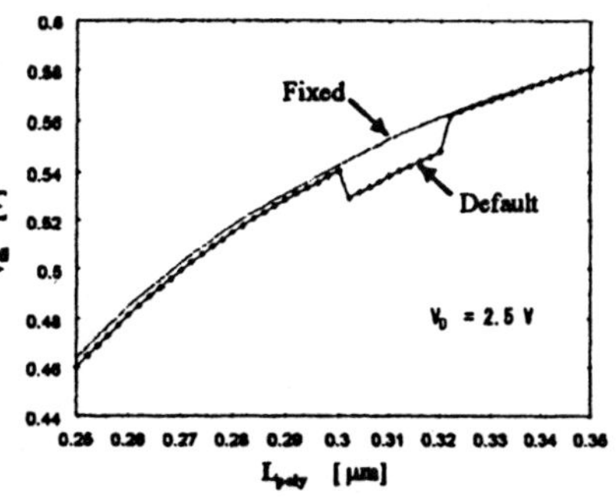

Fig. 7 shows the threshold voltage V_{th} as a function of the channel length L [9].

The mesh generation should be based on physics considering the relation between the mesh size and the physical quantities in order to reduce the meshing noise. Here, the mesh dependency of the inversion carrier density N_{inv} [10] and the surface roughness mobility μ_{surf} [11] are discussed, which are very important for MOSFET simulation.

The mesh dependency of N_{inv} was investigated by Tanimoto and Shigyo [10]. Fig. 8 shows the N_{inv}-V_G characteristics for different grid size, where one-dimension simulation was used with constant grid size. Here, Delaunay and Voronoi meshes were used for the comparison. Device simulators have usually used Delaunay mesh, where the grid locates at the surface of the silicon substrate. On the other hand, the boundary of the control volume (CV) is locates at the surface for Voronoi mesh. As show in the Fig. 8, Delaunay mesh overestimated N_{inv}. Voronoi mesh underestimated it. These can be explained using Fig. 9. Delaunay mesh overestimated the number of carriers in the control volume at the surface, hence N_{inv}.

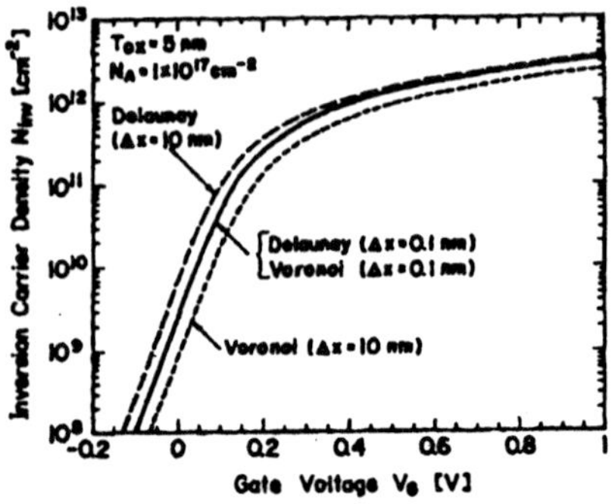

Fig. 8 Inversion carrier density N_{inv} as a function of gate voltage V_G [10].

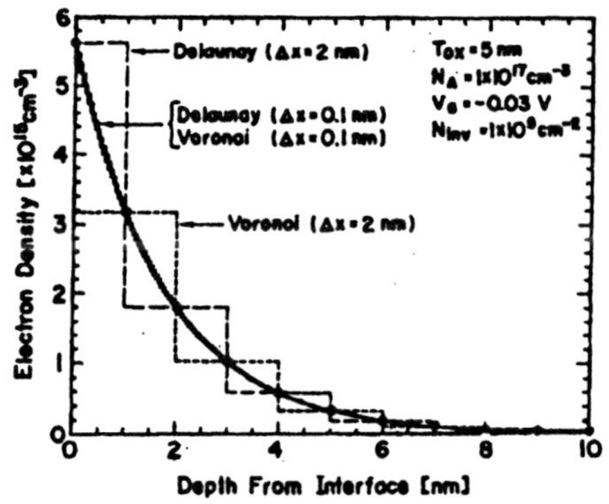

Fig. 9 Electron density as a function of depth [10]. Delaunay grid overestimated N_{inv}.

Fig. 10 shows minimum grid size to ensure the accuracy of N_{inv} for Delaunay mesh. Less than 1 nm grid is required to ensure 10% accuracy of N_{inv} in the subthreshold region for the substrate concentration N_A of 2×10^{17} cm^{-3}.

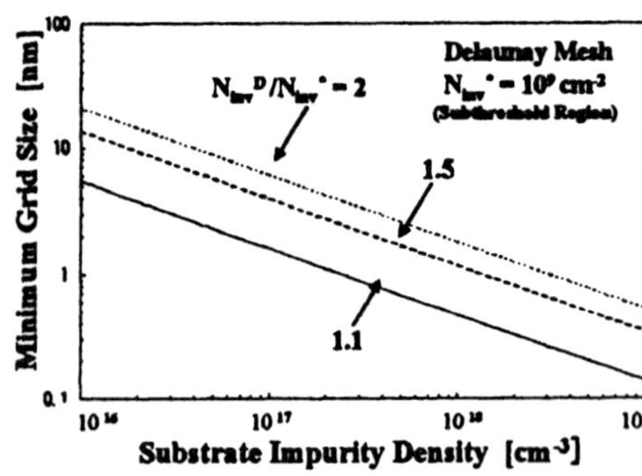

Fig. 10 Minimum grid size to ensure accuracy of N_{inv} for Delaunay mesh.

Fig. 11 shows the channel conductance as a function of grid size. Watt's model [12] considered the universality of μ_{surf} only by one girid at the surface. Thus, this model overestimated μ_{surf} with a decrease in the grid size, since the mobility at the next grid to the surface was considered to be bulk one. Shin et al. [13] proposed the method, which solved this problem. Their method provided a linkage between the *local* and *effective* mobility. Their method is mathematically correct, but it requires a fine mesh.

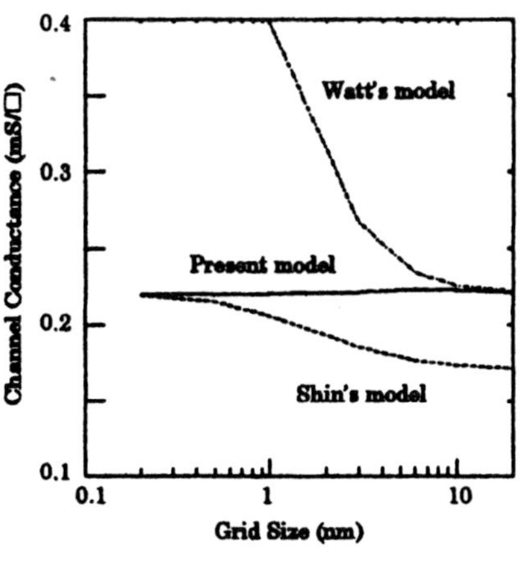

Fig. 11 Channel conductance as a function of grid size.

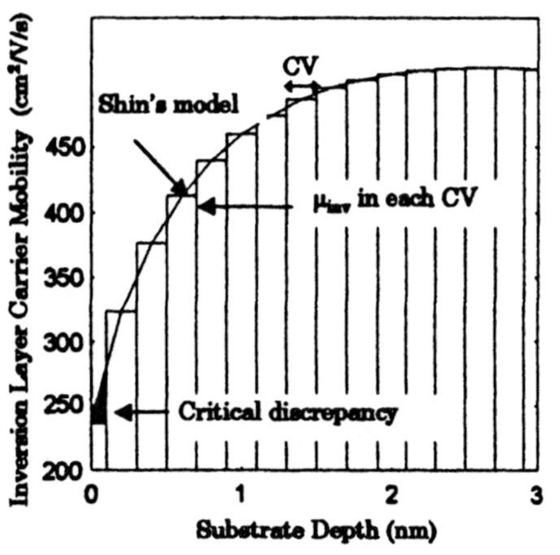

Fig. 12 Inversion carrier mobility as a function of depth [11].

Fig. 12 shows the cause of the grid size dependency. The major problem is the estimation of μ_{surf} at the surface. Enda and Shigyo [11] modified Shin's model with physical consideration. The vertical electric field E_V in the surface control volume was estimated using the values of E_V at the surface and next grids. Using this formulation, the grid size dependence on μ_{surf} could be reduced, as shown in Fig. 11.

3 Interconnect Variability

The importance of the interconnect RC delay is becoming larger with the advance in VLSI. Fig. 13 shows the interconnect structure used for the analysis. Periodic lines [14] were used. In the analysis, the vertical scaling for the interconnect height H and thickness T was not considered for the simplicity. A two-dimensional simulator [14] with 100 x 70 mesh points was used to calculate C.

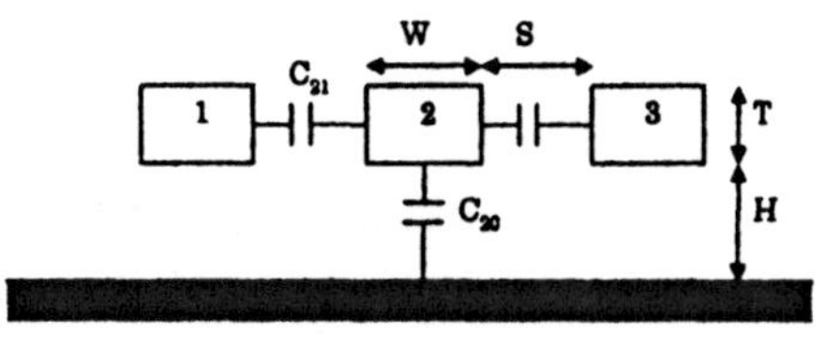

Fig. 13 Interconnect structure. T/H = 0.6.

Fig. 14 shows the relative variation of $\Delta C/C$ for +10% $\Delta W/W$. The space S was equal to $W - \Delta W$. The parallel plate capacitance C_P was calculated as $C_P/(\kappa_{ox}\varepsilon_0 L) \equiv W/H + 2\,T/S$, where κ_{ox} is dielectric constant of oxide, and ε_0 permittivity of free space. It is found that the simulated relative variation $\Delta C/C$ was *smallest* when $W/H \approx 2$, because of the fringing capacitance C_F, which is a parasitic component originated from the fringing

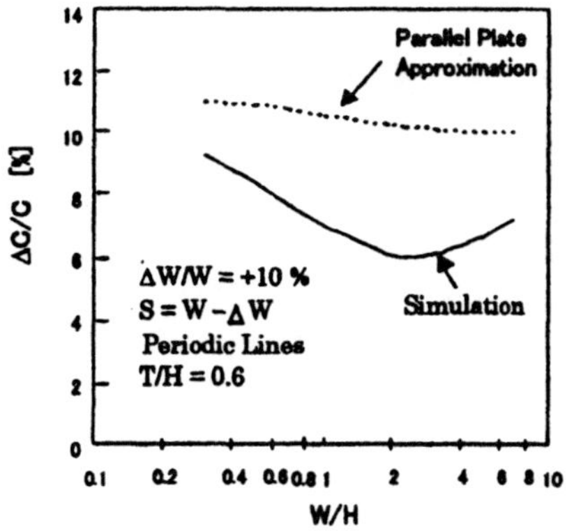

Fig. 14 Capacitance variation $\Delta C/C$ for +10 % CD variation, $\Delta W/W$ [15].

electric field. C_F reduces ΔC.

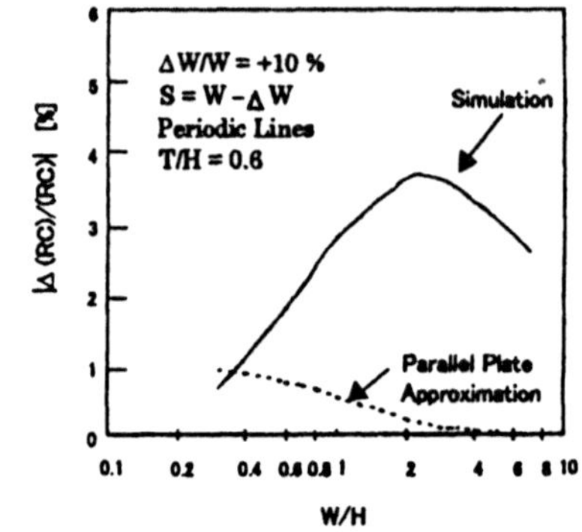

Fig. 15 RC delay variation $\Delta(RC)/(RC)$ for +10 % CD variation, $\Delta W/W$ [15].

Fig. 15 shows the relative variation of $|\Delta(RC)/(RC)|$ for +10% $\Delta W/W$. *Maximum* variation occurred when $W/H \approx 2$, which was opposite to the minimum case for $\Delta C/C$. There is a *tradeoff* between $\Delta C/C$ and $\Delta(RC)/(RC)$. The RC delay variation is decomposed as follows.

$$\frac{\Delta(RC)}{RC} = \frac{(R+\Delta R)(C+\Delta C)-RC}{RC} = \frac{\Delta R}{R}+\frac{\Delta C}{C}+\frac{\Delta R \Delta C}{RC}. \quad (1)$$

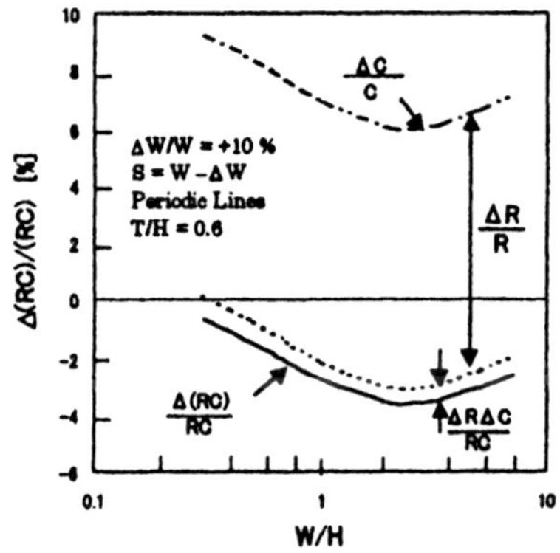

Fig. 16 Each component of (1).

In Fig. 15, $|\Delta(RC)/(RC)|$ was less than 1 % for the parallel plate approximation. This is because the resistance variation $\Delta R/R$ was cancelled by $\Delta C/C$, i.e., there is a *negative correlation* between the two. In the actual case, C_F reduced $\Delta C/C$, so that $\Delta R/R$ was *not* canceled by $\Delta C/C$. Fig. 16 shows each component of (1). The value of $\Delta R/R$ was constant so that the decrease in $\Delta C/C$ due to C_F resulted in the increase in $|\Delta(RC)/(RC)|$.

An interconnect design guideline to reduce C and/or RC delay variations is as follows. In the case of short interconnects, $R_{tr}C$ is crucial for the circuit delay, so that the fringing ratio F ($\equiv C_F/C_P$ where $C_F \equiv C - C_P$) should be *large* to reduce the C variation. On the contrary, F should be *small* for long interconnects to reduce the RC variation.

4 Conclusion

The robust designs of process, device and circuit are crucial for realize VLSI, since process fluctuations influence upon device and circuit performance. The statistical simulation using TCAD is one of methods to solve this problem. An example of the generation of MOSFET worst-case circuit models using TCAD was demonstrated. An analysis of the variability of the interconnect reveled that there is a *tradeoff* between $\Delta C/C$ and $\Delta(RC)/(RC)$ because of C_F. TCAD based statistical analysis become increasingly importance. The meshing noise should be solved with physical considerations.

Acknowledgements

The author expresses sincere thanks to Dr. J. K. Kibarian and the members in PDF Solutions for fruitful discussions. I thank all of my colleagues in Toshiba Corporation who have contributed to this work.

References

[1] Director, S.W., Maly, W. (Ed.) (1994): Statistical approach to VLSI. North-Holland.

[2] Jones, R.E., Mele, T.C. (1991): Use of screening and response surface measuremental designs for development of a 0.5-μm CMOS self-aligned titanium silicide process. IEEE Trans. Semi. Manuf., 4: 281.

[3] Bakoglu, H.B., Meindl, J.D. (1985): Optimum interconnection circuits for VLSI. IEEE Trans. Electron Devices, ED-32: 903.

[4] Walker, D.M.H. et al. (1993): A TCAD framework for development and manufacturing. Technology CAD Systems (Ed. F. Fashing et al.): 83.

[5] Shigyo, N., et al. (2000): Technology CAD based statistical simulation of MOSFET's. Solid-St. Electronics, 44: 1001.

[6] Greenwood, J.A., Sandomire, M.M. (1950): Sample size required for estimating the standard deviation as a per cent of its true value. J. American Stat. Assoc., 45: 257.

[7] pdFab version 3.1 User's Reference Manual (1997): PDF Solutions.

[8] Shigyo, N. (2000): An analysis of process fluctuation induced propagation delay variation using analytical model: Solid-St. Electronics, 44: 2183.

[9] Shigyo, N., Tanimoto, H., Enda, T. (2000): Mesh related problems in device simulation: Treatments of meshing noise and leakage current. Solid-St. Electronics, 44: p. 11.

[10] Tanimoto, H., Shigyo, N. (1992): Discretization error in MOSFET device simulation. IEEE Trans. Computer Aided Design, 11: 921.

[11] Enda T., Shigyo, N. (1997): Grid size independent model of inversion layer carrier mobility. Proc. SISPAD'97: 319.

[12] Watt J.T., Plummer, J.D. (1987): Universal mobility-field curve for electrons and holes in MOS inversion layers. Symp. VLSI Tech. Dig. Papers: 81.

[13] Shin, H. et al. (1989): A new approach to verify and derive a transverse-field-dependent mobility model for electrons in MOS inversion layers. IEEE Trans. Electron Devices, 36: 1117.

[14] Shigyo, N., Fukuda, S., Kato, K. (1989): The influence of boundary locations on wiring capacitance simulation. IEEE Trans. Electron Devices, ED-36: 1171.

[15] Shigyo, N. (2000): Tradeoff between interconnect capacitance and RC delay variations induced by process fluctuations. IEEE Trans. Electron Devices, 47: 1740.

Numerical Modeling of Impact-Ionization Effects on Gate-Lag Phenomena in GaAs MESFETs

A. Wakabayashi, Y. Mitani, D. Kasai and K. Horio

Faculty of Systems Engineering, Shibaura Institute of Technology
307 Fukasaku, Omiya 330-8570, Japan
(horio@sic.shibaura-it.ac.jp)

Abstract

Two-dimensional simulation of turn-on characteristics of GaAs MESFETs is performed in which surface states and impact ionization of carriers are considered. It is shown that the gate-lag (or the slow current transient) becomes weaker when including the impact ionization. This is attributed to the fact that the potential profiles along the surface is drastically changed when the surface states capture holes that are generated by impact ionization. The relation between the gate-lag and the so-called kink phenomenon is also discussed.

1 Introduction

The gate-lag in GaAs MESFETs is a phenomenon that the drain current shows slow transient even if the gate voltage is changed abruptly. This is a serious problem in digital radio systems that are pulsed on and off with short duty cycle. This phenomenon is also known to degrade the distortion characteristics. As for the mechanism of gate-lag, effects of surface states are suggested, but the detailed mechanism is not well understood [1],[2].

To understand high-voltage phenomena in GaAs MESFETs, such as drain-to-source breakdown, is also important for realizing high-performance microwave power devices and ICs. In relation to this, recently, an abnormal increase in output conductance with the drain voltage ("kink") was often observed at relatively low-voltage regions. Since the kink in GaAs MESFETs was firstly correlated to a sidegating effect, the kink was discussed in terms of substrate-related effects [3],[4]. However, the surface-related factors, such as the existence of surface states, should also affect the breakdown and the kink phenomena. Nevertheless, few works have been reported on how the surface properties affect the kink phenomena in GaAs MESFETs [5].

In this work, we have made two-dimensional simulation of GaAs MESFETs, in which surface states and impact ionization of carriers are considered, and studied how the surface states affect the gate-lag and the kink phenomena. Particularly, we discuss how the gate-lag is influenced by impact ionization of carriers.

2 Physical model

A simplified planar structure is mainly analyzed here, although a recessed-gate structure is also treated. The surface states are considered on the planes between source and gate and on the planes between gate and drain. As for a surface-state model, we adopt Spicer's unified defect model [6], and assume that the surface states consist of a pair of a deep donor and a deep acceptor. As to their energy levels, we consider the following case based on experiments, as in previous works [1],[5]: $E_{SD} = 0.87$ eV, $E_{SA} = 0.7$ eV [7]. Here E_{SD} is the energy difference between the bottom of conduction band and the deep donor's energy level, and E_{SA} is the energy difference between the deep acceptor's energy level and the top of valence band. The surface states are assumed to distribute uniformly with 5 Å from the surface and their densities (N_{SD}, N_{SA}) are typically set to $2\text{x}10^{20}$ cm^{-3} (10^{13} cm^{-2}). According to a previous work [1] where impact ionization is not included, the deep-acceptor surface state mainly determines the surface Fermi level, and it acts as a hole trap in this case.

Basic equations to be solved are expressed as follows.

(a) Poisson's equation

$$\nabla^2\psi = -\frac{q}{\varepsilon}(p - n + N_D - N_A + N_{SD}{}^+ - N_{SA}{}^-) \tag{1}$$

(b) Continuity equations for electrons and holes

$$\frac{\partial n}{\partial t} = \frac{1}{q}\nabla \bullet J_n + G - (R_{n,SD} + R_{n,SA}) \tag{2}$$

$$\frac{\partial p}{\partial t} = -\frac{1}{q}\nabla \bullet J_p + G - (R_{p,SD} + R_{p,SA}) \tag{3}$$

$$G = (\alpha_n |J_n| + \alpha_p |J_p|)/q \tag{4}$$

$$R_{n,SD} = C_{n,SD} N_{SD}{}^+ n - e_{n,SD}(N_{SD} - N_{SD}{}^+) \tag{5}$$

$$R_{n,SA} = C_{n,SA}(N_{SA} - N_{SA}{}^-)n - e_{n,SA} N_{SA}{}^- \tag{6}$$

$$R_{p,SD} = C_{p,SD}(N_{SD} - N_{SD}{}^+)p - e_{p,SD} N_{SD}{}^+ \tag{7}$$

$$R_{p,SA} = C_{p,SA} N_{SA}{}^- p - e_{p,SA}(N_{SA} - N_{SA}{}^-) \tag{8}$$

(c) Rate equations for the deep levels

$$\frac{\partial}{\partial t}(N_{SD} - N_{SD}{}^+) = R_{n,SD} - R_{p,SD} \tag{9}$$

$$\frac{\partial}{\partial t} N_{SA}{}^- = R_{n,SA} - R_{p,SA} \tag{10}$$

where $N_{SD}{}^+$ and $N_{SA}{}^-$ represent ionized densities of the deep donors and the deep acceptors, respectively. α_n and α_p are ionization rates for electrons and holes, respectively, and are taken from [8]. C_n and C_p are the electron and hole capture coefficients of the deep levels, respectively, e_n and e_p are the electron and hole emission rates of the deep levels, respectively, and the subscript (SD, SA) represents the corresponding deep level [1]. The other symbols have their normal meanings. These equations are put into discrete forms and are solved numerically.

3 Calculated results and discussions

Fig.1 shows a comparison of calculated turn-on characteristics of a GaAs MESFET with and without impact ionization. Even if the gate voltage is switched on, the drain current remains a low value for some periods and begins to increase slowly, showing large gate-lag. The gate-lag is smaller for the case with impact ionization. This can be understood from the potential profiles shown in Fig.2. As seen in Fig.2(a), without impact ionization, the drain voltage is entirely applied along the interface between drain electrode and surface-state layer in this hole-trap case [1], and the ionized deep-acceptor density N_{SA}^{-} increases in this region because of hole depletion. On the other hand, in the case with impact ionization, generated holes at the drain edge are captured by the deep acceptors, resulting in the decrease in N_{SA}^{-} there, and hence the potentials become applied along the surface-state layer between gate and drain electrodes (Fig.2(b)). Therefore, in the OFF state, the electron depletion under the surface-state layer is weaker for the case with impact ionization, resulting in the smaller gate-lag. It is experimentally suggested that the gate-lag becomes weakened in the high-voltage region where carrier generation becomes important [9].

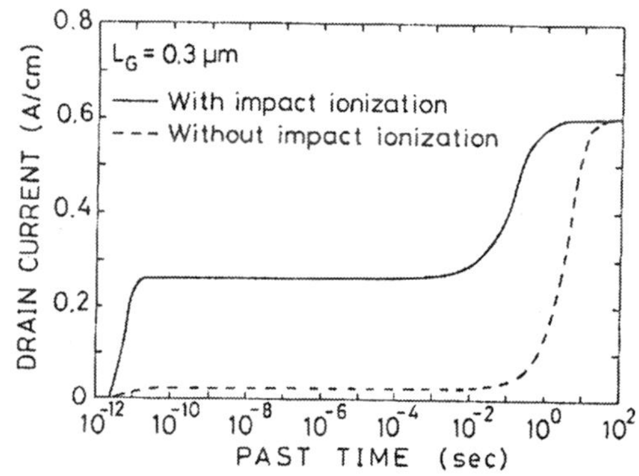

Fig.1. Comparison of calculated turn-on characteristics of GaAs MESFET with and without impact ionization. The surface-state densities are $2x10^{20}$ cm^{-3}.

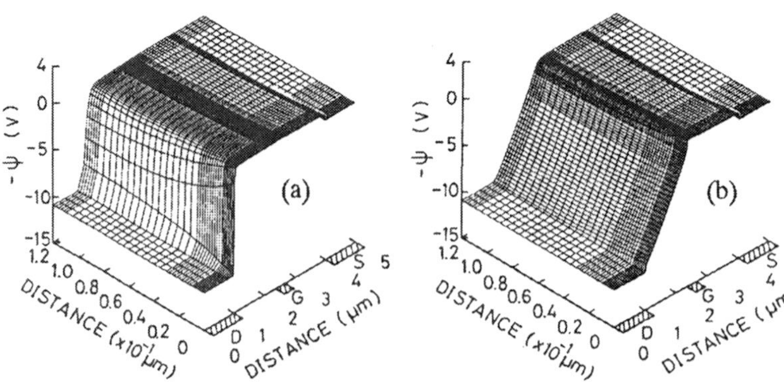

Fig.2. Comparison of potential profiles of GaAs MESFET (a) without and (b) with impact ionization.

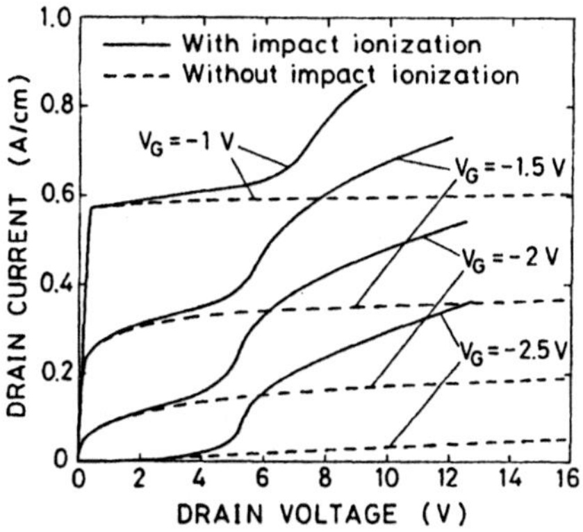

Fig.3. Calculated I_D-V_D characteristics of GaAs MESFET with surface-state densities of $4x10^{19}$ cm^{-3}.

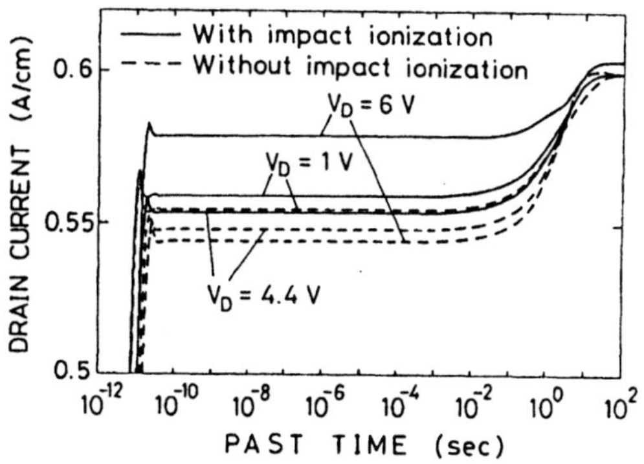

Fig.4. Calculated turn-on characteristics of GaAs MESFET as a parameter of the drain voltage V_D, corresponding to Fig.3.

Next, we describe a case with lower surface-state densities of $4x10^{19}$ cm^{-3} ($2x10^{12}$ cm^{-2}). Fig.3 shows a comparison of calculated drain characteristics of the GaAs MESFET with and without impact ionization. With impact ionization, the drain currents begin to increase steeply around V_D = 5 V, showing the kink behavior. It is understood that the kink occurs because holes are generated by impact ionization and are captured by the surface states, and hence the net negative charges at the surface decrease, leading to widening the channel thickness. Fig.4 shows the calculated turn-on characteristics as a parameter of the drain voltage V_D. Without impact ionization, the gate-lag becomes slightly larger for higher V_D. But, with impact ionization, the gate-lag becomes rather small at V_D = 6 V where the impact ionization becomes important. This is also understood from the fact that the potential profiles along the surface are influenced by impact ionization of holes and the following hole trapping by the surface states.

4 Conclusion

Effects of impact ionization of carriers (together with surface states) on the gate-lag and the kink phenomena in GaAs MESFETs are studied by two-dimensional simulation. It has been shown that the gate-lag becomes smaller when the impact ionization becomes important --- which is consistent with experiments, because the potential profiles along the surface are strongly influenced by the surface states' hole capturing.

References

[1] Horio K, Yamada T. IEEE Trans. Electron Devices 1999; 46: 648-655
[2] Horio K, Wakabayashi A, Yamada T. IEEE Trans. Electron Devices 2000; 47: 617-624
[3] Horio K, Satoh K. IEEE Trans. Electron Devices 1994; 41: 2256-2261
[4] Wilson MR et al. Proceedings of GaAs IC Symposium. IEEE, 1995, pp.109-112
[5] Horio K, Wakabayashi A. IEEE Trans. Electron Devices 2000; 47: 2270-2276
[6] Spicer WE et al. J. Vac. Sci. Technol. 1979; 16: 1422-1433
[7] Wieder HH. Surface Sci. 1983; 132: 390-405
[8] Bulman GE et al. IEEE Electron Device Lett. 1983; EDL-4: 181-185
[9] Parker AE, Root DE. Proceedings of ISSSE'98. URSI, 1998, pp.444-449

Monte Carlo simulation of multi-band carrier transport in semiconductor materials with complex unit cells

H-E. Nilsson[1], A. Martinez[2], M. Hjelm[1,2], E. Bellotti[3], K. Brennan[4]
[1]Dept. of Information Technology and Media, Mid Sweden University
S-851 70 Sundsvall, Sweden
Hans-Erik.Nilsson@ite.mh.se
[2]Dept. of Electronics, Kungl. Tekniska Högskolan, Electrum 229, S-164 40 Kista, Sweden
[3]Electrical & Computer Engineering Dept., Boston University, 8 Saint Mary's Street, MA 02215-2421, Boston, USA
[4]School of Electrical and Computer Engineering, Georgia Tech
Atlanta, GA 30332-0250, USA

Abstract

In a traditional Monte Carlo (MC) model the carrier preserves its band identity during the free flight between scatterings. However, this assumption may not be valid in semiconductor materials with complex unit cell. A new model is needed where the traditional way to use classical equations during the free flight between scattering is replaced by a fully quantum mechanical model of the Bloch carrier dynamics between scattering events. In this work we present such a model along with simulated results of the hole initiated impact ionization coefficients of 4H-SiC.

1 Introduction

Silicon Carbide is a semiconductor that recently has attracted a lot of attention due to its high breakdown field and high thermal conductivity. These characteristics make it a very promising semiconductor material for high field and high temperature applications. 4H-SiC is a hexagonal SiC polytype with 8 atoms in the unit cell. This level of complexity in the unit cell introduces new challenges in the area of full band Monte Carlo (MC) simulation. An important feature in the 4H-SiC band structure is the large number of bands within a limited energy range (see Fig. 1), which means that interband tunneling is significant at high electric fields. This is especially true for the valence band where the multi-band picture is very pronounced even at low energies. In order to model the hole transport in 4H-SiC a fully quantum mechanical transport model is needed. Krieger and Iafrate have proposed a theory for the time evolution of Bloch electrons in a homogeneous electric field [1]. This theory has been used to develop a fully quantum mechanical Monte Carlo simulator (QM-MC), where both scattering and drift are treated as quantum mechanical processes. In order to evaluate the importance of this type of model we have simulated the hole initiated impact ionization coefficient of 4H-SiC and compared the results with two independent measurements.

2 New numerical model for the carrier dynamics

In the theoretical framework developed by Krieger and Iafrate [1] the time development of the probability $|\alpha_n(t)|^2$ to be in band n for a Bloch electron moving in a homogenous electric field can be found by solving the following system of linear differential equations

$$\frac{d\alpha_n(t)}{dt} = \frac{1}{i\hbar}\sum_{n'} \alpha_{n'}(t) F X_{nn'}(\bar{k}(t)) \exp\left(\frac{i}{\hbar}\int_0^t \left[E_n(\bar{k}(t')) - E_{n'}(\bar{k}(t'))\right]dt'\right) \qquad (1)$$

F is the force due to the electric field and $E_n(\mathbf{k})$ is the energy dispersion for band n at wave vector **k**. We have dropped the vector notation for $X_{nn'}$ and F, since the motion in k-space is parallel to the force. If $\alpha_{n'}=\delta_{nn'}$ at time $t=0$ we can interpret $|\alpha_{n'}(t)|^2$ as the probability of changing band from band n to band n' after a drift t. Assuming an electric field applied in the c-axis direction (k_z direction) then the quantity $X_{nn'}(\mathbf{k})$ can be defined as

$$X_{nn'} \equiv \frac{1}{i}\int_{cell} U^*_{nk}\frac{\partial}{\partial k_z}U_{n'\bar{k}}\,dr^3 \approx \frac{1}{i|\Delta\bar{k}|}\begin{cases} I_{nn'}(\bar{k},\bar{k}+\Delta\bar{k}) & , \; n \neq n' \\ i\cdot imag(I_{nn'}(\bar{k},\bar{k}+\Delta\bar{k})) & , \; n = n' \end{cases} \qquad (2)$$

U_{nk} is the periodic part of the Bloch wave function and the integration is performed over the unit cell. It is possible to use the overlap integral of the wave functions obtained from the empirical pseudopotential band structure calculation to derive $X_{nn'}(\mathbf{k})$. The numerical approximation used to calculate $X_{nn'}(\mathbf{k})$ from the overlap integral $I_{nn'}(\mathbf{k}, \mathbf{k}')$ is also presented in equation (2).

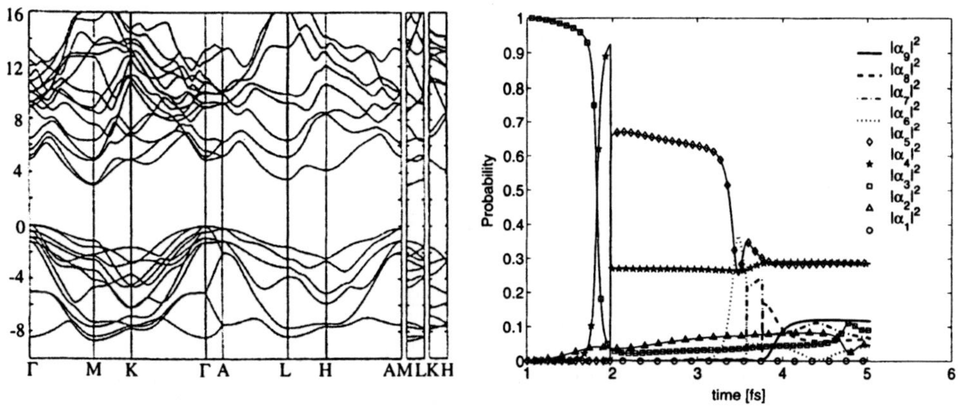

Fig. 1. Left: Empirical pseudopotential band structure in 4H-SiC. Right: Typical time evolution of the probability amplitudes.

The expression (2) in the case of n equal to n' ensures the condition that X_{nn} should be real [2].

There are some practical issues that need to be considered to extract $X_{nn'}$ from the band structure. $X_{nn'}$ should be used for integration, which means that all values need to have the same phase reference. Many eigenvalue solvers provide results with an arbitrary phase factor. The arbitrary phase factor is different at different bands and k-points and has to be removed before $X_{nn'}$ can be extracted using equation (2). The EISPACK eigenvalue solver [3] used in this work provides an arbitrary phase factor that is easy to predict and remove.

Equation (1) can be solved using a Runge-Kutta integration scheme. In Fig. 1, the probability density for a hole initially in band 3 is plotted as a function of time for a drift trajectory parallel to the c-axis. During the drift the probability to be in different bands vary drastically and at the end of the drift the hole can emerge into any one of the nine possible bands albeit with different probabilities.

The primary difference between the current technique and standard MC is that during the course of the drift, the carrier cannot be definitively assigned to a specific band. As such, at different time intervals during the drift, the magnitude of the physical quantities such as scattering time, etc. must be determined based on the probability densities. To this end, we use the simplest formulation given as,

$$A(\bar{k}(t)) = \sum_{n=1}^{N} |\alpha_n(t)|^2 A_n(\bar{k}(t)) \tag{3}$$

to determine $A(\mathbf{k}(t))$, the value of the desired quantity. $A(\mathbf{k}(t))$ is found by summing its value in each band, $A_n(\mathbf{k}(t))$ weighted by the probability density, $|\alpha_n(t)|^2$ over all N possible bands. The time between scatterings is thus determined terminating the free flight drift. The band index is then determined stochastically based on the probability densities and a subsequent scattering event and final state is selected for the carrier from this initial band.

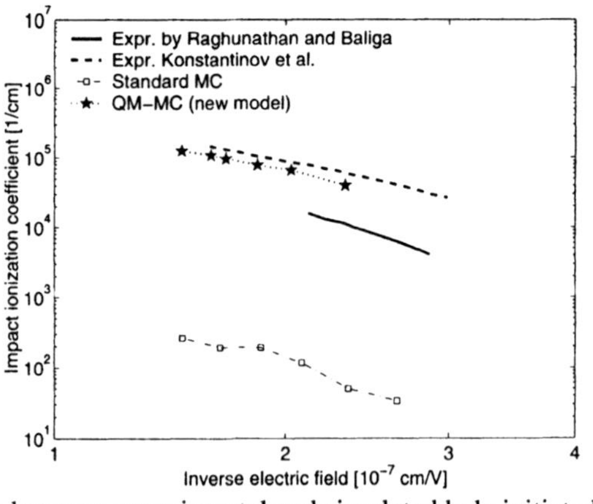

Fig. 2. Comparison between experimental and simulated hole initiated impact ionization coefficients in 4H-SiC

$X_{nn'}$ is a rapidly varying function of the k-vector and a very dense mesh has to be used in the discretization scheme. In this work we are considering transport in a homogenous electric field directed parallel to the c-axis in 4H-SiC. This allows us to reduce the storage requirements by considering only one component of the $X_{nn'}$ vector.

3 Monte Carlo simulation

The full band MC model uses numerically calculated k-dependent scattering rates. The scattering mechanisms are; acoustic phonon scattering, polar and nonpolar optical phonon scattering, ionized impurity scattering and impact ionization. The impact ionization transition rates have been calculated following the approach presented in Ref [4]. There are two independent measurements of the hole initiated impact ionization coefficients in 4H-SiC [5][6]. In Fig. 2 we have compared results from our new MC model with these two measurements. We have also included simulation results obtained using the standard full band MC approach with a classical model for the carrier dynamics. The new model provides results that are similar to the experimental data, while the standard MC model gives results that are almost three orders of magnitude lower. This demonstrates the importance of a fully quantum mechanical dynamic equation in MC simulation of charge transport in 4H-SiC. This conclusion can be extended to other semiconductors with a band structure containing a significant amount of band intersections.

4 Conclusion

A new approach to simulate the carrier transport in semiconductors with complex unit cell and band structure is presented. In the new MC model the traditional way of using classical equations during the free flight between scattering has been replaced by a fully quantum mechanical dynamic equation. The quantum mechanical model allows carriers to tunnel between bands during the drift. In the case of hole transport in 4H-SiC this tunneling is very strong and it enhances the carrier heating drastically. The simulated hole initiated impact ionization coefficients increase by three orders of magnitude when the new model is employed yielding results that are in good agreement with the presently available experimental data.

References

[1] J. B. Krieger, G. J. Iafrate (1986): Time evolution of Bloch electrons in a homogeneous electric field, Phys. Rev. B, 33, pp. 5494-5500

[2] G. J. Iafrate et al. (1998): Bloch electron dynamics in spatially homogeneous electric fields, Int. J. of High Speed Elec. and Sys., 9: pp. 223-250

[3] EISPACK numerical library, http://www.netlib.org/eispack

[4] J. Kolnik et al. (1996): Calculation of the wave-vector-dependent interband impact-ionization transition rate in wurtzite and zinc-blende phases of bulk GaN, J. Appl. Phys., 79: pp. 8838-8840

[5] A. Konstantinov et al. (1997): Ionization rates and critical fields in 4H silicon carbide, Appl. Phys. Lett., 71: pp. 90-92

[6] R. Raghunathan, B. J. Baliga (1999): Temperature dependence of hole impact ionization coefficients in 4H and 6H-SiC, Solid-State Electron., 43: pp. 199-211

Modeling Semiconductor Carbon Nanotube Rectifying Heterojunctions

Gary Pennington and Neil Goldsman
Department of Electrical Engineering, University of Maryland
College Park, MD 20742
garyp@eng.umd.edu, neil@eng.umd.edu

Abstract

We present a method of modeling the effective mass, band offset and current in a heterostructure electronic device constructed from two carbon nanotubes of different diameters.

1 Introduction

A carbon nanotube (NT) may be a single walled (SWNT) or a multi-walled (MWNT) tube, with length reaching up to 1mm (Saito et al. 1998). Each wall of the NT consists of a seamless cylindrical tubule formed from a 2D graphite sheet(graphene). The SWNT unit cell, shown in Fig. 1, is a cylindrical segment defined by two fundamental tube indices, n and m, which determine the tube's diameter and wrapping angle. A SWNT may be a metal or a semiconductor and typically ranges from 1 to 20 nm in diameter (Saito et al. 1998). SWNTs are further seperated into those which are optically active (chiral), and those which are not (achiral). These properties and others such as the bandgap (Saito et al. 1998), and, as found in this work, the effective mass, all are determined by the tube indices n and m. The same properties are exhibited by each wall of a MWNT, but typically larger diameters, up to 100nm, are observed. Since the lower conduction and upper valence bands of a MWNT will be determined by the outer wall, we will model their electrical transport properties considering the outer wall as an isolated SWNT.

Nanotubes of different diameter can be connected seamlessly by the incorporation of a small number of defects, which are usually pentagonal or heptagonal rings (Saito et al. 1998). The versatile electronic and structural properties and small diameter of carbon nanotubes offer great potential and interest to the nanoelectronics community (Martel et al. 1998, Deckker 1999). One structure of particular interest was a rectifying heterojunction that was recently fabricated (Papadopoulous et al. 2000) by the connection of two semiconducting MWNTs of different diameters.

We have developed a theory for explaining electronic properties of a heterojunction formed by nanotubes of different diameters. The model describes how to predict the effective mass, band offset and junction current based solely on the diameters, and thus the fundamental tube indices n and m, of the connected tubes.

2 Effective Mass and Band Structure

To get the dependence of the current on the choice of NT heterojunction, we will need to determine how the effective mass, m^*, of each tube depends on its diameter. For this we will use a π orbital tight-binding method (Saito et al. 2000), to first calculate the band structure. We find that this leads to a single analytical form for the hole and electron effective masses in any semiconducting NT which depends on the tube diameter

$$d(n,m) = \frac{a_{cc}\sqrt{3(n^2+m^2+nm)}}{\pi}. \tag{1}$$

The effective mass, $m^*(d)$, as a function of tube diameter is found to depend on the tube chirality according to:

$$m^*(d) = \begin{cases} \frac{29}{2} m_e \left(\frac{a_{cc}}{d}\right)^{\left[\frac{\pi}{2}+\frac{26\pi}{7}\left(\frac{a_{cc}}{d}\right)-\frac{37\pi}{3}\left(\frac{a_{cc}}{d}\right)^2\right]} & chiral \\ \frac{2}{3} m_e \left(\frac{a_{cc}}{d}\right) & achiral\ , \end{cases} \tag{2}$$

where m_e is the electron mass and $a_{cc} = 1.42\mathring{A}$ is the carbon-carbon bond length.

3 Band offset

We also present the first calculation of the band offset for a nanotube heterojunction using the average bond energy method of Wang et al. (1992). The average bond energy (E_b) can be used as a reference point to calculate the valence band offset between side 1 and 2 of a heterojunction according to

$$\Delta E_v = (E_{b1} - E_{v1}) - (E_{b2} - E_{v2})\,, \tag{3}$$

where E_v are the energies of the respective valence band maximums. We incorporate the wavefunction overlap between nearest neighbors and include the effects of σ-π mixing on the π-orbital band structure to calculate the difference in the average bonding energies. This difference, $E_{b1} - E_{b2}$, is found to be negligibly small. This indicates that very little charge moves across the junction in the formation of the heterojunction but since we are dealing with a very small interface, a significant amount of band bending will still most likely occur. Using the dependence of the bandgap on diameter, we arrive at the band offset which depends on the diameters of the tubes

$$\Delta E_v = E_{v2} - E_{v1} = \gamma a_{cc}\left(\frac{1}{d_1(n_1,m_1)} - \frac{1}{d_2(n_2,m_2)}\right). \tag{4}$$

Here $\gamma = 3.03$eV is the nearest-neighbor π transfer integral within the tight-binding bandstructure calculation.

Based on experimental evidence of large hole concentrations in nanotubes (Martel et al. 1998), the p-p nanotube heterojunction junction is shown in Fig. 2.

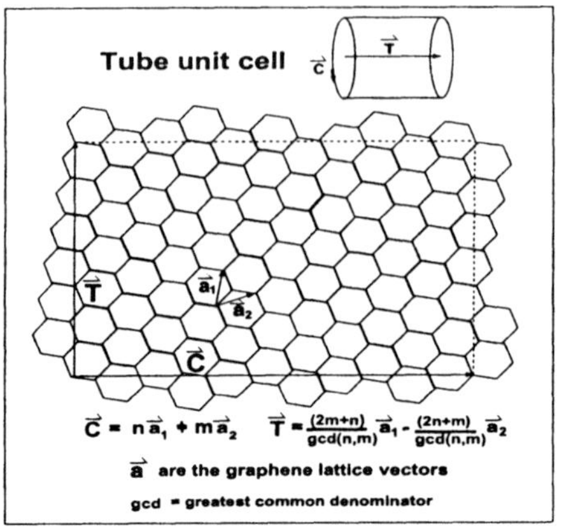

Fig. 1 Unit cell

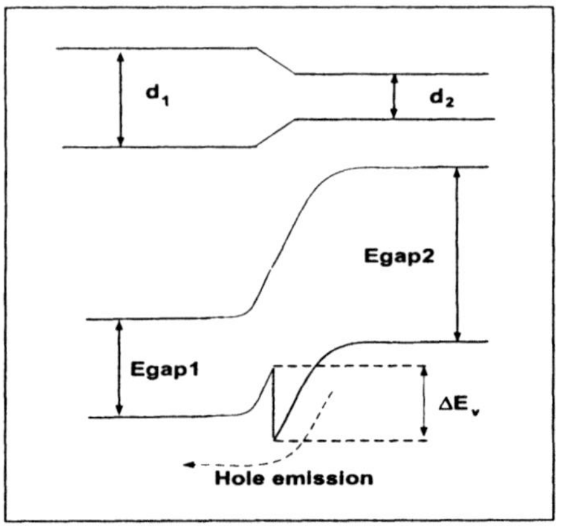

Fig. 2 Tube band-edge diagram

4 Current

To obtain the current across the NT junction with an applied forward bias, we adapt the heterojunction thermionic emission model (Milnes and Feucht 1972) to nanotubes. For the NT with large hole doping, the hole concentration is inversely proportional to the β power of the tube diameter, $N \propto \frac{1}{d^\beta}$,where $\beta \simeq 4$ (Papadopoulous et al. 2000). The dielectric constant will then scale as $\epsilon \propto N$ (Papadopoulous et al. 2000). Using this and equations we derived for effective mass (2) and band offset (4), we derive the following equation for the forward bias current across the junction, as a function of diameters, d_1 and d_2, of the connected tubes:

$$I_{d_1 d_2}(V) \propto I_o(d_1, d_2) \exp\left(\frac{qV + \frac{\gamma a_{cc}(d_1 - d_2)}{d_1 d_2}}{k_b T\left(1 + \left(\frac{d_1}{d_2}\right)^{2\beta}\right)}\right)\left(1 - \exp\left(\frac{-qV}{k_b T}\right)\right), \quad (5)$$

where

$$I_o(d_1, d_2) = \frac{1}{d_2^\beta \sqrt{m^*(d_2)}}\left(\left(\frac{d_2}{d_1}\right)^\beta \sqrt{\frac{m^*(d_2)}{m^*(d_1)}}\right)^{\frac{1}{\left(1+(\frac{d_1}{d_2})^{2\beta}\right)}}. \quad (6)$$

Using the experimental data of Papadopoulous et al. (2000) for a $\frac{d_1}{d_2} \simeq \frac{60}{40}$ NT junction, we find a good fit for β=3.3 as shown in Fig. 3(a). The dependence of the IV curve on the diameter ratio of three selected heterojuncions is illustrated in Fig. 3(b). As expected the current is clearly seen to turn on at lower forward bias voltages for heterojunctions with smaller diameter ratios and thus smaller band offsets.

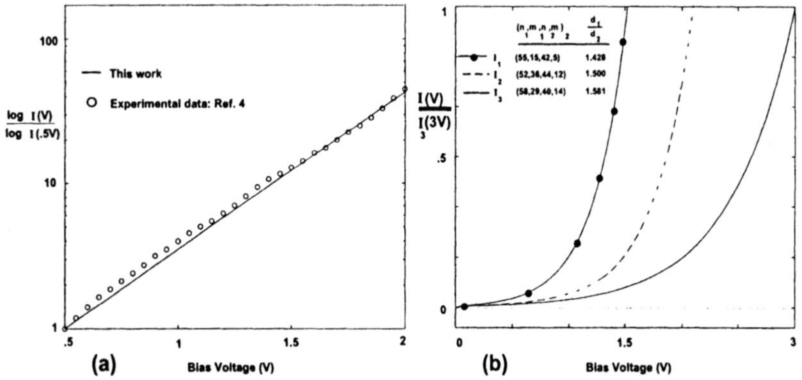

Fig. 3 Forward bias IV characteristics for SWNT junctions including , (a), the comparison of a $\simeq \frac{60}{40}$ diameter junction with experiment and , (b), the comparison of various junctions.

5 Conclusion

In this work we have developed expressions for the effective mass of any semiconducting carbon nanotube, and the band offset and forward bias current of a carbon nanotube heterojunction, in terms of the tube diameters. This allows a convenient prediction of how the electrical characteristics of a semiconductor-semiconductor NT junction vary as the component nanotubes are altered.

References

[1] Dekker C. (1999): Carbon nanotubes as molecular quantum wires. Phys. Today 52: 22-28

[2] Martel R. et al. (1998): Single- and multi-walled carbon nanotube field-effect transistors. Appl. Phy. Lett. 73: 2447-2449

[3] Milnes A. G. and Feucht K. L. (1972): Heterojunctions and metal-semiconducto junctions. Acad. Press, N.Y.

[4] Papadopoulos C. et al. (2000): Electronic transport in y-junction carbon nanotubes. Phy. Rev. Lett. 85: 3476-3479

[5] Saito R. et al. (1998): Physical properties of carbon nanotubes. Imperial College Press, London

[6] Saito R., Dresselhaus G., and Dresselhaus M. S. (2000): Trigonal warping effect in carbon nanotubes. Phy. Rev. B 61: 2981-2990

[7] Wang R., Ke S., and Huang M. (1992): Average bonding-antibonding energy at semiconductor heterojunctions and its applications to calculating the valence band offset. J. Phys. Condens. Matter 4: 8083-8090

Simulation of Vertical CEO-FETs by a Coupled Solution of the Schrödinger Equation with a Hydrodynamic Transport Model

J. Höntschel, R. Stenzel
University of Applied Sciences Dresden
Friedrich-List-Platz 1, D-01069 Dresden, GERMANY

W. Klix
University of Technology Dresden, SFB 358
Mommsenstr. 13, D-01062 Dresden, GERMANY

F. Ertl, T. Asperger, R.A. Deutschmann, M. Bichler, and G. Abstreiter
Walter Schottky Institute, TU Munich
Am Coulombwall, D-85748 Garching, GERMANY

Abstract

This paper describes the simulations of vertical cleaved-edge overgrowth field effect transistors (CEO-FETs). For the simulation the device simulator SIMBA is used, which is capable to handle complex device geometries as well as several physical models represented by certain sets of partial differential equations. With a multidimensional solution of the Poisson equation the Schrödinger equation is solved either in one or in two dimensions according to the confinement of the electrons in the area where quantum mechanical effects are expected. As a new feature the involvement of a hydrodynamic (HD) transport model is implemented to include non-equilibrium transport phenomena in extremely short channels. The experimental results are compared with the simulated data of this device.

1. Introduction

One intention of modern semiconductor technology is the reduction of the device length and width. With the realization of these nanometer structures several quantum mechanical effects appear. The electrons and the holes can not move freely in all of the three spatial directions. The movement of the charge carriers is confined in one, two or three directions, so that 2-, 1-, or 0-dimensional electron gases arise. In heterojunctions field-effect transistors a 2-dimensional electron gas (2DEG) is used as an active transistor channel. The cleaved-edge overgrowth technique [1] offers innovative capabilities for the design of novel quantum size field-effect transistors. The self-consistent solution of the Poisson and Schrödinger equation gives an accurate model to describe quantum mechanical effects in charge carrier gases. The coupling of this microscopic model with a hydrodynamic transport model, in which short-channel and overshoot effects are taken into account, can describe the device characteristics of modern quantum sized field effect transistors. First simulations of vertical CEO-FETs have been carried out to predict the device performance and to optimize the structure.

2. Simulation model

For the simulations a 2D-microscopic/macroscopic algorithm is used as described in [2], which consists of the solution of the effective mass Schrödinger equation

$$-\frac{\hbar^2}{2}\nabla\cdot\left(\frac{1}{m^*}\nabla\psi\right)+(V-E)\psi=0 \tag{1}$$

(m*: effective electron mass, E: discrete energy levels, ψ: wavefunctions, V: potential energy)
together with the Poisson equation

$$\nabla(\varepsilon\cdot\nabla\varphi)=-e\cdot\left(p-n+N_D^+-N_A^-+\rho_{ADD}\right) \tag{2}$$

(N_D^+, N_A^- ionized donor and acceptor density, ρ_{ADD} additionally fixed charge)
to calculate the exact electron density distribution within the two-dimensional electron gas (2DEG) channel. Schrödinger equation, Poisson equation, continuity equations

$$\nabla\cdot\mathbf{J}_p=-q\cdot\left(R-G+\frac{\partial p}{\partial t}\right) \tag{3}$$

$$\nabla\cdot\mathbf{J}_n=q\cdot\left(R-G+\frac{\partial n}{\partial t}\right) \tag{4}$$

transport equations

$$\mathbf{J}_p=-q\cdot\mu_p\cdot p\cdot\nabla(\varphi-\Theta_p)-k_B\cdot T_p\cdot\mu_p\cdot\nabla p-k_B\cdot p\cdot\mu_p\cdot\nabla T_p \tag{5}$$

$$\mathbf{J}_n=-q\cdot\mu_n\cdot n\cdot\nabla(\varphi+\Theta_n)+k_B\cdot T_n\cdot\mu_n\cdot\nabla n+k_B\cdot n\cdot\mu_n\cdot\nabla T_n \tag{6}$$

and energy balance equations (HD transport model),

$$\nabla\cdot\mathbf{S}_p=\mathbf{J}_p\cdot\mathbf{E}^*-\frac{3}{2}k_B\cdot p\cdot\frac{(T_p-T_L)}{\tau_{wp}}-\frac{3}{2}k_B\frac{\partial}{\partial t}(p\cdot T_p)-\frac{3}{2}k_B T_p(R-G) \tag{7}$$

$$\nabla\cdot\mathbf{S}_n=\mathbf{J}_n\cdot\mathbf{E}^*-\frac{3}{2}k_B\cdot n\cdot\frac{(T_n-T_L)}{\tau_{wn}}-\frac{3}{2}k_B\frac{\partial}{\partial t}(n\cdot T_n)-\frac{3}{2}k_B T_n(R-G) \tag{8}$$

both for electrons and holes are solved self-consistently in the Gummel algorithm to simulate the device characteristics at different bias conditions. R and G are the recombination and generation rate, Θ_p and Θ_n are so called band parameter for holes end electron respectively.

3. Results

The device structure used for the simulation is presented in Fig. 1 together with the doping densities. Fig. 2 and Fig. 3 show the calculated output and transfer characteristics, respectively. Additionally the results of a Schrödinger-Poisson solver coupled with a drift-diffusion (DD) model are inserted. The considerable differences show that short-channel and overshoot effects are taken into consideration by the HD transport model. Currently first experimental data are available for CEO-FETs operating at T = 4.2 K. The measured output characteristics (Fig. 4) correspond qualitatively to the calculated data at T = 300 K.

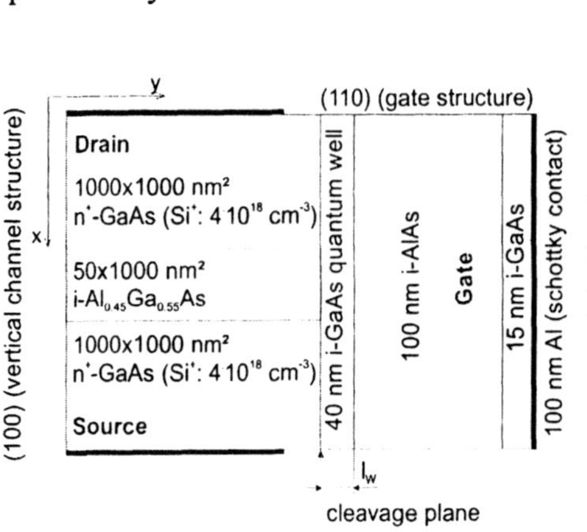

Fig. 1 Structure of the vertical CEO-FET

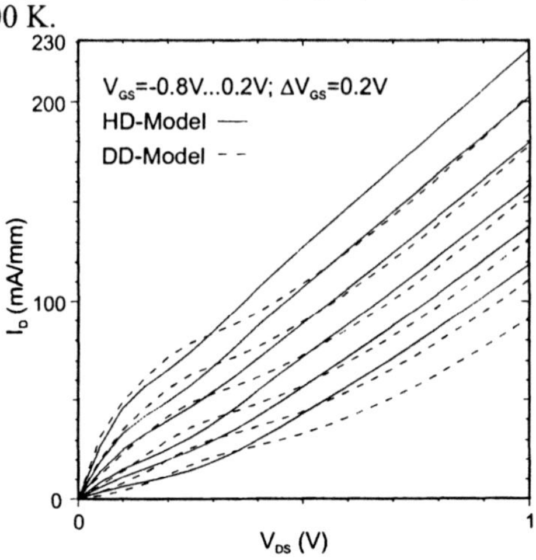

Fig. 2 Output characteristics for different simulation models

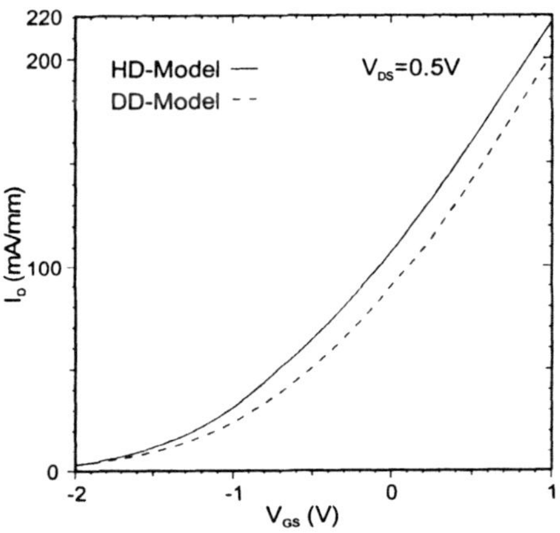

Fig. 3 Transfer characteristics for different simulation models

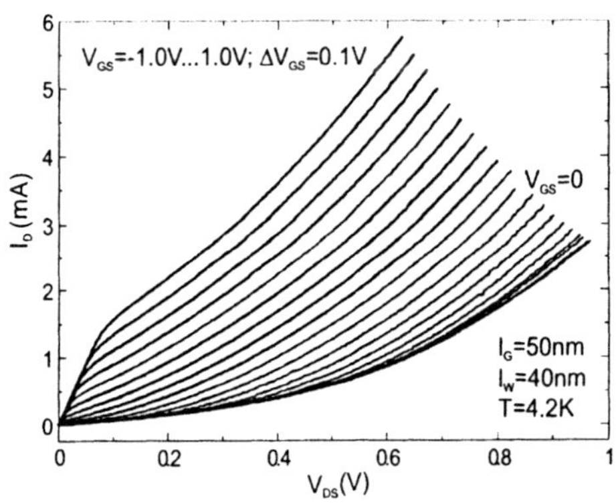

Fig. 4 Experimental CEO-FET output characteristics

The electron temperature in the channel region is depicted in Fig. 5. A strong increase of the temperature can be detected at the drain side near the channel edge. A plot of the conduction band around the quantum well formed in the middle of the gate shows Fig. 6. The discrete energy levels are marked within the quantum well. The transit frequency $f_T = 14$ GHz and the maximum frequency of oscillation $f_{max} = 31$ GHz are obtained at the bias point $V_{DS} = 1$ V, $V_{GS} = 0$. These insufficient values of the non-optimized structure result mostly from the overlap capacities in the gate-source and gate-drain region. The capability of device simulation contains the possibility to investigate scaling effects. The performance, especially the output and transfer characteristics, of CEO-FETs has been evaluated as a function of channel width (l_W) in the range from $l_W = 10$ nm to $l_W = 40$ nm. In Fig. 7 the output and in

Fig. 8 the transfer characteristics are illustrated for different channel widths at $V_{GS} = 0$.

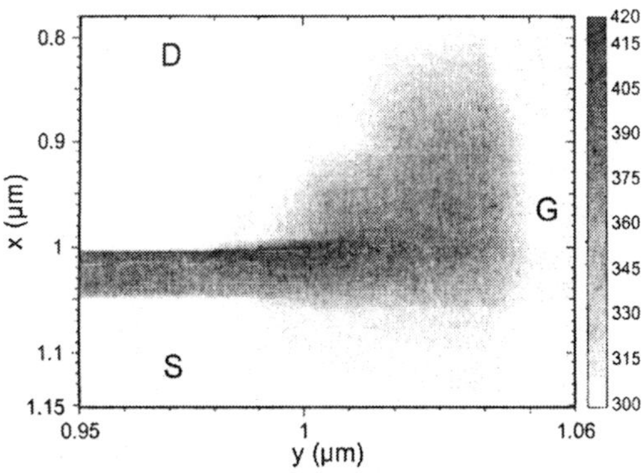

Fig. 5 Electron temperature T_e (K) in the channel region ($V_{GS} = 0$; $V_{DS} = 1.0$ V)

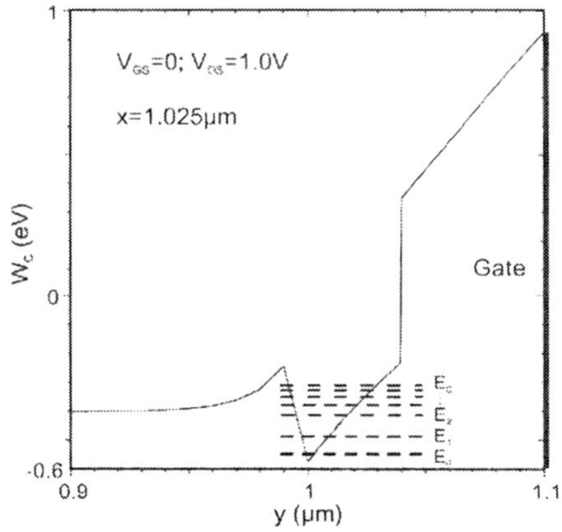

Fig. 6 Conduction band and discrete energies

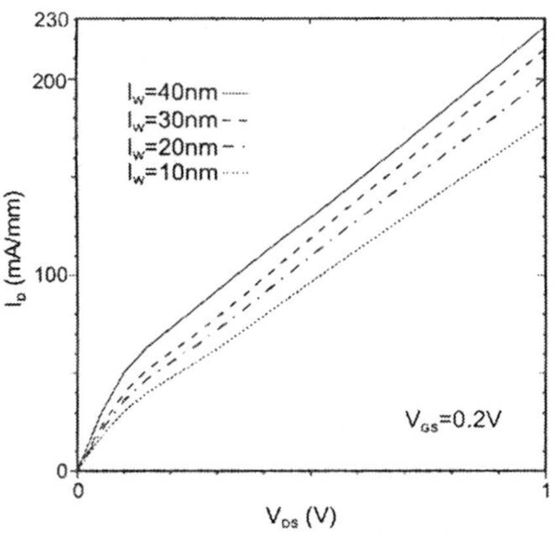

Fig. 7 Output characteristic for different gate widths

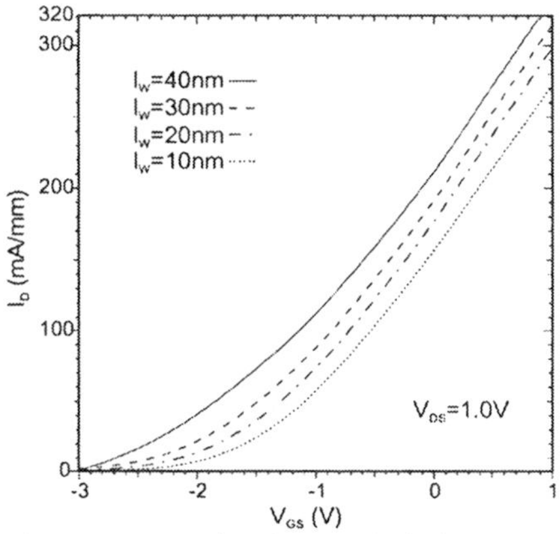

Fig. 8 Transfer characteristic for different gate widths

4. Conclusions

Numerical 2D-simulations of vertical CEO-FETs have been carried out. For the calculations, a coupled solution of Schrödinger equation with a hydrodynamic transport model is deployed. The non-optimized structure shows scaling effects, like short-channel and overshoot behaviour in the device characteristics. The variations of the channel width shows a increase the drain current with the channel width.

References

[1] Pfeiffer L.N., West K.W., Stormer H.L., Eisenstein J.P., Baldwin K.W., Gershoni G., Spector J. : Appl. Phys. Lett. 56, 1990, pp. 1697-1699.

[2] Pigorsch C., Klix W., Stenzel R.: Microelectronic Engineering 43-44, 1998, pp. 325-333.

A Computational Efficient Method for HBT Intermodulation Distortions and Two-Tone Characteristics Simulation

Kuen-Yu Huang, Yiming Li, C. P. Lee, and S. M. Sze
Department of Electronics Engineering, National Chiao Tung University
Hsinchu 300, Taiwan
ymli.ee87g@nctu.edu.tw

Abstract

In this paper, a new simulation method for two-tone characteristics calculations and the third-order intercept point (OIP3) of heterojunction bipolar transistor (HBT) in large-scale time domain is proposed. Base on waveform relaxation (WR) and monotone iterative (MI) methods, we solve a set of nonlinear ordinary differential equation (ODE) of equivalent circuit. With this approach, the two-tone characteristics in frequency domain for HBTs were directly computed from time domain result with fast Fourier transform (FFT). Simulation results on a realistic HBT are presented to show the accuracy and efficiency of the method.

1 Introduction

High power HBTs operated at high frequencies for power amplifiers have been of great interests for wireless applications in recent years [1-3]. One of the most favorite properties of the HBTs is their unusually high linearity [4-5]. The common approach to calculate the intermodulation distortion and two-tone characteristics for an HBT is to solve a set of equivalent circuit differential equations simulation in frequency domain. The harmonic balanced method is a standard approach for such problem. In circuit simulation, conventional approach to solve a set of nonlinear ODEs is the Newton's iterative method. Newton's iterative method is a local method that, in general, converges quadratically in a sufficiently small neighborhood of the exact solution.

In this paper, a novel approach for the calculation of two-tone characteristics and the OIP3 in large-scale time domain is proposed. Based on WR and MI methods, the HBT circuit equations are directly solved in time domain. The computed AC results are then analyzed with FFT to obtain the necessary information. Our new approach was using the MI method instead of the Newton's iterative method to solve the equations in time domain directly. The MI method has been successfully developed and applied to semiconductor device simulations by us earlier [6-8]. In comparison with Newton's iterative method, the major improvements of the MI method for circuit simulation are as follows: it converges globally for HBT, BJT, and MOSFET equivalent circuit models and it is readily for parallel computation. In Sec. 2, we present the model and method of calculation. Followed by results and conclusions.

2 A HBT Circuit Model and Method of Calculation

As shown in Fig. 1, the simulation model is firstly formulated with nodal equations, where the Gummel-Poon large signal Model is used in the simulation [9-10]. For HBT, where the Early effect can be ignored because high doping concentration in the Base of HBT. The circuit for DC simulation and two-tone characteristic are shown in Fig. 1. First of all, the DC circuit is simulated to find the DC solutions. The result shows the accuracy of the proposed method. The similar method is then applied to simulate the circuit with a two-tone input signal. As shown in Fig.1b, V_{IN} is the DC bias and the express of V_{in} is as follows:

$$V_{in} = V1 \times \sin(2\pi \times Freq1 \times t) + V2 \times \sin(2\pi \times Freq2 \times t), \quad (1)$$

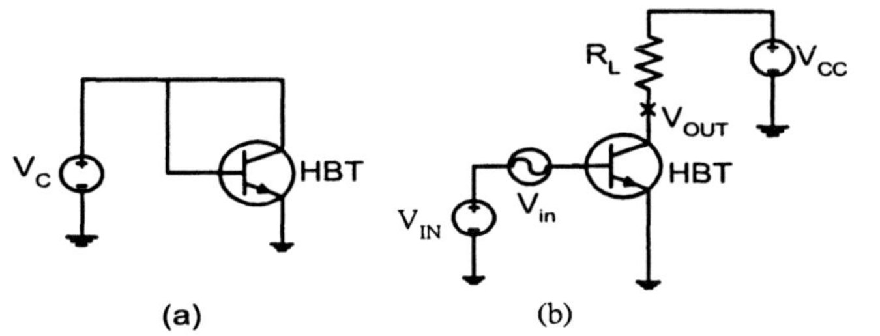

Fig. 1. (a) The circuit for DC simulation. (b) The circuit for two-tone simulation.

where the frequencies Freq1 and Freq2 are 1.05 and 0.95 GHz, respectively. For the non-linearity of equivalent circuit, we calculate the two-tone output signal in time domain with inter-modulation. More than twenty cycles are computed with the simulator and used in FFT to obtain frequency domain analysis.

In this work, As shown in Fig. 1b, there are six ODEs have to be solved. To solve these nonlinear ODEs, our approach is: (i) Use WR method to decoupled the equations. (ii) Each decoupled ODE is then calculated with MI method. The convergence criterion for outer and MI loops are 10^{-12} and 10^{-8} in maximum error, respectively.

3 Results and Discussions

As shown in Fig. 2, a Gummel plot of the simulated circuit as shown in Fig. 1a is presented to show the accuracy of the proposed method. The solid lines are our results with the new simulation approach and the dotted lines are results from HSPICE. Fig. 3 demonstrates the time domain simulation of the output voltage (V_{OUT}, as shown in Fig. 1b) with a two-tone input signal. The simulation was done directly in time domain scale that relies on the robustness of the developed circuit simulator in this work.

As shown in Fig. 4, the spectrum of V_{OUT} with an input voltage, V_{in}, is plotted, where the amplitude of the input is 0.01V. This result was calculated from FFT with computed time domain results. Output Power of the fundamental frequency and the third order inter-modulation (IM3) as a function of input power were calculated.

Furthermore, As shown in Fig. 5, the cross point of those two extrapolated lines is the third-order intercept point (OIP3). Higher OIP3 value presents better linearity of the device. Finally, Fig. 6 shows the OIP3 of the HBT device as a function of the collector current. The linearity of the same device varies with different biases.

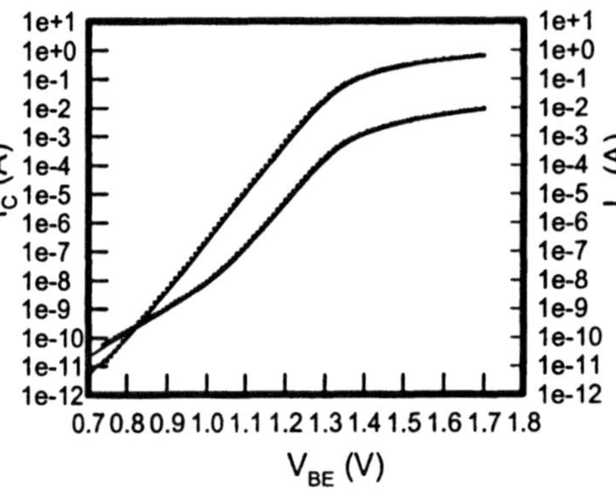

Fig. 2. Gummel plot of the HBT device.

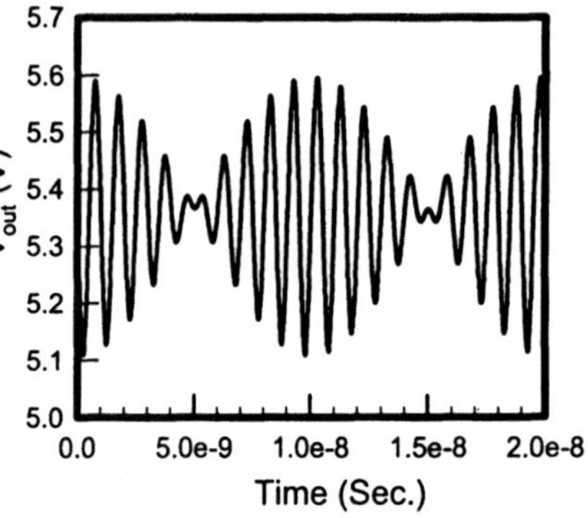

Fig. 3. Time domain simulation result of V_{OUT}.

4 Conclusions

A novel circuit simulation that bases on the WR and MI methods is proposed. With this approach, the intermodulation distortion and two-tone characteristics for a HBT device is directly computed from time domain results with FFT. Simulation results on a realistic HBT are presented to show the accuracy and efficiency of the method. This method provides a new alternative in large-scaled circuit simulation. Based on the constructive monotone iterative method, the developed circuit simulator is not only convergent globally but also inherent parallel.

References

[1] Guofu N., et al. Intermodulation characteristics of UHV/CVD SiGe HBTs. Proc. Bipolar/BiCMOS Circuits and Technology Meeting, 1999; 50-53

[2] Guofu N., et al. SiGe profile design tradeoffs for RF circuit applications. IEDM Tech. Dig., 1999; 573-576

[3] Sheu S., et al. A new approach for SPICE simulation of AlGaAs/GaAs HBT subjected to burn-in test. IEEE Trans. Electron Devices, 1998; 45: 326-329

[4] Sheu S., et al. Current gain long-term instability of AlGaAs/GaAs HBT: physical mechanism and SPICE simulation. Proc. IRPS, 1997; 248-252

[5] Chang M. F., et al. High efficiency and high linearity microwave power amplifiers based on ultra-high f/sub max/ selective buried sub-collector (SBSC) HBTs. Proc. IEEE

MILCOM, 1998; 704-707

[6] Li Y., et al. A novel approach for the two-dimensional simulation of submicron MOSFETs using monotone iterative methodProc. IEEE VLSI-TSA, 1999;27-30

[7] Mastorakis N. Recent Advances in Applied and Theoretical Mathematics, World Scientific, 2000

[8] Li Y., et al. A New Parallel Adaptive Finite Volume Method for the Numberical Simulation of Semiconductor Devices. Ad. Prog. Tech. Dig. CCP 2000; 138

[9] Getreu L E. Modeling the Bipolar Transistor, Elsevier, New York, 1984

[10] Liu W. Handbook of III-V Heterojunction Bipolar Transistors, John Wiley & Sons, 1998

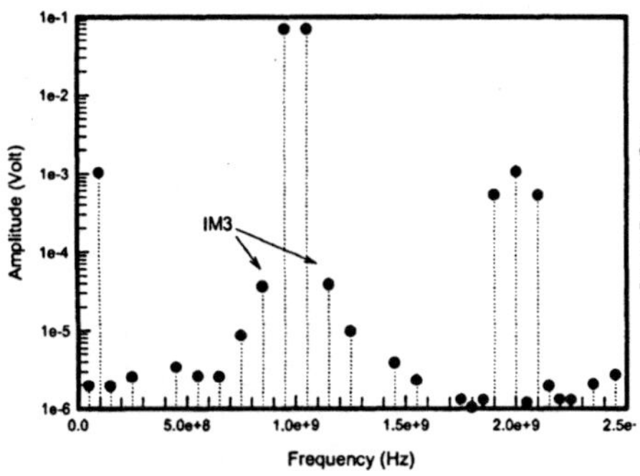

Fig. 4. The spectrum of V_{OUT}.

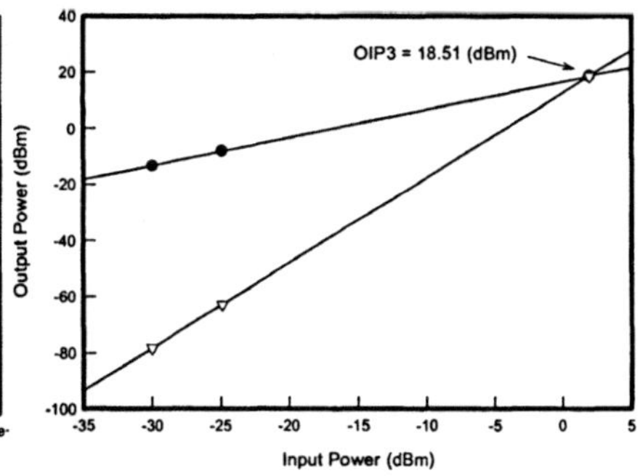

Fig. 5. Plot of output Power vs. input power.

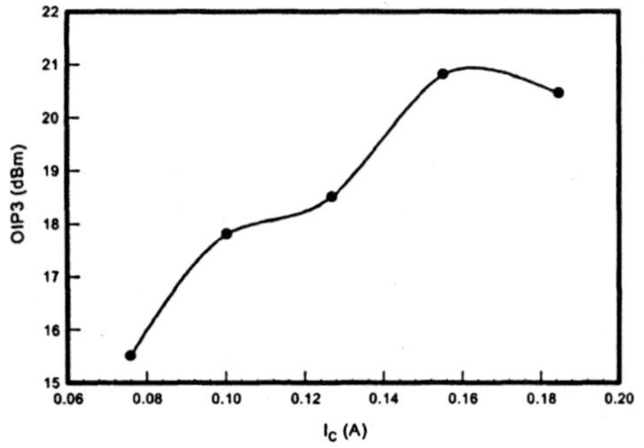

Fig. 6. OIP3 of the HBT device as a function of the I_C.

Modeling the Impact of Body-to-body Leakage in Partially-Depleted SOI CMOS Technology

MeiKei Ieong, Ralph Young, Heemyong Park, Werner Rausch, Isabel Yang, Samuel Fung,*Fariborz Assaderaghi, and *H-S Philip Wong

IBM SRDC, 2070 Route 52, Hopewell Junction, NY-12533
*IBM T.J. Watson Research Center, Yorktown Heights, NY 10598, USA
E-mail: mkieong@us.ibm.com Phone: (845) 892-4719 Fax: (845) 892-3039

Abstract

We study the impact of body-to-body leakage on the performance of partially-depleted silicon-on-insulator (SOI) CMOS technology using TCAD. The body-to-body leakage could be significant because of aggressive technology scaling as shown by process simulations. A fifteen percent degradation in overall delay time is observed by device and mixed-mode circuit simulations. Numerical junction leakage models for accurate body potential modeling are also discussed.

1. Introduction

As SOI CMOS technology becomes mainstream for high performance applications, accurate and efficient modeling is crucial for successful SOI technology development and optimization [1][2]. In addition to junction capacitance reduction, the elimination of body effects in stacked MOSFET (e.g. NAND) circuits can provide a significant performance advantage in partially depleted (PD) SOI technologies. The two MOSFETs in a stacked configuration usually share a common diffusion. The bodies of these MOSFETs are normally isolated by abutting the common diffusion against the buried oxide. However, as technology scales, both the poly-to-poly spacing and the junction depth are substantially reduced for density reduction and control of short-channel-effects, respectively. A leakage path between the bodies of the MOSFETs could be formed if the common diffusion does not abut the buried oxide due to scaling. In this paper, we evaluate the performance impact in such a situation. A complete TCAD flow from process, device, to circuit (mixed-mode) simulations has been established. We also discuss key numerical leakage models for accurate body potential modeling.

2. Junction Leakage Models

Because of the Vdd scaling and strong halo used in scaled MOSFETs, the body potential in a PD-SOI FET at and below threshold can be well characterized by two back-to-back diodes (see figure 1). The body potential at the off condition can be estimated by the continuity between the forward and reverse diode current. Figure 2 displays a typical diode current characteristic at forward and reverse-biases. Note that it is the balance of leakage current (not the magnitude) that determines the body potential. At high reverse bias, the weaker temperature dependence suggests that leakage is dominated by band-to-band tunneling.

Figure 3 shows that the band-to-band tunneling is mainly located near the halo and extension junction. At low bias, the leakage current is dependent on the trap density related to defects. This leakage can be accurately described by a field-enhanced carrier lifetime model in a SRH-like recombination formulation. However, the defect density is highly process dependent. The ion-implantation species and its implant energy and rapid thermal anneal (RTA) conditions can substantially affect the defect related leakage current. The defect related current is distributed along the source/body junction as shown in Figure 3. Thus, reducing the junction area can substantially increase the floating body effect.

Diode characteristics from a body-tied SOI MOSFET at different temperatures are needed to separate the defect-related current and the band-to-band tunneling current. Special attention must be paid to ensure diode current is not obscured by the gate induced drain leakage (GIDL). The formulations for junction leakage are derived from Hurkx [3] and Kane [4] and are summarized in Table 1. The models have been successfully used for silicon channel thickness optimization [5].

Table 1: Diode leakage models

Defect related leakage current
$U = \dfrac{(np-n_i^2)}{\tilde{\tau}_n(n+n_1)+\tilde{\tau}_p(p+p_1)}$ $n_1 = N_C \exp[\frac{E_t-E_C}{kT}], p_1 = N_V \exp[\frac{E_V-E_t}{kT}].$ $\tilde{T}_k = \dfrac{\tau_k(N_D,N_A,T)}{\aleph_k(F,T)},$ $\aleph_k(E,T) = \aleph_k(E,T)_{CL} + \aleph_k(E,T)_{QM}$
Band-to-band tunneling
$G^{BB} = D(\psi, E_{fn}, E_{fp}) \times A \times E^n \times \exp(-B/E)$ $D(\psi, E_{fn}, E_{fp}) = \frac{1}{\exp[(-E_{fp}-q\psi)/kT]+1} - \frac{1}{\exp[(-E_{fn}-q\psi)/kT]+1}$ $A = A_0 \times \left[\frac{m_r}{E_g}\right]^{\frac{1}{2}}, m_r = 0.198 \times m_0 \times \frac{m_T}{m_{300}}$ $B = B_0 \times \left[\frac{m_r}{0.198 \times m_0}\right]^{\frac{1}{2}} \times \left[\frac{E_g}{1.1eV}\right]^{1.5}$

3. Impact of Body-to-body Leakage

When the poly-to-poly spacing is scaled with technology, the common diffusion between two SOI MOSFETs in a stacked structure (e.g. NAND gate) can become shallower due to shadowing effects. A schematic device structure is shown in Figure 4. The body potential can be very different depending on whether the common diffusion junction is abutted to the BOX or not. The body potential could be strongly linked when there is a direct body-to-body leakage path. When the two bodies are strongly linked, the potentials in the two bodies are identical. The potential is self-limited by the forward biased diode of transistor (1). The body potential is pinned at around 0.4V. If the two bodies are isolated very well, the body for transistor (2) should be charged up to 1.8V (Vdd) when the gate in transistor (2) is turned "on". The large difference in body potential of transistor (2) can substantially degrade circuit performance. Detailed process simulations using IBM's T-Suprem were used to obtain the junction profile for the common diffusion. A small reduction of the poly-to-poly spacing resulted in a shallower common diffusion due to shadowing effects. The body-to-body leakage is increased which reduces the body potential of transistor (2). The junction capacitance is also increased due to a larger junction area. To study the body-to-body leakage impact on circuit performance, mixed-mode capability in Fielday3 [6][7] was used to simulate a SOI two-way NAND-gate circuit. Figure 5 displays the simulated input and output wave forms. More than 10-15% degradation in overall delay time is observed. The absence of the body effect in a NAND gate is one of the performance advantages for SOI. To gain this benefit, body-to-body leakage must be minimized by careful device design. Further scaling of the silicon channel could alleviate this body-to-body leakage problem. In summary, accurate numerical leakage models have been developed and calibrated. The models have been used for body-to-body leakage minimization.

References

[1] H. Park *et al., IEDM*, p.337, 1999.

[2] S. Fung *et al., VLSI Tech. Symposium*, p.206, 2000.

[3] G.A.M. Hurkx *et al., IEDM*, p.307-10, 1989.

[4] E.O. Kane, *J. Appl. Phys.*, 32(1):83-91, 1961.

[5] I Yang *et al.*, p.431, *IEDM* 1999.

[6]E.M. Buturla *et al.*, (1981): Finite-element analysis of semiconductor devices: the FIELDAY program. IBM J. Res. Develop., **25**: pp. 218 *ff*

[7] Ieong *et al., SISPAD in Seattle*, p.147, 2000.

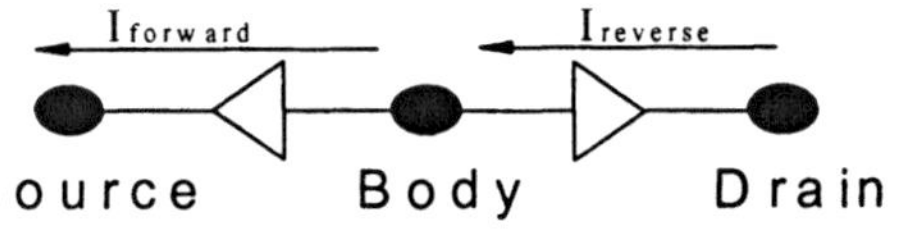

Figure 1: Equivalent circuit model for the floating body potential at and below threshold. The body potential can be determined by balancing the reverse and forward biased diode current.

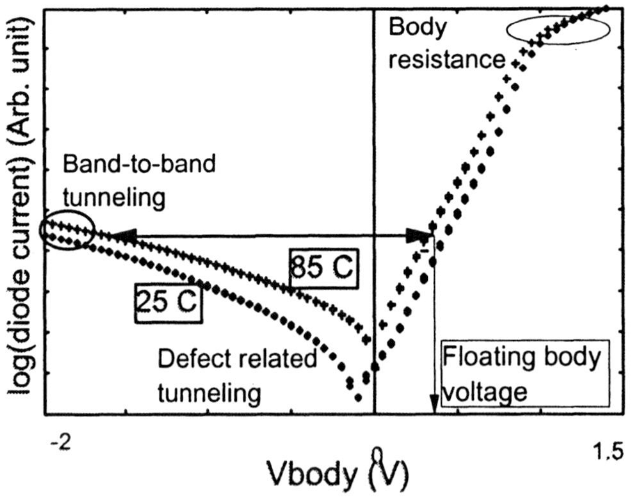

Figure 2: Measured diode leakage current at 25C and 85C. The floating body potential of a scaled PD-SOI MOSFET can be determined by equating the forward and reverse biased diode currents.

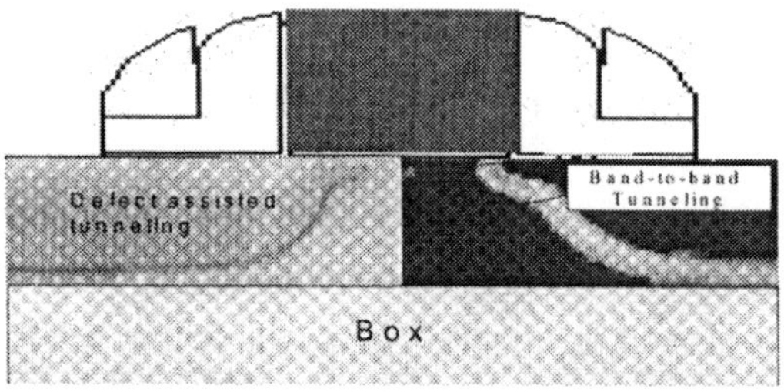

Figure 3: Simulated diode leakage current density at Vds=1.8V, Vs=Vg=0V. The maximum band-to-band tunneling current is located near the drain extension and halo junction. The defect-related current is distributed along the source and body junction.

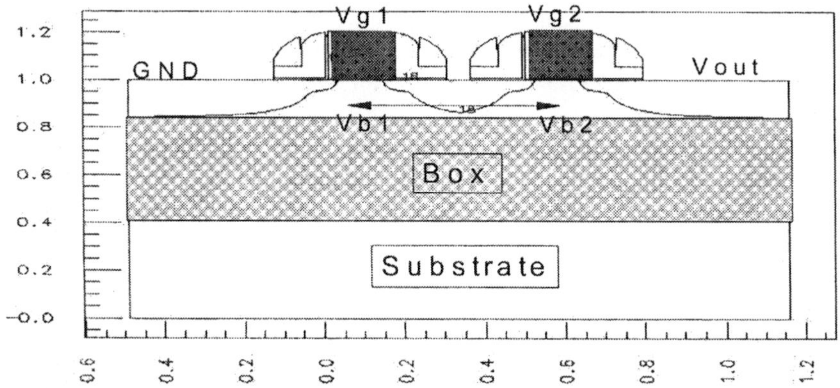

Figure 4: Schematic device structure and terminal voltages definition.

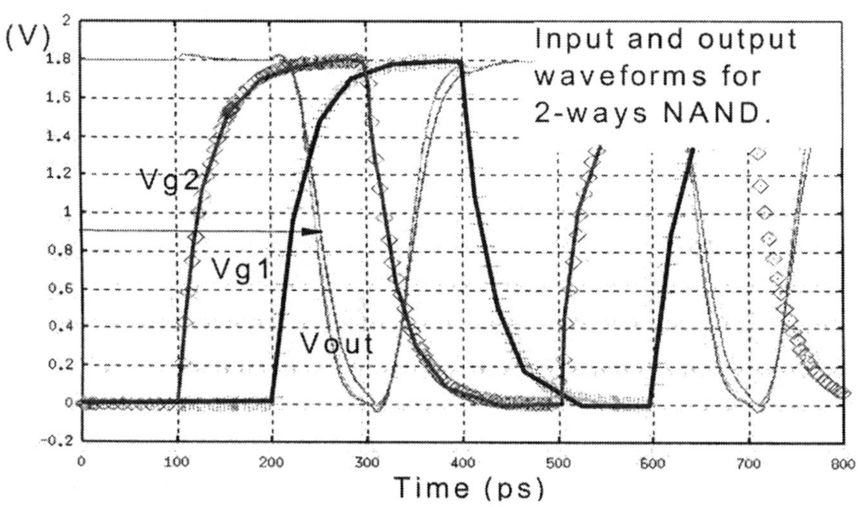

Figure 5: Mixed-mode simulation comparing strongly body-linked and weakly body-linked NAND gate on a 0.25 um SOI CMOS technology. The strongly body-linked NAND gate is about 10-15% slower.

Compact device model for partially depleted SOI-MOSFETs

---For simulation of transient drain current arising from the floating body effects---

Y.Fujii, R.Yoshimura, T.Matsuoka and K.Taniguchi
Department of Electronic and Information Systems, Graduate School of Engineering, Osaka University, Suita, Osaka, 565-0871 Japan
TEL +81-6-6879-7791, FAX +81-6-6879-7792, E-mail: fujii@eie.eng.osaka-u.ac.jp

1 INTRODUCTION

An accurate simulation of integrated circuits consisting of partially depleted (PD) SOI MOSFETs requires a compact MOS model including parasitic phenomena arising from the floating body. Floating body effects include the occurrence of kink in the saturation current and anomalous sub-threshold slope where impact ionization becomes relevant. In addition, the drain current overshoot [1,2] at “switch on” has been reported even in the absence of impact ionization. The drain current overshoot in PD SOI MOSFETs has a significant history dependence: during switching the body potential proportional to the remaining body charge is determined by both carrier generation/recombination in the body and leakage current through the drain/source-to-body junctions.

The aim of this paper is to develop a complete SPICE model including the floating body effects to accurately estimate history dependent gate delay for reducing the risk of inadequate circuit designs.

2 EXPERIMENTAL

Partially depleted NMOS transistors with gate oxide thickness of 5.7 *n*m were fabricated on 150 *n*m Si film on the buried oxide with 400 *n*m thickness. Figure 1 shows a schematic of the apparatus for pulsed time-current measurements for which NMOS transistors with W/L=10/1 (μm) were used. A pulse voltage was applied to the gate of the MOS devices at a given drain voltage of 0.5V. Time dependence of the drain current was derived from the evolution of the voltage across the resistor connected to the drain which was monitored with the oscilloscope. In order to make such pulse measurements on a microsecond time scale possible, we used 1kΩ resistor.

Figure 2(a) shows the time dependence of the measured drain currents after a pulse voltage of 1.0 V was applied to gate electrodes at t=0*sec* for MOSFETs being subjected to different "off" duration. The measurements have been carried out at a drain voltage of 0.5 V. Figure 2(a) demonstrates that the drain current significantly overshoots at "turn-on" of the MOSFETs. This arises from accumulated minority carriers in the body as shown in Fig.3 which give rise to the increase of the bulk potential, resulting in the increase of the drain current due to positive substrate bias effect.

3 SUB-CIRCUIT MODEL FOR FLOATING BODY POTENTIAL

To model the floating body potential which significantly affects the electrical characteristics of PD SOI MOSFETs, a sub-circuit shown in Fig.4 is proposed for the body voltage based on the equations given by

$$\frac{d\Delta Q_b}{dt} = \alpha\left(\sqrt{V_D - \Delta V_b + \phi_{diff}} - \sqrt{\phi_{diff}}\right) - \frac{\Delta Q_b}{\tau}$$

$$- I_s\left[\exp\left(\frac{q\Delta V_b}{nk_B T}\right)\right] + \sum_i C_{ib}\frac{d(V_i - \Delta V_b)}{dt}$$

$$\Delta Q_b = \sum_i C_{ib}\Delta V_b$$

where ΔQ_b and ΔV_b are excess charge and incremental body potential in the floating body, respectively.

The first two terms of RHS represent the generation current at the drain/source-to-body junction and the recombination current in the body. The following two terms are forward biased PN junction current at the body-to-source and the displacement currents due to capacitive coupling to other electrodes, *i*= source, drain, gate and Si-substrate. The emission coefficient *n* is introduced into the general diode equation covering the recombination current for the forward biased junction at the body-to-source. A compact device model implemented into commercial circuit simulators[3] includes the latter two terms. The first two terms are, however, left untouched so we try to include those terms into a compact device model. In the drain voltage range of our interest, those terms are approximated as

$$\alpha\left(\sqrt{V_D - \Delta V_b + \phi_{diff}} - \sqrt{\phi_{diff}}\right) \approx g_D(V_D - \Delta V_b)$$

$$and \quad \frac{\Delta Q_b}{\tau} = \frac{\sum_i C_{ib}}{\tau}\Delta V_b = g_B\Delta V_b.$$

As shown in Fig.5, the nominal conductance g_D and g_B are substituted by the resistance depending on width of the depletion region. Once the body potential, ΔV_b, is found, the drain current overshoots can be calculated dynamically through an accurate threshold voltage model in the forward body-bias regime.

4 COMPARISON WITH EXPERIMENTAL AND SIMULATION RESULTS

Note that the decay time of the drain current overshoot covers in a quite wide range from several μsec up to 100 *m*sec. There are two physical mechanisms involved: (1) the rapid decay associated with forward biased leakage current at the body-to-source junction and (2) the slow decay due to the counter-balance of carrier generation at the body-to-drain junction[4] and carrier recombination in the body. Fig.2(b) demonstrates that the device model presented above well predicts floating-body effects resulting from majority carrier charging in the floating body for switching time of 10 *μ*sec up to 100 *m*sec, over four orders of magnitude. Fig.6 shows the normalized peak current as a function of exponential "off" time duration.

REFERENCES

[1] K.Kato and K.Taniguchi, IEEE Trans., Electron Devices, ED-33, 133(1986)

[2] H.C.Shin et al., IEEE Trans., Electron Devices, ED-43, 318(1996)

[3] P.Su et al., Proc. of IEEE CICC, Orlando, p.197, 2000

[4] A.L.Lacaita and L.M.Perron, Microelectronic Engineering, 48, 319(1999)

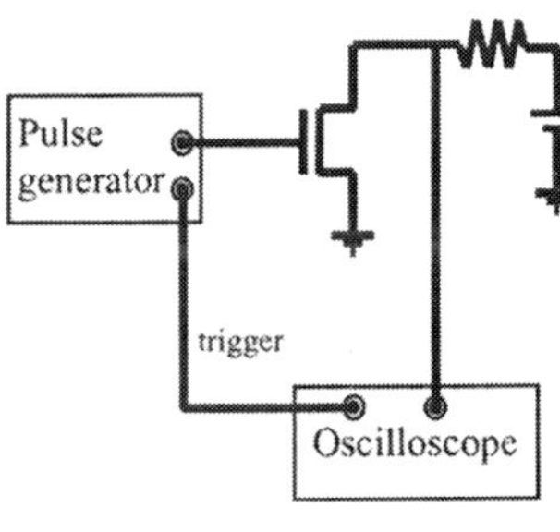

Fig.1 diagram of apparatus

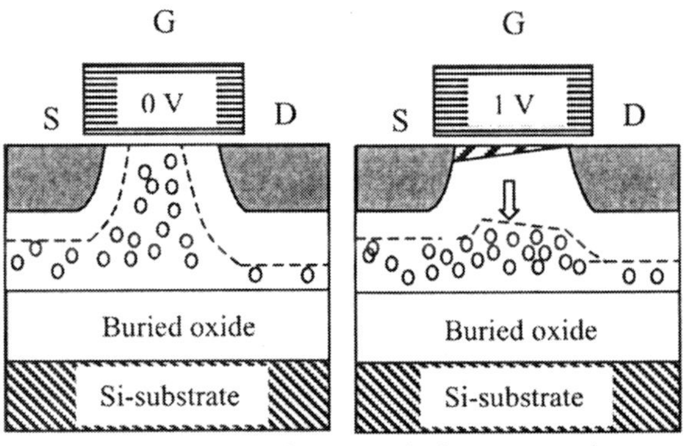

Fig.3. At "switch on", holes near the surface are squeezed into the floating body, leading to positive charged body.

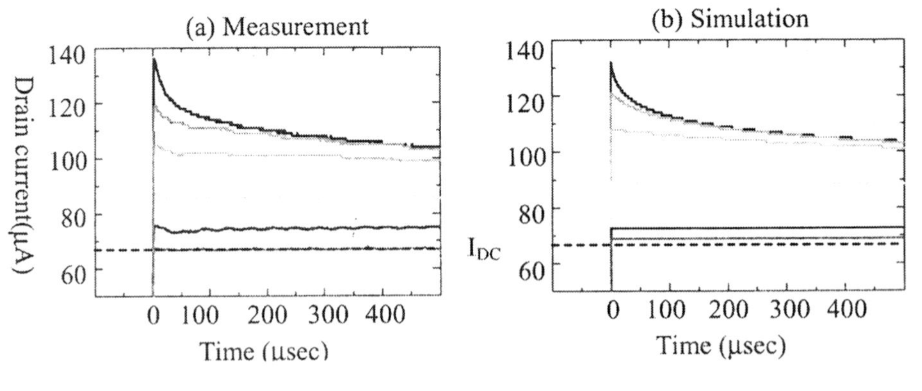

Fig.2. Transient drain current as a parameter "off" duration: t_{off}=10μsec(bottom) to 100msec(top). I_{DC} represents the steady state drain current at DC bias condition.

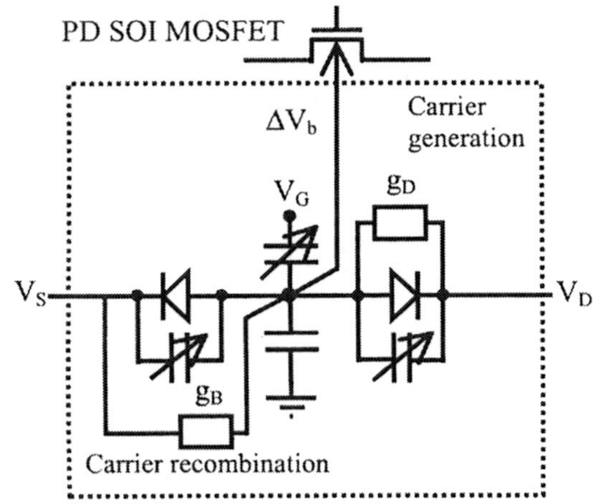

Fig.4. Dashed rectangle: a sub-circuit to track transient behavior of floating body potential, ΔV_d, which controls the threshold voltage of PD SOI MOSFET.

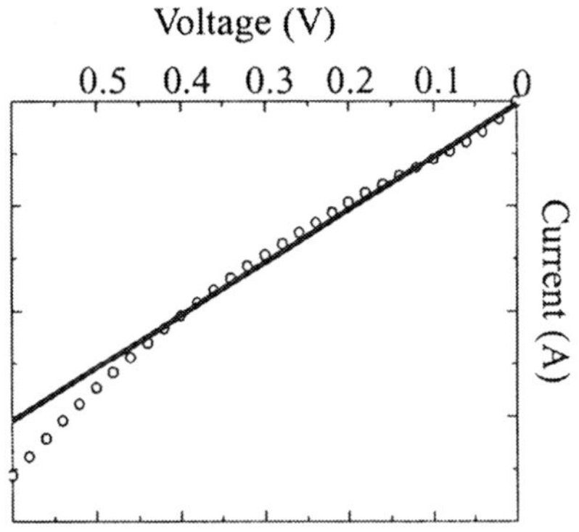

Fig.5. The reverse current characteristic of diode and nominal conductance.

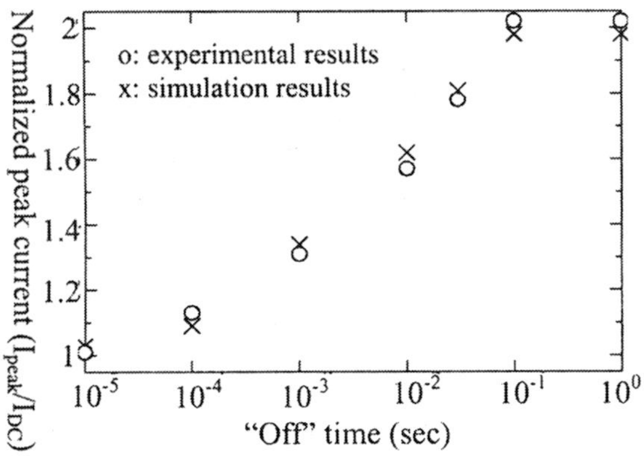

Fig.6. Normalized peak current as a function of "off" duration. Open circles: measured results and Crossed are simulated results

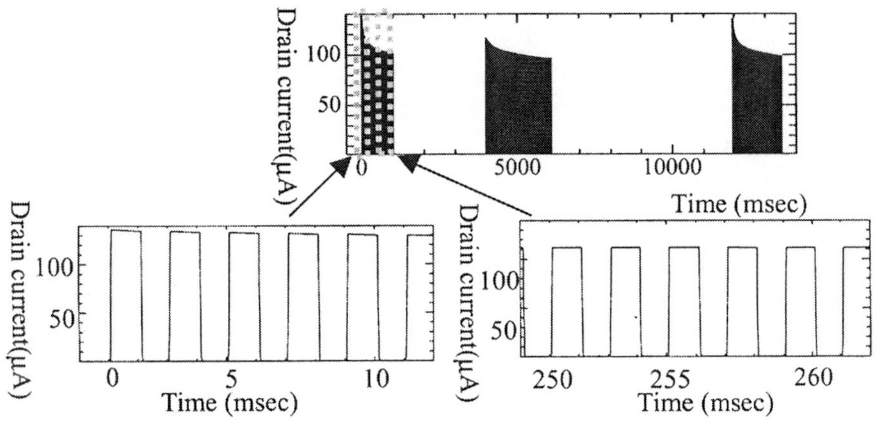

Fig.7. History dependence of drain current under 1msec pulse sequence

Two-Dimensional Model for the Subthreshold Slope in Deep-Submicron Fully-Depleted SOI MOSFET's

Hans van Meer and Kristin De Meyer
STDI division, IMEC, Kapeldreef 75, B-3001 Leuven, Belgium
ESAT, K.U. Leuven, Kasteelpark Arenberg 10, B-3001 Leuven, Belgium
vanmeer@imec.be

Abstract

A 2D analytical model for the calculation of the subthreshold slope has been derived for deep-submicron Fully-Depleted SOI MOSFET's using a Green's function solution technique. The accuracy of the equations has been verified by a 2D numerical device simulator. It is shown that the analytically derived model for the subthreshold slope is in good agreement with 2D numerical simulation data.

1 Introduction

Fully-Depleted (FD) Silicon-On-Insulator (SOI) CMOS has shown to be a promising technology for deep-submicron circuit applications. The analytical modelling of the subthreshold slope of FD SOI MOSFET's has already been reported by several authors [1]-[3]. However, since the coupling effects between front- and back-gate become more complicated for short-channel FD SOI devices, the subsequent analytical models for the subthreshold slope have become more complicated and inaccurate as well. This paper presents a two-dimensional analytical approach which is able to model the subthreshold slope of deep-submicron FD SOI MOS transistors accurately.

2 Analytical model

Since for deep-submicron MOSFET's the short-channel effects can not be neglected, the calculation of the subthreshold slope S should take into account the *drift component* of the *drift-diffusion equation* [4]

$$J_{DS} = \underbrace{q\mu_{eff} V_{th} \frac{\partial n}{\partial x}}_{\text{diffusion component}} - \underbrace{q\mu_{eff}\, n \frac{\partial \psi}{\partial x}}_{\text{drift component}} \tag{1}$$

Normally, only the diffusion component is analysed, which results in accurate predictions for the subthreshold slope of long-channel FD SOI MOS devices only [1]. If the drift-diffusion equation is combined with the continuity equation, $\partial J/\partial x = 0$, a general expression for the free carrier concentration, which relates the electrical potential $\psi(x,y)$ in the 2D device architecture to the carrier density, can be found. If this expression is introduced into the drift-diffusion equation again, an integral equation form for the current density can be obtained, namely

$$J_{DS} = q\mu_{eff} V_{th} N_{SD} \frac{\exp\left(-\frac{V_{DS}}{V_{th}}\right)-1}{\int_0^L \exp\left(-\frac{\psi(x,y)-\psi(0,y)}{V_{th}}\right)dx} \tag{2}$$

which is valid for any bias regime and for any device architecture. From Eq. (2), it follows that the subthreshold slope S, which is defined as $S = \partial V_{GS}/\partial \log(I_{DS})$, can be obtained for deep-submicron FD SOI MOS transistors if an accurate 2D model for the electrical potential $\psi(x,y)$ is available. In our calculations, we used a 2D analytical model for the electrical potential, which is based on an exact solution of the 2D Poisson's equation for FD SOI MOSFET's (Fig. 1).

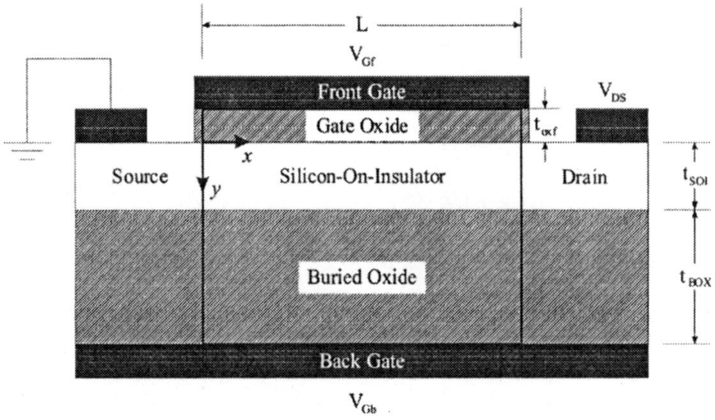

Fig. 1. Schematic diagram of a Fully-Depleted SOI MOSFET. The bold lines indicate the domain over which the Poisson's equation is solved.

The solution has been obtained by using the Green's function solution technique [5]. By combining the results on the Green's functions for the 2D Poisson's equation, shown in Fig. 2a, and Green's theorem, presented in Fig. 2b, a 2D model for the electrical potential in an FD SOI MOSFET can be derived. The final result is shown in Fig. 3.

3 Results and discussion

By combining the obtained integral equation for the carrier density and the 2D analytical model for the electrical potential, the carrier concentration can be calculated as function of the lateral position along the front Si/SiO_2 interface of a FD SOI MOSFET. The results, which are presented in Fig. 4, are obtained 1) by using the 2D device simulator MEDICI [6]; 2) by using a 1D model which only evaluates the *diffusion component* in the *drift-diffusion equation* [1] and 3) by using our proposed 2D analytical model. Figure 4 clearly shows that the carrier concentration

$$G_x(x,y;x',y') = \frac{2}{L}\sum_{m=1}^{\infty} \sin k_m x \,\sin k_m x' \times \frac{\cosh k_m y\,\cosh k_m (t_{SOI} - y')}{k_m \sinh k_m t_{SOI}}$$

$$G_y(x,y;x',y') = \frac{c}{t_{SOI}}\sum_{n=0}^{\infty} \cos k_n y \,\cos k_n y' \times \frac{\sinh k_n x\,\sinh k_n (L - x')}{k_n \sinh k_n L}$$

$c = 1$ for $n = 0$; $c = 2$, for $n > 0$

with $k_m = \frac{m\pi}{L}$; $k_n = \frac{n\pi}{t_{SOI}}$

(a)

$$\psi(x,y) = \iint_V \frac{\rho(x',y')}{\varepsilon} G(x,y;x',y')\,dx'dy' + \oint_S \frac{\partial\psi(x',y')}{\partial n'} G(x,y;x',y')\,dS' - \oint_S \psi(x',y') \frac{\partial G(x,y;x',y')}{\partial n'}\,dS'$$

(b)

Fig. 2. (a) The Green's functions for the silicon film of the Fully-Depleted SOI MOS transistor. (b) Green's theorem, used for the calculation of the 2D analytical solution of the potential.

$$\psi(x,y) = -\frac{qN_{chan}}{2\varepsilon_{Si}} x(L-x) + V_{bi} + \frac{V_{DS}}{L}x + \sum_{m=1}^{\infty} \frac{D_{sf}^m \cosh k_m (t_{SOI} - y) - D_{sb}^m \cosh k_m y}{\varepsilon_{Si} k_m \sinh k_m t_{SOI}} \sin k_m x$$

Fig. 3. 2D analytical solution of the electrical potential in the Fully-Depleted SOI MOS transistor shown in Fig. 1.

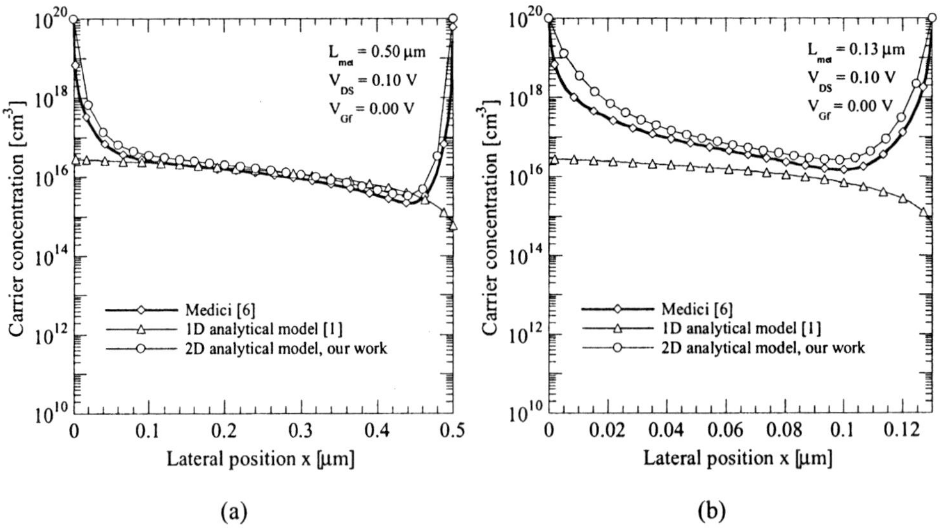

Fig. 4. Carrier concentration as function of lateral position x at the front Si/SiO$_2$ interface for a (a) 0.5 µm and a (b) 0.13 µm long FD SOI transistor. The silicon thickness t_{SOI} = 25 nm; the front gate oxide t_{oxf} = 3 nm; the buried oxide t_{BOX} = 400 nm. The substrate doping $N_{chan} = 5\cdot10^{17}$ cm^{-3}.

should indeed be calculated by taking into account the *drift component*. The approaching S/D areas result in a significant increase of the carrier concentration in short-channel FD SOI MOSFET's, resulting in a large error in the subthreshold slope model of the 1D model [1].
Finally, the results obtained on the subthreshold slope S are summarised in Fig. 5. The subthreshold slope has been calculated for both a 'long'- (0.5 μm) and a 'short'- (0.13 μm) channel SOI device, as well as at 'low' (0.1 V) and 'high' (1.5 V) drain bias. Besides the data acquired from our proposed 2D model, also the data obtained from the 1D model and the 2D device simulator MEDICI are shown in Fig. 5. It can be concluded that the proposed 2D analytical model agrees well with the 2D device simulator MEDICI, while the 1D model shows clearly to be unusable for deep-submicron dimensions.

	$S(V_{DS} = 0.1$ V) [mV/dec]			$S(V_{DS} = 1.5$ V) [mV/dec]		
Gate length	MEDICI [6]	1D model [1]	2D model (our work)	MEDICI [6]	1D model [1]	2D model (our work)
0.50 μm	60.7	60.0	60.9	60.7	60.0	60.9
0.13 μm	67.7	60.0	67.1	72.1	60.0	68.8

Fig. 5. Subthreshold slope for a 'long' (0.5 μm) and a 'short' (0.13 μm) Fully-Depleted SOI MOSFET at different drain biases: comparison between the 1D, 2D analytical models and the 2D device simulator MEDICI.

4 Conclusions

We have presented a 2D analytical model for the calculation of the subthreshold slope for deep-submicron Fully-Depleted SOI MOSFET's using a Green's function solution technique. The accuracy of the equations has been verified by a 2D numerical device simulator. From the results presented, it can be concluded that the analytically derived model for the subthreshold slope is in good agreement with 2D numerical simulation data.

References

[1] Wouters, D.J., Colinge, J.P., Maes, H.E., "Subthreshold slope in thin-film SOI MOSFET's", IEEE Trans. Electron Devices, Vol. 37, No. 9, pp. 2022-2033, 1990.

[2] Balestra, F., Benachir, M., Brini, J., Ghibaudo, G., "Analytical models of subthreshold swing and threshold voltage for thin- and ultra-thin-film SOI MOSFETs", IEEE Trans. Electron Devices, Vol. 37, No. 11, pp. 2303-2311, 1990.

[3] Joachim, H.-O., Yamaguchi, Y., Ishikawa, K., Inoue, Y., Nishimura, T., "Simulation and two-dimensional analytical modeling of subthreshold slope in ultrathin-film SOI MOSFETs down to 0.1 μm gate length", IEEE Trans. on Electron. Devices, Vol. 40, No. 10, pp. 1812-1817, 1993.

[4] Biesemans, S., De Meyer, K., "Analytical calculation of subthreshold slope increase in short-channel MOSFET's by taking drift component into account", Jpn. J. Appl. Phys, Vol. 34, No. 2B, pp. 917-920, 1995.

[5] Guo, J.-Y., Wu, C.-Y., "A new 2-D analytic threshold-voltage model for fully depleted short-channel SOI MOSFET's", IEEE Trans. on Electron. Devices, Vol. 40, No. 9, pp. 1653-1661, 1993.

[6] MEDICI, V4.0, Technology Modeling Associates, Inc., Palo Alto, 1998.

3D Thermal Analysis for SOI and its impact on Circuit Performance

R. V. Joshi, [1]S. S. Kang, and C. T. Chuang
[1]IBM, Rochester, MN - 55901
IBM, T. J. Watson Res. Center, Yorktown Heights, NY - 10598
(rvjoshi@us.ibm.com, 914-945-1118)

Abstract

3D-thermal analysis for multifinger devices is developed for partially depleted SOI and Cu interconnects. Using this technique the temperature rise for multifinger devices is at least a factor of 3 higher than predicted by 1-D thermal resistance models.
Also increase in temperature degrades the performance incrementally.

1. Introduction

As silicon process technology, driven by improved performance requirements, migrates to SOI, low resistivity metals and low-k dielectrics, the thermal properties of these materials play a significant role in determining the temperature build-up, power and performance of the chip. In this work we have evaluated the temperature rise in high duty cycle critical circuits using 3-D thermal analysis for the first time.

2. Model/Simulation

The steady state temperature distribution of a multi-finger NFET with a realistic physical geometry in a clock buffer was studied using a commercial 3-D finite-difference thermal code. The geometry of a four-finger FET on a chip including the body, source and drain MCBAR as a local interconnect, gate polysilicon, five layers of metal, vias and dielectric was modeled in cartesian coordinates.

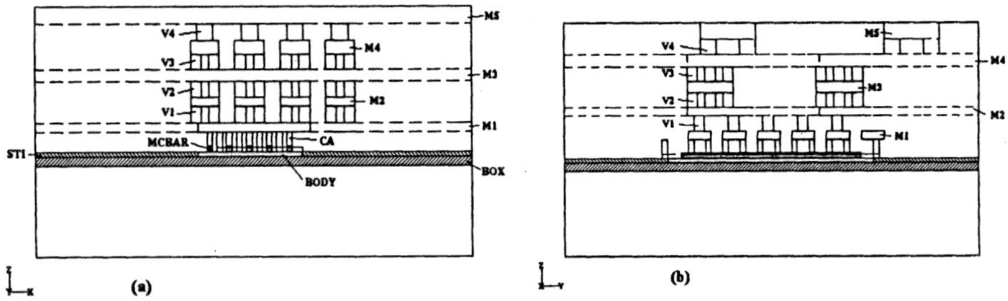

Figure 1 (a) and (b) show the geometry of the 20x20x11 μm region modeled.

The temperature field was determined by numerically solving the three dimensional Poisson equation for temperature. Heat dissipation was specified in the volume occupied by the channel region of the FET and in the five levels of metal lines based on circuit simulations for a clock buffer. Adiabatic (or symmetry) boundary conditions were specified on all the x and y boundary planes. A fixed chip temperature T=1 °C was specified on the z boundary 4 μm below the body of the FET while a fixed thermal conductance ($k=T|_z = C(T-T_o)$) to a surface (e.g. a chip carrier) at T_o =1 °C was specified on the z boundary at the last metal to mimic thermal conduction through a solder bump with a thermal resistance of 700 °C/W. A nine-finger FET was modeled by starting the FET body at the x=0 boundary and adding a half wide channel at that boundary.

---

[1] Presently with Aavid Themalloy LLC, Concord, NH 03301.

The calculations for the bulk technology were done by eliminating the BOX layer in the model. The physical design dimensions and materials considered are summarised in Table 1.

Layer	Thickness (μm)	Width (μm)	Material	Thermal Conductivity (W/m-°C)	Volumetric heating (mW/μm³)
BOX	0.4		SiO_2	1.4	
STI	0.2		SiO_2	1.4	
Body	0.2		Si	120	
Channel	0.1	0.15	Si	120	0.06**
PC	0.2	0.26	Silicide	40	
MCBAR	0.3	0.25	Tungsten	170	
CA	0.6	0.25	Tungsten	170	
M1	0.4	1.1	Cu / Al	390 / 210	0.0136
V1	0.7	0.44	Cu / Al	390 / 210	
M2	0.4	1.35	Cu / Al	390 / 210	0.0218
V2	0.7	0.5	Cu / Al	390 / 210	
M3	0.5	2.24	Cu / Al	390 / 210	0.0268
V3	0.7	0.5	Cu / Al	390 / 210	
M4	0.6	1.35	Cu / Al	390 / 210	0.0444
V4	0.7	0.75	Cu / Al	390 / 210	
M5	0.8	2.6	Cu / Al	390 / 210	0.0154
**heat dissipation per μm of channel width.					

Table 1. Physical design used in the thermal model.

Thermal simulations were carried out for a range of one to nine active fingers, SOI and bulk device, copper metalization, and silicon dioxide back end dielectric. Circuit performance simulations were likewise carried out using SPICE models with p(pfet) *Leff*=0.15 μm and n(nfet) *Leff*=0.12 μm.

3. Results and Discussions

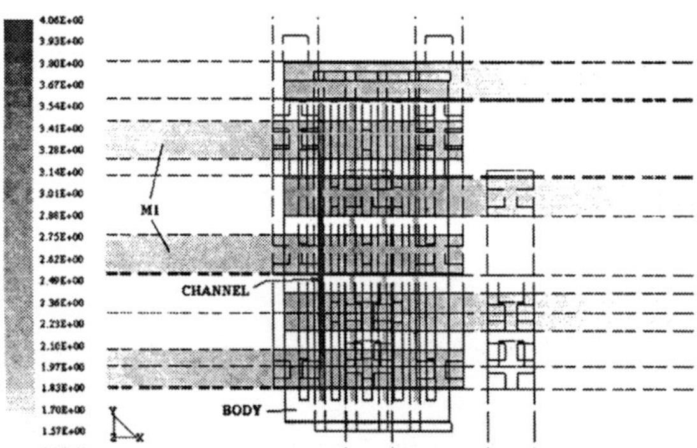

Figure 2. Junction and level 1 metal (M1) peak temperature rise in the SOI device with one active finger.

Only the leftmost channel is active and shows the highest temperature rise. The normalized junction-to-chip temperature rise or thermal resistance is 50 °C-μm/mW (3 °C rise for 0.06 mW/μm heat dissipation). Values reported in the literature range from 50-90 °C-μm/mW for single-finger devices with similar dimensions [1-2].

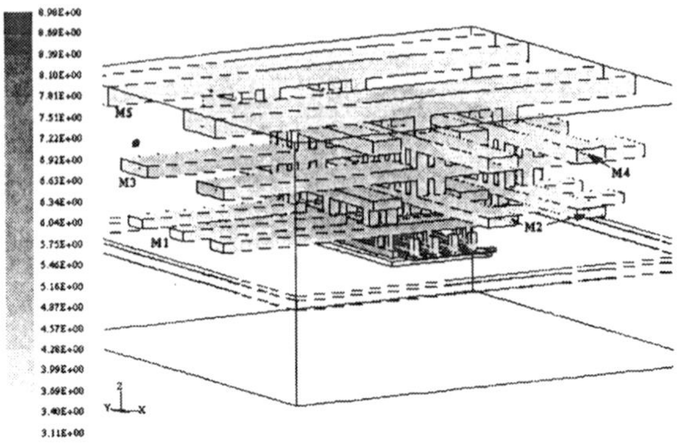

Figure 3. Illustrates the temperature distribution in all the metal interconnect (MCBAR to M5) with four fingers active.

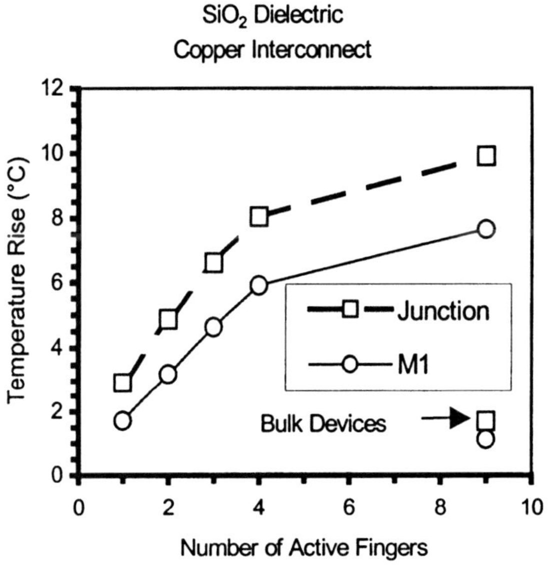

Figure 4. Effect of multiple active fingers on Junction and M1 temperature rise.

With multiple fingers active, the junction temperature rise increases as shown in Figure 4 even though the heat dissipation per unit channel width remains the same (0.06 mW/μm). Because of the close thermal coupling between fingers within the same body, our results clearly show that the junction-to-chip thermal resistance increases by more than a factor of three as more fingers become active. These results show that the conventional use of a single thermal resistance in SPICE type models for a SOI device without differentiating between a single-finger or a multi-finger structure is erroneous, in particular, the use of a thermal resistance value derived or measured from a single-finger isolated device would fail to properly account for the temperature rise of multi-finger devices due to self-heating. Also the SPICE models can not account for both types of geometries simultaneously for thermal calculations. By comparison, the temperature rise in a similar bulk device is small (Fig. 4) and the circuit simulation errors are negligible.

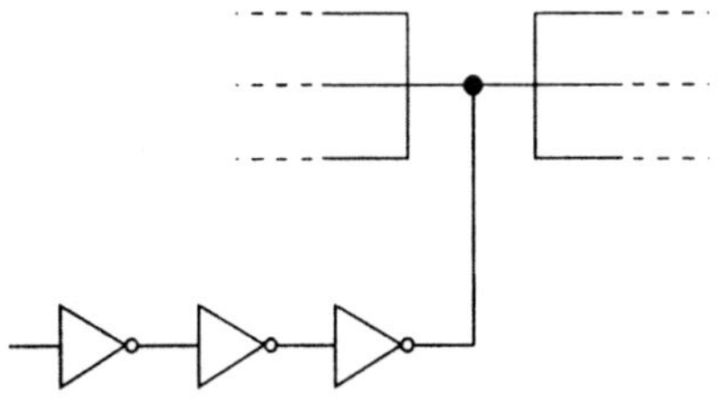

Figure 5. Clock tree used for performance analysis.

Circuit performance was evaluated using the clock tree from a real microprocessor shown in Figure 5.

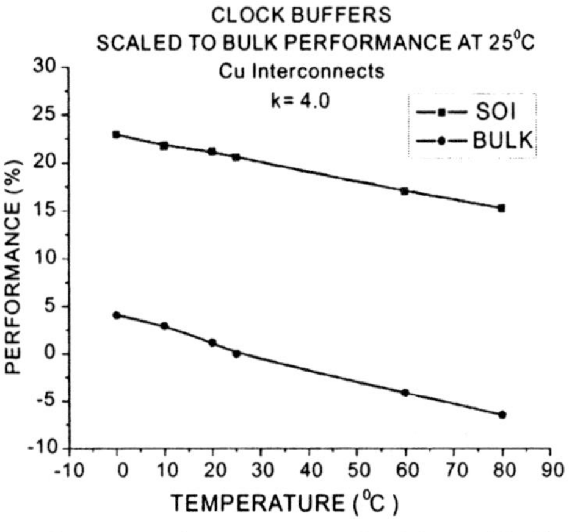

Figure 6. Performance v/s temperature for SOI and bulk clock buffer.

Performance changes with respect to the bulk CMOS performance at 25°C for SOI and bulk devices at a dielectric constant of 4.0 are shown in Fig. 6. For the chosen *Leff* and VDD (1.8 V) the performance gain is over 20% for SOI in the temperature range of 30 - 80 °C and reduces at lower temperatures. The performance change for SOI and bulk per 10 °C change in junction temperature is 1.2% and 1.32% respectively. Also the increased temperature adds to the degradation in reliability of SOI (not shown here).

4. Conclusions

1. 3-D thermal analysis shows significant coupling between adjacent fingers of multi-finger SOI devices. The junction temperature rise in multi-finger devices was a factor of three higher than predicted by 1-D thermal resistance models of single finger devices. 2. Junction temperature rise in SOI due to self-heating in high power devices like clock buffers can be 10 °C. Conduction from the channel can raise the M1 temperature 10 °C above that of the silicon. 3. Increase in temperature degrades the performance incrementally.

References:

1. Su, Lisa T. et. al., IEEE Transactions on Electron Devices, Vol. 41, No. 1, Jan 1994, pp. 69-75.
2. Workman, Glen O. et. al., IEEE Transactions on Electron Devices, Vol. 45, No. 1, Jan 1998, pp.125-133.

Modelling of High-Voltage SOI-LDMOS Transistors including Self-Heating

A.C.T. Aarts, M.J. Swanenberg* and W.J. Kloosterman*

Philips Research Laboratories, Eindhoven, The Netherlands

* Philips Semiconductors, Nijmegen, The Netherlands

Abstract

This paper presents (i) the compact modelling and (ii) the parameter extraction strategy of a 12 V SOI-LDMOS transistor. The LDMOS transistor is characterized by a macro model consisting of the physics based Philips' MOS Model 9 (MM9) for the channel and MOS Model 31 (MM31) for the drift region. Incorporation of the effect of self-heating during parameter extraction shows that the DC-characteristics can be described consistently and accurately over a wide range of biases, temperatures and device dimensions.

1 Introduction

The use of SOI (Silicon-On-Insulator) in a high-voltage process provides a new technology for consumer and automotive applications [4]. The advantage of SOI comprises, amongst others, latchup-free operation, a high packing density and less leakage current at high temperature operation. Optimal design of these high-voltage IC-applications requires accurate models for circuit simulation, describing the characteristics of the high-voltage devices over a wide range of biases, temperatures and device dimensions. Due to the combination of SOI and high-power densities, the self-heating of the device is significant. Therefore, a model that accounts for the temperature rise due to self-heating is essential for reliable results. In contrast to [2] where the effect of self-heating has been limited to the drift region, we incorporate self-heating in the drift region as well as in the channel. In addition to [1] where self-heating has also been included in an SOI-LDMOS model, this paper presents the parameter extraction strategy as well as the full scalability of an SOI-LDMOS model over a range of temperatures and device dimensions.

2 SOI-LDMOS Technology

Fig. 1 shows the cross-section of a 12 V n-channel LDMOS transistor in SOI (gate-oxide thickness is 38 nm). The p-well is diffused under the gate to about 1 μm from the source. The drift region of lightly doped n-type Si is necessary to sustain the high voltages. For characterization purposes the p-well body contact is separated from the source contact. The transistor is isolated by LOCOS as well as by a thick buried oxide (box) layer (of thickness t_{box} = 1 μm) underneath the SOI-layer. The length L_{box} of the SOI-layer is approximately 27 μm.

In forward operation electrons from the inversion channel in the p-well body are transported through the drift region to the drain contact. The formation of an accumulation layer under the gate in the drift region decreases the resistance, whereas the depletion layer at the p-well/drift region junction increases this resistance. For higher drain voltages the current is also decreased by depletion at the surface of the drift region from the drain side.

3 LDMOS Modelling

Fig. 2 shows the macro model used for DC-characterization of the 12 V SOI-LDMOS transistor. The channel in the p-well is modelled by MM9, whereas the n^- drift region is modelled by MM31 [3]. The latter is a depletion and accumulation mode MOSFET model, including velocity saturation, pinch-off and temperature effects. It describes the MOS-action of the gate field plate above the n^- drift region, as well as junction-depletion from the p-well body. In the macro model, the small influence of a bias at the Si-substrate is neglected. Notice that in most bias regimes this macro model can also be used for AC-characterization.

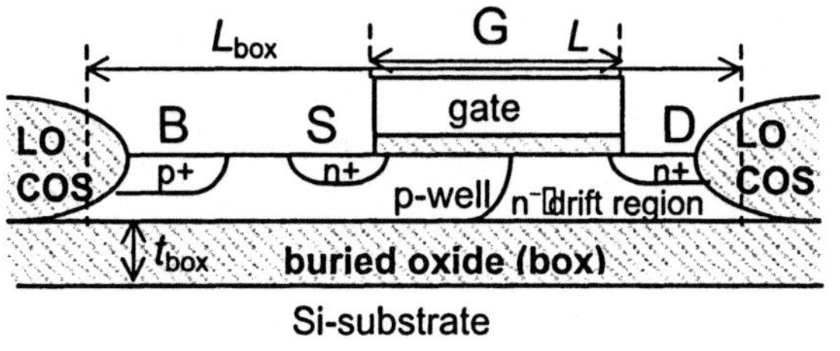

Fig. 1. Cross-section of the 12 V n-channel SOI-LDMOS transistor.

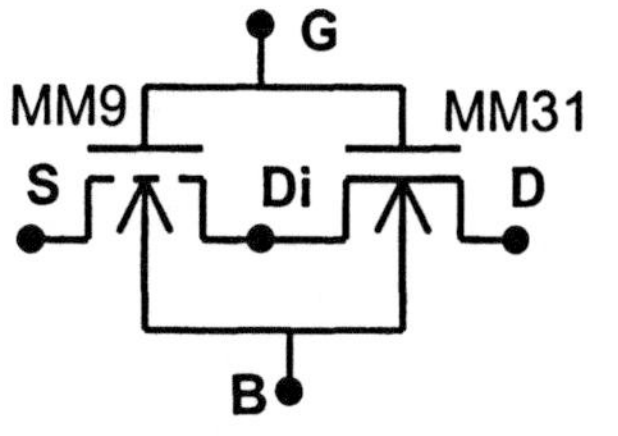

Fig. 2. The macro model used for DC–characterization.

4 Self-Heating

Due to considerable power dissipation and the low thermal conductive box layer, the temperature in the device rises significantly. Analysis of the one-dimensional heat flow shows that the temperature rise ΔT in the device due to self-heating is given by

$$\Delta T = R_{\text{th}} I_{\text{DS}} V_{\text{DS}}, \qquad R_{\text{th}} = \frac{t_{\text{box}}}{k_{\text{box}} A_{\text{box}}}, \tag{1}$$

provided that heat loss in the Si-substrate is neglected. Here, k_{box} = 1.4 Wm/K is the thermal conductivity of oxide, and $A_{\text{box}} = W_{\text{box}} L_{\text{box}}$ is the area of the box layer. The thermal resistance R_{th} is extracted from the DC-conductance, by incorporation of temperature scaling rules according to

$$X|_{T+\Delta T} = X|_T + S_X \Delta T \qquad \text{or} \qquad X|_{T+\Delta T} = X|_T \left(\frac{T}{T+\Delta T}\right)^{\eta_X}, \tag{2}$$

for a temperature-dependent parameter $X|_T$ at a given ambient temperature T. Here, $X|_{T+\Delta T}$ denotes the parameter corrected for self-heating, and η_X, S_X are the temperature coefficients.

5 LDMOS Extraction Strategy

For a set of devices of widths W varying from 5 µm to 114 µm, gate lengths L between 1 and 2 µm (the channel length is fixed by the p-well diffusion) and temperatures T between 10 and 125 °C, the model parameters are extracted according to the following strategy:

1. Assuming η_X and S_X, one evaluates at temperature T the model including self-heating, and extracts the parameters $X|_T$.
2. By executing step 1 at various T-values, the temperature coefficients η_X and S_X can be determined.
3. Repeat step 1 and 2 in an iterative procedure, until the temperature coefficients found in step 2 correspond to those used in step 1.

4. Extract for a reference temperature the parameters for various widths W, and determine the width scaling rules.
5. Extract the parameters for various gate lengths L, and determine the length scaling rules for the drift region.

These temperature and device scaling rules are subsequently used to determine the current at a certain temperature and device dimension.

6 Results

In Fig. 3 the effect of self-heating is shown by a reduction of the conductance in the saturation regime. For a lower temperature this effect is more pronounced. Fig. 4 shows the increase of the thermal conductance $1/R_{th}$ with box width W_{box}. We observe that, due to two-dimensional effects, the extracted values deviate slightly from the linear trendline predicted by (1) with $A_{box} = W_{box}L_{box}$. The slope of this trendline predicts a thermal conductance k_{box} of 1.3 Wm/K, which corresponds reasonably to the theoretical value of 1.4 Wm/K.

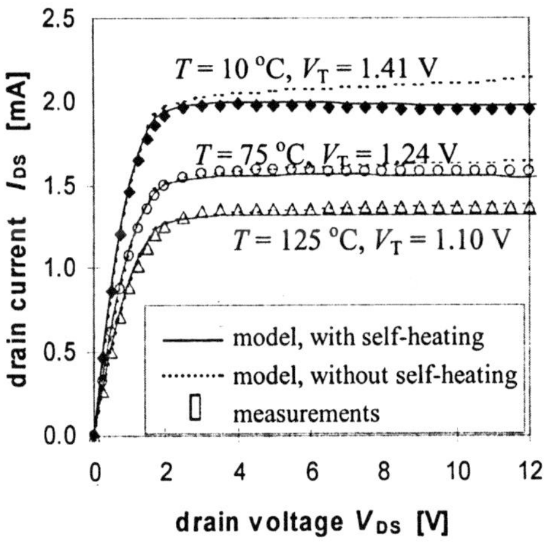

Fig. 3. The effect of self-heating for $W = 17$ μm, $L = 1.6$ μm and $V_{GS} = V_T + 3.1$ V.

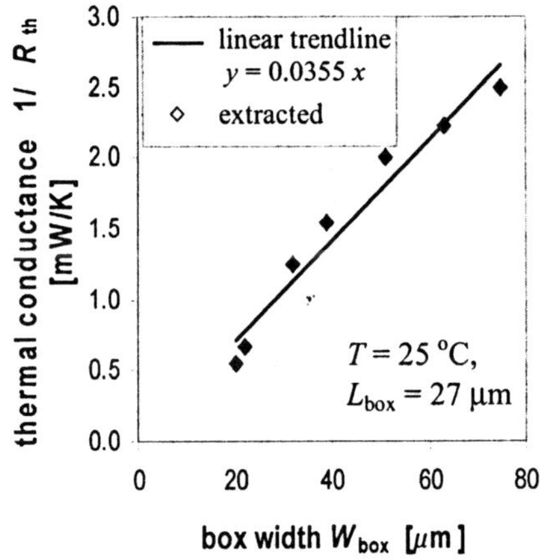

Fig. 4. The extracted thermal conductance $1/R_{th}$ and the linear trendline corresponding to (1).

Figs. 5 and 6 show, for grounded Si-substrate and p-well contact, measured DC-data and modelled results for larger device widths. Fig. 5 shows the drain current and transconductance g_m versus gate voltage. Here, a temperature rise ΔT due to self-heating of up to 110 °C is reached. Fig. 6 shows the drain current and conductance $|g_{DS}|$ versus drain voltage. The self-heating effect is clearly shown by a negative output conductance. In contrast to [2], where the effect of self-heating is due to a decrease in the conductance of the drift region, here the effect of self-heating is mainly caused by a decrease in the conductance of the *channel.*

7 Conclusions

By use of the macro model consisting of MM9 for the channel and MM31 for the drift region, an LDMOS transistor in SOI has been characterized accurately over a wide range of dimensions, temperatures and bias conditions. The effect of self-heating has been found to be significant, and needs to be incorporated during parameter extraction. Incorporation of self-heating has led to a physics based model that accurately predicts the SOI-LDMOS characteristics. The model is successfully used in high-voltage IC-design.

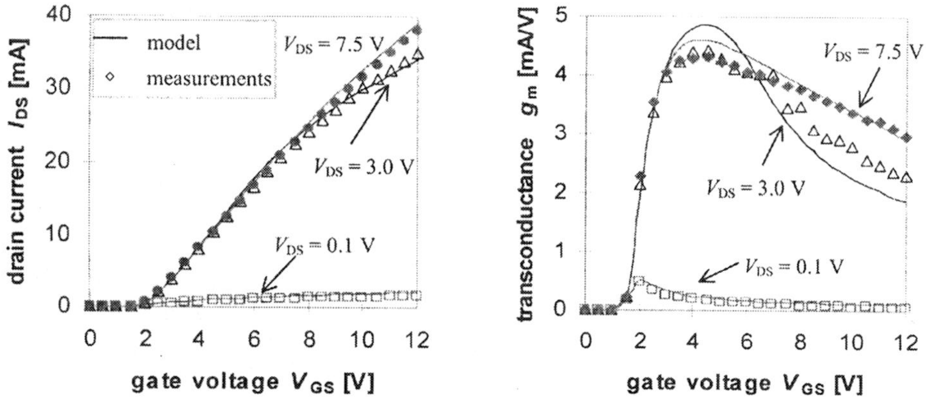

Fig. 5. The measured and simulated drain current and transconductance, for $W = 114$ μm, $L = 1.6$ μm and $T = 25$ °C.

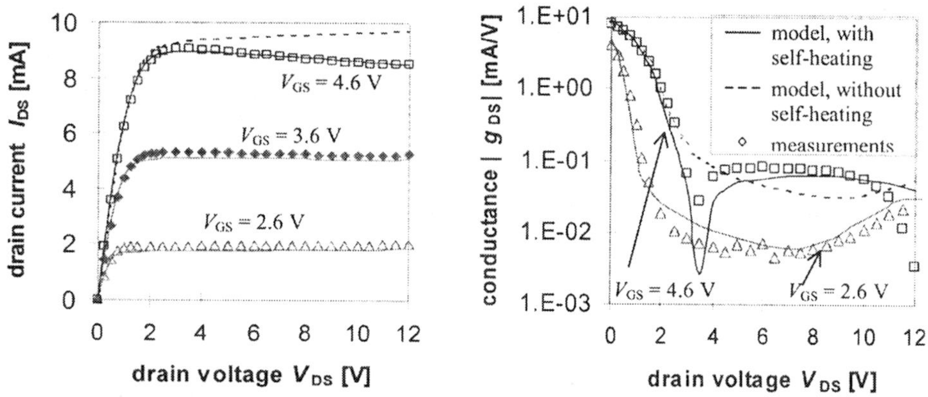

Fig. 6. The measured and simulated drain current and conductance, for $W = 90$ μm, $L = 1.6$ μm and $T = 25$ °C.

References

[1] D' Halleweyn, N. et al. (1999), A Compact Model for SOI-LDMOST, Including Accumulation, Lateral Doping Gradient and High Side Behaviour. In: Proc. ISDRS '99, pp. 195-199.

[2] Liu, C.M., et al. (1995), A Closed-form Physical Back-Gate-Bias Dependent Quasi-Saturation Model for SOI Lateral DMOS Device with Self-Heating for Circuit Simulation. In: Proc. ISPSD '95, pp. 321-324.

[3] Philips (2001), http://www.semiconductors.philips.com/Philips_Models/

[4] van der Pol, J. A., et al. (2000): A-BCD: An Economic 100 V RESURF Silicon-On-Insulator BCD Technology for Consumer and Automotive Applications. In: Proc. ISPSD '00, pp. 327-330.

An Efficient Tool For Extraction of Interconnect Models in Submicron Layouts

P. Maffezzoni, A. Brambilla, A. L. Lacaita

DEI, Politecnico di Milano, p.za L. Da Vinci 32, 20133 Milano, Italy

1 Introduction

During these last years the width of interconnects in silicon layouts has been reduced to less than $0.25\mu m$, the number of metal levels has been increased up to five and the contributions to parasitic capacitances has become dominant. This is why a renewed interest has been triggered on the development of improved extraction methods and recent literature reports a large number of proposals. Capacitance evaluation is a well known and studied problem which requires the solution of the Laplace's equation for the electrostatic potential. Many codes have been developed so far, based on finite element or finite difference methods, solving the Laplace's equation even in a 3D geometry; however they can be seldom applied to real life silicon layouts, since due to the layout geometrical complexity they easily run out of computer resources. The bottleneck is usually circumvented by avoiding the solution of Laplace's equation for the entire layout and trying to break the problem into many elementary geometries (sub-problems) or by not considering the real 3D geometry. Those are the so called "quasi 3D" or "2.5D" extractors. In this paper we present an alternative approach for parasitic extraction which solves the Laplace's equation considering the entire layout without making any geometrical simplification or breaking. It is based on an extended version of the Floating Random Walk algorithm (FRW) [2]. The use of FRW is not novel, but so far it has been limited to the extraction of the total capacitance of the interconnect. Here we show how to use FRW together with the Picard-Carson iterative procedure [1] in order to efficiently get a compact model of an interconnect.

2 The Picard-Carson method

Given the interconnect shown in fig. 1 of total length L, it is possible to represent its behavior at the input ($x = 0$) and output ($x = L$) terminals in the Laplace domain by means of the transmission matrix T relating the voltage $V(L)$ and the $I(L)$ current, measured at the end of the interconnect, with the corresponding values $V(0)$, $I(0)$ at the input

$$\begin{bmatrix} V(N) \\ I(N) \end{bmatrix} = \begin{bmatrix} A & B \\ C & D \end{bmatrix} \cdot \begin{bmatrix} V(0) \\ I(0) \end{bmatrix}$$

The A, B, C, D entries can be computed using closed forms [1]:

$$A(s) = I + A_1 s + \ldots + A_N s^N + \ldots \qquad B(s) = B_1 + B_2 s + \ldots + B_N s^{N-1} + \ldots$$
$$C(s) = C_1 s + C_2 s^2 + \ldots + C_N s^N + \ldots \qquad D(s) = I + D_1 s + \ldots + D_{N-1} s^{N-1} + \ldots$$

where $A_i = \zeta^{(2i)}$, $B_i = \zeta^{(2i-1)}$, $C_i = \psi^{(2i-1)}$, $D_i = \psi^{(2i)}$. Saying $c(x)$ and $r(x)$ the line capacitance and resistance per unit length, the above terms are derived by means of the iterated integrals:

$$\zeta^{(h)}(x) = \int_0^x \psi^{(h-1)}(\tau)r(\tau)d\tau \quad ; \quad \psi^{(h)}(x) = s \cdot \int_0^x \zeta^{(h-1)}(\tau)c(\tau)d\tau \quad x \in [0, L]$$

where $\zeta^{(0)}(x) = \psi^{(0)}(x) = 1$ represent the initial guess. The $r(x)$ function is known, while $c(x)$ is evaluated using the FRW algorithm, as summarized in the following section.

3 Parasitic capacitance extraction and FRW

The self capacitance of an electrode is equal to the electrical charge induced on it when its potential is raised to 1V and the other electrodes and ground plane are at 0V. The charge on the electrode is in turn equal to the flux of the electrical displacement vector through the Σ Gaussian surface contouring the interconnect. In this way the capacitance estimation is recasted in terms of evaluation of flux. The same approach can be adopted for the estimate of $c(x)$, which, in FRW, is done in a statistical way. Following a Monte Carlo procedure pick a point **p** on the Σ Gaussian surface as shown in fig. 1 and start a random walk from it by choosing hops on maximum spheres that do not intersect other interconnects or ground plane. Selection of **B, C, D, E** points (see fig. 2) is randomly done according to a probability density function related to the Green function that links potential or electric field displacement in the center of each sphere to potentials on its surface. A possible random walk that end on an electrode in the **E** point is shown in fig. 2. The average value of the potentials collected on electrodes gives an estimation of the $c(p)$ value.

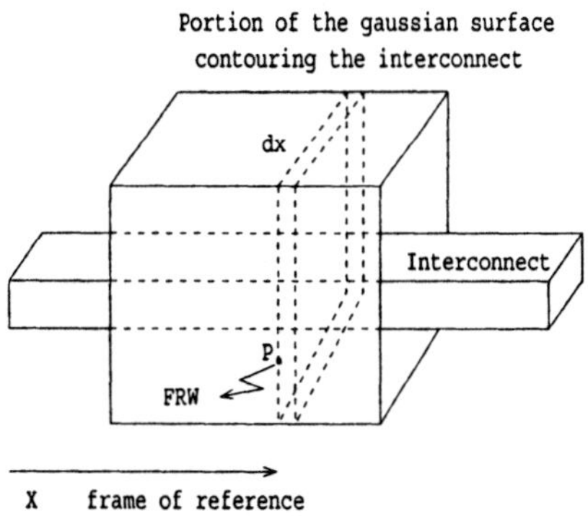

Fig. 1. A portion of the Gaussian surface through which the average value of the electric field E_η is estimated. Generic FRW is started from **P** point.

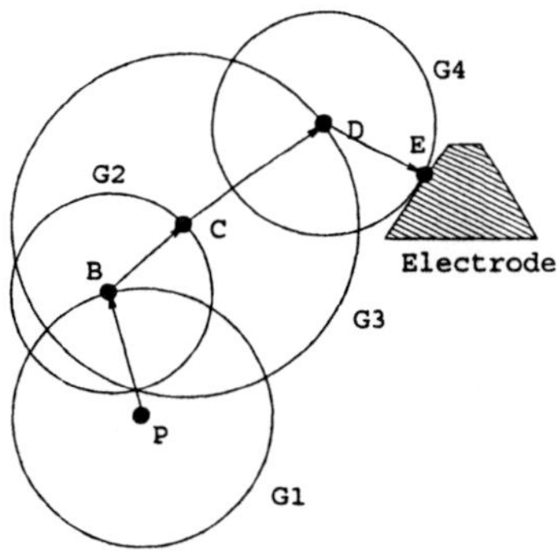

Fig. 2. A FRW walk composed of discrete hops; a walks ends when point (for example **E**) lies on the ground plane or on an interconnect.

It can be shown that *the recursive computation of the entries of the T matrix in done through this statistical sorting approaches, the exact values as the number of floating random walks goes to infinity.* In practice a high but finite number of iterations (e.g. 10000) are performed till the estimated statistical error falls below a suitable relative value (e.g. 5%).

4 Simulation results

The efficient extraction of the lumped model and the statistical evaluation of the above integrals make the new algorithm very efficient and suitable for applications to industrial 3D layouts. Figs. 3, 4 show a performance comparison between SIMPLEX [3] and our approach, denoted to as FASTNET, in terms of CPU time and memory allocation. In both Figs. bars on the left refer to a simple cell embedded in the layout of Fig. 5. During parasitic extraction of this cell, FASTNET allocates more memory with respect to SIMPLEX due to the overhead for storing the layout structure. Bars on the right in the same Figs. show the performance in the analysis of the more complex layout in Fig. 5. In this case FASTNET memory allocation and CPU time remain almost the same, while SIMPLEX requires much more resources: the CPU time grows exponentially and also the memory allocation is about three time larger. Note that in both cases FASTNET is even more accurate (< 5%) than SIMPLEX (< 10%). *In general FASTNET CPU time remains almost independent of the dimensions of the circuit layout thus becoming extremely efficient in handling parasitic extraction in realistic industrial layouts.*

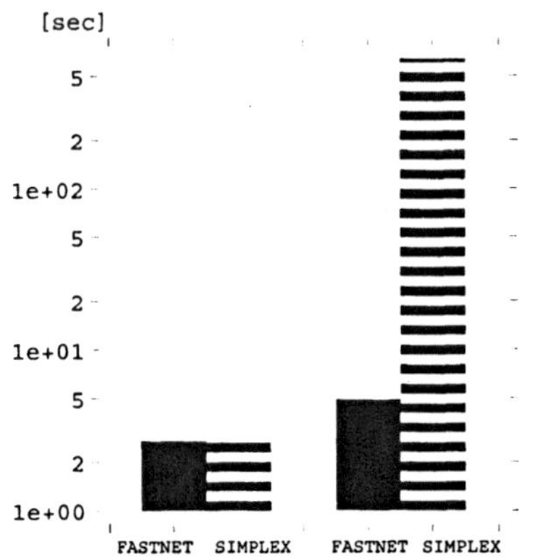

Fig. 3. CPU time required by FASTNET and SIMPLEX for computing the equivalent lumped models of the interconnects of a simple cell embedded in the layout of Fig. 5 (left two bars) and of the more complex layout of Fig. 5 (right two bars).

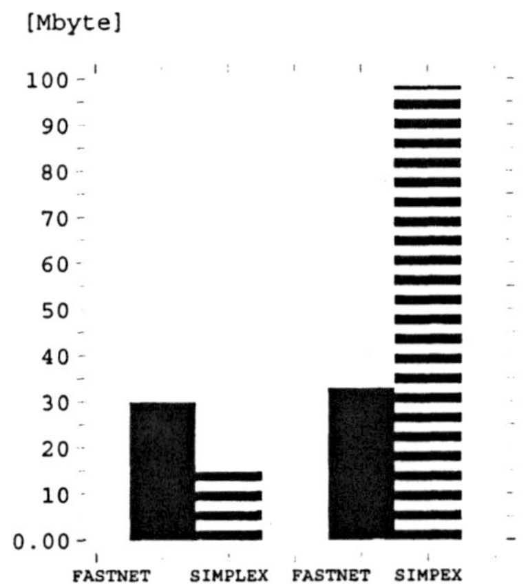

Fig. 4. Memory required by FASTNET and SIMPLEX for extracting the capacitance of nets of a simple cell embedded in the layout of Fig. 5 (left two bars) and of the layout in Fig. 5 (right two bars).

Fig. 6 compares the number of RC cells needed by SIMPLEX and FASTNET to model an interconnect and to reach the same accuracy in the delay evaluation. Note that since geometry of layout has not been partitioned by FASTNET the equivalent RC network is always composed of a few number of cells independently from line length L. Despite compactness this equivalent model has an accuracy comparable to that composed of 100 cells of SIMPLEX as shown in fig. 7.

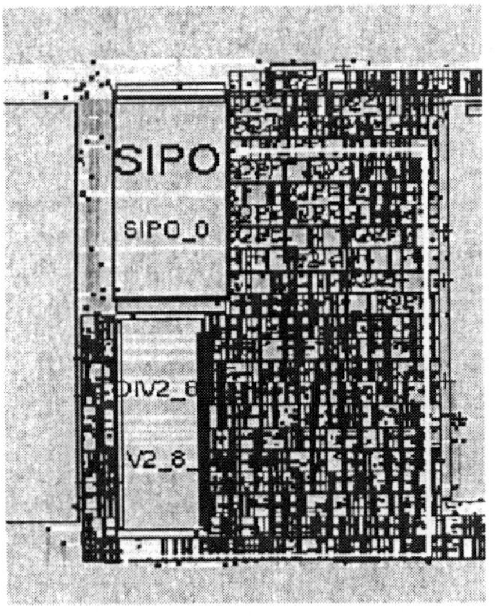

Fig. 5. Typical interconnect (highlighted in white) in a 0.25μm 6 metal level technology layout.

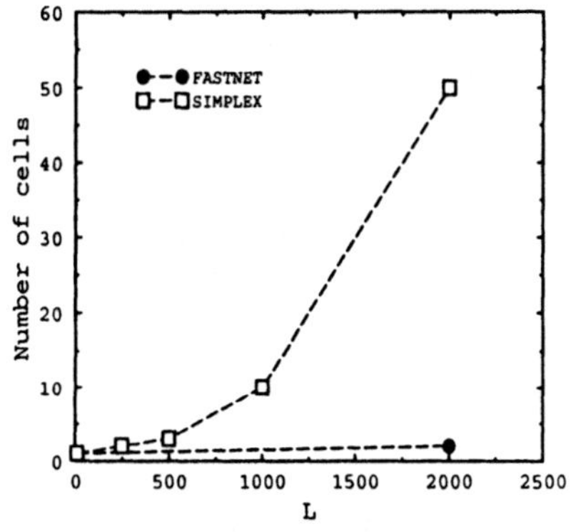

Fig. 6. Number of lumped cells required by FASTNET and SIMPLEX to get an accuracy < 1% versus the L line length.

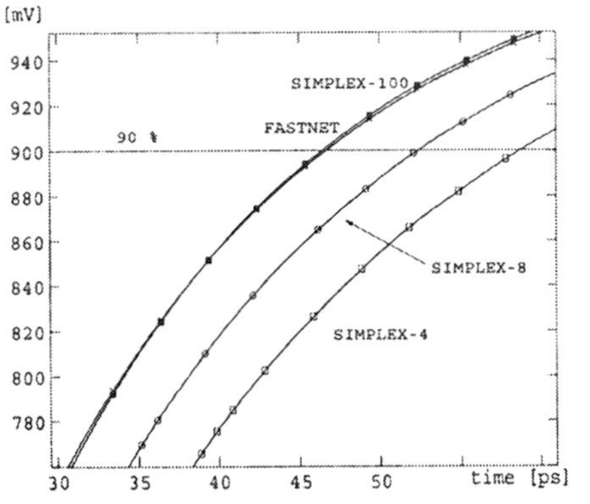

Fig. 7. Leading edge of the voltage pulse at the end of the line in Fig. 5, as derived by SIMPLEX when using 100 RC cells (crosses) and by forcing SIMPLEX to use only 8 (circles) and 4 cells (empty squares). Result obtained with the two cell equivalent network (filled squares) extracted by (FASTNET).

5 Conclusions

A fast and accurate method to compute and reduce the lumped RC model of a distributed net has been presented. It is based on an extension of the Floating Random Walk method and demonstrate that statistical methods are attractive to these applications since they need limited memory resources and are able to deal with large 3D industrial layouts.

6 Acknowledgments

The work has been supported by CNR (Italian National Research Council) in the frame of the project MADESS-II.

References

[1] M. S. Ghausi, J.J. Kelly, "*Introduction to Distributed-Parameter Networks with Application to Integrated Circuits* ", Holt, Rinehart and Winston, New York 1968.

[2] G. M. Royer, "*A Monte Carlo Procedure for Potential Theory Problems*", IEEE Trans. on MTT, vol. MTT-19. No. 10, pp. 813-818, October 1971.

[3] *SIMPLEX reference manual*, (1999).

A Comparative Study of Two Numerical Techniques for Inductance Calculation in Interconnect Structures

C. Harlander, R. Sabelka, and S. Selberherr

Institute for Microelectronics, TU Vienna, Gusshausstrasse 27–29, A-1040 Vienna, Austria
Phone: +43-1-58801/36028, Fax: +43-1-58801/36099, Email: Harlander@iue.tuwien.ac.at

Abstract

We present an advanced algorithm for an extraction tool that computes inductances of interconnect structures. As already pointed out in [1] the pursued energy concept leads to a 6-fold integral which can also be evaluated by use of the Monte Carlo method. Classical implementation of the Monte Carlo method, where the whole geometry has to be hunted for the associated element loses efficiency. Our approach is applied without time consuming element location for the random point coordinates to compute this integral.

1 Introduction

One consequence of technology scaling by shrinking feature sizes and increasing clock frequencies is the growing importance of interconnect lines. The performance of interconnects is limited by various parasitic effects (eg. signal delay, capacitive and inductive crosstalk, attenuation). The utilization of new materials (Copper and low-k dielectrics) reduces the RC time constant. Thereby decreased resistance and capacitance bring out inductive effects more intensively, requiring consideration in circuit simulation. Thus, inductance extraction becomes necessary for critical nets.

2 Physical approach

We compare two stationary inductance calculation methods both based on a numerical solution of Neumann's formula [2] for a precalculated current density distribution:

$$L_{ik} = \frac{\mu}{4\pi}\frac{1}{I_i I_k}\int\limits_{V_i}\int\limits_{V_k'}\frac{\vec{J}_i(\vec{r})\cdot\vec{J}_k(\vec{r}\,')}{|\vec{r}-\vec{r}\,'|}\,\mathrm{d}V\,\mathrm{d}V'\ . \tag{1}$$

The integration is carried out numerically, where special attention has to be paid on the singularities of the integrand, or with the Monte Carlo method. For both methods the stationary current density is calculated with the finite element method. The first method [1] employs a summation of the contributions of all pairs of finite elements to solve the integral (1), where different kinds of approximation are used, depending on the term $|\vec{r}-\vec{r}\,'|$. For a large distance (compared to the tetrahedron diameter) simple integration formulae are sufficient. The evaluation for small distances demand special formulae with certain integration points, published by Stroud [3] who presented various integration formulae which are applicable for various n-simplexes (e.g. the unit triangle, the unit tetrahedron) as integration region. If $\vec{r}$ and $\vec{r}\,'$ are in the same tetrahedron, a partially analytic integration scheme is used to increase the accuracy of the integration.

3 The Program Package

The SMART ANALYSIS PROGRAMS [4] use the finite element method, because of it's advantages, as numerical robustness, the ability to solve nonlinear systems, high obtained accuracy, and general applicability.

The geometry can be defined either directly from the layout by specifying layer thicknesses, or by a rigorous topography simulation [5, 6]. The layout of the interconnect structure can also be imported from CIF or GDSII files, or created interactively with a graphical layout editor [7]. Furthermore, the program package includes three preprocessors, one for two-dimensional applications (CUTGRID) the other for three-dimensional applications. The preprocessor LAYGRID allows a layer-based input of the simulation geometry and the specification of the boundary conditions on the borders of each subdomain. The fully unstructured three-dimensional Delaunay grid generator DELINK [8] utilize an advanced-front algorithm, whereby the mesh generation starts from the initial front to fill up the solids with tetrahedrons.

A preconditioned conjugate gradient solver (ICCG), which has been optimized specifically for the discretized Laplace operator, is used to solve the linear systems for domains of conducting materials [9]. By applying Ohm's law to the derivative of the electrostatic potential the distribution of the electric current density is obtained. The simulation is performed with the module STAP (Smart Thermal Analysis Program), where both inductance extraction methods have been implemented.

Two postprocessors complete the program package, whereby the visualization tool SV is based on VTK [10], a flexible and powerful visualization library. Both postprocessors can be used to verify the grid quality, and for the visualization of several distributions (e.g. electric potential, temperature, current density), whereby SV provides numerous features, as eg. cutting plains, volume rendering and contour faces representation of distributions. Fig. 1 gives an overview about the SMART ANALYSIS PROGRAMS.

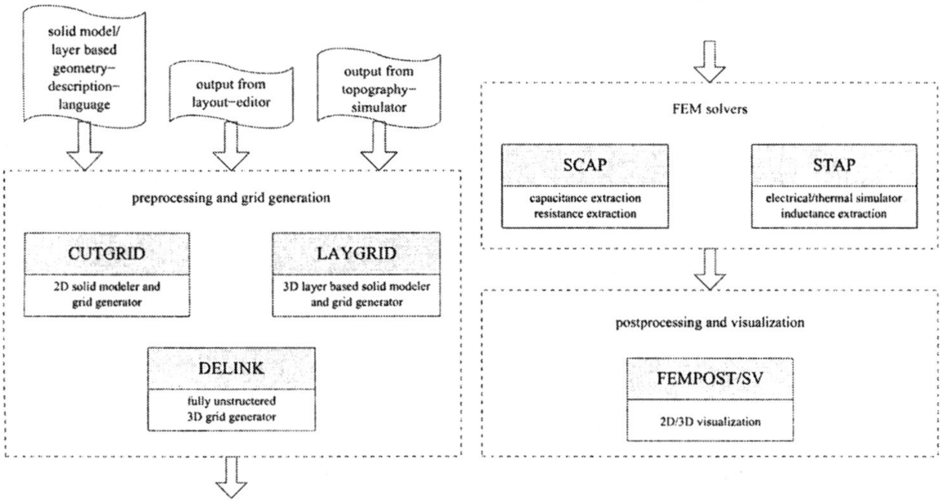

Fig. 1: The SMART ANALYSIS PROGRAMS: tools and dataflow

4 The Monte Carlo Implementation

A well-known choice for the evaluation of multiple integrals is to apply the Monte Carlo method. Associated with this method, where by random the point coordinates are chosen, is a fairly high effort on CPU-time, because of the time consuming search for the associated element of the random point coordinates. To reduce the error a high number of function evaluations has to be carried out, whereby for each evaluation the aligned element with the precalculated current density must be found. To improve the convergence during the Monte Carlo sampling several variance reduction schemes (e.g. importance sampling, control variates) are known to accelerate the computation procedure [11].

One big advantage of our implementation is to bypass the high computational effort for the element location. We first determine the associated element to the evaluated probability function, and then locate the point inside the tetrahedron. For this purpose we take two arrays for every conductive segment. In the first one is the volume of each element, whereby the sum of all entries is scaled to one. In the second one is the probability function already evaluated for each conductor element by adding up all entries from the beginning to the current index of the first array. Then the random generator chooses a number between zero and one. The associated element complying to the probability function is found by a binary search.

To ensure a uniform probability the local coordinates of the integration points are found by shooting into the unit cube. The first point inside the registered unit tetrahedron is taken. For the interpolation of the current density inside each element quadratic shape functions are used.

5 Application Example

Figure 2 and Fig. 3 show the current density of two planar transformers. These transformers are build of two interwound spirals each of 3, respectively, 5-turns metal with 5 μm width, a spacing of 15 μm, and an inner length of 54 μm.

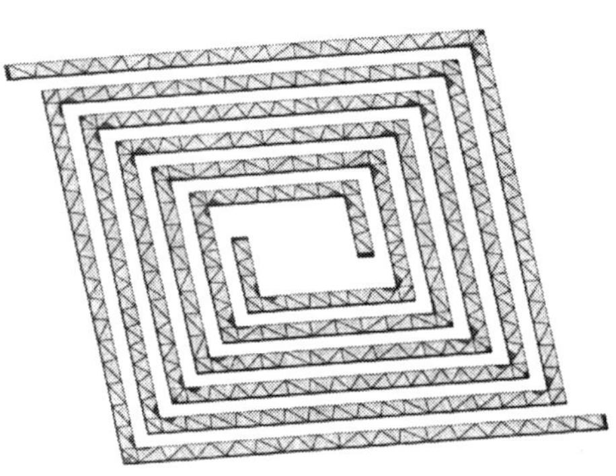

Fig. 2: Current density distribution of the planar transformer with 3-turns metal

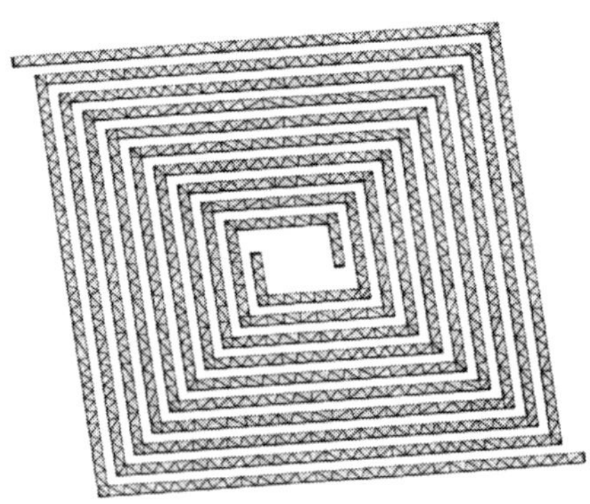

Fig. 3: Current density distribution of the planar transformer with 5-turns metal

By utilizing the preprocessor LAYGRID three different grids were made. In Table 1 the simulation times for the current density and the Monte Carlo method, respectively, the first numeric method as accomplished above, and the calculated inductances are listed. The simulations were performed on a Digital Alpha workstation (DEC600/333 MHz).

The number of samples for the Monte Carlo method was 1 million. The first column of Table 1 implies all elements of the conductive segments, whereby tetrahedral grid elements with quadratic shape functions were used. The analysis time for the Monte Carlo method is not so strongly influenced by the number of elements (n), because the computational effort for the binary search grows with ln(n). The simple integration formulae for the mutual inductances demand with increasing n almost the same time. Table 1 emphasize the advantages of the Monte Carlo method explicitly.

	Number of Elements	Time [s]				Results [nH]			
		MC		Method [1]		MC		Method [1]	
		M	L	M	L	M	L	M	L
3-turns	1800	17	33	1	327	0.67	1.04	0.67	1.06
	1968	17	34	1	627	0.67	1.05	0.67	1.06
	2648	18	34	3	1764	0.67	1.06	0.67	1.06
5-turns	4383	18	35	7	1945	2.71	3.58	2.70	3.62
	4653	19	35	8	2088	2.71	3.60	2.70	3.62
	5697	19	36	15	7885	2.70	3.60	2.69	3.63

Table 1: Analysis time and results of the planar transformers

6 Conclusion

We have presented a comparative study of two numerical techniques for inductance calculation in interconnect structures. Both methods are implemented into the package SMART ANALYSIS PROGRAMS, which allows simultaneous extraction of three-dimensional effective parameters of VLSI circuits.

Acknowledgment

Our work is partly supported by Compaq Corporation, Shrewsbury, USA and by Sony Corporation, Atsugi, Japan.

[1] C. Harlander, R. Sabelka, and S. Selberherr, "Inductance Calculation in Interconnect Structures," in *Proc. 3rd Intl. Conf. on Modeling and Simulation of Microsystems*, (San Diego, California, USA), pp. 416–419, Mar. 2000.

[2] F. W. Grover, *Inductance Calculations: Working Formulas and Tables*. D. Van Nostrand Company, New York, 1946.

[3] A. H. Stroud, *Approximate Calculation of Multiple Integrals*. Prentice-Hall, Englewood Cliffs, N.J., 1971.

[4] R. Sabelka and S. Selberherr, "A Finite Element Simulator for Three-Dimensional Analysis of Interconnect Structures," *Microelectronics Journal*, vol. 32, pp. 163–171, Jan. 2001.

[5] W. Pyka, R. Martins, and S. Selberherr, "Optimized Algorithms for Three-Dimensional Cellular Topography Simulation," *IEEE J.Technology Computer Aided Design*, 2000. `http://www.ieee.org/products/online/journal/tcad/accepted/Pyka-March00/`.

[6] R. Martins, W. Pyka, R. Sabelka, and S. Selberherr, "Modeling Integrated Circuit Interconnections," in *Proc. Intl. Conf. on Microelectronics and Packaging*, (Curitiba, Brazil), pp. 144–151, Aug. 1998.

[7] R. Martins and S. Selberherr, "Layout Data in TCAD Frameworks," in *Modelling and Simulation*, pp. 1122–1126, Society for Computer Simulation International, 1996.

[8] P. Fleischmann, W. Pyka, and S. Selberherr, "Mesh Generation for Application in Technology CAD," *IEICE Trans.Electron.*, vol. E82-C, no. 6, pp. 937–947, 1999.

[9] R. Bauer and S. Selberherr, "Preconditioned CG-Solvers and Finite Element Grids," in *Proc. CCIM*, vol. 2, (Breckenridge, USA), Apr. 1994.

[10] W. Schroeder, K. Martin, and B. Lorensen, *The Visualization Toolkit: An Object-Oriented Approach to 3D Graphics*. Prentice-Hall, 1996.

[11] G. Leonhardt and W. Fichtner, "Acceleration of Inductance Extraction by Means of the Monte Carlo Method," Tech. Rep. 99/8, Integrated Systems Laboratory, ETH Zürich, 1999.

Comparison of finite element stress simulation with X-ray measurement for the aluminum conductors with different passivation topography

Tai-Kyung Kim,Young-Pil Kim[1],Won-Young Chung,Young-Kwan Park,

and Jeong-Taek Kong

Semiconductor R&D Center, Samsung Electronics Co., Ltd.

San #24, Nongseo-Ri, Kiheung-Eup, Yongin-City, Kyunggi-Do, 449-711, Korea

taikyung@samsung.co.kr

Abstract

This paper evaluates the dependence of the thermal stresses of aluminum conductors and SiO_2 layers deposited by several different processes using the finite element method(FEM) and topography simulation. The results of topography simulation for four different deposition processes agree well with scanning electron microscopes and subsequent FEM stress simulations are compared with X-ray diffraction measurement data. Simulation results show that the different stresses are created in aluminum lines from different passivation processes. Especially, the stresses of aluminum lines decrease as the encapsulation of aluminum conductors by the oxide layer with void profiles decrease. Through this topography simulation followed by stress simulations for different deposition processes, we can systematically define the failure mechanisms of aluminum lines for various passivation processes.

1 Introduction

Today's VLSI technology requires multi-level metals and dense interconnection schemes, which can be realized by a suitable process to fill spaces between metal lines with a high aspect ratio without voids. The mechanical stresses generated by different thermal expansion between the aluminum and the SiO_2 layer must be controlled for immunity of the hillock formation and voids of metal lines. The formation and growth of voids from mechanical stresses of aluminum lines have been studied, especially in terms of the dependence of storage temperature, line width, topology, and thickness of a passivation layer[1]. In addition, the intrinsic stress of a passivation layer influences to the stresses in the aluminum lines. Previous work has reported the comparison of FEM stress simulation with X-ray diffraction measurement[2]. In this paper, we first reports the simulation results on the stresses of aluminum lines in several different deposition processes for a passivation layer that has different intrinsic stresses and different topographies including voids in the oxide layer.

To investigate the stress dependency of the aluminum on the passivation topography, an internally developed topography simulator is used to generate the void containing profiles for four deposition processes. In addition, the subsequent mechanical stress simulations are performed using ABAQUS for each profile obtained from those topography simulations. The results of topography simulation agree well with the vertical SEM micrographs and the stresses are also compared with X-ray measurements for all the deposition processes including the PECVD (plasma enhanced chemical vapor deposition), HDP (high density plasma) CVD and flowable oxide(FOX) by spin coating. Especially, aluminum lines which embedded void profiles in the dielectric layer have a lower stress than those processed by a void-free IMD process.

2 Topography simulation: details and results

In this work, we use an in-house topography simulator for investigating the oxide deposition process. This topography simulator has been integrated with a plasma equipment simulator based on the two dimensional hybrid model, HPEM(Hybrid Plasma Equipment Model), to obtain the density of each chemical and ion species[3,4]. Patterned aluminum lines were passivated by four deposition processes, HDPCVD oxide, two PECVD oxides using the SiH_4 source and the TEOS source, and FOX. These deposition processes are described by well-known surface reaction models[5].

Fig. 1 shows the results of profile simulation. The final profiles for different deposition processes match well with vertical SEM pictures. The PECVD oxides using the SiH_4 source and the TEOS source both create the voids between metal lines. On the other hands, FOX and HDPCVD oxides do not create voids. These simulation profiles with void structures are fed back to the finite element stress simulation. Thus, we can characterize the effect of the void profile of an IMD layer on the stress distribution of metal lines.

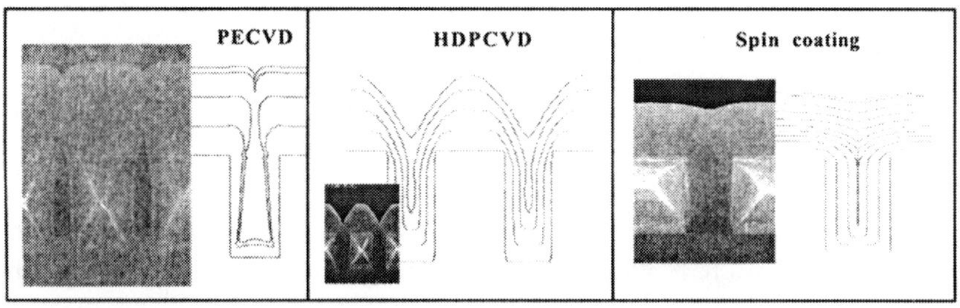

Fig. 1. Comparison between simulation profiles and SEM micrographs for different deposition processes of SiO_2

3 FEM stress simulation: details and results

We investigate the influence of the thermal stress distribution in a composite passivation layer with a real topography. ABAQUS· is used to model the stress distribution of the two dimensional profile obtained from topography simulation. We adopted the frozen view model, in which the maximum stress arising in the manufacturing process is usually evaluated during the temperature ramping down process. The stress simulation of a non-patterned wafer is performed to evaluate the material behavior according to the deposition process, and then the aluminum line passivated by SiO_2 is simulated to calculate the detailed stress distribution with respect to the pattern size and the passivation profile.

The stress simulation results of the non-patterned wafers are compared with the wafer warpage measurement, as shown in Fig. 2. We can extract the material behavior as a function of temperature for the different deposition processes. Fig-2(a) shows that different deposition processes and temperature cycles produce different intrinsic stresses and material behaviors of SiO_2 layers. The SiO_2 passivation layer of the patterned wafer must be described by the different modulus, thermal expansion coefficient, and viscous relaxation behavior to generate the experimental data. From these stress hysteresis simulations, we can describe the exact viscous relaxation effect and intrinsic stresses on the operating temperature. In addition, we simulated the non-passivated aluminum layer in the sputtered deposition process, as shown Fig. 2-(b). Especially, we can also describe the plastic deformation of the aluminum layer and extract the yield stresses for special deposition conditions.

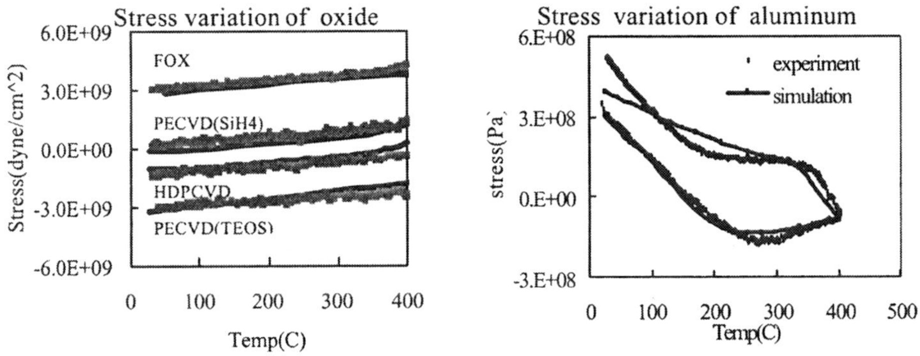

Fig. 2. Comparison of stress hysteresis simulation of a non-patterned wafer with the wafer warpage measurement.

We show the results of stress simulations starting from a real topography where Ti 30nm\Al 800nm\Ti 10nm\TiN 35nm metal stack is patterned and followed by four different passivation processes. Table 1. shows the comparison of FEM stress simulations to X-ray data that measure the residual stresses of the passivated metal lines, with monochromatic CuKa beam, at the 40kV 100mA beam condition. The most important result is that the trend of averaged tensile stress of an aluminum line from simulation correspond to the stresses from experiments according to the deposition process of the SiO_2 passivation layer.

Aluminum lines passivated by PECVD oxide using TEOS have a higher tensile stress, about 80MPa higher than that of the PECVD oxide using SiH_4. The differences of stresses in aluminum come only from the different intrinsic stresses of the SiO_2 layer. Some Researchers have argued that the high intrinsic compressive stress in the passivated layer is responsible for the high tensile stress in the metal[1]. However, simulation and experimental results show that the stresses of aluminum lines are dependent only on the intrinsic stresses of the SiO_2 layer. The different stresses between the PECVD and HDPCVD oxides are turned out to be originated from two reasons, intrinsic stresses and the existence of void profiles. In spite of low compressive intrinsic stresses of the HDPCVD oxide, stresses of aluminum passivated with the HDPCVD oxide has a high tensile stress, while that of PECVD passivation is relaxed by void profiles of SiO_2. Simulational stresses of aluminum passivated by the PECVD without void profiles has a higher tensile than that of the HDPCVD oxide. Thus, finite element stress modeling of aluminum must consider the void profiles in the passivation layer. In spite of voidless profile in the FOX layer, the stress relaxation of aluminum is due to the tensile intrinsic stresses and the large viscous relaxation properties of the FOX layer.

From this type of stress simulation with a real topography, we distinguish the evolution mechanism of thermal stresses for passivated aluminum lines with different deposition processes.

Table 1. Comparison of FEM stress modeling with X-ray measurement.

		PECVD (SiH4)	PECVD (TEOS)	FOX	HDPCVD (SiH4)	PECVD w/o void
XRD	S_x	242	338	132	457	x
(MPa)	S_y	98	197	89	363	x
FEM	S_x	245	328	175	417	420
(MPa)	S_y	198	275	111	258	355

4 Conclusions

We investigated the causes of stress differences in aluminum lines deposited by different processes. The level-set based topography simulator reproduced real topographies for different deposition process. Through the FEM stress simulation with the real topographies, we can systematically define why the deposition process of the passivation layer creates the different stresses and failures of aluminum lines.

Such a systematic FEM stress simulation with a real topography and intrinsic stresses is turned out to be useful in finding out the failure mechanism of integrated devices[6]

References

[1] R.E.Jones, Jr., et al, Appl. Phys. Lett, Vol. 50, pp725, 1987.

[2] B.Greenebaum, et al, Appl. Phys. Lett, Vol. 58, pp1845, 1987.

[3] W.Y.Chung, et al, SISPAD, pp340, 2000.

[4] R. J. Hoekstra, et al, J. Vac. Sci. Technol. A, vol. 15, pp. 1913-1921, 1997.

[5] J. Li, et al, J. Vac. Sci. Technol. B. Vol. 13, pp10, 1995.

[6] T.K.Kim, et al, IEDM, pp725, 1998.

A New Compact Spice-like Model of E^2PROM Memory Cells Suitable for DC and Transient Simulations

L. Larcher[1], P. Pavan[1], M. Cuozzo[2], A. Marmiroli[2]
[1] D.S.I. and INFM, Università di Modena e Reggio Emilia, 41100 Modena, Italy
[2] STMicroelectronics, 20041 Agrate Brianza (MI), Italy

Abstract

This paper presents for the first time a new compact Spice-like model of a E^2PROM Memory Cells suitable for both DC and transient circuit simulations. This model is based on a new Floating Gate voltage calculation procedure that improves strongly the accuracy of the modeling of the cell. Moreover, this model features many advantages compared to the previous ones: i) it is simple to implement and scale; ii) its computational time is not critical; iii) its parameter extraction procedure is the same of a MOS transistor; iv) it can be easily upgraded to take into account leakage current contributions (SILC).

1 The Model

Models of Floating Gate devices proposed in the literature are mostly based on the "classic" lumped element description of the FG cell shown in Fig.1(a). Memory operations are described starting from the exact knowledge of FG voltage, V_{FG}, which is calculated through fixed capacitive coupling coefficients, as reported in the paper of Pavan et al. (1997).

$$V_{FG} = \alpha_{CG} V_{CG} + \alpha_D V_D + \alpha_S V_S + \alpha_B V_B$$

However, these coefficients are not easy to measure, since the FG node cannot be directly accessed, and they depend on the applied terminal voltages, as stated by Duane et al. (1998). Therefore, considering them as constants introduces errors in the FG voltage calculation, thus compromising the correct model of the cell. For this reason, we have developed a new V_{FG} calculation procedure, which does not use fixed coefficients. As shown in Fig.1(b), the model we propose is composed by four elements: a MOS transistor whose Source, Body, Drain are S, B, D, of the cell, and Gate is the FG of the cell; a capacitor connected between FG and Control Gate; a voltage-controlled voltage source, V_{FG}, between FG and ground, which implements the V_{FG} calculation procedure and has been necessary to overcome the problem of simulating a capacitive net in DC conditions; a voltage-controlled current source, $I_{W/E}$, connected between FG and D, to reproduce write/erase transient currents. The new V_{FG} calculation procedure represents the core of our model and has been implemented in the C code routine controlling the voltage controlled voltage source. It is based on the solution of the charge balance equation at the FG node. That is, the sum of the charge on the MOS gate, Q_G, plus the charge on the bottom plate of C_{CG}, has to be equal to the charge forced in/out the floating gate during cell program/erase operations, Q_{FG}, which in DC condition is a constant and depends on the state of the cell (for example,

Q_{FG}≈11.5fC and Q_{FG}≈-4fC, for an erased and written cell, respectively). Q_G is a complex function of the S, B, D, FG voltages, whose analytical expression has been obtained from the charge equations of the MOS compact transistor model adopted (in our case: Philips MOS Model 9, www.semiconductors.philips.com/Philips_Models/).

$$Q_G(V_{FG}, V_S, V_D, V_B) + C_{CG}(V_{FG} - V_{CG}) = Q_{FG}$$

The voltage-controlled current source, $I_{W/E}$, has been implemented by a C code routine to reproduce write and erase currents. As known, write and erase of a E^2PROM cell occur by Fowler-Nordheim tunnel across the thin oxide between D and FG of the cell. The oxide field has been calculated by means of a self consistent model taking into account poly-depletion as well as charge quantization effects, which has been developed by Larcher et al. (2001). Results obtained by this model has been parameterized on the oxide thickness, the p-Si substrate and the n-well D doping. In this way, the model can successfully evaluate the correct oxide field for different device parameters and voltage combinations.

2 Results

The model proposed in this paper allows to reproduce accurately the whole electrical behavior of E^2PROM memory cells (W=0.3μm, L=0.75μm and C_{GC}=3fF) manufactured in the existing technology (0.25μm) by ST-Microelectronics. In Figs.2-3, examples of fitting capabilities of our model are shown in DC conditions compared to experimental data. The agreement between measurements and simulations is excellent without the need of any free parameter to adjust fitting quality. Only a small amount of charge has been assumed on the Floating Gate node, which is probably due to charging effects during the manufacturing process of the cell. Moreover, this model is able to simulate very accurately also the transient behavior of E^2PROM (see Figs.4(a)-(d)). Particularly, threshold voltage shifts occurring during both write and erase operations are excellently fitted without calibrating any additional parameters (see Figs.4(a)-(b)), and the use of this model allows to inspect and monitor physical quantities like the FG voltage and the erase current which cannot be directly measured (see Figs.4(c)-(d)).

3 Conclusions

In this paper a new Spice-like compact model of a E^2PROM cell has been proposed. It allows the reproduction of the whole electrical behavior (DC and transient conditions) of the memory cell with great accuracy, and it features new great advantages compared to previous models. 1) It is based on a new procedure which improves strongly the FG voltage estimate, as well as the whole modeling of the cell. 2) It is very simple to implement, since it uses standard circuit elements whose parameters can be determined by applying the MOS parameter extraction procedure to the dummy cell (FG and Control Gate are short-circuited). The only additional parameter is the capacitance between FG and CG, C_{CG}. 3) This model is easily scalable, since the compact MOS model already takes the scaling rules into account and they do not influence the V_{FG}

calculation procedure. 4) The write/erase current source can be changed or upgraded independently of the other parts of the model, thus enabling future development the description of W/E mechanisms as well as the inclusion of leakage current contributions (SILC). 5) The accuracy of the model depends on the compact MOS model adopted, thus taking advantage of the many efforts to improve and scale the MOS compact models.

References

[1] R. Duane, A. Concannon, P. O'Sullivan, A. Mathewson (1998): Advanced numerical modelling of non-volatile memory cells. In Proceedings of the European Solid-State Device Research Conference, ESSDERC'98, Bordeaux, France. Edition Forntieres, pp. 304-307

[2] L. Larcher, P. Pavan, F. Pellizzer, G. Ghidini (2001): A new model of gate capacitance as a simple tool to extract MOS parameters. IEEE Trans. Electron Dev. 48: 935-945

[3] P. Pavan, R. Bez, P. Olivo, E. Zanoni (1997): Flash memory cells – An overview". In Proceedings of the IEEE, 85, pp.1248-1271

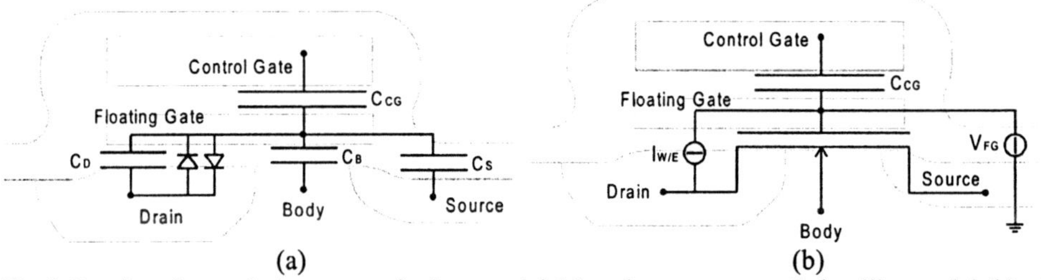

Fig.1. Previous lumped element equivalent model (a) and new compact Spice-like model (b) of a E^2PROM memory cell.

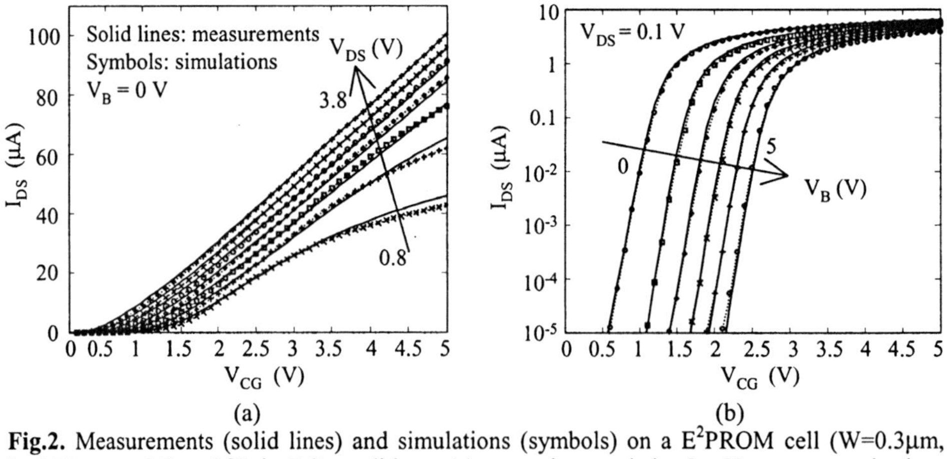

Fig.2. Measurements (solid lines) and simulations (symbols) on a E^2PROM cell (W=0.3μm, L=0.75μm and C_{GC}=3fF) in DC conditions: (a) trans-characteristics I_{DS}-V_{CG} parameterized on V_{DS}; (b) sub-threshold trans-characteristics I_{DS}-V_{CG} parameterized on V_B.

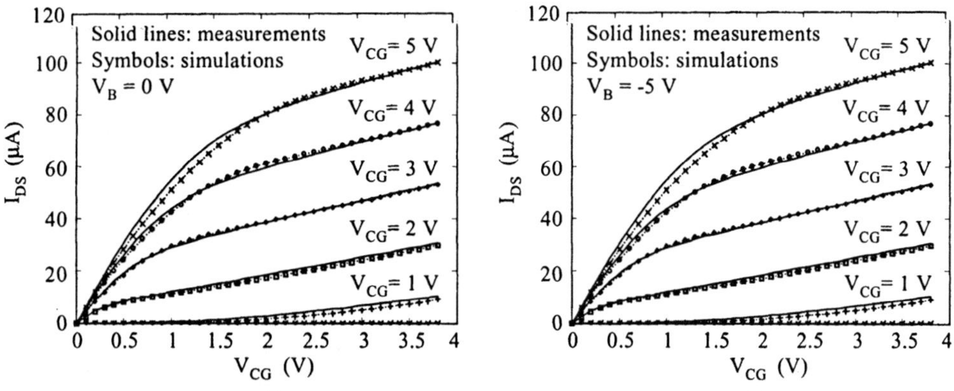

(a) (b)

Fig.3. Measured (solid lines) and simulated (symbols) I_{DS}-V_{DS} characteristics for V_B=0V (a) and V_B=-5V (b) of a E^2PROM cell (W=0.3μm, L=0.75μm and C_{GC}=3fF) in DC conditions.

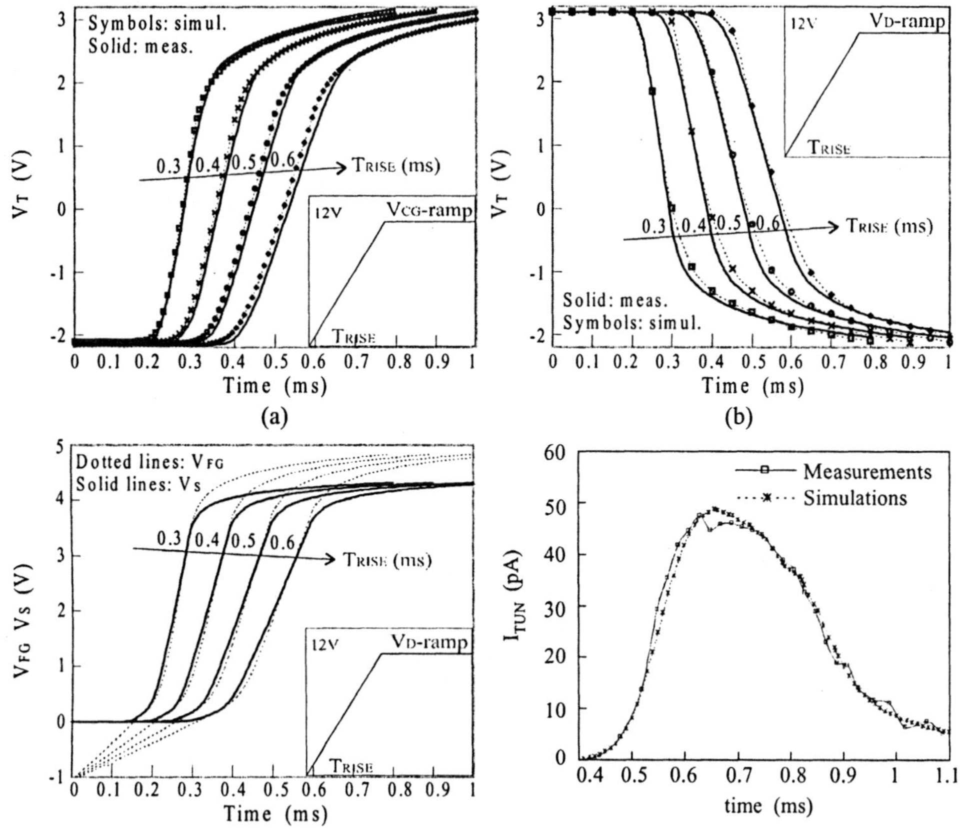

(c) (d)

Fig.4. Measurements (lines) and simulations (symbols) of the threshold voltage shift in erase (a) (V_D=V_S=V_B=0V, V_{CG}-ramps) and write (b) transients (V_{CG}=V_B=0V, V_D-ramps, S floating); (c) simulations of V_S (solid lines) and V_{FG} (dotted lines) during a write transients and of the Fowler-Nordheim current, I_{TUN}, during the erase of the cell.

Simulation of Flash Memory Programming Characteristics

K. Matsuzawa and T. Ishihara
Advanced LSI Technology Laboratory, Toshiba Corporation
8, Shinsugita-cho, Isogo-ku, Yokohama 235-8522, Japan
kazuya.matsuzawa@toshiba.co.jp

Abstract

A practical approach is presented to simulate programming characteristics of Flash memories. The current continuity equation for carriers injected in SiO_2 is embedded in the Poisson equation. The amount of hot electrons at the interface of substrate and SiO_2 is used as the fixed boundary condition for NOR-Flash, whereas spatial distribution of the FN tunnelling probability in SiO_2 is considered for NAND-Flash.

1 Introduction

In many cases of simulations for programming characteristics of NOR- and NAND-Flash memories, the capacitance value around the floating gate (FG) has been necessary (Lorenzini et al. 1999, Kim et al. 1998). However, it is sometimes troublesome to determine the value. In the present approach, the current continuity equation for carriers injected in SiO_2 is embedded in the Poisson equation around FG instead of specifying the capacitance value. The programming characteristics by hot electrons and FN tunneling are demonstrated.

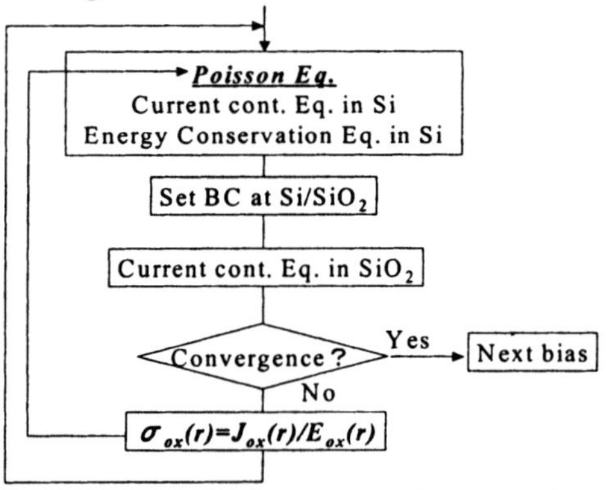

Fig. 1. A flowchart for simulation of programming characteristics of NOR-Flash.

2 Programming for NOR-Flash

Fig. 1 shows a flowchart for simulation of the hot electron programming of NOR-Flash in the present approach. First, the amount of hot electrons is estimated by solving the energy conservation equation in the silicon region. Then, distribution of current density around FG is obtained by solving the current continuity equation in the SiO_2 region. Distribution of electric conductivity around FG is calculated and the current continuity equation composed of current density J_{OX} in the SiO_2, J_{FG} in FG, and displacement current density J_{DISP} is set up around FG as shown in Fig. 2. The continuity equation can be solved in the framework of the Poisson solver, since all the current densities are expressed by gradient of the potential ψ. Fig. 3 shows a schematic of behavior of hot electrons injected in SiO_2. The amount of the hot electrons that surmount the energy barrier is used as the fixed boundary condition for the current continuity equation in SiO_2 and then the distribution of hot electrons injected into SiO_2 is obtained as shown in Fig. 4.

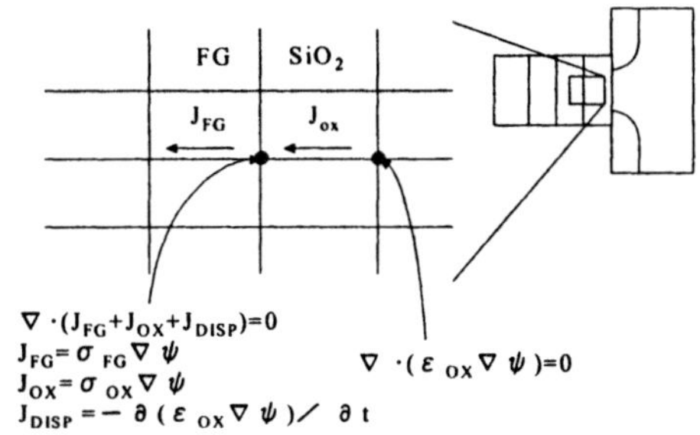

Fig. 2. Current continuity equation at the interface of FG and SiO_2.

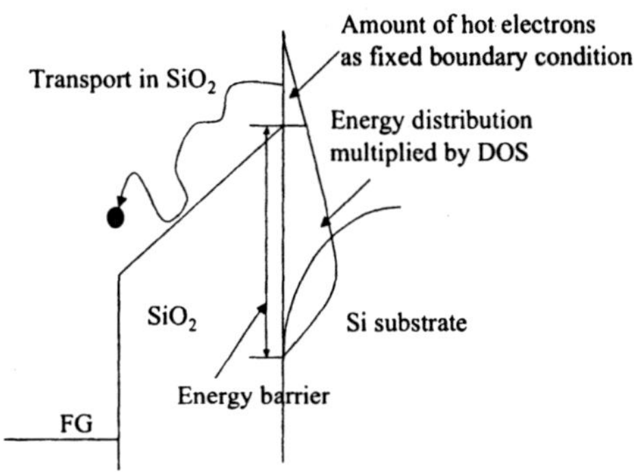

Fig. 3. Schematic of hot electron behavior in SiO_2.

Fig. 5 shows influences of the drain voltage V_D and the control gate voltage on the transient behavior of FG voltage V_{FG}. Ultimately, V_{FG} becomes lower as V_D is higher, since more hot electrons are generated. On the other hand, the ultimate value of V_{FG} tends to become the same value under the same V_D condition even if V_{CG} is different. This is because the combination of the gate current and V_{FG} tends to balance under the constant energy supply, namely the drain voltage.

3 Programming for NAND- Flash

Fig. 6 shows a schematic of FN tunneling. The tunneling probability was calculated based on the WKB approximation, considering the energy band gap in the floating gate and the substrate. Fig.7 shows the generation rate distribution of carriers in SiO_2 caused by the FN tunneling from the substrate under a programming condition. The gap of the distribution is due to the energy band gap of the substrate. It was found that the energy band gap affects the transient of V_{FG} as shown in Fig. 8.

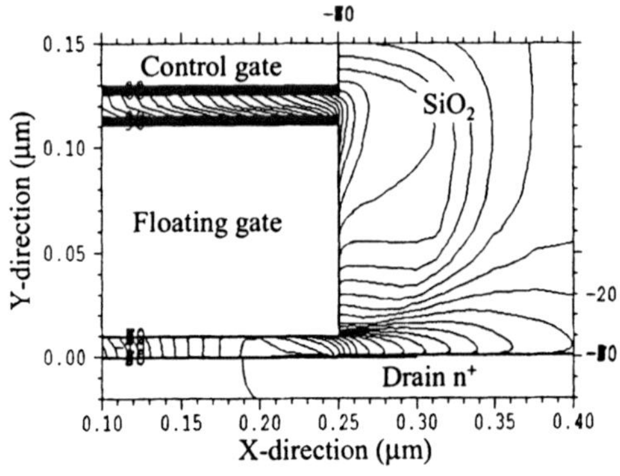

Fig. 4. Electron distribution injected into SiO_2.

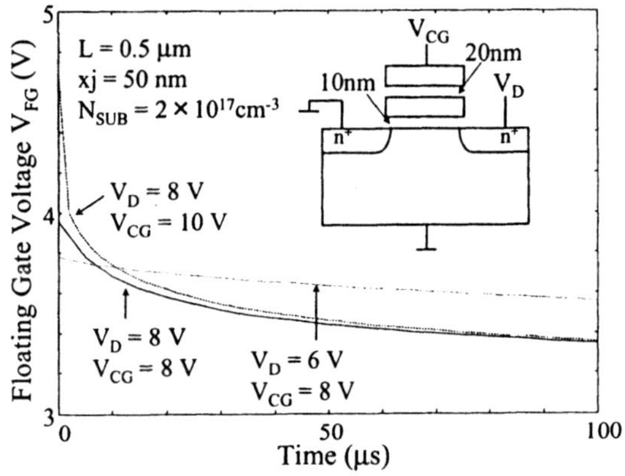

Fig. 5. Writing characteristics of NOR-EEPROM.

4 Conclusion

It is demonstrated that it is possible to simulate the programming characteristics of NOR- and NAND-Flash memories by embedding the current continuity equation for carriers injected into SiO_2 in the Poisson equation. The present approach allows us to simulate the characteristics self-consistently without specifying the capacitance value. Therefore, we expect that it can become a practical TCAD tool.

Acknowledgement

Authors would like to thank Mr. Hazama for his helpful information.

References

[1] Lorenzini, M., Vissarion, R., Rudan, M. Three-dimensional modeling of the erasing operation in a submicron Flash-EEPROM memory cell. IEEE Trans. Electron Devices 1999; 46: 975-983

[2] Kim, D.M., Cho, M.K., Kwon, W.HStacked gate mid-channel injection Flash EEPROM cell- Part II: Analysis of gate current and modeling of programming characteristics. IEEE Trans. Electron Devices 1998; 45: 1703-1709

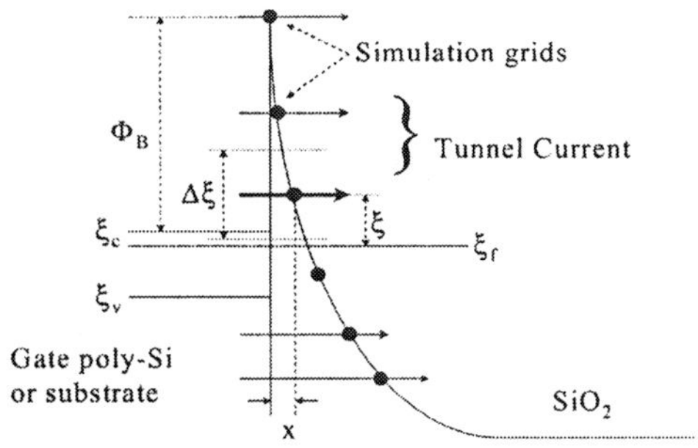

Fig. 6. Schematic of FN tunneling.

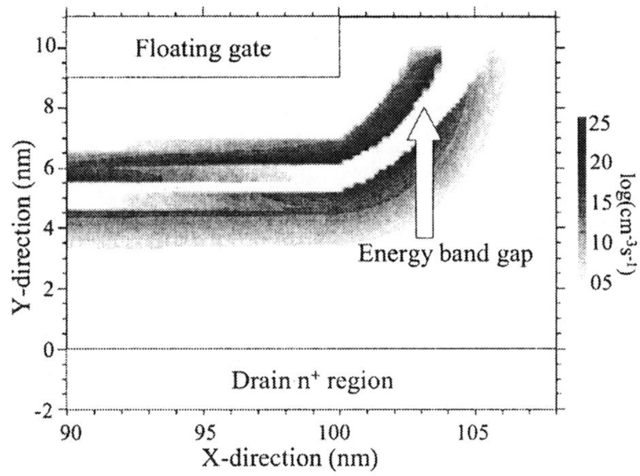

Fig. 7. Generation rate distribution by FN tunneling.

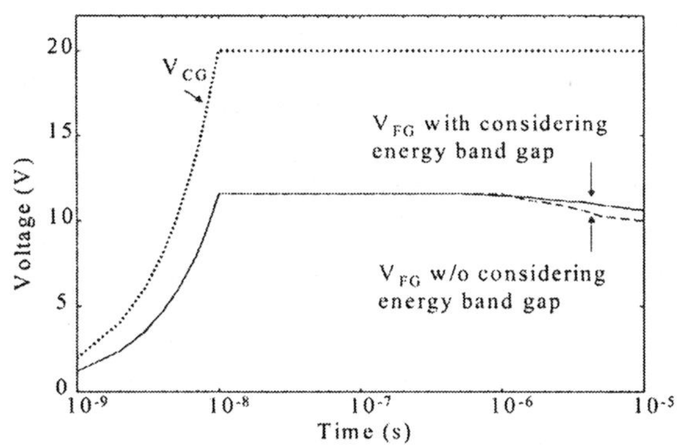

Fig. 8. Writing characteristics of NAND-Flash with similar parameters to those in Fig. 5.

A Figure of Merit for Flash Memory Multi-Layer Tunnel Dielectrics

B. Govoreanu[a,b], P. Blomme[a,b], M. Rosmeulen[a,b], J. Van Houdt[a] and K. De Meyer[a,b]

[a]STDI Division, IMEC Leuven
Kapeldreef 75, 3001 Leuven, Belgium

[b]ESAT, KU Leuven
Kasteelpark Arenberg 10, 3001 Leuven, Belgium

E-mail: Bogdan.Govoreanu@imec.be

Abstract

In this paper a figure of merit for evaluating the performance of multi-layer tunnel dielectrics for Flash memory applications is proposed. Further analysis provides an in-depth understanding of the multi-layer stacks and allows to select the most suitable stack for memory application.

1 Introduction

Scaling down the tunnel oxide of Flash memories faces major problems. The stress induced leakage current (SILC) strongly degrades the charge retention capability of the memory cell, once the tunnel oxide thickness is scaled down below 7-8 nm. Even in the absence of SILC, reducing the oxide thickness to 4-5 nm seems to be very difficult because of the direct tunneling current. Recent work [1] proposes to use high-k dielectric materials for replacing tunnel oxides and it has been shown [2] that by using a stack of several materials, enhancement of the Fowler-Nordheim tunneling is, in principle, possible.

In some cases, deposition of high-k dielectrics requires that a thin interfacial layer of SiO_2 exists between the silicon and the high-k dielectric. Hence, considering stacks with SiO_2 being one of the layers is of practical importance. This study shows that the existence of the oxide does not necessarily involve a decrease of the performance, and that careful analysis allows for selecting an optimal thickness of the oxide w.r.t. some specific performance indicators. For the sake of simplicity, in this paper only results of two-layer stacks are presented. They are compared with an oxide layer of the same electrical thickness (Equivalent Oxide Thickness - EOT). This choice has a major practical implication: since the capacitance of the stack is the same as that of the reference oxide layer, the capacitive coupling of the floating gate will remain unchanged and therefore the performance of the stack relative to that of the reference layer is a direct measure for the device performance improvement. Moreover, an easy translation of the time scale to the current density scale is possible, which simplifies the programming speed/retention time analysis.

2 Models

The tunnel current density was calculated using the independent electron approximation [3] and a WKB approach with Taylor expansion around the Fermi level in the injecting electrode, and taking into account the variation of the effective mass of the dielectrics across the stack. Although simplistic, this method gives results similar to the Airy implementation [4] and is still valid, at least within an order of magnitude, for the range of thicknesses used in Flash memory devices.

Fig.1 shows typical tunneling curves for the reference layer and for a stack of the same EOT. It is assumed that the dielectric constant of the reference layer (for this analysis, SiO_2 layer) equals the lowest dielectric constant within the stack. At high voltages, the energy barrier shifts downward so that only the first layer of the stack still shows a barrier to the tunneling particle. The electric field is the same in the reference layer and in the first layer of the stack and it therefore follows that in the high voltage range both I-V curves coincide. Maximum performance of the stack w.r.t. the reference layer is obtained when programming at a voltage V_P for which the current ratio J_{stack}/J_{ref} is maximum. Moreover, the retention performance of the stack is better as long as the maximum disturb voltage does not exceed the V_R value, where the stack and the reference curves cross. To get a narrow transition region between the two voltage levels, and therefore a steep programming curve, the ratio V_P/V_R should be relatively small.

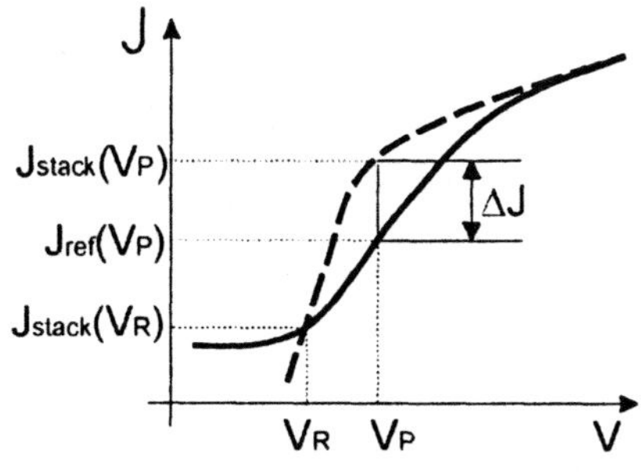

Fig.1. Tunneling curves for the reference layer (solid line) and for the stack of the same electrical thickness

The corresponding ratio of the currents $J_{stack}(V_P)/J_{stack}(V_R)$, should be of the order of several decades, according to the programming speed and charge retention requirements, expressed in current levels. The programming speed is related to the current level in a voltage range around V_P and is traded off for low voltage operation. *The voltage factor* $F_V = V_P / V_R$ *and the current factor* $F_J = \log(J_{stack}(V_P)/J_{ref}(V_P))$ *subject to the constraints* $J_{stack}(V_R) \leq J_{max}$ *and* $J_{stack}(V_P) \geq J_{min}$ *are proposed as a figure of merit for comparing stacks of identical EOT.* The current levels J_{max} and J_{min} should be fixed according to the retention and speed requirements, being related to the coupling capacitances of the cell.

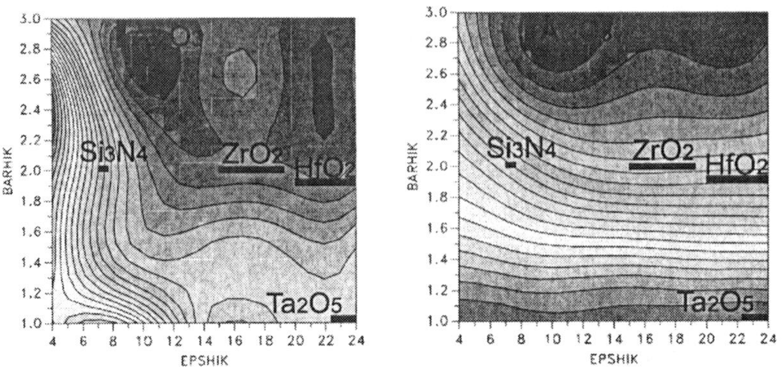

Fig.2. Contour plots of F_V (left) and F_J (right) as functions of k_{hk} (EPSHIK) and $\Phi_{B0,hk}$ (BARHIK) at $t_{ox,stack}$ = 3 nm, for 7 nm EOT.

3 Results and Discussion

The dielectrics that are currently being investigated for SiO_2 replacement usually have lower energy barriers and higher dielectric constants (Tab. 1). Without loss of generality we limit ourselves here to a range from 4 to 24 for the dielectric constant and 1 to 3 eV for the barrier height relative to the bottom of the Si conduction band.

Material	SiO_2	Si_3N_4	Al_2O_3	ZrO_2	HfO_2	Ta_2O_5
Dielectric constant k[-]	3.9	7.5	9-10	15-20	20-25	~25
Barrier Height Φ_{B0}[eV]	3.15	2.0	~3.0	~2	~1.5	~1.0

Table 1: Dielectric constants and barrier heights of some dielectrics

Simulations were carried out for stacks of different EOT and showed similar results to those presented hereafter, where an EOT of 7 nm is considered. A multilevel design of experiment (DOE) approach was used. The independent variables are the dielectric constant of the high-k material k_{hk}, the barrier height of the high-k material ($\Phi_{B0,hk}$) and the physical thickness of the oxide in the stack ($t_{ox,stack}$). A statistical interpolation method [5] was used to construct the models. Contour plots for a fixed thickness of 3 nm oxide are shown in Fig. 2, where the black strips correspond to the considered dielectrics. From the analysis of the models, the following conclusions are drawn: (a) better F_V are obtained for the Al_2O_3, ZrO_2 and HfO_2 stacks; (b) F_J is, however, more favorable in case of either HfO_2 or Ta_2O_5 and it is also less sensitive to the dielectric constant; this is consistent with the fact that such barriers have less "tunneling area" as compared to the stacks with higher barrier height; (c) high programming currents indicate stacks with ZrO_2 or Al_2O_3 or layers whereas lower retention current is achieved for stacks with higher k and lower Φ_{B0} if the oxide is not very thin and for stacks with higher k and higher Φ_{B0} for very thin oxide layers. Overall, it can be derived that the SiO_2/ZrO_2 stack is among the most promising combinations. By fixing k_{hk} and ($\Phi_{B0,hk}$) an optimal thickness of the oxide layer can be found by minimizing the objective function $FO = \alpha(F_V - F_{V0})^2 + (1-\alpha)/F_J^2$, with α a parameter which trades off the voltage range operation for the programming speed. For the case of a stack of 5 nm EOT and assuming a transition region so that Fv = 2.0, the optimal thickness of the SiO_2 layer in the stack is 1.8 nm, if no requirement is placed on the current factor. However, maximizing the programming speed leads to an oxide thickness of 3 nm (Fig.3), if no constraint is placed on Fv.

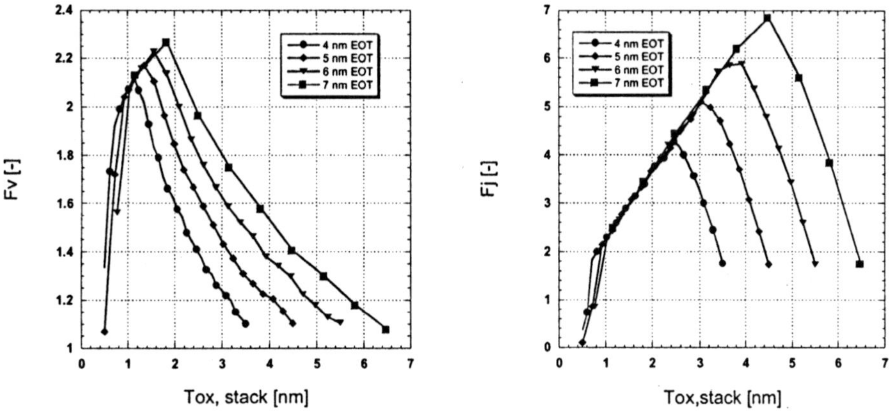

Fig. 3. Voltage and current factors for a SiO_2/ZrO_2 stack, as a function of the oxide thickness in the stack, at different equivalent oxide thicknesses

4 Conclusion

The tunneling current through multilayer stacks was compared to the current through a reference monolayer tunnel dielectric. This allowed for the definition of a figure of merit to be used for evaluating several stacks of a given EOT as replacement for the tunnel dielectric in Flash memories. SiO_2/ZrO_2 stack is proposed as one of the most promising candidates, which allows for higher speed or lower voltage programming.

Acknowledgement – The authors wish to thank Dr. M.Lorenzini, Dr. W.Magnus, Dr. W.Schoenmaker (IMEC Leuven) for useful discussions and comments.

References

[1] Manchanda L. High-K dielectrics for CMOS and Flash. Proc. ICSSDM, 1999, 150-151

[2] Likharev K.K. Layered tunnel barriers for nonvolatile memory devices. Appl Phys Lett 1998, 73(15):2137-2139

[3] Harrison W.A. Tunneling from an independent particle point of view. Phys Rev 1961, 123(1): 85-89

[4] Schenk A, Heiser G. Modeling and simulation of tunneling through ultra-thin gate dielectrics. J Appl Phys 1997, 81(12):7900-7908

[5] Sacks J, Welch W.J, Mitchell T.J, Winn H.P. Design and Analysis of Computer Experiments. Stat Science, 1989, 4:409-435.

Enhanced Diffusion of Phosphorus due to BPSG layer in SEG-MOSFETs

Jaehee Lee, Woo-Seock Cheong*, Jae-Hoon Choi, and Jae-Chul Om

Device Physics and Modeling Team and *Advanced Process Team 3
Memory R&D Division, Hynix Semiconductor Inc.
San 136-1, Ami-ri, Bubal-eub, Ichon-si, Kyoungki-do, 467-701 Korea
TEL: 82-31-639-1887 FAX: 82-31-630-4545 E-mail:jaelee@hynix.com

Abstract

We report on the enhanced diffusion characteristics of phosphorus in SEG(selective epitaxial growth)-MOSFETs due to in-diffusion from BPSG (borophosphosilicate glass) layer. Because of structural complexity of SEG-MOSFETs, three-dimensional simulation was implemented and our results successfully show that a great deal of impurity in-diffusion from BPSG to silicon layer generates a lot of interstitial. Therefore, the diffusivity of phosphorus is increased dramatically by TED effect, causing the characteristics of SEG-MOSFETs is degraded severely.

1. Introduction

As the dimensions of MOS devices are shrunk to the 0.1μm and below, ultra-shallow source and drain junctions are strongly required to prevent short channel effects. MOSFETs with ESD(elevated source and drain) realized by SEG of silicon have been proposed as one of the alternative approaches to very shallow S/D junctions for improving short-channel effects[1]. In the case of NMOS devices of DRAM cell, as shown in Figure 1, ESD is faced by BPSG layer that is dielectric material for cell transistor of DRAM. It is known that BSG, PSG, or BPSG yields solid-phase diffusion of boron or phosphorus to the silicon bulk layer [2,3]. As the formation of SEG needs high temperature (>800℃) for a few minutes, the enormous dopant diffusion by highly doped silicate glasses can cause degradation of MOS characteristics. In this paper, we report on the enhanced diffusion kinetics and degradation characteristics of SEG-MOSFETs due to in-diffusion from BPSG layer.

2. In-diffusion characteristics

Figure 2 shows in-diffusion characteristics of BPSG. Epitaxial silicon layer with phosphorus of $4\times10^{19}cm^{-3}$ was grown to 2500 Å, followed by deposition of BPSG and by thermal anneal of 900℃ for 6 minutes. This result shows that BPSG is primarily a source of phosphorus, and the phosphorus out diffusion is enhanced at higher boron concentration. A huge amount of phosphorus in BPSG diffuses into silicon bulk layer, while boron diffusion is retarded. It is because that concentration of phosphorus in BPSG is about 4 wt% and has the segregation coefficient from silicon to oxide of 10, on the contrary boron has the segregation coefficient from silicon to oxide of 0.1.

Figure 3 shows the comparison of diffusion distance between TSUPREM4 default values and that used in this simulation. A huge amount of phosphorus diffusion at high concentrations was successfully observed by our diffusion parameters considering diffusion barrier lowering effect[4,5].

3. SEG-MOSFET degradation

Figure 4 shows phosphorus diffusion results for epitaxial growth layer. Epi growth was performed at 900℃ for 6 minutes and BPSG was additionally deposited on the epi layer in Figure 4(b). After thermal treatment of 900℃, experimental results were compared to simulation, which showed good agreement. In Figure 4(b), the diffusion barrier lowering effect causes a huge amount diffusion of phosphorus into silicon. This in-diffused dopant substitutes the silicon on lattice site, which generates interstitial and causes transient enhanced diffusion of phosphorus in silicon bulk. Diffusion length of phosphorus in bulk layer is increased 15 times greater than that in Figure 4(a). It is found that in-diffusion from BPSG not only increases the doping concentration at the interface area but also causes the dopant in bulk area to be diffused by TED effect.

Based on the above simulation results, degradation effect of 0.15μm NMOS cell device was investigated. Because of structural complexity of SEG-MOSFET as shown in Figure 1, it is hard to get the exact solution by two-dimensional simulation. Therefore, three-dimensional simulation was implemented as shown in Figure 5.

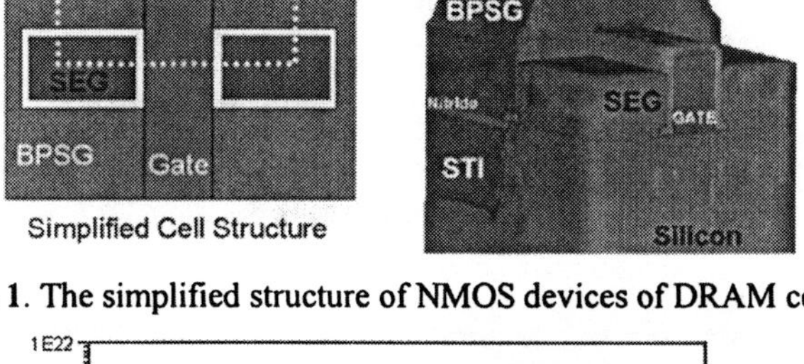

Fig. 1. The simplified structure of NMOS devices of DRAM cell.

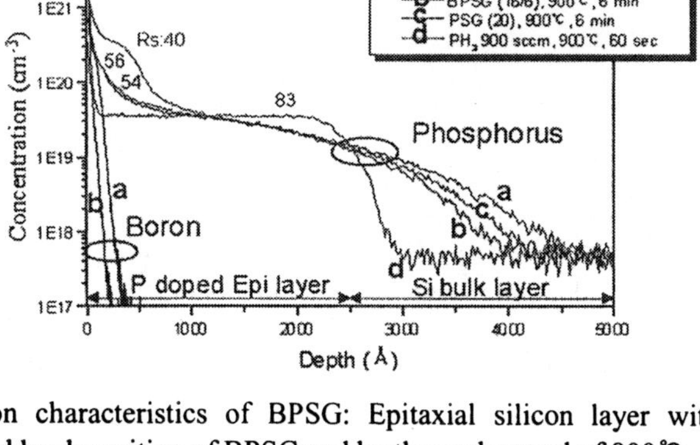

Fig. 2. In-diffusion characteristics of BPSG: Epitaxial silicon layer with phosphorus of $4x10^{19}cm^{-3}$ followed by deposition of BPSG and by thermal anneal of 900℃.

Phosphorus diffused out of BPSG, as mentioned before, not only increases the concentration of N-type dopant at the interface but generates interstitial of which diffusivity is so high that interstitial broadens out into silicon bulk layer immediately. Interstitials kick out bulk phosphorus to diffuse. Y-Z plane of Figure 5 shows that abnormally diffused phosphorus changes the desired doping profile under the gate. Therefore, net doping of channel near the STI interface is decreased or inverted from P-type to N-type, increasing narrow width effect and leakage current.

To prevent such in-diffusion from BPSG, plasma treatments were implemented after BPSG patterning process. Hardening the surface of BPSG layer by plasma treatment effectively prohibits silicon bulk layer from in-diffusion of BPSG layer. Finally, the characteristics of the NMOSFET having elevated source and drain for cell device is shown in Figure 6. Because the source and drain were formed by in-situ doping in epitaxial growth process instead of ion implantation, reverse short channel effect of threshold voltage is not found, which is usually shown in short channel device. Our simulation results show good agreement with experimental data without any fitting manipulation.

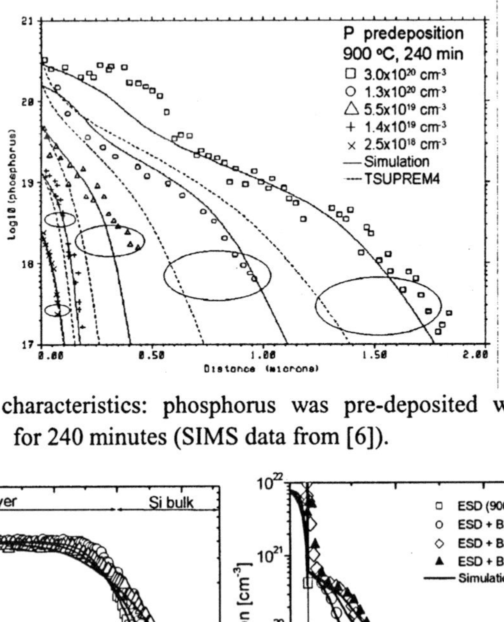

Figure 3. Diffusion characteristics: phosphorus was pre-deposited with various ambient concentration at 900℃ for 240 minutes (SIMS data from [6]).

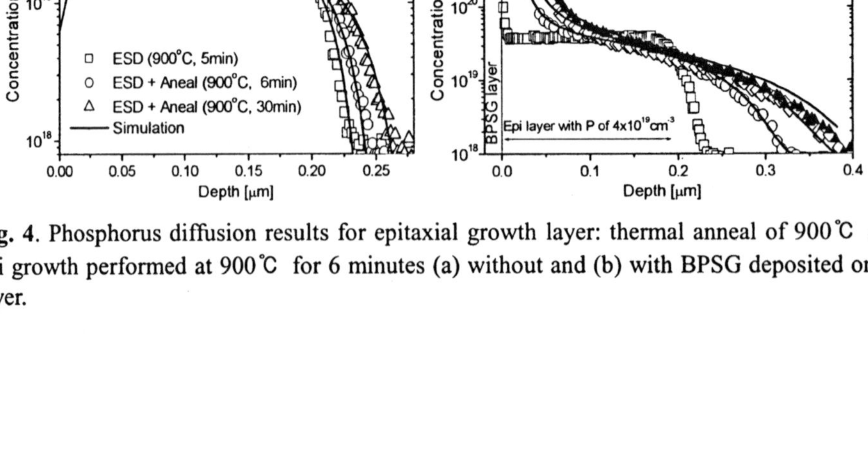

Fig. 4. Phosphorus diffusion results for epitaxial growth layer: thermal anneal of 900℃ after epi growth performed at 900℃ for 6 minutes (a) without and (b) with BPSG deposited on epi layer.

4. Conclusions

In this paper, the degradation of SEG-MOSFETs due to enhanced diffusion of phosphorus is reported. Even though our device has a structural complexity, three-dimensional simulation successfully shows that abnormally diffused phosphorus by BPSG changes the desired doping profile under the gate. Therefore, net doping of channel near the STI interface is decreased or inverted from P-type to N-type.

References

[1] Nakahara, Y. et al. (1996) In: *Sym. VLSI Technology* pp. 174-175.
[2] Ono, M. et al. (1993) In: *IEDM 1993*, pp. 119-122.
[3] Packan, P. et al. (1998) In: *IEDM 1998*, pp. 505-508.
[4] Mathiod, D. et al. (1984): J. Appl. Phys., Vol. 55, p. 3518.
[5] Wittel, F. (1996): *Ph. D. thesis*, Boston University.
[6] Yoshida, M. (1974): J. Appl. Phys., Vol. 45, p. 1498.

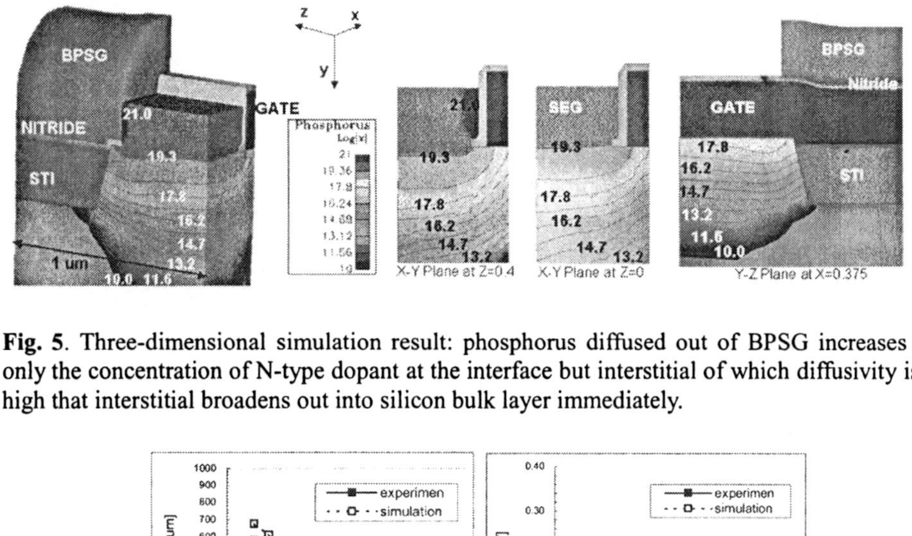

Fig. 5. Three-dimensional simulation result: phosphorus diffused out of BPSG increases not only the concentration of N-type dopant at the interface but interstitial of which diffusivity is so high that interstitial broadens out into silicon bulk layer immediately.

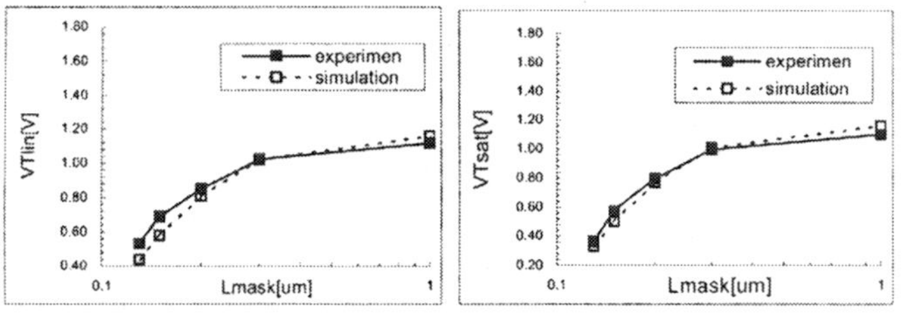

Fig. 6. The electrical characteristics of the cell NMOSFET with ESD.

Neutron-SER Modeling & Simulation for 0.18μm CMOS Technology

Changhong Dai, Nagib Hakim, Steve Walstra, Scott Hareland, Jose Maiz, Scott Yu, and Shiuh-Wuu Lee
Intel Corporation, Santa Clara, California, USA
changhong.dai@intel.com

Abstract

This paper presents a new and physical modeling approach for neutron SER with excellent accuracy demonstrated on SRAMs fabricated using 0.18μm CMOS technology. The SER contribution of each type of recoil ion and a fast roll-off behavior of neutron SER for high Q_{CRIT} nodes are reported for the first time.

1 Introduction

Neutrons from terrestrial cosmic rays and alpha particles from radioactive impurities in packaging materials are the primary radiation sources that contribute to the soft-error rate (SER) of electronic products in non-space applications. [1-2] While the alpha particle contribution increases at a faster rate as processing technology scales to smaller feature sizes,[3] its impact can be minimized using advanced packaging techniques, such as low-emissivity solder materials. Neutron contributions, however, cannot be easily reduced due to the relatively fixed environment and will continue to dominate the logic circuit SER. In the past two decades, many experimental and theoretical studies have been performed to analyze neutron SER, but most of them have used the charge-burst model, which treats all recoil ions collectively, thus preventing detailed understanding of each recoil ion's contribution. [4-8] The charge-burst model also relies on experimental determination of charge collection efficiency, which could be more accurately analyzed using numerical 3D device simulations that provide superior predictability to future technology generations and enable S/D doping engineering for SER reduction.

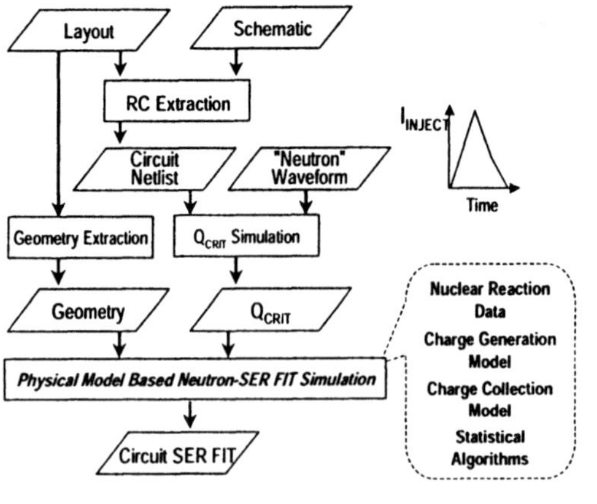

Fig. 1. Neutron-SER modeling and simulation methodology and flow

This paper presents a new and physical modeling and simulation approach for neutron SER, shown in Fig.1, including (1) neutron-silicon-nuclear-reaction-based recoil ion charge generation model, (2) numerical-device-simulation-based charge collection model, and (3) statistical algorithms for FIT (Failure-In-Time) rate simulation. We also present the simulated and experimental neutron-SER results of SRAMs fabricated using 0.18µm CMOS technology.[9] This paper for the first time reports the SER contribution of each type of recoil ion as well as the fast roll-off behavior of neutron-SER for high Q_{CRIT} nodes. The physical models and simulation capabilities discussed in this paper have also been used to study the impact of CMOS process scaling and SOI on soft error rates of logic processes. [10]

2 Modeling and Simulation Methodology

Q_{CRIT} for a given node is the critical amount of charge required for a given circuit to switch when a current pulse representing a recoil ion's charge is injected at that node. It is obtained via circuit simulations from the minimum pulse height that logically fails the circuit. Compared to alpha strikes, recoil ions typically have a much faster transient process, resulting in a sharper collection waveform and smaller Q_{CRIT} because the circuit node has less time to respond. The neutron SER can then be calculated by computing the total probability of all recoil ions with a collected charge exceeding Q_{CRIT}.

3 Charge Generation and Collection Modeling

As an electrically neutral particle, a neutron can only interact with silicon nuclei through strong interaction, which can produce recoil ions with Zs ranging from 2 (He) to 15 (P). Fig. 2a shows the probabilities of each type of ion produced within a unit volume and time as a function of the recoil ion's kinetic energy. These probabilities are based upon the nuclear reaction cross-sections between monotonic neutrons and silicon nuclei and weighted by the sea-level neutron energy spectrum, as shown in Fig. 2b. [2, 11-12] Secondary neutrons and protons can also be produced, but are not included in the modeling because of their negligible contribution.

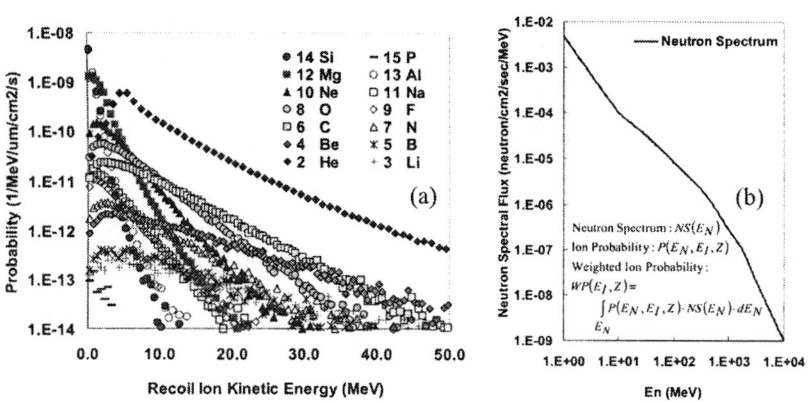

Fig. 2. (a) Energy transfer probability of recoil ions in neutron-silicon interaction, weighted by the sea level neutron spectrum; (b) Sea level neutron spectrum

The charge generation model is based on the ion energy loss rate in silicon (shown in Fig. 3) and the recoil path location relative to the S/D-to-well junction and trench isolations (shown in Fig. 4a).

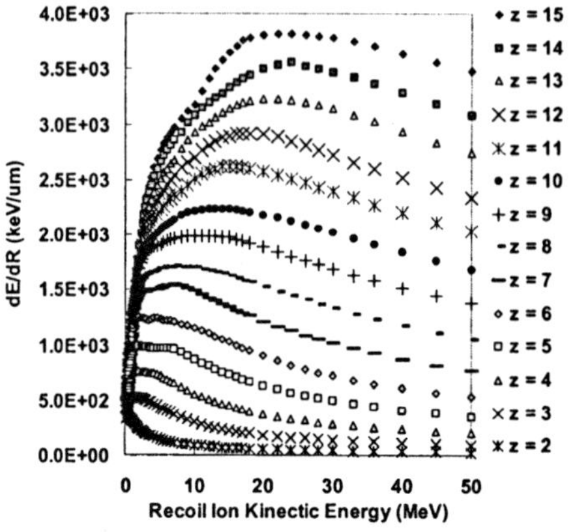

Fig. 3. Recoil ion energy loss rate in silicon, simulated using TRIM [13]

The charge collection model is obtained using Intel's proprietary numerical device simulator, by placing a point charge source at various locations near the S/D-to-well junction during the transient simulation and measuring the proportion of charge collected. As shown by Fig. 4b, the charge collection efficiency is near 1 inside depletion region and decreases almost linearly outside this region. By combining this collection efficiency with the charge generation function along the recoil path, the total charge collection can be obtained for a given recoil event.

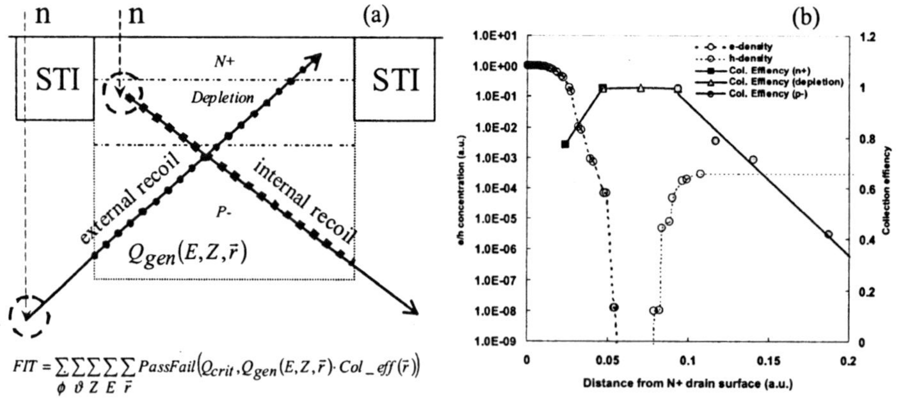

Fig. 4 (a) Charge generation model and the general equation for charge collection and FIT calculation; (b) Charge collection efficiency derived from 3D device simulations and its correspondence to the S/D-to-well junction doping profile

4 Statistical Algorithms for FIT Simulation

A random recoil ion event consists of seven variables: recoil angles (ϕ, θ), ion nuclear number Z, ion kinetic energy E, and recoil starting location (x, y, z). The distribution associated with ϕ and θ is typically assumed to be uniform for the non-elastic scattering events and is sampled linearly. The distribution associated with Z and E can be obtained from Fig. 2a and is sampled with a prioritized order. The distribution associated with the recoil starting location is uniform but not limited to the charge collection box underneath the S/D area, thus making it difficult to capture the contribution of "external" recoil ions. As shown by Fig. 5, a novel technique was developed to use segmented surface areas to count the cumulative probability of external recoil ions entering the charge collection box and the internal cubes to count the probability of "internal" recoil ions.

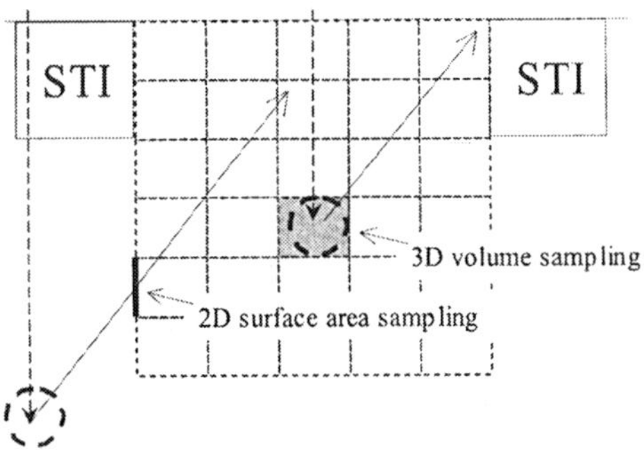

Total Ion Probability = $TP(E_I, Z)$

For 3D volume sampling of internal recoils

$$TP(E_I, Z) = V_{sampling}\, WP(E_I, Z)$$

For 2D surface sampling of external recoils

$$TP(E_I, Z) = S_{sampling} \cdot \int_{E_I}^{E_{max}} \frac{WP(E, Z) \cdot (\hat{n} \cdot \hat{r})}{(dE/dR)} \cdot dE$$

$\hat{n}$ and $\hat{r}$ are the unit vectors of sampling surface and ion path

Fig. 5. The novel technique for effective sampling of recoil ions initiated inside and outside the charge collection box underneath S/D area. The surface area sampling comprehends all recoil ions initiated outside but eventually entering the box.

5 Results and Discussion

Fig. 6a shows the simulated and experimentally measured neutron-SER FIT rates of SRAMs fabricated using 0.18µm technology. While the models and simulation algorithms are developed for logic products, SRAMs are usually the most convenient vehicle for verification and calibration. The physical-model-based statistical simulation of neutron SER agrees with the experimental data very well without using any fitting parameters. Both simulated and measured results show a power-law dependence between the FIT rate and Q_{CRIT}, which is different from the exponential relationship observed from alpha SER. [3] By examining the SER contribution of each type of recoil ion, as shown in Fig. 6b, we find that each ion type does obey the exponential dependence on Q_{CRIT}, just as alpha particles do, revealing the same fundamental SER physics masked by the collective behavior of recoil ions. The simulated results also show a saturation trend at very low Q_{CRIT}, similar to what has been reported before for alpha-SER because of the dominant contribution from secondary alpha particles. [3] At very high Q_{CRIT}, the simulation results show a fast roll-off behavior, which can be physically explained by the exclusive contribution from heavy recoil ions that bear the exponential dependence to make neutron-SER roll off faster than what a power-law predicts. Both trends are extremely critical for accurate SER modeling as well as technology and design planning.

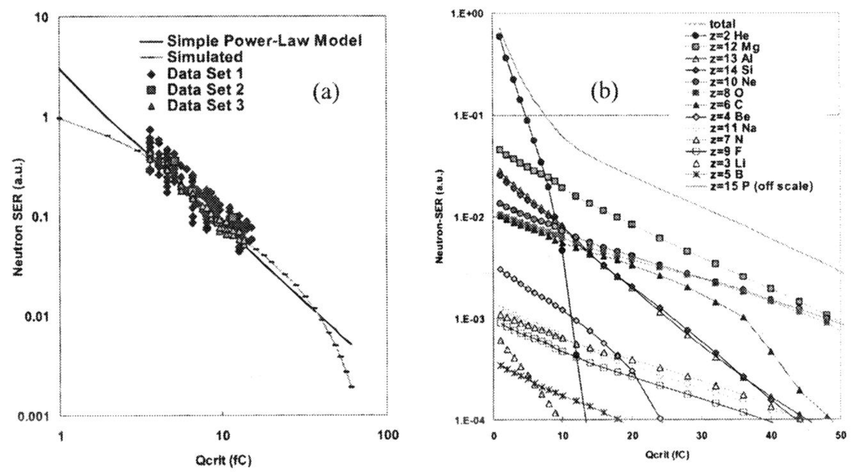

Fig. 6. (a) The simulated and experimentally measured neutron-SER FIT rates of the SRAMs fabricated using 0.18mm logic technology. As a collective behavior of neutron-SER, a power-law dependence on Q_{CRIT} is observed; (b) SER contribution of each type of recoil ion that obeys an exponential dependence on Q_{CRIT}. The total neutron-SER is dominated by secondary alpha particles at low Q_{CRIT} region and heavy ions at high Q_{CRIT} region.

6 Conclusions

A new and physical neutron-SER modeling and simulation approach has been developed with excellent accuracy demonstrated on SRAMs fabricated using 0.18μm CMOS technology. The SER contribution of each type of recoil ion and a fast roll-off behavior at high Q_{CRIT} are reported for the first time.

Acknowledgements

The authors would like to thank H.H.K. Tang of IBM, S. Maston, J. Bordelon and N. Mielke for their critical contribution to this work.

References

[1] T. C. May et al., "Alpha particle-induced soft errors in dynamic memories", *IEEE TED*, vol. ED-26, p. 2, 1979.

[2] J. F. Ziegler et al., "The effect of sea level cosmic rays on electronic devices", *J. Appl. Phys.*, vol. 52, p. 4305, 1981.

[3] C. Dai et al., "Alpha-SER modeling & simulation for sub-0.25um CMOS technology", *Symposium on VLSI Technology*, p. 81, 1999.

[4] J. R. Letaw et al., "Guidelines for predicting single-event upsets in neutron environments", *IEEE Trans. Nucl. Sci.*, vol. 36, p. 2348, 1991.

[5] E. Normand, "Single event effects in avionics", *IEEE Trans. Nucl. Sci.*, vol. 43, p. 461, 1996.

[6] G. R. Srinivasan et al., "Accurate, predictive modeling of soft error rate due to cosmic rays and chip alpha radiation", *IEEE/IRPS Proceeding*, p. 12, 1994.

[7] Y. Tosaka et al., "Measurements & analysis of neutron-reaction-induced charges in a silicon surface region", *IEEE Trans. Nucl. Sci.*, vol.44, p. 173, 1997.

[8] Y. Tosaka et al., "Measurement and analysis of neutron-induced soft errors in sub-half-micron CMOS circuits", *IEEE TED*, vol. ED-45, p. 1453, 1998.

[9] S. Yang et al., "A high performance 180 nm generation logic technology", *IEEE IEDM Tech. Digest*, p. 197, 1998.

[10] S. Hareland et al., "Impact of CMOS process scaling and SOI on the soft error rates of logic processes", *Symposium on VLSI Technology*, 2001.

[11] H. H. K. Tang et al., "Cascade statistical model for nucleon-induced reactions on light nuclei in the energy range 50 MeV – 1 GeV", *Phys. Rev.*, vol. C42, p1598, 1990.

[12] H. H. K. Tang et al., "Nuclear physics of cosmic ray interaction with semiconductor materials: Particle-induced soft errors from a physicist's perspective", *IBM Journal of R&D*, 1996.

[13] J. F. Żiegler et al., *TRIM User Manual*, January 1995.

TCAD Driven Process Design of 0.15μm Fully-Depleted SOI Transistor for Low Power Applications

N. Miura, H. Hayashi, H. Komatsubara, M. Mochizuki,
H. Matsuhashi, Y. Kajita and K. Fukuda
VLSI Research Center, Oki Electric Industry Co., Ltd.
550-1 Higashiasakawa, Hachioji, Tokyo 193-8550, Japan
Email : miura288@oki.co.jp

Abstract

We presented a TCAD-driven total design methodology of FD-SOI MOSFETs, starting from 0.35μm/2.5V shrinking to 0.15μm/1.5V. Jumping from 0.35μm to 0.15μm, two-phase experiments are performed effectively supported by exhaustive applications of TCAD local models. SOI specific consideration of SOI film thickness variations (σTsoi) and floating-body effects are the key points for the TCAD driven strategy.

1 Introduction

Silicon-on-insulator(SOI) is a promising technology for low power applications. Especially, fully-depleted (FD) SOI MOSFET has advantages in short channel effects and subthreshold characteristics. In deep sub-micron region, however, trade-off relationships of FD-SOI MOSFETs are more complicated than those of bulk transistors. Therefore TCAD-driven process design would be more efficient in FDSOI cases. In this paper, TCAD-driven total design methodology of SOI MOSFETs, starting from 0.35μm/2.5V shrinking to 0.15μm/1.5V, is presented. SOI specific consideration of SOI film thickness variations (σTsoi) and floating-body effects are the key points for the TCAD driven strategy.

2 Fabrication Process

The commercially available SIMOX wafers with 100nm-thick buried oxide are used as the starting material. The starting Tsoi is about 60nm and the final Tsoi is thinned down to 40nm. Gate oxide (electrical thickness Tgox:4nm) is grown after a LOCOS isolation. After gate pattering and spacer formation, a cobalt-silicide process is applied to reduce the source-drain sheet resistance.

3 TCAD driven development of 0.15μm FD-SOI transistor

In 0.15μm FDSOI transistor development, we focus on the power consumption in view of supply voltage reduction (requirement: Vdd=1.5V) maintaining the Ion-Ioff(@Vd=2.5V) of the preceding generation 0.35μm FD-SOI. In order to keep process compatibility for quick development, the basic well process sequences of

0.35μm FD-SOI transistor are inherited. Our TCAD driven process design methodology of 0.15μm FD-SOI transistor consists of the following 3 steps (Fig.1).

3.1 Certifying overall process strategy with 0.35um SOI models (phase-0)

Fig.2 shows the calibration results of 0.35μm SOI. Using this calibration model, we estimate short channel effects and Ion-Vdd trends before developing 0.15μm SOI. Fig.3 shows the simulated Ion trends having the same Ioff for FD-SOI devices. On the basis of this outlook for device scaling and Vdd reduction, basic scaling directions (Tgox=4nm, Tsoi=40nm) are determined.

3.2 Building basic Vth window considering σVth effects (phase-1)

In first-step experiment, we construct the response surface functions(RSFs). The RSFs can be used for the following 3 purposes: (A) Making rough process windows (B) Estimating device performance variation (C) Obtaining experimental data for local model refinement [1] to optimize the process conditions. TCAD calibration model of the preceding generation is effective for estimating process control parameters range (Fig.4: as for (A)). Also the variation of device performance due to process variations (considering σTsoi, σTgox, σLgate and σVtdose) is predictable (Fig.5: as for (B)). In order to clarify the effect of σTsoi, we obtain σVth extracted from RSFs (Fig.6). As Fig.6 shows, the behavior of σVth-Tsoi varies significantly as Lgate is varied. The behavior specific to FD-SOI devices are caused by the combination of Tsoi variations and short channel effects. In this case (Tsoi=40nm), it is revealed that the influence of σTsoi is smaller than that of short channel effect. Process optimization based on (C) is discussed in 3.3 .

3.3 Optimization of extension/halo considering floating-body effects (phase-2)

In this optimization step, the refined TCAD calibration model is used for more precise device performance prediction even if process parameters are extrapolated. In 0.15μm SOI process, LDD structure, which is adopted in 0.35μm SOI for suppressing floating-body effects, cannot satisfy the Ion-Ioff requirements for 1.5V Vdd. Furthermore, it is difficult to keep a fully-depleted state at long Lgate, especially in low-Ioff requirement (<0.01nA/μm). In order to overcome this situation, extension/HALO process is adopted in 0.15μm SOI process. As refined-model predicted, S/D extension increases the Ion compared to LDD (Fig.7) because of suppressing parasitic source-drain resistance, and HALO implant can effectively suppress to be partially-depleted state at long channel (Fig.8) without thinning Tsoi and/or heavier channel doping. On the other hand, as Ion are increased, BVsd is decreased because the impact-ionization and floating-body effects are enhanced. Especially in FD-SOI devices, such phenomena are also influenced by the change of Vt-rolloffs which are due to the variations of Tsoi and Lgate. In order to maximize the Ion under the specifications of BVsd, we optimized the process considering the effect of σTsoi and σLgate by using TCAD (Fig.9).

4 Device Performance

Fig.10 shows that 0.15μm FD-SOI devices have the advantage of bulk devices having the same Tgox and Ioff. Fig.11 shows the fabricated IdVd characteristics of 0.35μm SOI and 0.15μm SOI transistor (having the same Ioff). As we expected, the supply voltage of 0.15μm SOI can be successfully decreased while maintaining the Ion-Ioff of 0.35μm SOI.

5 Summary

We presented a TCAD-driven total design methodology of FD-SOI MOSFETs, starting from 0.35μm/2.5V shrinking to 0.15μm/1.5V. 0.15μm FD-SOI MOSFETs are successfully optimized in short time by the TCAD-driven process design. Jumping from 0.35μm to 0.15μm, two-phase experiments are performed effectively supported by exhaustive applications of TCAD local models. TCAD advantages specific to FDSOI MOSFETs design are as follows:

a) The effects of σTsoi on σVth are clarified in the 1st Vth window step.

b) Considering σTsoi, BVsd limited by floating-body effects are optimized in the 2nd extension/HALO optimization step.

References

[1] Mar, J. (1996): The Application of TCAD in Industry. In: Proceedings of International Conference on Simulation of Semiconductor Process and Devices, SISPAD' 96, Tokyo. Japan Society of Applied Physics, Tokyo, pp.139-145

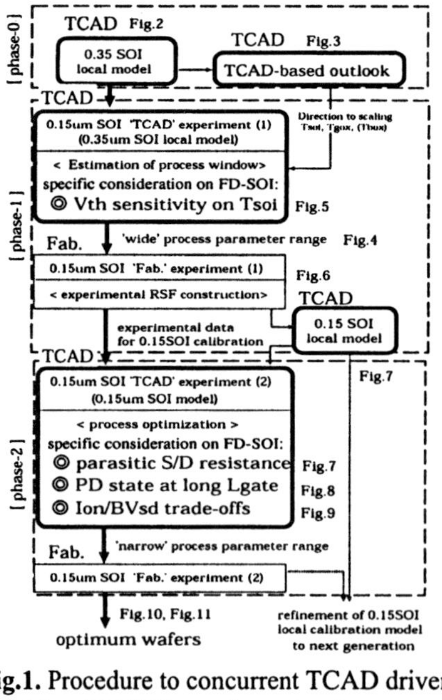

Fig.1. Procedure to concurrent TCAD driven 0.15um FD-SOI transistor development.

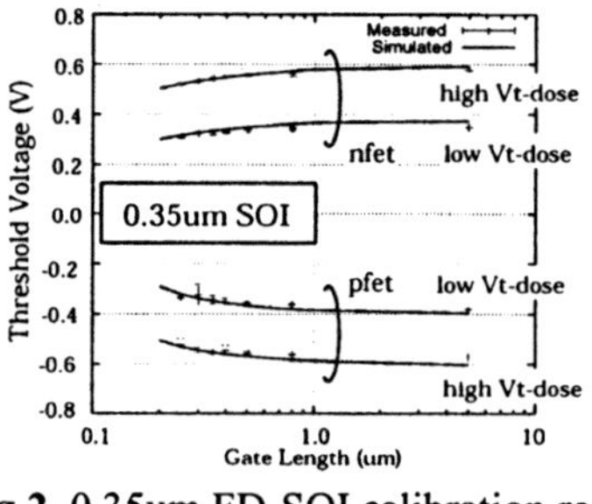

Fig.2. 0.35um FD-SOI calibration results.

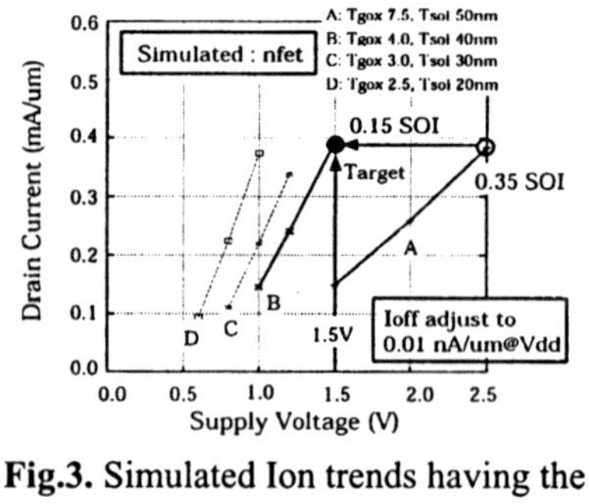

Fig.3. Simulated Ion trends having the same Ioff FD-SOI devices.

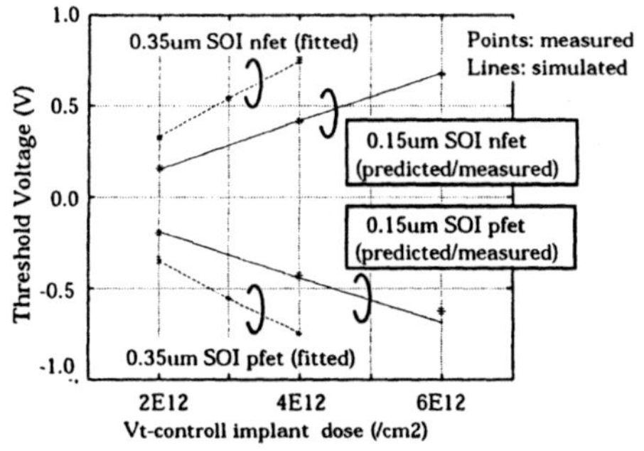

Fig.4. Vt-Implant dose dependence of Vth.

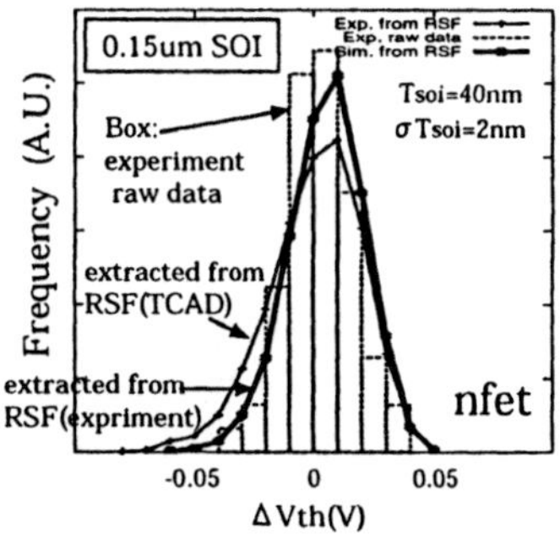

Fig.5. 0.15um SOI Vth distributions.

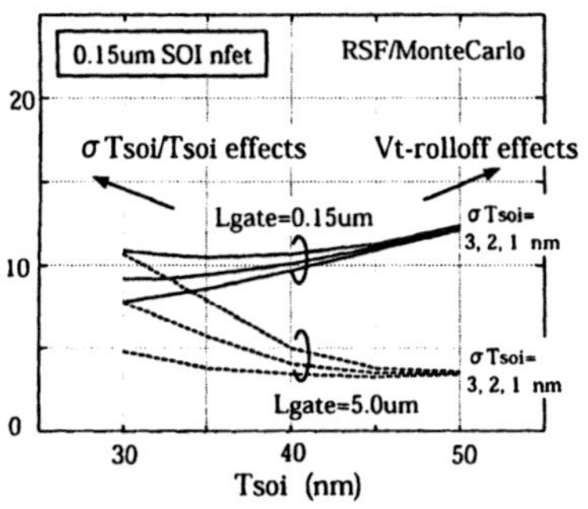

Fig.6. σVth v.s. Tsoi for σTsoi=1,2,3nm

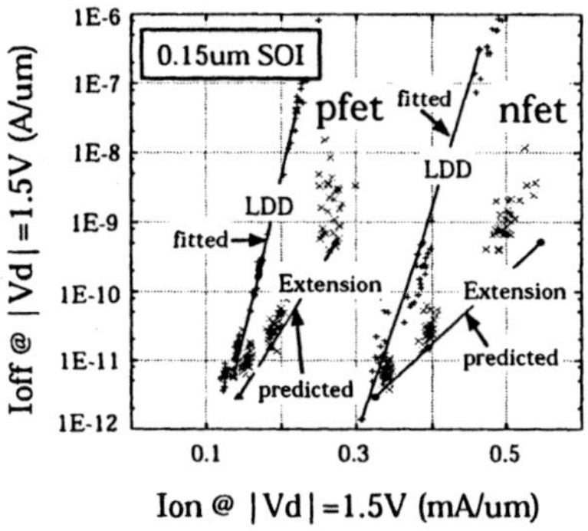

Fig.7. 0.15um SOI: Ion-Ioff characteristics.

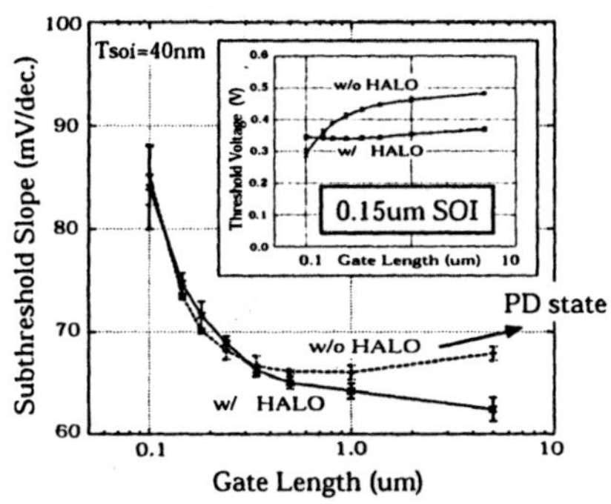

Fig.8. Subthreshold slope for nfet.

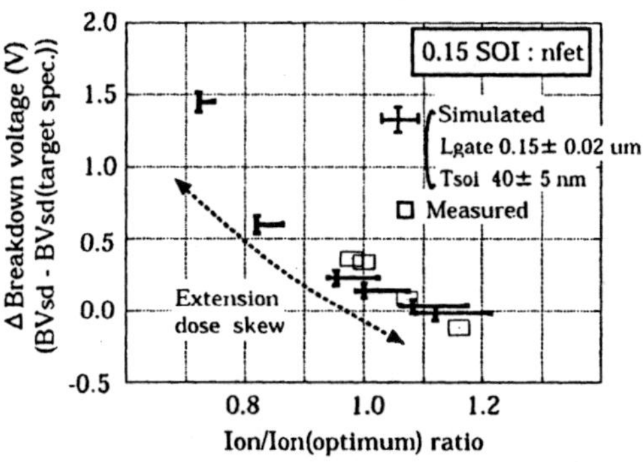

Fig.9. BVsd v.s. Ion considering σTsoi and σLg.

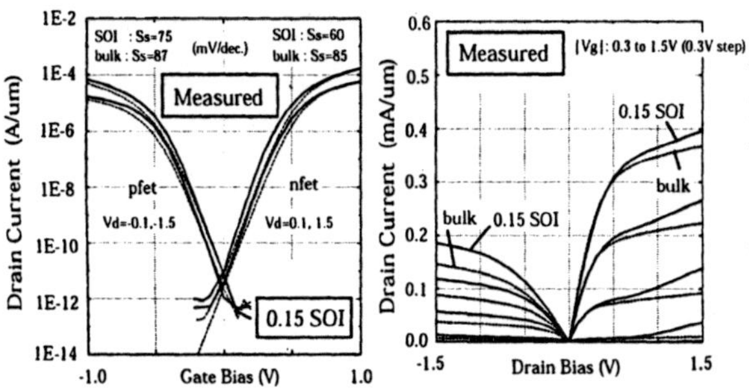

Fig.10. IdVg and IdVd characteristics of fabricated 0.15um FD-SOI and bulk (Tox_inv=4nm).

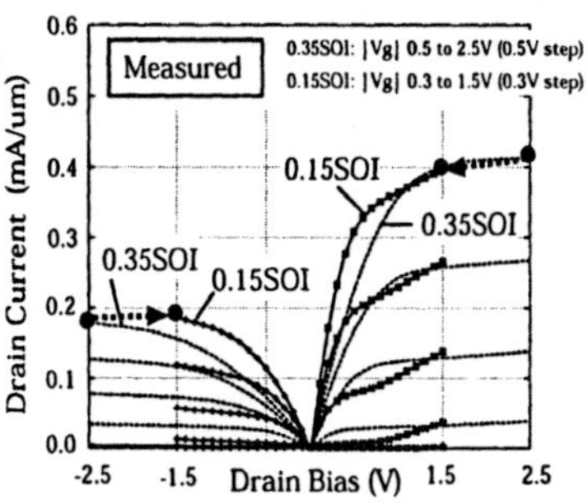

Fig.11. IdVd characteristics of 0.35um SOI and 0.15um SOI.

A Simulation Evaluation of 100nm CMOS Device Performance

S K Jones, D J Bazley
Caswell Technology, Marconi Optical Components,
Caswell, Towcester, Northants NN12 8EQ, UK
stephen.k.jones@marconi.com

E Augendre, G Badenes, A de Keersgieter
IMEC,
Kapeldreef 75, B-3001 Leuven, Belgium

T Skotnicki,
ST Microelectronics,
850 rue Jean Monnet, 38926 Crolles, France

Abstract

This paper presents a simulation evaluation of the performance of 100nm MOSFET devices. Calibration of 2D physical device models against measured 0.13 micron NMOS and PMOS devices forms the basis for a study of different MOSFET device options. The sensitivity of individual design factors is assessed on the device Ion/Ioff performance. The feasibility of achieving the ITRS-2000 specifications is addressed and limitations to the benefit of gate dielectric thickness scaling are identified.

1 Introduction

The continued applicability of 2D drift-diffusion transport modelling to MOSFETs, albeit with careful calibration of velocity-field characteristic and transverse field dependence, is a huge benefit to the TCAD support of technology development. Models can be quickly calibrated to a sufficient accuracy and deployed in large-scale simulation studies to provide an initial feasibility analysis and guidance of device design and performance specifications for new CMOS technology generations. This paper presents an evaluation of the Ion/Ioff performance of 100nm MOSFET devices based upon predictive simulation from device models calibrated on 0.13μm CMOS.

2 Models

In this work, models were calibrated to data from 0.13 micron CMOS development with 2.5nm gate dielectric[1]. This included assessment of the doping activation in the gate, as shown in fig. 1, and the mobility model, as shown in fig 2. A parameterised

---

[1] S Kubicek et al "Investigation of intrinsic transistor performance of advanced CMOS devices with 2.5nm NO gate oxides" Proc IEDM-99 p823 (1999)

MOSFET template was then established in the 2D-device simulator, allowing variation of important doping and geometry variables. These include physical gate dielectric (equivalent SiO_2, EOT) thickness (Tox), gate activation, source/drain extension (MDD) dose and junction depth (X_J), inner spacer (1) and outer spacer (2) widths, channel (VTA) and halo dose, gate length (L_G) and supply voltage (V_{DD}). Design-of-experiments methods were used to evaluate the effects of these variables and to build response-surface models for the device parameters, which were then exported into a spreadsheet tool.

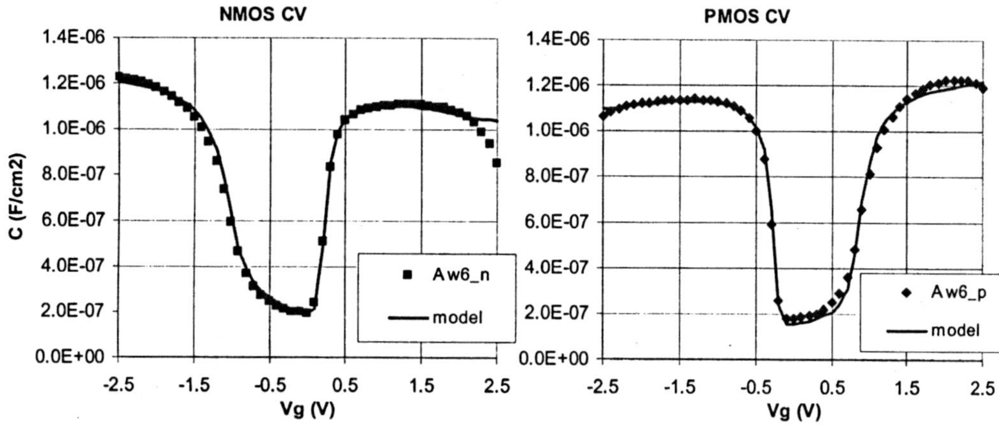

Fig. 1. C(V) for the N+ and P+ gate capacitors comparing model and measurements.

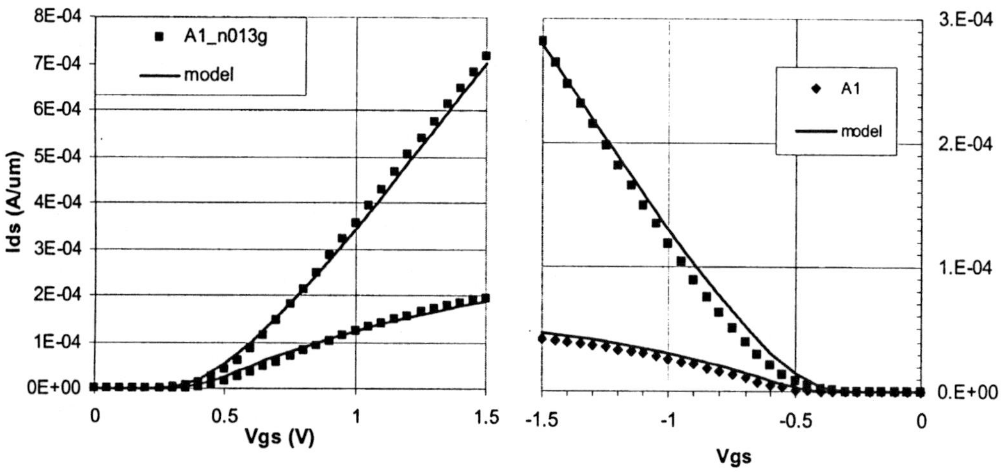

Fig. 2. Id(Vg) plots for 130nm NMOS device at Vds=0.1 and 1.5V comparing calibrated Avant! MEDICI™ model and measurements.

3 Device Sensitivity

The main effects in the on/off current (Ion/Ioff) space are shown as vector plots in fig. 3 for NMOS and PMOS conventional MOSFETs with halo doping to counteract short-channel effects. The centre-point device structure variables are shown and use Tox of 1.85nm. The effect of modifying the channel dose, shown as a dashed line, represents the neutral behaviour of the device (VTA tunes the desired off-current for the technology). Design directions to the right of this line increase performance.

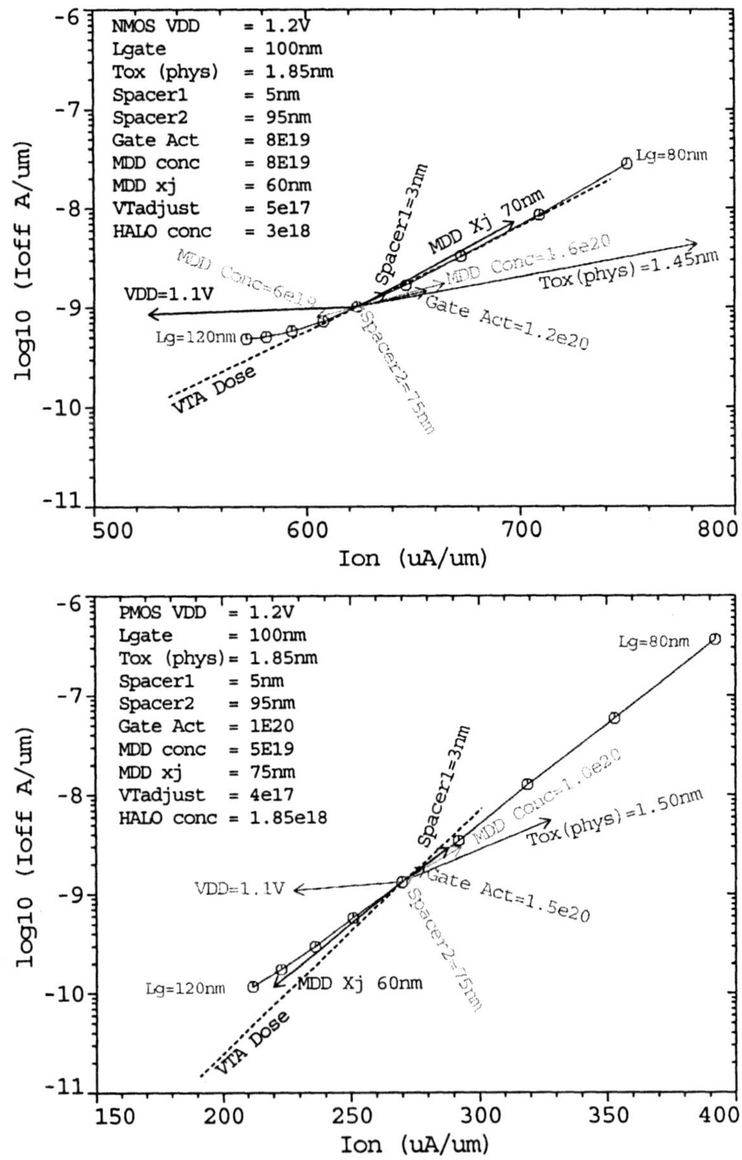

Fig. 3. Effects on Ion/Ioff performance of NMOS (upper) and PMOS (lower). $I_{on} = I_{ds}(@V_{ds}=V_{gs}=V_{DD}, L_G^{nom})$, $I_{off}=I_{ds}(@V_{ds}=V_{DD}, V_{gs}=0, L_G^{nom})$. L_G^{nom} is the nominal gate-length.

The strongest effect is from gate oxide thickness: reducing to 1.45nm gives a 25% increase to the NMOS Ion with only x4 increase in Ioff (assuming that gate leakage is controlled to <1A/cm^2). There are also benefits in increasing the extension and gate doping activation. Interestingly, reducing L_G does not give a strong increase in performance, due to worsening short-channel effects at this X_J: a 15% decrease in L_G gives x9 increase in Ioff. V_{DD} has a significant effect: reducing from 1.2 to 1.1V gives a 16% reduction in Ion. Aggressive scaling of the equivalent gate dielectric thickness leads to a high-field mobility reduction which outweighs the increased gate capacitance, ultimately producing a net degradation in the drive-current normalised to a given off-current (figure 4). For 1.2V the limit is 0.85nm increasing to 1.2nm for low leakage devices.

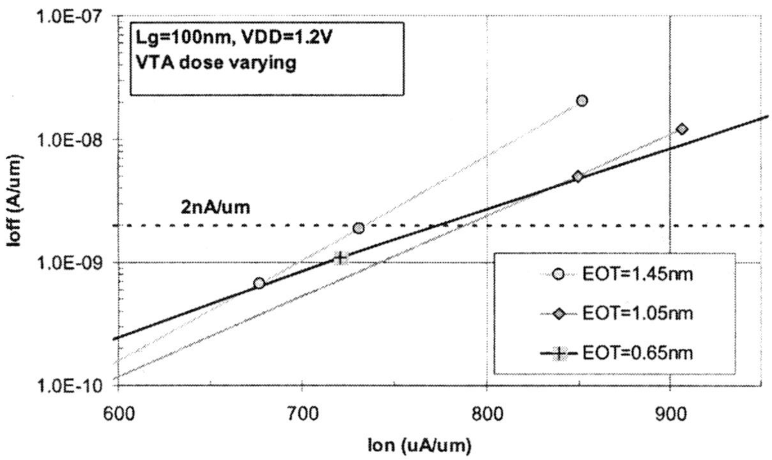

Fig. 4. Predicted I_{on}/I_{off} behaviour of L_G=100nm NMOS devices for different gate dielectric thickness (equivalent SiO_2) at V_{DD}=1.2V. The lines represent the effect of adjusting channel dose. For I_{off} = 2nA/µm, scaling the gate dielectric below 0.85nm gives no benefit.

4 Conclusions

We estimate that the ITRS[2] high performance targets of N/PMOS I_{on}=750/350µA/µm can just be achieved with Tox=1.5nm, V_{DD}=1.2V at I_{off} of 2nA/µm for nominal L_G (100nm), which is x10 lower than the ITRS I_{off} specification at worst-case gate length. The drive-current benefit from future scaling of gate dielectric thickness (and gate length) is compromised by mobility degradation at high fields and high doping.

Acknowledgements: This work was partly funded by the EC Project IST-1999-11599 HUNT (Hundred-Nanometre CMOS Transistors).

[2] ITRS-2000 Update, http://public.itrs.net

A Practical Approach to Modeling Strained Silicon NMOS Devices

Phil Oldiges, Xinlin Wang, MeiKei Ieong, Stephen Fischer, and Ken Rim[†]
IBM SRDC, 2070 Rte. 52, Hopewell Jcn., NY 12533
[†]IBM T.J. Watson Research Center, Yorktown Heights, NY 10598
E-mail: poldiges@us.ibm.com Phone: (845)892-2511 fax: (845)892-3039

Abstract

Parameters for a generalized mobility model are extracted from hardware measurements of strained silicon NMOS devices. Only the phone limited mobility parameter was shown to vary with amount of strain in the silicon.
Using results of the mobility model fitting, simulations of strained and unstrained silicon devices were performed. These simulations indicated that strained silicon devices should have improved short channel control as well as yield a minimum of 50% improvement in on current.

1. Introduction

It has been known for a long time that biaxial strain in silicon leads to higher mobility for in-plane transport of electrons [1,2]. Both the low and high field mobility are increased due to a splitting of the conduction band, resulting in a higher population of electrons in the lower energy valleys where they exhibit a lower effective mass. The most common and perhaps best developed method of creating a thin strained silicon layer is by growing a silicon layer on unstrained $Si_{1-x}Ge_x$ as demonstrated in [3]. In this work, we show a practical method for simulating strained silicon NMOS devices using standard drift-diffusion equations and a commonly used silicon mobility model that has been calibrated to low field (linear) Id-Vg hardware data. Using this calibration, we then simulate hypothetical sub 100 nm gate length strained silicon and conventional devices to compare expected DC performance, showing better short channel effect, Idlin and Ion for the strained silicon devices.

2. Low Field Mobility Calibration

Strained silicon and conventional long channel devices were fabricated and the linear I-V characteristics were measured. For the strained silicon devices, Ge content was fixed at either 15% or 20%. An approximate device structure was defined based on the process used to manufacture the devices, and Fielday [4] simulations were then performed. For both the control device and strained silicon devices, the Mujtaba mobility model [5] was used. The Ge content in the SiGe layer and strain in the silicon were taken into account by using appropriate band offsets in the defined device.

A Levenberg-Marquardt optimization program was used to adjust mobility model parameters to fit the Id-Vg hardware data. Since strained silicon should not necessarily exhibit the same mobility-field relationship as conventional silicon, we added two more adjustable parameters in the Mujtaba model. Within the phonon limited mobility portion of the model, there are two terms dependent upon transverse field. One of these terms has the functional form: $C/(E_t^{\gamma 1})$, with $\gamma 1=1/3$. The surface roughness limited mobility portion of the model has a term which shows the functional form: $\delta/(E_t^{\gamma 2})$, with $\gamma 2 = 2$. The two parameters that we used as adjustable ones were $\gamma 1$ and $\gamma 2$. In addition to $\gamma 1$ and $\gamma 2$, we also fit the δ and C parameters. During the optimization procedure, we found that $\gamma 1$ and $\gamma 2$ changed by less than 5% from their original values and that the surface roughness mobility term, δ also changed by less than 5%. This tells us that the low field mobility improvement is due to changes in phonon scattering and that surface roughness scattering is not affected much by strain. For higher transverse electric fields than for those shown here it is expected that the surface roughness scattering term will have a large impact [7], and for strained silicon devices the surface roughness limited mobility is expected to be larger than that for conventional silicon devices.

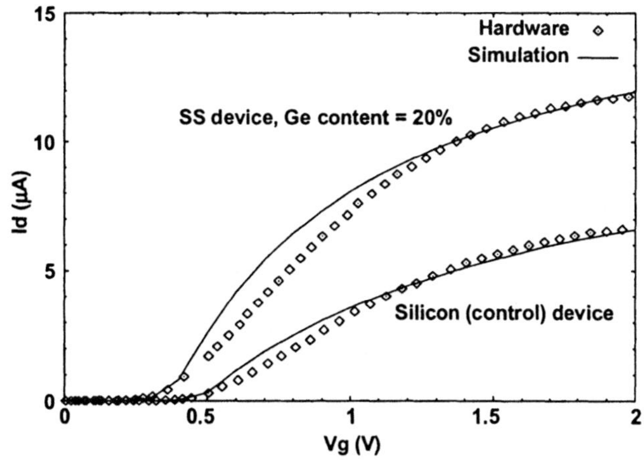

Fig. 1. Comparison of measured and simulated linear Id-Vg characteristics of 10X10 strained and unstrained silicon devices.

Table 1. Normalized values of C extracted from optimization.

Device	Ge content	C (normalized)
Control	0%	1.0
SS	15%	2.7
SS	20%	4.0

The only parameter that had a significant impact on the fit was the phonon limited mobility parameter, C. Figure 1 shows the fit compared to hardware data for the silicon control device and the strained silicon device with a Ge content of 20%. A similar fit was performed to device characteristics with 15% Ge content. The values of C (normalized to the silicon control device) are shown in Table 1. To a reasonable approximation, the phonon limited mobility is proportional to the Ge content of the unstrained SiGe layer. Since the biaxial strain in the thin silicon layer is linearly proportional to the Ge content, the phonon limited mobility can be seen to be proportional to strain. The fit of the simulations to hardware is certainly not as good as it can be. Taking into account a more detailed understanding of the device structure including S/D resistance and poly depletion effects and using additional mobility parameters would improve the match, but our intent here is to show feasibility using a conventional silicon mobility model and its application to device design.

3. DC Device Performance Improvements

We define a generic sub 100 nm process such that the off current is 10 nA/μm for the shortest channel length device (80nm). Although Ge content and strain will affect the dopant distribution in actual devices [6], we assume that the doping profiles in both sets of devices follow the same shape. Power supply was set to 1.5 V. We compare a control (silicon) device to a strained silicon device with Ge content of 20%. For high field mobility considerations, [1] shows that velocity saturation in silicon is 1.07×10^7 cm/sec and in $Si_{0.8}Ge_{0.2}$, Vsat = 1.3×10^7 cm/sec. Figure 2 shows the Vt roll off characteristics of these prototype devices. There is a lot of roll up in the characteristics due to the relatively heavy halo that was defined. Figure 3 shows the ratio of Idlin for the various gate lengths. For long channel lengths, we see an improvement in Idlin of greater than the mobility improvement of 75% due to the strain. The additional improvement in Idlin is due to Vtlin of the strained silicon devices being less than that of the control devices. For shorter channel lengths, the difference in Vtlin is less and the halo doping in the strained silicon devices is starting to affect the channel mobility. Figure 4 shows the Ion/Ioff characteristics. A significant improvement in on current can be obtained using strained silicon devices. Higher off currents for long channel devices are due to higher junction leakage in the strained silicon devices.

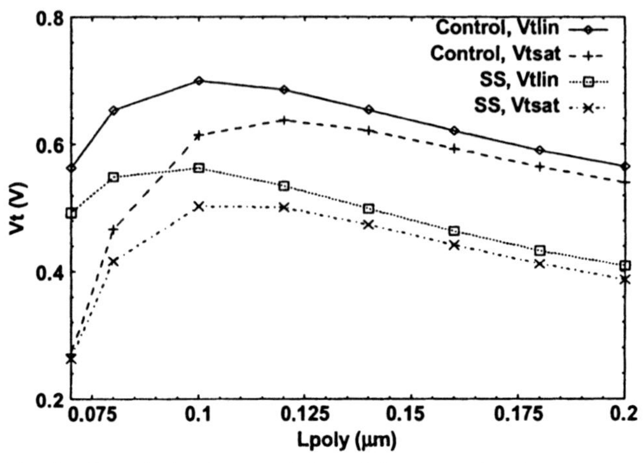

Fig. 2. Vt roll off characteristics of the control and strained silicon devices. Vtsat for the 70 nm gate length devices are similar due to the constraint of off current being 10 nA/μm.

Devices designed around nominal gate length should be compared to look at important differences between strained silicon and conventional device design. Because we constrain the off current at minimum channel length to be 10 nA/μm, Vtsat for minimum device length are the same. Due to band

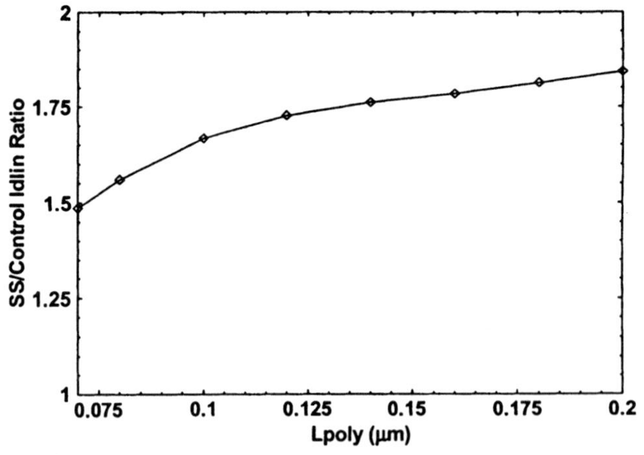

Fig. 3. Ratio of linear current for strained silicon and control devices as a function of gate length.

offsets in the strained silicon layer, Vtsat is shifted to a lower value than a conventional device with the same doping profile. To compensate for this, the background doping in strained silicon devices must increase, yielding better control over short channel effects. For a nominal gate length of 0.1 μ m, Vtsat of the strained silicon device is approximately 100 mV lower than the conventional device. Because of the lower Vt and higher saturation velocity we see an Ion improvement of ~50%. We can also see that short channel effect is improved significantly for strained silicon devices.

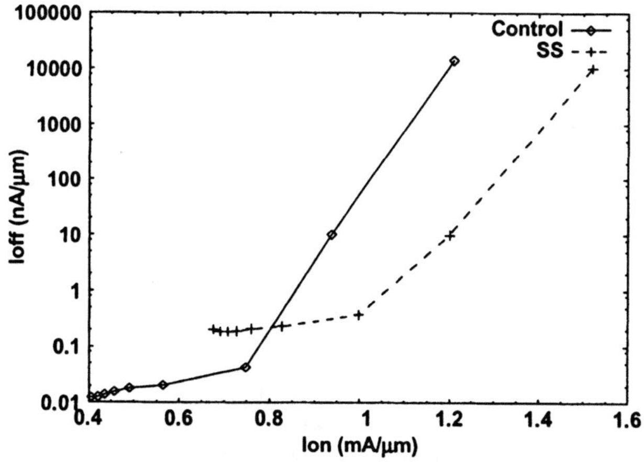

Fig. 4. Ion/Ioff characteristics of the control and strained silicon devices. Significantly higher on currents can be obtained using strained silicon. Off currents for long channel length strained silicon devices are larger than the control devices due to higher junction leakage.

4. Conclusions

A simple and effective method to simulate strained silicon devices was shown. A simple calibration showed that phonon limited mobility in strained silicon will be significantly better than conventional silicon devices. It was also seen that the field dependence of the mobility is comparable in strained silicon devices to that of conventional devices. Even though typical low field mobility improvements of 75% for 20% Ge content SS devices have been seen, greater than 75% improvement in Idlin can be obtained due to a lower Vtlin of the strained silicon devices. Better short channel effects and on currents can be obtained due to effects of band offsets and doping profiles in the strained silicon devices.

References

[1] Vogelsang, T. and Hofmann, K.R., Proc. of Dev. Res. Conf., pp. 34-35, 1992.

[2] Welser, J., *et al.*, IEEE Elect. Dev. Lett. **15**, 3: pp. 100-102, 1994.

[3] Rim, K., *et al.*, IEEE Trans. Elect. Dev. **47**, 7: pp. 1406-1415, 2000.

[4] Buturla, E.M. *et al.*, IBM J. Res. Develop., **25**: pp. 218*ff.*, 1981.

[5] Mujtaba, S. , Ph.D. dissertation, Stanford Univ., 1995.

[6] Laudon, M., *et al.*, App. Phys. Lett. **78**, 2: pp. 201-203, 2001.

[7] Rim, K., *et al.*, VLSI Symp. Tech. Dig., 2001.

Accounting for Quantum Effects and Polysilicon Depletion in an Analytical Design-Oriented MOSFET Model

Matthias Bucher *mbucher@elab.ntua.gr*
National Technical University of Athens (NTUA), Athens, Greece
Jean-Michel Sallese *jean-michel.sallese@epfl.ch*
Swiss Federal Institute of Technology (EPFL), Lausanne, Switzerland
Christophe Lallement *lallem@erm1.u-strasbg.fr*
ERM-PHASE / ENSPS, Strasbourg, Illkirch, France

Abstract

An analytical MOSFET model is presented that accounts for energy quantization in inversion charge and depletion in the poly gate. The model consistently describes effects on charges, transcapacitances, drain current and transconductances in all regions of operation, depending on five physical device parameters and bias conditions. Comparison to experimental data is provided and parameter extraction briefly discussed. The model offers manageable equations providing insight into the physical phenomena, thereby supporting analog circuit design practice as well as efficient circuit simulation.

1 Introduction

Depletion in the polysilicon gate and energy quantization of the mobile carriers drastically reduce the performance of deep submicron CMOS technology. Quantum mechanical (QM) effects and polydepletion (PD) delay the formation of the inversion layer with applied gate bias. The most commonly observed changes in device characteristics are increased threshold voltage [1], decreased slope of inversion charge versus gate voltage resulting in decreased gate capacitance, as well as reduced drain current (e.g. [2]-[4]).

The surface-potential based charge-sheet models offer a suitable basis for MOSFET modelling. Depletion in the polysilicon gate can be included using the depletion approximation [3]. A suitable description of quantum mechanical effects according to Van Dort e.a. [1] has been used in the context of surface-potential based models [2][5]. Even if approximations to relieve the computational burden are made [5], the surface-potential based models do usually not offer an easy insight into the physical phenomena and their interactions.

Manageable expressions for device characteristics in terms of process parameters and bias conditions are very desirable for analog IC design. An alternative to purely surface potential based models is offered by the "charge linearization" models [6]-[9], following the "EKV" [6] model approach. The charge linearization models are charge-sheet models that offer an analytical approximation to the surface-potential based model, retaining the essence of the physical phenomena required for precise analog IC design while offering increased analytical versatility and computational efficiency.

We present the essentials of simultaneously accounting for PD and QM effects in the context of the charge linearization following [9] and using the Van Dort model to account for QM effects. Once the local charges, now including QM and PD effects, are known, a full model for charges, transcapacitances, drain current and transconductances can be formulated for all regions of operation of the MOSFET, from weak to moderate and strong inversion as well as from conduction to saturation [8]. The new analytical model is compared to experimental data. Parameter determination is briefly discussed.

2 Analytical model formulation

The relationship between voltages and charges across the MOS structure is established supposing uniform doping concentration and complete depletion in gate and substrate [9],

$$V_G = V_{FB} + \Psi_s - \frac{Q_C'}{C_{ox}'} + \frac{1}{\gamma_p^2}\left(\frac{Q_C'}{C_{ox}'}\right)^2 \tag{1}$$

where V_G, V_{FB} and Ψ_s are the gate voltage, flat-band voltage and surface potential, Q_C' is the total charge in the semiconductor, $C_{ox}' = \varepsilon_{ox}/T_{ox}$ is the oxide capacitance. The gate and substrate factors are defined as $\gamma_{p(s)} = \sqrt{2q_e\varepsilon_{si}N_{p(s)}}/C'_{ox}$ related to the poly and substrate doping levels $N_{p(s)}$, and other symbols have their usual meanings. Solving (1), the inversion charge $Q_i' = Q_C' - Q_B'$ (where $Q_B' = -\gamma_s C_{ox}'\sqrt{\Psi_s}$ is the depletion charge) is obtained in terms of V_G and Ψ_s. Linearizing Q_i' in terms of Ψ_s leads to a versatile analytical model accounting for PD effects. Even if this linearization is performed under strong inversion conditions – where PD effects are most notable – the model remains applicable at all levels of inversion over a wide range of device parameters with excellent accuracy.

However, QM effects lead to a considerably increased surface potential compared to the classical solution [2], particularly in strong inversion. The strong inversion condition $\Psi_s \cong 2\phi_F + V_{ch}$, where ϕ_F is the bulk Fermi potential and V_{ch} the channel potential, is no longer directly applicable. Therefore, a modification of (1) is necessary to account for QM effects, considering the increase of the surface potential $\Psi_s^{qm} = \Psi_s + \Delta\Psi_s$. According to Van Dort [1], energy quantization increases the energy gap,

$$\Delta E_g \cong \kappa \cdot \left|\frac{Q_C'}{C_{ox}'}\right|^{2/3} \qquad \kappa = \frac{13}{9}\cdot\kappa_Q\cdot\left[\frac{C_{ox}'^2}{4kT\varepsilon_{si}}\right]^{1/3} \tag{2}$$

where ΔE_g is expressed in Volt and κ_Q is a constant [1]. The corresponding increase of the surface potential in strong inversion may be estimated as $\Delta\Psi_s \cong \Delta E_q$ [5][8]. Expanding $\Delta\Psi_s$ to second order in terms of Q_C' around $Q_{B0}' \cong -\gamma_s C_{ox}'\sqrt{2\phi_F + V_{ch}}$, we have

$$\Delta\Psi_s \cong p + q\cdot\left|\frac{Q_C'}{C_{ox}'}\right| - r\cdot\left(\frac{Q_C'}{C_{ox}'}\right)^2 \qquad \text{where} \begin{cases} p = (2\kappa/9)[Q_{B0}'/C_{ox}']^{2/3} \\ q = (8\kappa/9)[Q_{B0}'/C_{ox}']^{-1/3} \\ r = (\kappa/9)[Q_{B0}'/C_{ox}']^{-4/3} \end{cases} \tag{3}$$

where p, q and r are the expansion factors depending on κ. Eqn. (1) is now solved for Q_C' using Ψ_s^{qm} containing the additional charge terms from the expansion (3), as

$$\frac{Q_C'}{C_{ox}'} = \frac{1 + q - \sqrt{(1+q)^2 + 4(\gamma_p^{-2} - r)(V_G - V_{FB} - \Psi_s - p)}}{2(\gamma_p^{-2} - r)}. \tag{4}$$

The inversion charge density Q_i' is then obtained as a function of V_G, Ψ_s and the other auxiliary variables and process parameters. In the above, p simply is a shift in flatband voltage $\Delta V_{FB} = p$ due to QM effects. Linearizing the inversion charge versus surface potential, the charge linearization factor $n_Q \equiv -[dQ_i'/d\Psi_s]/C_{ox}'$ is obtained. The factor n_Q does directly affect inversion charge and drain current [8][9]. Modified expressions for the pinch-off voltage V_P, and the slope factor $n_v \equiv [dV_P/dV_G]^{-1}$, both related to the substrate and gate effects, are obtained. While n_v is not directly used in the analytical model, it is employed for parameter extraction and has further practical importance; a useful approximation is $V_P \cong (V_G - V_{TO})/n_v$, where V_{TO} is the threshold voltage. These expressions are not detailed here due to limited space. Note that all model expressions correctly revert to their counterparts in [9] if QM effects are negligible.

As a consequence, the new model for charges and drain current is obtained through integration of inversion charge density along the channel. Transcapacitances and transconductances are obtained as their partial derivatives [8][9], where QM and PD effects are now accounted for consistently in all model expressions. In particular, the charges model is then further exploited to model mobility dependence on charges [4][8].

3 Experiment and discussion

The model parameters to be determined are the following: V_{FB}, $2\phi_F$, C_{ox}', γ_s and γ_p, or equivalently T_{ox}, N_s, N_p for the latter three. Parameter determination uses an original scheme involving DC measurements at constant current as well as CV characteristics. Devices are from a standard $0.25\mu m$ CMOS technology with a physical T_{ox} of $5nm$.

In Fig. 1, the analytical expressions for the pinch-off voltage V_P and the slope factor n_v are compared to the measurements obtained from the constant current technique [8][10][11], where devices are biased in moderate inversion at a constant level of inversion. Model parameters are adjusted so that V_P, n_v and the transcapacitances characteristics in Fig. 2 are simultaneously and correctly fitted. The slope factor n_v as well as the transcapacitances, all involving derivatives, are particularly sensitive to bias conditions and parameter values, thus helping to obtain the best set of physical device parameters. Note that the slope factor n_v increases due to both QM and PD effects; the weak inversion slope $S = 2.3 \cdot n_v \cdot U_T$ [8] is degraded. The transconductance-to-current ratio, an important quantity in analog IC design, is degraded in weak inversion due to both QM and PD effects.

The gate capacitance of an n-channel device from depletion to strong inversion is shown in Fig. 2 a). The simulated curves correspond to the analytical model, including QM and PD effects individually, and the final result including both contributions. The value of the gate capacitance corresponds to the physical oxide thickness.

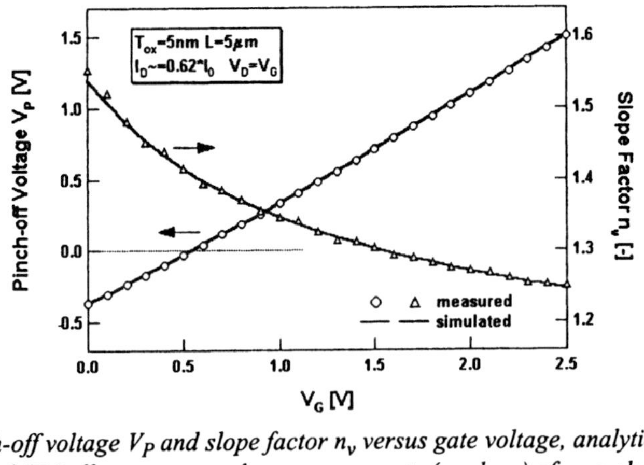

Fig. 1: *Pinch-off voltage V_P and slope factor n_v versus gate voltage, analytical model (lines) including PD and QM effects, compared to measurements (markers) of an n-channel device. The measurement is performed at constant current in moderate inversion.*

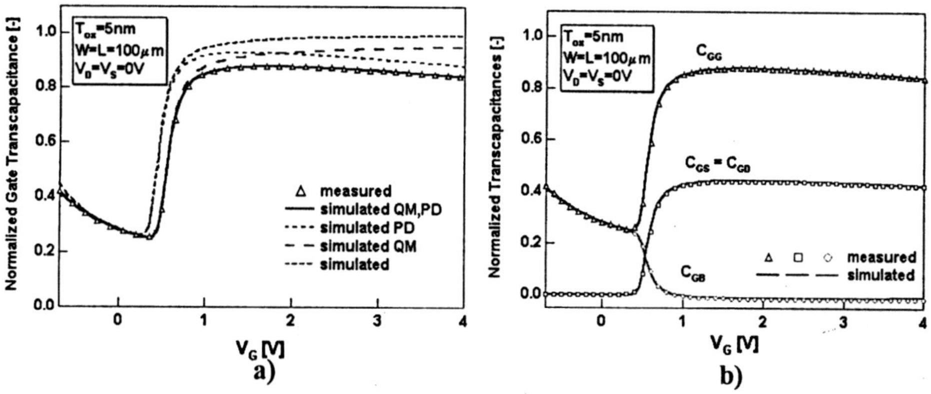

Fig. 2: *Transcapacitances versus gate voltage compared to measurements (markers); a) gate capacitance, model with QM/PD effects (line); QM and PD effects separately; model without QM/PD effects (dashed), b) transcapacitances using the full QM/PD model.*

In Fig. 2 b), other transcapacitances for the same device show excellent agreement between measurement and simulation. The linearization factor n_Q is responsible for the reduction of gate capacitance (and transconductance) observed mainly in strong inversion.

4 Conclusions

In summary, the extension of an analytical circuit simulation model to account for the combined effects of polydepletion and energy quantization has been presented. The charge linearization scheme leads to manageable analytical expressions offering insight into the physical phenomena. All device characteristics are affected in a consistent manner, depending simply on five device parameters, and bias conditions. Model parameter extraction is thus considerably simplified, while the accuracy of the model fits is excellent for all bias conditions. The new charges model now accounting for polydepletion and quantum effects is then used as the basis for a complete compact MOSFET model for IC design and simulation using deep submicron CMOS technologies.

Acknowledgment

The first author wishes to acknowledge W. Grabinski for performing the CV measurements, F. Krummenacher and C. Enz for helpful discussions and Y. Papananos for his support.

References

[1] M. Van Dort, P. Woerlee, A. Walker, *Solid State Electronics*, Vol. 37, N° 3, pp. 411-414, 1994.

[2] R. Rios, N. Arora, C.-L. Huang, N. Khalil, J. Faricelli, L. Gruber, *IEDM Tech. Dig.*, 1995, pp. 937-940.

[3] C.-L. Huang, N. D. Arora, *IEEE Trans. ED*, Vol. 40, N° 12, pp. 2330-2337, Dec. 1993.

[4] T. Janik, B. Majkusiak, *IEEE Trans. ED*, Vol. 45, N° 6, pp. 1263-1271, June, 1998.

[5] G. Gildenblat, T. L. Chen, P. Bendix, *Electron. Lett.*, Vol. 36, N° 12, pp. 1072-1073, June 2000.

[6] C. C. Enz, F. Krummenacher, E. A. Vittoz, *Analog Int. Circ. Signal Proc. J.*, Vol. 8, pp. 83-114, 1995.

[7] A. I. A. Cunha, M. C. Schneider, C. Galup-Montoro, *IEEE J. Solid-State Circuits*, Vol. 33, N° 10, pp. 1510-1519, 1998.

[8] M. Bucher, Ph. D. Thesis No. 2114 (1999), Swiss Federal Institute of Technology (EPFL).

[9] J.-M. Sallese, M. Bucher, C. Lallement, *Solid State Electronics*, Vol. 44, N° 6, pp. 905-912, 2000.

[10] M. Bucher, C. Lallement, C. Enz, *Proc. IEEE Int. Conf. on Test Structures* (ICMTS'96), Vol. 9, pp. 145-150, March 1996.

[11] C. Lallement, M. Bucher, C. Enz, *Solid State Electronics*, Vol. 41, N° 12, pp. 1857-1861, 1997.

Investigations of Salicided and Salicide-Blocked MOSFETs for ESD Including ESD Simulation.

V. Axelrad*, Y. Huh, J.W. Chen, P. Bendix
*SEQUOIA Design Systems, Woodside, CA, USA
LSI Logic Corp., Santa Clara, CA, USA

Abstract

Standard salicided MOSFETs have been repeatedly shown to have inferior ESD protection properties in comparison to salicide-blocked MOSFETs. Standard explanations typically attribute this to shallower current flow and higher peak current density in salicided devices due to the higher conductivity of salicides. In this work we present a numerical analysis of the phenomenon using physical mixed-mode circuit-device simulation. Our results show that the inherent lack of thickness uniformity known to exist in salicide layers can lead to local concentration of current flow and thus local failure of the device.

1. Introduction

Many authors have discussed the merits of using salicide-blocked grounded gate MOSFETs as primary protection elements in I/O circuits [1]-[3]. The usual explanation (offered with a great deal of handwaving) is that the blocked salicide device allows current to flow deeper into the silicon during an ESD event, thereby dispersing heat more effectively, in turn leading to better ESD performance [4]-[5]. For salicided transistors, the current flow is restricted closer to the surface (where the salicide resistance is low), leading to higher temperatures and lower failure voltages compared to salicide-blocked devices [6]-[7].

Our investigations show that the above picture is misleading and does not represent the major effect. We show here that the major improvement of salicide-blocked devices comes from a more uniform turn-on of each grounded gate NMOS finger in a multi-fingered structure (many devices in parallel). Due to salicide thickness variation in salicided MOSFETs, the performance of each finger can be different, leading to non-uniform current density in the fingers which can lead to catastrophic failure of the finger with the thickest salicide. For salicide-blocked devices, a more uniform finger turn-on is more likely.

The interpretation of this result is owing largely to a novel simulation software [8] which we use for characterizing an ESD event. In comparison to other device simulation tools (PISCES, Medici, etc.) the main differences are:

1) It does not require a process simulator to generate the input device structure. One can inverse model a transistor's profiles from desired electrical characteristics that are specified by the user (Vth's, Idsat, breakdown voltage, etc.).
2) Manual gridding is not required. The simulator constructs an efficient grid without user intervention.
3) An I/O circuit with multiple elements (mosfets, resistors, diodes, etc.) can be entered using built-in schematic capture and simulated without exhausting the grid node count limit of more traditional device simulators.

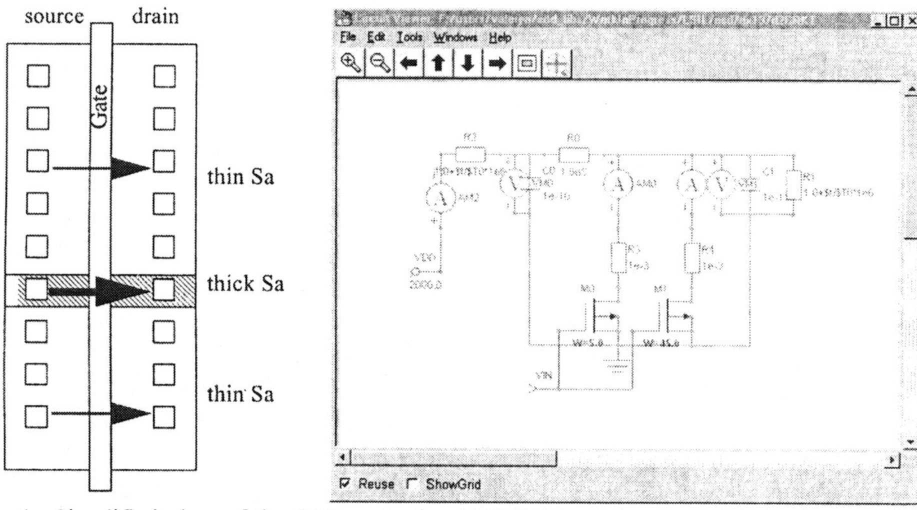

Fig. 1. Simplified view of the ESD protection MOSFET and the simulated circuit: to model salicide thickness variation, two grounded MOS slices (fingers) in parallel with different salicide thicknesses are used. The total device width is 50μm. Thin salicide fingers account for 45μm (M1), the remaining 5μm are modeled as a thick salicide MOSFET finger M0.

And finally, compared to ESD simulators based on Spice-like circuit simulation [9]-[10], a device simulator is capable of dealing with geometry and doping variables (such as salicide thickness) which a circuit simulator cannot do.

2. Mixed-Level ESD Analysis

In Fig. 1, we show the circuit used to simulate an HBM event. The protection device consists of a grounded gate MOSFET comprised of two parallel MOSFETs, a thin salicide one with a width of 45μm and a thick salicide one with a width of 5μm.

Fig. 2 (left) shows a cross section of the MOSFET indicating the salicide thickness parameter. MOSFET doping profiles were specified according to SIMS measurements. Simulated breakdown curves (Fig. 2, right) demonstrate good agreement with measured data without any additional calibration. The differences between the two before breakdown are insignificant. After breakdown, however, the thin salicide device has a higher on resistance, once again showing that salicide thickness variation will lead to variation in finger performance.

Fig. 3 shows plots of electric potential along a horizontal cross-section under the salicided drain of the two MOSFETs. Higher voltage drop and therefore higher on-resistance are evident for the thin-salicide case (crosses). Since all transistor slices (fingers) are connected in parallel, lower on-resistance of a slice results in higher dissipated power and stronger heating there.

In Fig. 4 (left), we show the current density per micron of MOSFET width through the two parallel protection GGNMOSs, and in Fig. 4, right we show the output voltage VM1 as well as peak temperatures for the two devices. The thicker salicide device (5μm width) has higher current density leading to a higher temperature in this device. Note that this is the effect we expect based on standard thinking but for the wrong reason. The standard claim is that the thicker salicide device should heat up more because

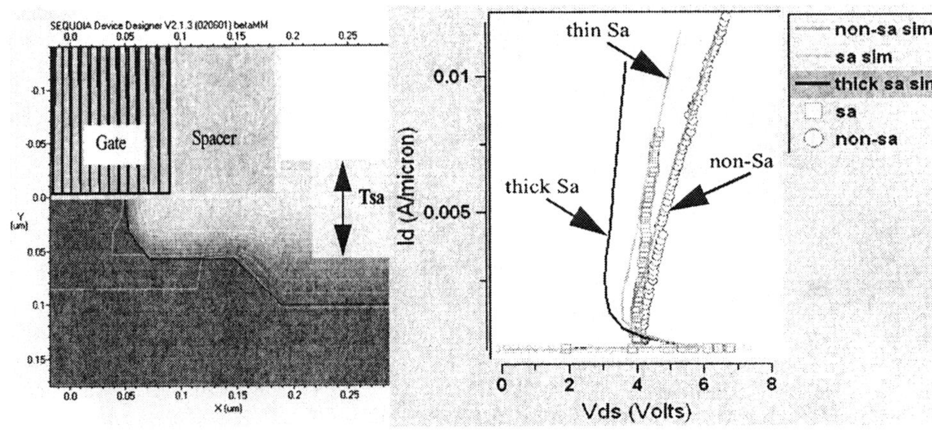

Fig. 2. Left: Parametrized MOSFET with the salicide thickness parameter Tsa indicated (thick salicide case is shown). Right: measured (symbols) and simulated (lines) breakdown curves for the non-salicided, thick- and thin-salicide devices with Lpoly=0.18μm. Good agreement between simulation and experimental data is observed. Salicide thickness has a significant effect on post-breakdown on-resistance.

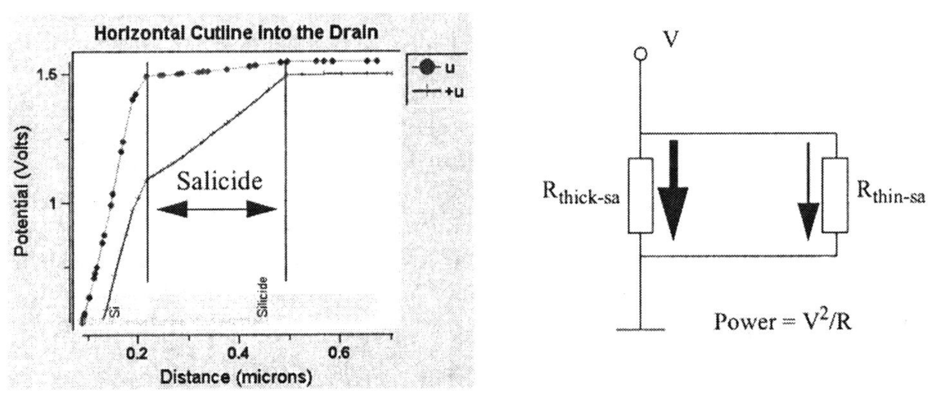

Fig. 3. Electric potential during the ESD pulse along a horizontal cutline close to the surface of the thick-sa MOS (circles) and thin-sa MOS (crosses). Much higher voltage drop in the salicide region for the thin-sa device is evident, indicating its higher on-resistance. Since all MOS slices (fingers) are connected in parallel, the slice with the lowest resistance pulls the highest power and therefore heats up the most. In our case it is the thick-sa device.

its current doesnt flow deep enough into the silicon to benefit from better heat dissipation of surrounding silicon. Here we see that it heats up more because it has lower resistance and draws more current. In addition, this clearly shows that salicide thickness variation can lead to large differences in current drawing and heating of the device.

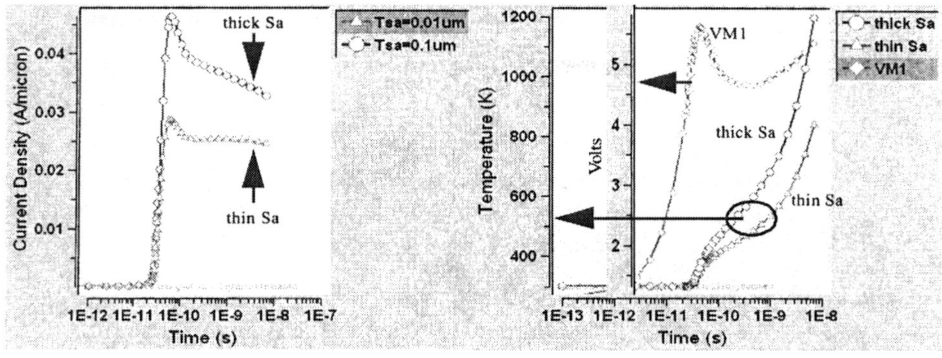

Fig. 4. Left: Current per unit width, the thicker salicide device pulls a higher current density and consequently heats up faster. Right: Output waveform VM1 (see Fig. 1, blue curve) and peak temperatures in the thick salicide device (red) and thin salicide device (green). Permanent device failure occurs between 800K (aluminum failure) and 1200K (silicon melts).

3. Conclusions

We demonstrated a novel approach to the analysis of ESD protection circuits. The approach is based on a mixed-level simulation system, which combines physical depth of analysis with complex circuit effects important for ESD events. The simulation system was applied to study salicided versus non-salicided grounded gate MOSFETs.

Our analysis suggests that the experimentally well-documented ESD-weakness of salicided MOSFETs is caused by the significant intrinsic variability of salicide layers. Locally varying salicide thickness can lead to locally reduced device on resistance, thus increased current density, local heating and possible local failure at the thick salicide locations in the device.

4. References

[1] A. Palella, et al., A Design Methodology for ESD Protection Networks, EOS/ESD 1985, pp. 24-40.

[2] C. Duvvury, et al., ESD Design Considerations for ULSI, EOS/ESD 1985, pp. 45-48.

[3] C. Diaz, et al., Studies of EOS Susceptibility in 0.6um nMOS EOS I/O Protection Structures, EOS/ESD 1993, pp. 83-91.

[4] A. Amerasekera et al., ESD Failure Modes: Characteristics, Mechanisms and Process Influences, IEEE TED-39 1992, p.2.

[5] S. Beebe, Methodology for Layout Design and Optimization of ESD Protection Transistors, ESD/EOS 1996, pp. 265-275.

[6] T. Polgreen, et al., Improving the ESD Failure Threshold of Silicided n-MOS Output Transistors by Ensuring Uniform Current Flow, IEEE TED-39, pp. 379-388.

[7] K. Verhaege, et al., Wafer Cost Reduction through Design of High Performance Fully Silicided ESD Devices, ESD/EOS 2000, pp. 18-28.

[8] SEQUOIA *Device Designer Users' Guide* , SEQUOIA Design Systems, 1998-2001

[9] S. Beebe, Simulation of Complete CMOS I/O Circuit Response to CDM Stress, ESD/EOS 1998, pp. 259-270.

[10] M. Mergens, et al., ESD-level Circuit Simulation - Impact of Gate RC-Delay on HBM and CDM Behavior, ESD/EOS 2000, pp. 446-155.

Bipolar Transistor's Intrinsic and Extrinsic Capacitance Determination

B. Ardouin, T. Zimmer, H. Mnif, P. Fouillat
Laboratoire de Microélectronique IXL,UMR 5818, University of Bordeaux I
33405 Talence Cedex, France
email: ardouin@ixl.u-bordeaux.fr

D. Berger, D. Céli
ST Microelectronics, 850 rue Jean Monnet, 38 926 Crolles, France

Abstract

We present a new method to determine the intrinsic and extrinsic base-emitter and base-collector junction capacitances parameters of bipolar transistors as a function of junction bias. The capacitance specific compact model parameters (C_{j0}, V_j, γ_j) can be extracted for the intrinsic as well as for extrinsic part. The results can be introduced in recent compact bipolar models.

1 Introduction

Parasitics in bipolar transistors are becoming of great importance due to the reduction in the device dimensions. Therefore, the junction capacitances distributed nature has to be studied, because it affects the device HF behaviour, but also the DC behaviour via the direct and inverse Early effect. The recent bipolar compact models (Hicum [1], Mextram [2], VBIC [3]) take this effect into account by using a distributed R-C network (junction capacitance splitting across the base resistance). All compact models describe the junction capacitance behaviour under reverse and low forward bias with the relation:

$$Cj = C_{j0} \Big/ \left(1 - \frac{V}{Vj}\right)^{\gamma j} \tag{1}$$

The capacitance splitting implies the need of two different equations (1) and the corresponding junction parameter sets (C_{j0}, V_j, γ_j) for the intrinsic and extrinsic capacitances. To determine the intrinsic and extrinsic part of the junction capacitances, several methods exist:

1) The first one consists in evaluating the capacitances parameters using a device simulator. Nevertheless, a direct method would be preferred since parameter extraction in real world compact modelling is always based on electrical measurements.

2) The second method is described in [2]. It allows an approximated partitioning of the zero bias capacitance only for the BC junction, considering that the intrinsic and extrinsic junctions have the same characteristics, and consequently the same junction parameters, which is not true, especially in the case of SIC implant. Additionally, this method is particularly sensitive to previous parameter extraction errors.

3) The third method is described in [3]. It consists in a geometrical analysis of the junction capacitances: The intrinsic and extrinsic capacitances are considered to be proportional to an effective junction area and an effective perimeter, respectively. Therefore, the effective geometry

has to be known precisely, which is not always possible. Moreover, for the B-C junction, especially if a SIC implant exists, this method looses its physical meaning and is no more valid. Therefore, in the following section, we present a new self consistent method, based on HF measurements, that allows to split the capacitances values into an intrinsic and an extrinsic part.

2 Direct Extraction Procedure

Under reverse and low forward bias conditions for all junctions, the following assumptions can be made:

(a) the transistor is supposed to have no transfer current, which implies infinite values for $r\pi$ and $r\mu$, and that the diffusion capacitances C_{de} and C_{dc} can be neglected, giving the equivalent circuit of figure 1.

(b) since the transistor has no current gain, the influence of the series resistance re, and rc are further neglected, and rbx is lumped with rb. The resulting simplifications give the Y-parameters of interest:

$$\mathrm{Re}(Y_{11})=\frac{1}{rb}\frac{\left(\dfrac{\omega^2}{1/rb\,(Cjei+Cjci)}\right)^2}{1+\left(\dfrac{\omega}{1/rb\,(Cjei+Cjci)}\right)^2}\qquad \mathrm{Re}(Y_{11}+Y_{12})=C_{jei}\frac{\left(\dfrac{\omega^2}{1/rb\,(Cjei+Cjci)}\right)}{1+\left(\dfrac{\omega}{1/rb\,(Cjei+Cjci)}\right)^2}\tag{2}$$

The first step of the method is a classical junction capacitance determination using S-parameters. The total junction capacitances $C_{jeT}(V_{be})$ and $C_{jcT}(V_{bc})$ are determined using well known relations:

$$C_{jcT}=\frac{\mathrm{Im}(-Y_{12})}{\omega}\qquad C_{jeT}=\frac{\mathrm{Im}(Y_{11}+Y_{12})}{\omega}\tag{3}$$

with $C_{jeT}= C_{jei} + C_{jex}$ and $C_{jcT}= C_{jci} + C_{jcx}$. We can note that the result depends greatly on the frequency used for measurements, as show figure 2 where $C_{jcT}(f)$ and $C_{jeT}(f)$ are plotted.

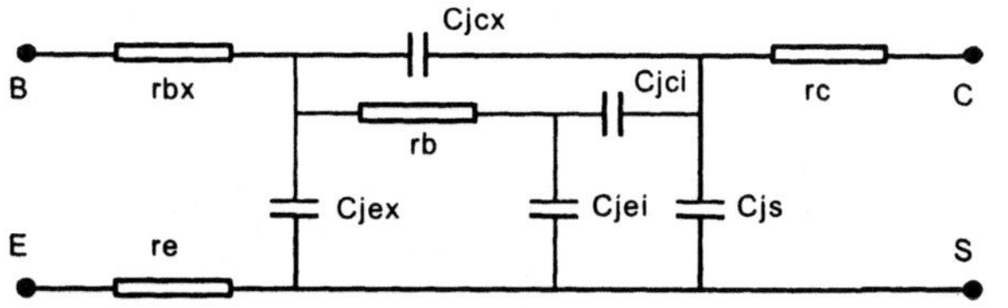

Figure 1: equivalent circuit for "Y-parameter determination"

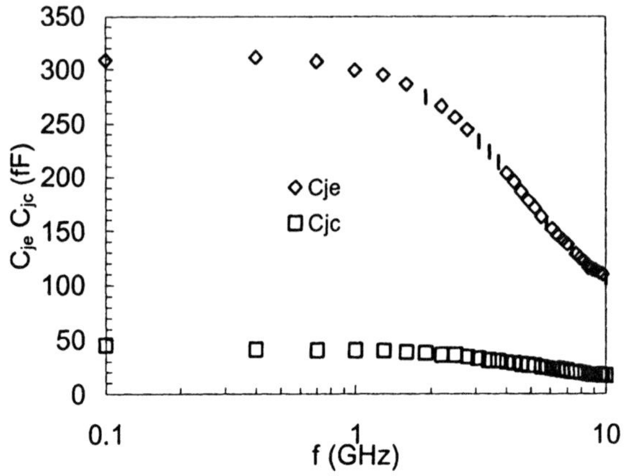

Figure 2: $C_{jcT}(f)$ and $C_{jeT}(f)$ equation 3

The second step of the method is based on the same measurements, using a higher frequency range. Moreover, (2) can be rearranged as:

$$\frac{\omega^2}{\mathrm{Re}(Y_{11})} = rb\,\omega_T^2 + rb\,\omega_T\,\omega^2 \qquad \frac{\omega^2}{\mathrm{Re}(Y_{11}+Y_{12})} = \frac{\omega_T}{Cjei} + \frac{\omega^2}{\omega_T\,Cjei} \qquad (4)$$

$$\text{with } \omega_T = 1/\,r_b\left(C_{jei} + C_{jci}\right)$$

Plotting $\omega^2/\mathrm{Re}\,(Y_{11})$ and $\omega^2/\mathrm{Re}(Y_{11}+Y_{12})$ versus $\omega^2=(2\pi f)^2$ and applying a linear regression allows to determine rb, Cjei and Cjci. As an example, figure 3 (a) shows the plots of $\omega^2/\mathrm{Re}(Y_{11})$ and $\omega^2/\mathrm{Re}\,(Y_{11}+Y_{12})$ versus ω^2 for $V_{bc}=0$ and $V_{be}=0$. We can note that the curves are nearly straight lines, which confirms the validity of the equivalent circuit used to derive (2) and (4).
The determination of Cjei and Cjci can be performed for each bias point, then allowing the determination of Cjei and Cjci voltage variation and the related junction parameters extraction (C_{j0}, V_j, γ_j) by using a standard method.
The extrinsic capacitances are determined using: $C_{jex}=C_{jeT}-C_{jei}$ and $C_{jcx}=C_{jcT}-C_{jci}$.

3 Results

The method was applied to advanced SiGe heterojunction bipolar transistors fabricated by ST Microelectronics, having a 45 GHz transit frequency and a current gain of 150. The transistors are realised in a self aligned technology with SIC implant and 0.4µm minimum emitter width.
Figure 3 (b), shows the extracted C_{jei}, C_{jex}, C_{jci} and C_{jcx} versus junction voltage for a 1.6x25.6 µm² emitter bipolar transistor. The extracted base resistance rb=105 Ω at zero bias. Moreover, the comparison of the results for transistors of different sizes have shown the scaling coherence of this approach.

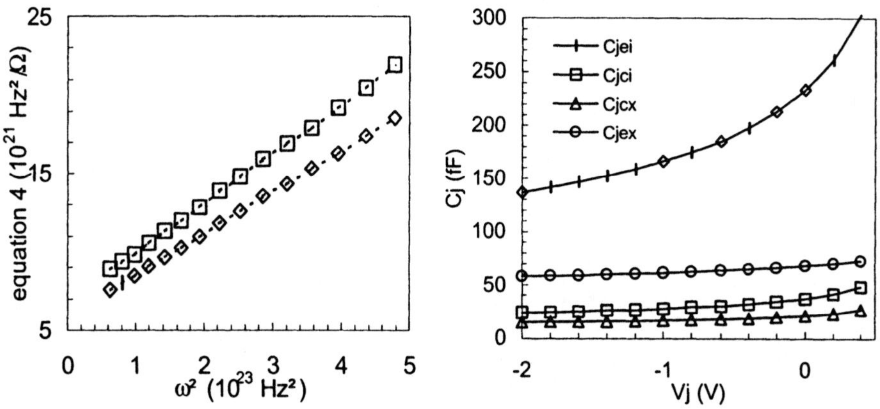

Figure 3: **(a)** $\omega^2/\mathrm{Re}(Y_{11})$ and $\omega^2/\mathrm{Re}(Y_{11})$ vs ω^2 **(b)** C_{jei} and C_{jci} versus V_{be} and V_{bc}

4 Conclusion

A new method for bipolar transistor's intrinsic and extrinsic capacitances parameter determination is presented. These parameters are essential for recent bipolar compact models. This splitting is an important improvement for the high frequency behaviour modelling and for the output conductance modelling (in the Hicum and Mextram model the Early effect depends on the intrinsic capacitances). This method has the advantage to be a direct method, based on measurements, which is physically more accurate for a given device than methods based on test structures and geometry considerations. Furthermore, a full electrical device modelling has been performed using the Hicum model, and we got excellent results for both, DC and AC characteristics.

References

[1] Schroter M. Hicum bipolar transistor model, available on the web: http://www.iee.et.tu-dresden.de/iee/eb/comp_mod.html.

[2] de Graaf H. C, Klaassen F.M, Compact transistor modelling for circuit design, Springer Verlag, Wien New York.

[3] Mc Andrew C, Seitchik J, Bowers D, Dunn M, Foisy M, Getreu I, McSwain M, Moinian S, Parker J, Roulston D, Schröter M, Wijnen P, Wagner L, VBIC95, the vertical bipolar Inter-company model, IEEE J. of sol. state cir., vol 31, N° 10, October 1996.

[4] Berger D, Gambetta N, Céli D, Dufaza C, Extraction of the base-collector capacitance splitting along the base resistance, BCTM, Minneapolis, September 24-26, 2000.

[5] Lee T. Z, Schroter M, presentation to the compact modelling council –The Hicum bipolar model: methodology for bipolar model parameter extraction, available on the web: http://www.eigroup.org/cmc/bipolar/default.htm.

Varying Characteristics of Bipolar Transistors with Emitter Contact Window Width

J. Fu, S. Mijalkovic, W. J. Eysenga, H. W. van Zeijl and W. Crans
Delft Institute of Microelectronics and Submicron Technology (DIMES)
Feldmannweg 17, P. O. Box 5053, 2600 GB, Delft, The Netherlands
E-mail: fujun@ectmuxs1.et.tudelft.nl, w.crans@its.tudelft.nl

Abstract

On the basis of careful selection and calibration of model parameters we performed process and subsequent device simulation on WEB bipolar transistors. As a result, we concluded, that the mechanical stresses may be responsible for the measured varying device DC characteristics with emitter contact window width.

1 Introduction

Fully-implanted washed-emitter-base (WEB) NPN bipolar transistors were optimized for operation in the 10-30 GHz range with low-stress silicon nitride film as surface isolation layer [6] (Nanver et al. 1996). The DC characteristics of the transistors were observed to vary with the emitter contact window width in an unexpected way. We investigated such a phenomenon with the aid of process and device simulators, i.e., TSUPERM-4 and MEDICI, respectively, based on careful selection and calibration of the corresponding model parameters. As a result, the observed variations in device characteristics with the emitter contact window width may reasonably be ascribed to the influence of the mechanical stresses induced by the silicon nitride surface isolation layer.

2 Processing sequence

The main steps of WEB bipolar process used for manufacturing the transistors can be described as follows. After the formation of an n^+ arsenic buried layer on top of (100) p type Si substrate (2-5Ωcm), a 0.9μm epitaxial layer doped with 1×10^{16} cm^{-3} arsenic was grown, followed by thermal growth of 30 nm silicon oxide and the subsequent implantation of p^+ junction isolation and n^+ collector plug. Next low ohmic extrinsic base and link base were formed by boron implantation with doses of 3×10^{15} cm^{-2} and 5×10^{13} cm^{-2}, respectively, both at an energy of 15keV. Low stress silicon nitride SiN_x layer with a thickness of 300 nm was successively deposited. All contact windows were opened through plasma etching of SiN_x and pad oxide in one masking step, followed by emitter (arsenic, 40keV, 7.5×10^{15} cm^{-2}) and intrinsic base (boron, 20keV, 1.75×10^{14} cm^{-2}) implants. All dopants were activated by a single thermal anneal for 30 min in an argon ambient at 950°C. The cross-sectional view of the final device structure is shown in Fig. 1.

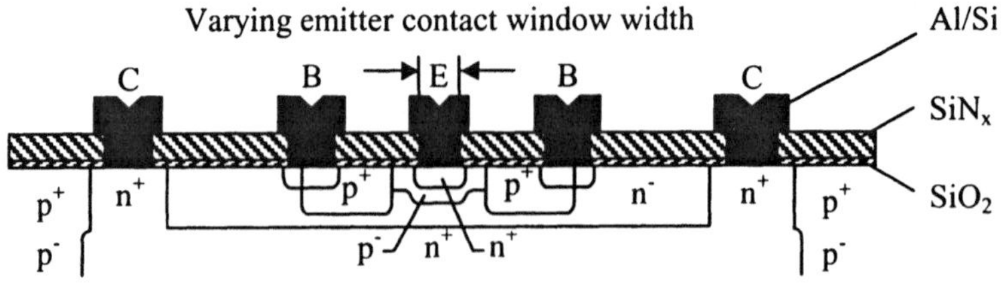

Fig. 1. Schematic cross section of WEB bipolar transistor

3 Simulation

The WEB bipolar process and DC characteristics of the fabricated transistors were simulated by using TSUPREM-4 (version 6.6) and MEDICI (version 4.1), respectively. In order to improve the accuracy of process simulation, the fully coupled point defect model was selected to predict post-implant dopant diffusions during some of the critical process steps and the corresponding model parameters (e.g., dopant diffusivity, equilibrium concentration and diffusivity of silicon interstitials, and so on, see reference [2] (Fu et al. 2001)) were calibrated by using Inverse modelling driver PROFILE. It is worth noting that such a calibration procedure was based on comparisons of simulated dopant profiles given by TSUPREM-4 with the corresponding SIMS data detected from the accompanying wafers of the WEB bipolar process published in reference [6] (Nanver et al. 1996). On the other hand, Phillips unified mobility model and field dependent mobility model, which are thought to be suitable for bipolar transistor simulation, were chosen to perform the subsequent MEDICI device simulation. In accordance with the default values of the model parameters used in Phillips unified mobility model, the values of the parameters employed to describe the bandgap narrowing effect followed reference [3] (Klaassen et al. 1992). As for the recombination of carriers doping concentration dependent SRH model and Auger model were utilized with the corresponding model parameters given by reference [4] (Law et al. 1991).

4 Results and discussions

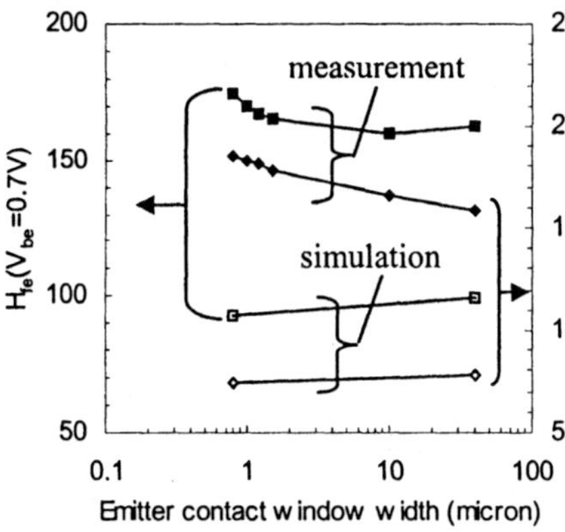

Fig. 2. Common emitter current gain H_{fe} and Early voltage V_A versus emitter contact window width for WEB bipolar transistors

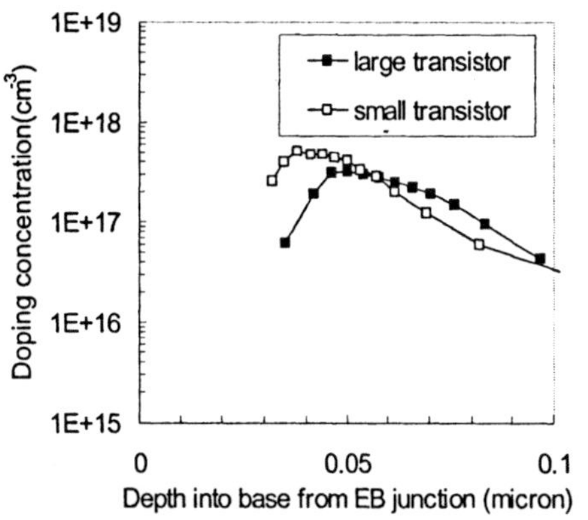

Fig. 3. Base doping profiles in the middle of emitter contact window for large and small transistors extracted by using differential capacitance method

Fig. 2 exhibits the measured common emitter current gain H_{fe} (V_{be}=0.7V) and Early voltage V_A for WEB bipolar transistors as a function of emitter contact window width with a fixed emitter contact window length of 40μm. For comparison the simulated H_{fe} and V_A are plotted as well in Fig. 2 for large emitter contact window width (40×40μm^2) and small one (40×0.8μm^2), respectively. It should be noted that the process conditions for fabricating the set of transistors differ somewhat in the dosages of link base and intrinsic base boron implants from those in reference [6] (Nanver et al. 1996) used for model parameter calibration. In fact, the nominal dosages of boron implantation used for link base and intrinsic base formation corresponding to Fig. 2 are 5×10^{13} cm^{-2} and 1.75×10^{14} cm^{-2}, instead of $1.1\times10^{14}cm^{-2}$ and $2.4\times10^{14}cm^{-2}$ in reference [6] (Nanver et al. 1996), respectively. For this reason the boron implant dosages involved in process simulation here were scaled accordingly. It can be clearly found that the simulated H_{fe} and V_A are significant less than those from measurement for both the large device and the small device. In addition, the simulation can not predict the trend of either H_{fe} or V_A versus the emitter contact window width. For the time being we don't discuss the current gain and its complicated mechanisms. In contrast to current gain, the magnitude of Early voltage is dominantly determined by the integrated quantity of base dopants per unit area, i.e., Gummel number. Consequently the higher measured V_A's compared to the simulated ones can reasonably be explained as relatively higher Gummel numbers (averaged over the emitter contact window) in the actual devices than those in the unpatterned accompanying samples prepared for SIMS detection. Accordingly, the increase in measured V_A with decreasing emitter contact window width as shown in Fig. 2 may be attributed to the increasing Gummel number with shrinkage of emitter contact window width. This assumption has been confirmed by the base doping profiles in the middle of emitter contact window extracted by means of so-called differential capacitance technique [1] (Fu et al. 2000) for the large and the small transistor, respectively, as shown in Fig. 3.

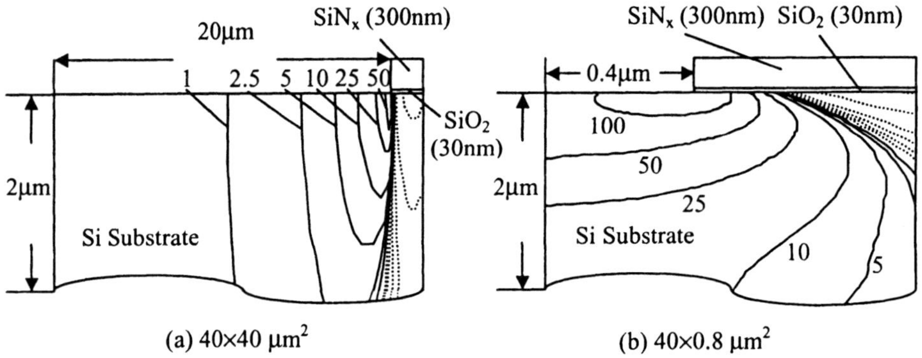

Fig. 4. Mechanical stress contours inside device structure for the large (a) and the small (b) transistor simulated by FLOOPS (only half of the emitter contact window is shown in consideration of symmetry of the structure)
unit: 10^7 dyn/cm^2; solid lines: tensile stress; dotted lines: compressive stress

A possible explanation for the emitter contact window width dependent dopant profiles is the influence of mechanical stresses induced by silicon nitride surface isolation layers on the dopant diffusions. The stress contours inside device structure for large and small transistor are obtained by FLOOPS simulation based on S. Mijalkovic's work [5] (Mijalkovic 2000) as plotted in Fig. 4 (a) and (b), respectively (only half of the emitter contact window is shown in consideration of symmetry of the structure). As we can clearly see from Fig. 4 that localized stresses are created around the edges of emitter contact windows as a result of patterning of silicon nitride layers. Moreover, the magnitude of the stresses near the center of emitter contact windows would be increased with decreasing window width as predicted by the FLOOPS stress analysis. It should be emphasized that we did not take any of the mechanical stress related effects into consideration during the TSUPREM-4 and MEDICI simulation. So the observed deviation of simulation from measurement regarding the DC characteristics of bipolar transistors may reasonably be ascribed to the effects of stresses. As a result, varying base doping profiles with emitter contact window width, and thus the elevated V_A with decreasing emitter contact window width can be accounted for in terms of perturbation of dopant diffusions arising from influence of mechanical stresses.

5 Conclusions

The varying DC characteristics of WEB bipolar transistors with emitter contact window width were investigated through process and subsequent device simulation. The observed deviation of simulation from the corresponding measurement and variations in measured V_A as a function of emitter contact window width led us to the conclusion that the Gummel number and hence Early voltage can be increased as a result of perturbed dopant diffusions in the presence of mechanical stresses induced by the patterned silicon nitride layer. The extracted base doping profiles inside the device structure provided further evidence in favor of our conclusion. In order to interpret varying current gain versus emitter contact window width, however, further investigations are required to gain more insight into the influence of the stress and strain states.

Acknowledgement

The authors wish to thank Dr. L. K. Nanver et al. for providing the measurement data of dopant profiles and device characteristics for WEB bipolar process. The authors also would like to thank the NWO-Technology Foundation-STW who is sponsoring this project under grant DEL. 4415.

References

[1] Fu J. , S. Mijalkovic, W. J. Eysenga, H. W. van Zeijl, and W. Crans (2000): The Influence of Mechanical Stresses on the Characteristics of Bipolar Transistors. In: Proceedings of Semiconductor Advances for Future Electronics 2000 (SAFE 2000), Veldhoven, the Netherlands

[2] Fu J. , W. J. Eysenga, and W. Crans (2001): Calibration of model parameters for process simulation on dopant ion implant and diffusion by using Inverse Modelling method. In: Proceedings of Semiconductor Advances for Future Electronics 2001 (SAFE 2001), Veldhoven, the Netherlands (to be published)

[3] Klaassen D. B. M. , J. W. Slotboom, and H. C. de Graaff (1992): Unified apparent bandgap narrowing in n- and p-type silicon. Solid-State Electronics. 35: 125-129

[4] Law M. E. , E. Solley, M. Liang, and Dorothea E. Burk (1991): Self-consistent model of minority-carrier lifetime, diffusion length, and mobility. IEEE Electron device letters. 12: 401-403

[5] Mijalkovic S. (2000): A piecewise linear Galerkin approach to stress analysis of nearly incompressible materials. Communications in Numerical Methods in Engineering. 16: 537-543

[6] Nanver, L. K. , E. J. G. Goudena, and H. W. van Zeijl (1996): Optimization of fully-implanted NPN's for high-frequency operation. IEEE transactions on electron devices. 43: 1038-1040

Ensemble Monte Carlo Particle Modeling of InGaAs/InP Uni-Traveling-Carrier Photodiodes

M. Ryzhii and V. Ryzhii
Computer Solid State Physics Laboratory
University of Aizu, Aizu-Wakamatsu 965-8580, Japan
E-mail: m-ryzhii@u-aizu.ac.jp

Abstract

An ensemble Monte Carlo particle model for uni-traveling-carrier (UTC) photodiodes is developed. This model is used to study electron and hole nonequilibrium phenomena and transient response of UTC photodiodes.

1 Introduction

A novel heterostructure optoelectronic device - the uni-traveling-carrier (UTC) photodiode proposed and studied by Ishibashi et al. (1997) and Shimizu et al. (1998) - combines ultra-fast operation and high power output. Such a photodiode includes a doped narrow-gap absorption region (p^+–InGaAs) and an undoped or lightly doped wide-gap collector region (N^-–InP). To eliminate the effect of the spike in the conduction band at the InGaAs/InP heterointerface preventing the electron escape from the absorption region into the collector region, a thin heavily doped cliff layer is inserted near the heterojunction. This structure is sandwiched between P^+- and N^+- layers serving as the contacts. The band structure of a UTC photodiode is shown in Fig. 1.

In this communication, we report the study of the response of InGaAs/InP UTC photodiodes to optical pulses with different energies generating electrons and holes in the absorption region. For this purpose we use an ensemble Monte Carlo (MC) particle technique similar to that used previously for the studies of the transient processes in bipolar heterostructure transistors and quantum well infrared photodetectors (M. Ryzhii 1997, M. Ryzhii and V. Ryzhii 2000) and adjusted to the structural features of UTC photodiodes. Our simulator takes into account the band structures of InGaAs and InP layers, the mechanisms of the electron and hole scattering including carrier-carrier interactions, and the self-consistent electric field in the diode active region (between the contacts).

2 Model

The parameters of materials of the UTC photodiode structure used in the MC calculations were extracted from the book by Adachi (1992). The model implemented neglects the influence of a narrow spike formed in the conduction band at the heterojunction on the electron transport due to relatively high tunnel transparency of this spike. For the MC simulations, ensembles of up to 5×10^5 particles (electrons and holes) were protracted. The time-step and the mesh size correspond to 10 fs and 2 nm, respectively. The structural parameters of

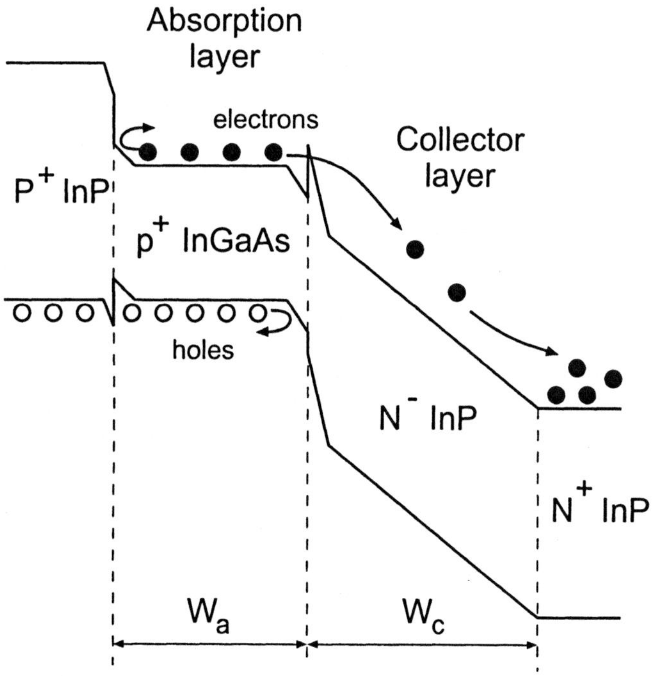

Fig. 1. Schematic view of the UTC photodiode band diagram.

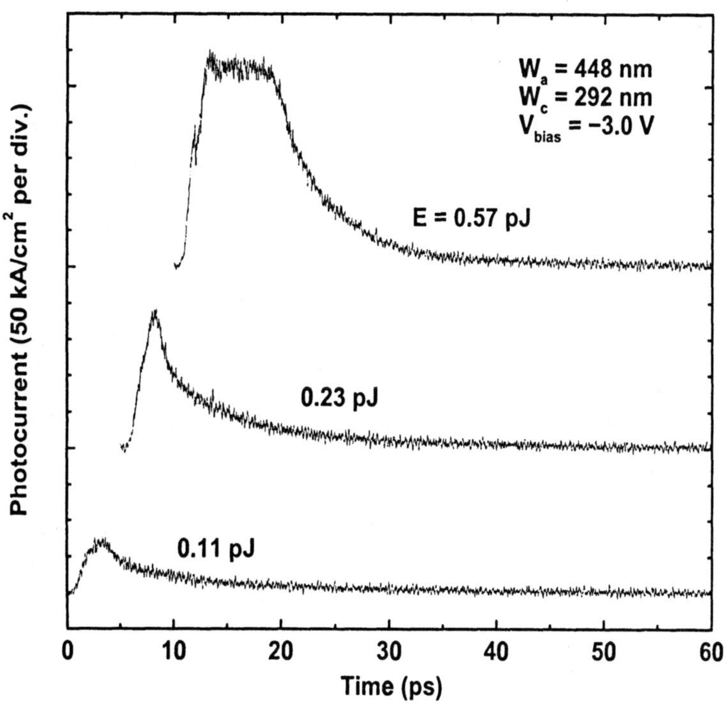

Fig. 2. Transient photocurrent for different absorbed optical energies.

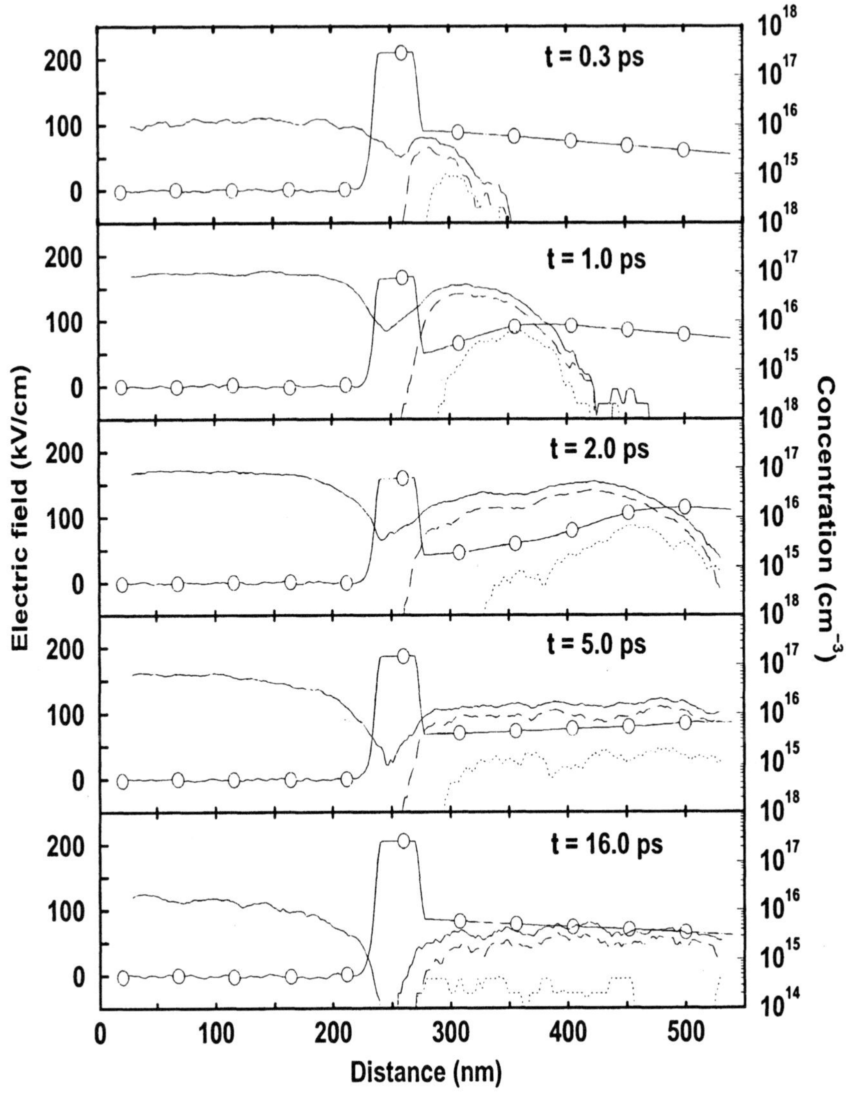

Fig. 3. Snapshots of electric-field (circles) and electron concentration spatial distributions. Solid line corresponds to net electron concentration, dashed line - L-electron concentration, and dotted line - X-electron concentration.

UTC photodiodes used in the simulations were assumed to be similar to that in the experimental studies, namely, the thicknesses of the absorption and collector layers $W_a = 228 - 888$ nm and $W_c = 292$ nm, respectively, the acceptor concentration in the main part of the absorption layer (except a thin undoped spacer adjacent to the heterojunction) $N_a = (0.25 - 1.0) \times 10^{18}$ cm^{-3}, and the device area $A = 20$ μm^2.

3 Results

Using the developed model, we calculated the spatio-temporal distributions of the electric field and carrier concentrations and the transient photocurrent induced in the external circuit. The obtained temporal dependences of the photocurrent were used to calculate the UTC photodiode responsivity as a function of signal frequency.

Figure 2 shows the temporal dependences of photocurrents in a UTC photodiode with $W_a = 448$ nm, and $W_c = 292$ nm at $V_{bias} = -3.0$ V in response to optical pulses ($\lambda = 1.55$ μm) having the Gaussian shape (with FWHM of 1000 fs) and different energies. As seen from Fig. 2, the height of the photocurrent peak increases with increasing absorbed optical energy E while the width of this peak varies insignificantly. At higher optical energies, the photocurrent saturates (so that its top becomes nearly flat) and the duration of the photocurrent pulses substantially rises. This is due to an essential limitation of the current extracted from the absorption region caused by the electron space charge in the collector region. This effect and its details are in agreement with experimental data.

Figure 3 shows the spatial distributions of the electric-field and electron concentration (both in the absorption and collector layers) at different moments for a UTC photodiode with $W_a = 228$ nm, and $W_c = 292$ nm exposed to an optical pulse with FWHM of 390 fs at $V_{bias} = -2.0$ V.

The authors are thankful to Dr. T Ishibashi for valuable discussions.

References

[1] Adachi S. Physical Properties of III-V Semiconductor Compounds: InP, InAs, GaAs, GaP, InGaAs, and InGaAsP. John Wiley & Sons, New York, 1992

[2] Ishibashi T, Kodama S, Shimizu N, Furuta T. High-speed response of uni-traveling-carrier photodiodes. Jpn. J. Appl. Phys. 1997; 36:6263-6268

[3] Ryzhii M. Monte Carlo modeling of transient effects in resonant-tunneling bipolar transistors. Jpn. J. Appl. Phys. 1997; 36:5060-5062

[4] Ryzhii M, Ryzhii V. Monte Carlo modeling of transient recharging processes in quantum-well infrared photodetectors. IEEE Trans. Electron Devices 2000; 47:1935-1942

[5] Shimizu N, Watanabe N, Furuta T, Ishibashi T. Improved response of uni-traveling-carrier photodiodes by carrier injection. Jpn. J. Appl. Phys. 1998; 37:1424-1426

An Effective Methodology for Predicting the Distribution of MOSFET Device Characteristics Using Statistical TCAD Simulations

Takaaki Tatsumi
LSI Technology Development Division,
Semiconductor Network Company, Sony Corporation
4-14-1, Asahi-cho, Atsugi-shi, Kanagawa-ken, Japan
tatsumi@ulsi.sony.co.jp Tel:+81-46-230-5719 Fax:+81-46-230-5945

Abstract

We have developed a novel and effective method for predicting the distribution of MOSFET device characteristics, which also enables us to specify the most typical process conditions for any device characteristics. In our approach, the distribution of the device characteristic caused by the fluctuation of every single process are calculated and then merged. Comparison with the result of the measured data has established that our method is accurate and practical.

1. Background

Predicting accurate distributions of device characteristics using TCAD at the device development phase is indispensable for the concurrent designing of cell libraries and circuits with appropriate performance margins. While designing, the saturated drain current of NMOS and PMOS (IdsN and IdsP) pair is used as the reference of the best/worst performance conditions. By measuring the IdsN-IdsP pairs of manufactured chips, it has been found that their distribution assumes an oval shape, as shown in Fig.1. Process conditions which lead to particular device characteristics, the FF and SS points in Fig.1 for example, should be specified for accurate prediction of device characteristics such as Id-Vg, Id-Vd and C-V. Monte Carlo simulations (Kunitomo et al. 1999) and response surface methods (Felt et al. 1996), which are well-known methodologies for predicting the distribution, have a problem in that they cannot specify the typical process conditions which lead to particular device characteristics.

We have developed a new methodology for predicting the distribution of device characteristics, which also enables us to specify the most typical process conditions for any device characteristics with only a few calculations. In our approach, to obtain the total distribution and process conditions, arrays which represent the distribution of the device characteristic caused by the fluctuation of every single process are prepared and then merged.

2. Methodology for Predicting the Distribution of Device Characteristics

Before applying our method, well-calibrated process and device simulators should be prepared. The method consists of the following four steps:

(step 1) Processes which have significant effects on the device characteristic, saturated drain current IdsN and IdsP in this case, are selected by calculating

sensitivities of every process fluctuation to the characteristic. From this result, the distribution of every selected process is mapped onto the distribution of the device characteristic.
(step 2) For every process selected, a two dimensional array with IdsN and IdsP for row and column is prepared. Elements of the arrays are filled with probability of occurrence of the corresponding IdsN-IdsP value pairs calculated from the distribution of the device characteristic.
(step 3) Arrays for every significant process specified in step 2 are merged. Elements of the merged array are calculated using the following equation:

$$P_{p1,p2}(i,j) = \sum_{m,n}(P_{p1}(m,n) * P_{p2}(i-m, j-n))$$

where $P_p(i,j)$ is the probability of occurrence of the (i,j) element, with the origin specified at nominal values of IdsN and IdsP, caused by the fluctuation of process p. This operation represents that assuming process p1 and p2 are independent, every element of the array caused by process p2 is distributed following the distribution of another process p1. In the computational implementation of this operation, every array element has an attribute list, where the corresponding process name, value and probability of occurrence are described. These lists are then combined and passed to the resultant array during the operation, which enables us to predict the most typical process conditions for the specified IdsN-IdsP value pair.
(step 4) To specify an appropriate 3σ or any distribution area from the resultant array, elements with a higher probability of occurrence are selected from the array and added until the sum of the probability reaches the specified rate. When the 3σ area is specified, the values of the array elements are added until the sum reaches 99.7%.

3. Application Example

We have applied this method to predicting the distribution of the IdsN-IdsP characteristics of a 0.18μm generation MOSFET and then evaluated our method by comparing the resultant distribution with the measured data and data calculated using Monte Carlo simulation.

3.1. Specification of Device Characteristics Distribution

We have analyzed the sensitivity of a device characteristic of saturated drain current (IdsN and IdsP) to every process fluctuation, and three major processes whose fluctuations have significant effect on the characteristic are selected. The nominal values and standard deviation (3σ) of the processes and their effect on the characteristic are shown in Table 1. Though every process fluctuation is assumed to follow Gaussian distribution, distribution of the characteristic caused by Lg fluctuation does not because of the non-linearity of the drain current vs. gate length relation. We specified two separate Gaussian distributions of the characteristic, one for the case gate length is manufactured longer than the nominal, and the other for shorter. Gate length mismatch represents the difference between the gate lengths of paired NMOS and PMOS; the distribution of the characteristic caused by the mismatch depends on the gate lengths of the pair because of short channel effects. We specified the distribution of the characteristic caused by gate length mismatch as a function of Lg.

3.2 Calculation Results and Their Accuracy in Re-calculation

Calculation results are shown in Fig.2 with 1σ, 2σ and 3σ distributions indicated. Most typical process conditions that lead to FF, SS, FS and SF points on the 3σ line are extracted and shown in Table 2. With these process conditions, device characteristics are calculated again and compared with the original prediction (Fig.3 and Table 3). All differences are less than 2%, which indicates the degree of accuracy of our method for the extraction of process conditions which lead to particular device characteristics.

3.3 Comparison with Measured Data and Monte Carlo Simulation

The 3σ line derived using our method is shown with the calculated points using Monte Carlo simulation (1,000 points, Fig.4) and the measured points of manufactured chips (11,649 points, Fig.5). Compared with Monte Carlo simulation, 99.3% of calculated points are within our 3σ line and its distribution indicates good agreement with our method. As for the measured data, 99.3% of the measured points are within our 3σ line and its distribution tends to shift slightly in the direction of less Ids, compared with our method. These comparison results suggest that our method provides sufficient accuracy in the development phase.

4. Conclusion

We have developed an efficient method to predict the distribution of device characteristics. Using this method, not only the distribution but also the typical process conditions for any particular device characteristics can be extracted from the result. Comparison with the result of Monte Carlo simulation and with the measured data has established that our method is accurate and practical.

References

Kunitomo,H., Sato,H., Tsuneno,K., Ikematsu,R., Masuda,H. (1999): "TCAD Prototyping with New Accurate Worst-Case Definition for a 0.2 Micron CMOS ASIC Process," 4th IWSM, pp.38-41, 1999 4th IWSM, pp.38-41

Felt,E., Zanella,S., Guardiani,C., Vincentelli,A.S. (1996): "Hierarchical Statistical Characterization of Mixed-Signal Circuits Using Behavioral Modeling," Proc. IC-CAD, pp.374-380, 1996 Proc. IC-CAD, pp.374-380

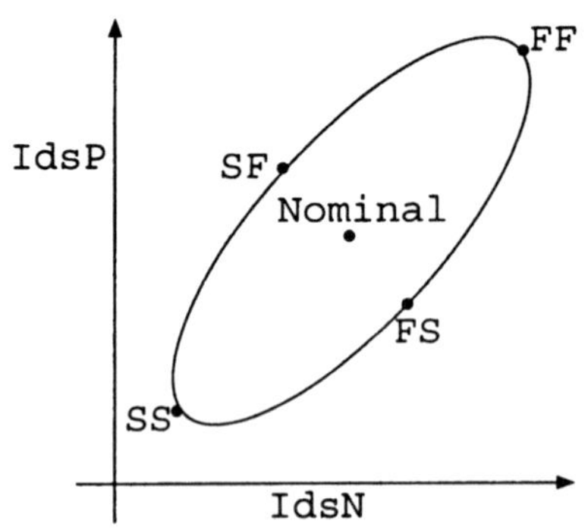

Fig.1 Measured distribution outline of IdsN-IdsP pairs. Nominal point is for center values of IdsN and IdsP. FF stands for NMOS Fast and PMOS Fast, SS for NMOS Slow and PMOS Slow, and so on.

Table 1 Fluctuations of three major processes, and their effect on IdsN and IdsP.

Process fluctuations	nominal	standard deviation (3σ)
Lg	0.15μm	0.015μm
Lg Mismatch	0μm	0.0056μm
Tox	3.47nm	0.069nm (NMOS)
	3.67nm	0.073nm (PMOS)
Nominal values of IdsN and IdsP(mA/μm)		
NMOS	5.60	
PMOS	1.97	
Calculated deviations(3σ) caused by single process fluctuations(mA/μm)		
Lg	IdsN 1.22(shorter),0.836(longer)	
	IdsP 0.389(shorter),0.27(longer)	
Lg Mismatch	$IdsN(Lg)=158Lg^2-55Lg+5.0$	
	$IdsP(Lg)=-17Lg^2+3.2Lg$	
Tox	IdsN 0.158	
	IdsP 0.101	

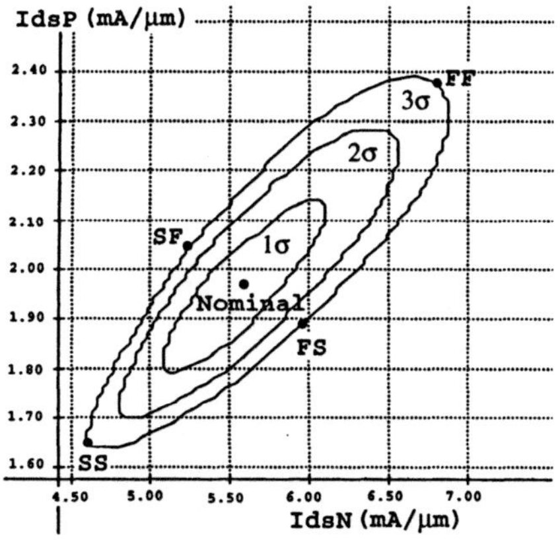

Fig.2 Calculated distribution of IdsN-IdsP pairs and FF,SS,FS and SF points on 3σ outline.

Table 2 Most typical process conditions predicted with our method, that lead to FF, SS, FS and SF points in Fig.2.

```
FF: IdsN:6.8-6.835, IdsP:2.37-2.38(mA/µm)
 Lg(µm)        N:0.1356 , P:0.1356
 Lg Mis(µm)    N:0, P:0
 Tox(nm)       N:3.451 , P:3.65
SS: IdsN:4.665-4.7, IdsP:1.64-1.65(mA/µm)
 Lg(µm)        N:0.1654 , P:0.1654
 Lg Mis(µm)    N:-0.0003774 , P:0.0003774
 Tox(nm)       N:3.493 , P:3.694
FS: IdsN:5.89-5.925, IdsP:1.87-1.88(mA/µm)
 Lg(µm)        N:0.1490 , P:0.1490
 Lg Mis(µm)    N:-0.005358 , P:0.005358
 Tox(nm)       N:3.483 , P:3.684
SF: IdsN:5.26-5.295, IdsP:2.05-2.06(mA/µm)
 Lg(µm)        N:0.1512 , P:0.1512
 Lg Mis(µm)    N:0.005585 , P:-0.005585
 Tox(nm)       N:3.460, P:3.660
```

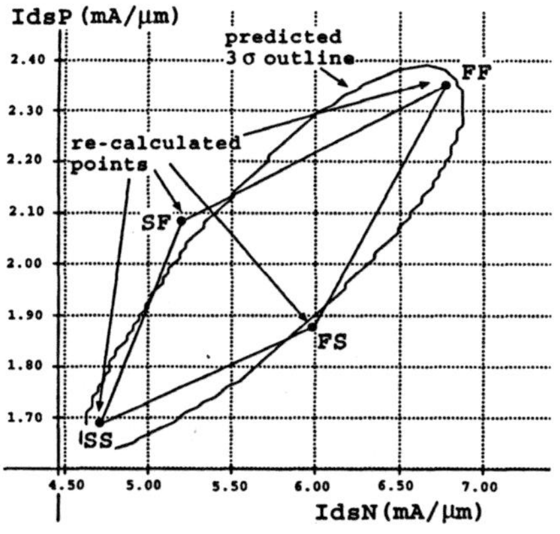

Fig.3 Comparison of re-calculated device characteristics with the distribution of the original prediction.

Table 3 Difference of re-calculated saturated current (IdsN and IdsP) compared with original prediction at FF,SS,FS and SF points.

	FF	SS	FS	SF
IdsN	-0.6%	0.8%	1.4%	-1.1%
IdsP	0.4%	1.0%	0.4%	1.5%

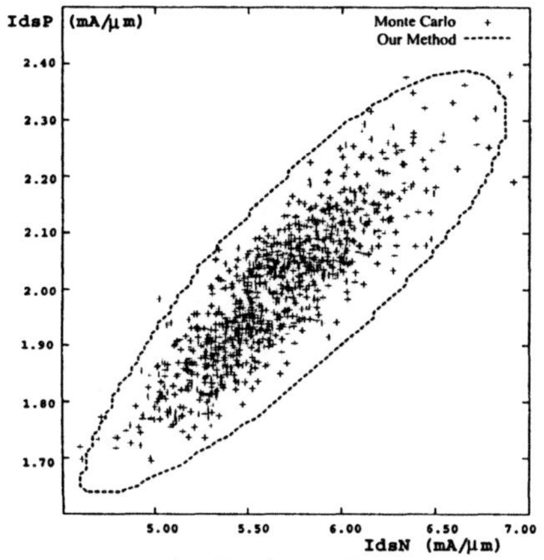

Fig.4 Distribution calculated by Monte Carlo Simulation (cross) and 3σ outline by our method (line).

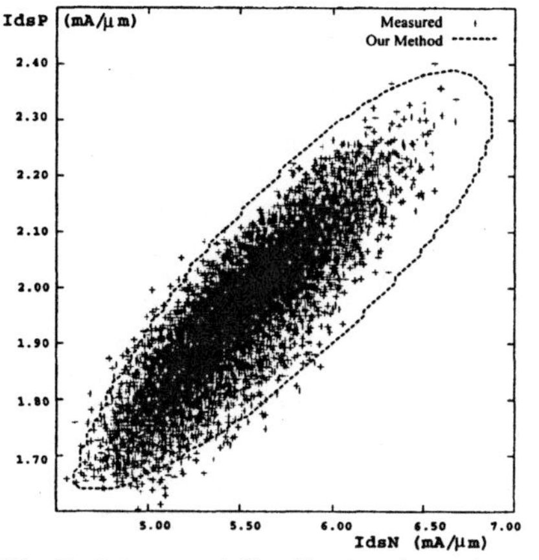

Fig.5 Measured distribution (cross) and 3σ outline predicted by our method (line).

Quantum mechanical balance equations for modeling transport in closed electric circuits.

Bart Sorée, Wim Magnus, Wim Schoenmaker
IMEC, Kapeldreef 75
B-3001 Leuven, Belgium
Bart.Soree@imec.be, Wim.Magnus@imec.be, Wim.Schoenmaker@imec.be

Abstract

We derived a set of quantum mechanical balance equations for a closed electric circuit represented by a closed region Ω. These equations were numerically simulated to obtain information about global and local transport properties. In particular we studied a pure quantum wire at low temperature containing only acoustical phonons. We also investigated the influence of a single elastic scatterer present in the quantum wire.

1. Introduction

In ref. [1], it was argued that the reservoir concept is not suitable to set up a quantum transport formalism. An alternative approach was proposed based on the non-trivial topology of closed electric circuits.

2. Derivation of quantum mechanical balance equations

The electric circuit is represented by a closed region Ω in which the electrons are confined. The region Ω also contains a battery that is responsible for the electromotive force (EMF) caused by an externally applied voltage. In general electromagnetic fields $(\vec{E}, \vec{B})$ will be present. The one-electron wave functions and the field operators $\hat{\Psi}$ vanish at the circuit boundary $\hat{\Psi}(\vec{r}, t) = 0$ at $\partial\Omega$. The electromagnetic fields $\vec{E}$ and $\vec{B}$ are related to a scalar and vector potential $(\Phi, \vec{A})$. The total Hamiltonian can be written as $\hat{H} = \hat{H}_e + \hat{H}_P + \hat{H}'$ where

$$\hat{H}_e = \int_\Omega d\tau \left(\frac{1}{2} m \vec{V}^* \hat{\Psi}^+ \cdot \vec{V} \hat{\Psi} + (U - e\Phi) \hat{\Psi}^+ \hat{\Psi} \right). \tag{1}$$

Eq. (1) represents the electron energy in the presence of internal potentials U and electromagnetic fields, where $\vec{V} = (-i\hbar\vec{\nabla} + e\vec{A})/m$ is the velocity operator. $\hat{H}_P$ and $\hat{H}'$ respectively represent the phonon bath energy and the electron-phonon interaction energy. The time evolution of operators in the Heisenberg picture is governed by the Heisenberg equation of motion $i\hbar \frac{\partial \hat{\Psi}}{\partial t} = [\hat{\Psi}, \hat{H}]$. This equation of motion is used to derive a set of *local* balance equations for energy and momentum. Integration of these *local* balance equations over the region Ω and taking the steady-state ensemble average then results in the following set of global balance equations in the steady-state regime :

$$IV_\epsilon = \frac{i}{\hbar} \langle [\hat{H}_e, \hat{H}'] \rangle = P(\Delta\mu, T_e) \tag{2}$$

$$\int_\Omega d\tau \, \langle \hat{\rho} \rangle \, \vec{E} = \frac{i}{\hbar} \langle [\hat{\vec{P}}, \hat{H}'] \rangle = \vec{F}(\Delta\mu, T_e) \tag{3}$$

where IV_ϵ is the power delivered by the battery, while $\vec{F}(\Delta\mu, T_e)$ is the friction force arising from the phonons. The product IV_ϵ in Eq. (2) arises due to an integral theorem for closed electric circuits [2].

3. Solving the balance equations

In order to solve the balance equations (2)-(3) in the steady-state regime we use first order perturbation theory to calculate the ensemble average of an arbitrary Hermitian operator $\hat{A}(t)$ in the Heisenberg picture

$$\langle \hat{A}(t) \rangle = \langle \hat{A}^{(0)}(t) \rangle_0 - \frac{i}{\hbar} \int_0^t dt' \left\langle \left[\hat{A}^{(0)}(t), \hat{H}_{int}^{(0)}(t') \right] \right\rangle_0 . \tag{4}$$

where the superscript (0) denotes the interaction picture for operators. The total Hamiltonian is divided into an unperturbed part $\hat{H}_0$ and an interaction part $\hat{H}_{int}$. The initial density operator $\hat{\rho}_0$ is not chosen to be the Gibbs-ensemble. Instead, it is taken to be a "boosted" initial density operator, i.e.

$$\hat{\rho}_0^B = \frac{1}{Z} \exp\left(-\beta_e(\hat{H}_e^B - \mu\hat{N})\right) \exp\left(-\beta\hat{H}_P\right), \tag{5}$$

$$\hat{H}_e^B = \hat{H}_e - \sum_q \hat{c}_q^+ \hat{c}_q \frac{1}{2} sgn(I_{qq}) \Delta\mu. \tag{6}$$

The initial density operator $\hat{\rho}_0^B$ is parameterized by $\beta_e = 1/k_B T_e$ and $\Delta\mu$, where I_{qq} is the current carried by the eigenstate $|q>$. Physically, it represents a moving and non-interacting electron system and a free phonon system in thermal equilibrium. The line integral of the electric field in the region Ω must be equal to the EMF, i.e.

$$V_\epsilon = \oint_C \vec{E} \cdot d\vec{l}. \tag{7}$$

The independent variables $\Delta\mu$, T_e and V_ϵ are found by solving Eqs. (2), (3) and (7) in a self-consistent way. A similar approach was also used in hot electron transport [3, 4].

4. Results

First, we have applied the balance equation approach to a pure quantum wire without impurities. The confinement is taken to be very strong and only sufficiently low temperatures are considered, so that only the lowest subband has to be taken into consideration. The dissipative medium consists of acoustical phonons only. The physical parameters that were taken for the quantum wire are listed in table 1. In fig. 1 we show the results of a simulation where the current I and the electron temperature T_e are plotted versus the voltage V_ϵ ranging between 0.1 mV and 1 V. Apparently there exists a threshold of approximately $V_\epsilon \approx 0.1V$ where the electron temperature T_e starts to increase and where I shows a non-linear behavior. Beyond this threshold the simulation results are to be doubted, because for higher electron temperature we need to take into account more than one subband and optical phonons in order to ensure energy relaxation. Next, we have studied the same quantum wire, but now containing one impurity. For the sake of simplicity we have modeled the impurity potential by a delta-function, i.e. $U(x) = U_0\delta(x)$. Accordingly, the electric field consists of a homogeneous part E_H

Fermi energy	ϵ_F	1 meV
Lattice temperature	T	1 K
Effective mass	m_e^*	$0.1\ m_e$
Lateral dimensions	L_y, L_z	10 nm
Wire length	L_x	1 mm
Sound velocity	v_s	10^3 m/s
Deformation potential	D	9 eV
Mass density	ρ_0	10^3 kg/m^3

Table 1: Physical parameters

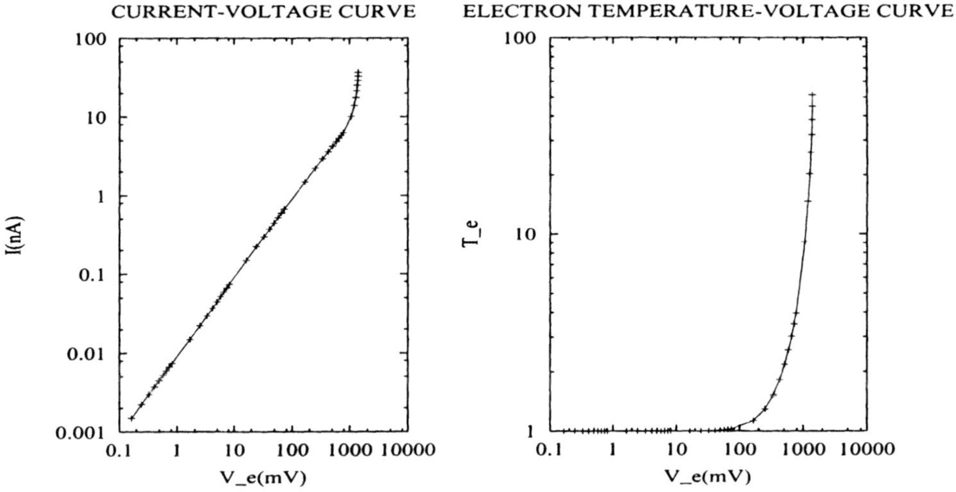

Fig. 1: Pure quantum wire

and a localized part $E_\delta(x) = V_\delta \delta(x)$. Due to Eq. (7) we have to require that the total voltage be equal to the sum of the homogeneous and localized voltage drop, i.e.

$$\oint_C \vec{E} \cdot d\vec{l} = E_H L_x + V_\delta = V_H + V_\delta = V_\epsilon. \tag{8}$$

The results of the simulations are shown in fig. 2 where both the localized and the homogeneous voltage drop are shown as a function of the transmission coefficient $T(k_F)$ at Fermi-level. The total voltage drop is taken to be 1 mV. For a weak impurity potential, i.e. $T \approx 1.0$ we obtain the same results as the pure quantum wire. The total voltage drop V_ϵ tends to the homogeneous voltage drop V_H, while $V_\delta \to 0$. As the barrier height U_0 increases or the transmission coefficient T becomes smaller than 1, the homogeneous voltage drop decreases while the localized voltage drop increases. For a very large barrier height $U_0 >> 1$, or $T << 1.0$, the total voltage drop gets equal to the localized voltage drop V_δ, while no voltage drop goes into the homogeneous part. If we look at the resistance for decreasing values of T it seems that an "anomaly" occurs. Classically one expects a monotonic increase of the resistance when the barrier height increases. However, fig. 2 clearly shows that the resistance first decreases and subsequently increases for $T < 0.1$. This phenomenon is due to the functional dependence of the transmission coefficient T on the electron energy.

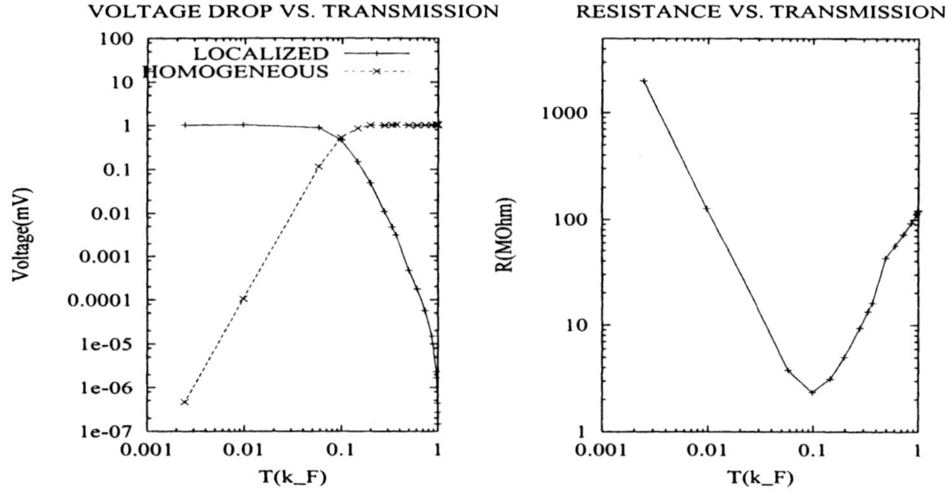

Fig. 2: Quantum wire with the presence of a delta-potential

5. Conclusions

The results we have obtained clearly show that the quantum mechanical balance equation approach is able to investigate global as well as local transport properties. In particular we demonstrated that the presence of an impurity in a quantum wire with strong confinement at low temperatures directly affects the spatial localization of energy dissipation. Moreover, this method resolves a longstanding problem of determining the spatial localization of energy dissipation in the presence of a potential barrier. In the future we will further apply the balance equation method to investigate spatially inhomogeneous structures such as quantum point contacts and carbon nanotubes.

Acknowledgment Part of this work is funded by the Flemish Institute for Science and Technology (IWT), Flanders (Belgium).

References

[1] W. Magnus and W. Schoenmaker. Quantized conductance, circuit topology and flux quantization. *Phys. Rev. B*, 61:10883, 2000.

[2] W. Magnus and W. Schoenmaker. On the use of a new integral theorem for the quantum mechanical treatment of electric circuits. *J. Math. Phys.*, 39:6715, 1998.

[3] X.L. Lei and C.S. Ting A new approach to non-linear transport for an electron-impurity system in a static electric field. *Journ. of Phys. C: Solid State Physics*, 18:77-92, 1985.

[4] X.L. Lei and N.J.M. Horing Balance-equation approach to hot-carrier transport in semiconductors. *Int. Journ. Mod. Phys. B*, 6:801, 1992.

A Nonlinear Iterative Method for InAs/GaAs Semiconductor Quantum Dots Simulation

Yiming Li[1], O. Voskoboynikov[1,2], C. P. Lee[1], and S. M. Sze[1]
[1]Department of Electronics Engineering, National Chiao Tung University
Hsinchu 300, Taiwan
Email: ymli.ee87g@nctu.edu.tw
[2]Kiev Tarans Shevchenko University
Kiev, Ukraine

Abstract

A computational efficient nonlinear iterative method for computing the ground and excited state energies of an electron confined by an InAs quantum dot embedded in GaAs matrix is proposed. We treat the problem with the effective one electronic band Hamiltonian, the energy and position dependent electron effective mass approximation, and the Ben Daniel-Duke boundary conditions. Computational results show that the parabolic band approximation is valid only for the dots with large volume. For excited states the nonparabolic effect has also been found to be stronger than for ground states.

1 Introduction

During the last decade the experimental development and theoretical study of semiconductor quantum dots (QDs) have been of great interests. For a well-fabricated QD the electron spectrum consists of a set of discrete levels and the density of levels becomes a set of δ-functions. This makes the semiconductor QDs very attractive for various applications in micro and nano optoelectronics [1-2].

In the modelling and simulation of semiconductor quantum dots, one needs a physical assumption about the electronic confinement potential in the QD system. Among those applied confinement potential models (the parabolic lateral potential and the infinite wall potential) (see [3] and references therein), a finite hard wall boundary potential model is the most realistic approach. In this case the problem cannot be solved analytically and it requires a computer modelling for the numerical simulation of such problems. In this paper, a computational efficient nonlinear iterative method for computing the ground and excited state energies and the corresponding wave functions of an electron confined by an InAs QD embedded in GaAs semiconductor matrix is presented for the first time. The model is formulated with the effective one electronic band Hamiltonian, the energy and position dependent electron effective mass approximation (so-called nonparabolic band approximation), and the Ben Daniel-Duke boundary conditions. In contrast to most of other calculations, our model includes a hard-wall (of finite height) 3D confinement potential that is induced by real discontinuity of the conduction band at the edge of the InAs/GaAs QD. Based on a nonlinear iterative, the balanced and shifted QR, and inverse iteration methods, a novel solution algorithm for InAs/GaAs QD simulation is developed and successfully implemented. With the developed QD simulator, the effects of parabolic and nonparabilic band approximation in energy levels are studied for different shapes of QD. Our results show that the parabolic band approximation (effective mass is independent on energy) is applicable only for the dots with large volume. For excited electron states the nonparabolic effect has also been found to be stronger than for ground states. In Sec. 2, the QD model and computational method are stated. Results and discussions are presented in Sec. 3. Conclusions are given in Sec. 4.

2 A Quantum Dot Model and Computational Algorithm

We consider electrons confined in QD and use one-band effective Hamiltonian H [4].

$$H = -\frac{\hbar^2}{2}\nabla_r\left(\frac{1}{m(E,r)}\right)\nabla_r + V(r), \tag{1}$$

where $m(E, r)$ is the electron effective mass that depends on both energy and position

$$\frac{1}{m(E,r)} = \frac{P^2}{\hbar^2}\left[\frac{2}{E + E_g(r) - E_c(r)} + \frac{1}{E + E_g(r) + \Delta(r) - E_c(r)}\right], \tag{2}$$

and $V(r) = E_c(r)$ is the QD confinement potential. The $E_c(r)$, $E_g(r)$, $\Delta(r)$, and P are the position dependent electron band edge, band gap, spin-orbit splitting in the valance band, and momentum matrix element, respectively [9]. When the QD has a disk, ellipsoid, or conical shape of radius R_0 and of height z_0 we can solve the problem with cylindrical coordinate (R, ϕ, z). The QD system is cylindrical symmetry, so the wave function can be written as: $\Phi(r) = \Phi(R,z)\exp(il\phi)$, where $l = 0, \pm1, \pm2, \ldots$ is the electron orbital quantum number and the model is reduced to

$$-\frac{\hbar^2}{2m_i(E)}\left(\frac{\partial^2}{\partial R^2} + \frac{\partial}{R\partial R} + \frac{\partial^2}{\partial z^2} - \frac{l^2}{R^2}\right)\Phi_i(R,z) + V_i(R,z)\Phi_i(R,z) = E\Phi_i(R,z) \tag{3}$$

where $V_{i=1}(R, z) = 0$ is inside the dot and $V_{i=2}(R, z) = V_0$ is outside the dot. The boundary conditions are

$$\Phi_1 = \Phi_2 \text{ and } \frac{1}{m_1(E)}\left\{\frac{\partial\Phi_1}{\partial R} + \frac{df_s}{dR}\frac{\partial\Phi_1}{\partial z}\right\} = \frac{1}{m_2(E)}\left\{\frac{\partial\Phi_2}{\partial R} + \frac{df_s}{dR}\frac{\partial\Phi_2}{\partial z}\right\}, \tag{4}$$

where $z = f_s(R, z)$ (s is disk, ellipsoid, or conical shape QD, see Fig. 1) presents a contour generator of the QD structure in (R, z) plane. To obtain a "self-consistent" solution of the model, we propose here a nonlinear iterative algorithm as shown in

Fig. 2. The feedback nonlinear iteration scheme is: (a) Set an initial energy E_0, (b) Compute effective mass m, (c) Solve Schrödinger equation with boundary conditions for energy E, (d) Back to (b). The iteration loop will be terminated when a specified stopping criterion on energy is reached. To solve the Schrödinger equation in step (c) a finite difference method with nonuniform mesh, balanced and shifted QR algorithm, and inverse iteration technique are applied in this work. The dominant method for solving matrix eigenvalue problem is still the QR algorithm, in the year 2000 [5].

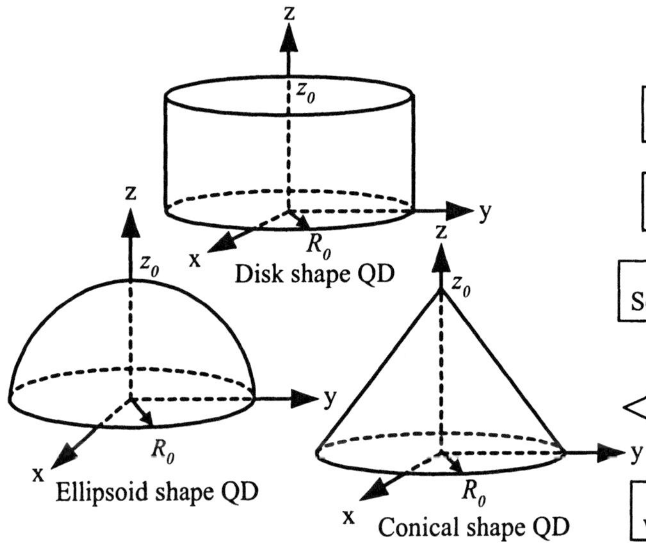

Fig. 1. Schematic diagram for various QDs.

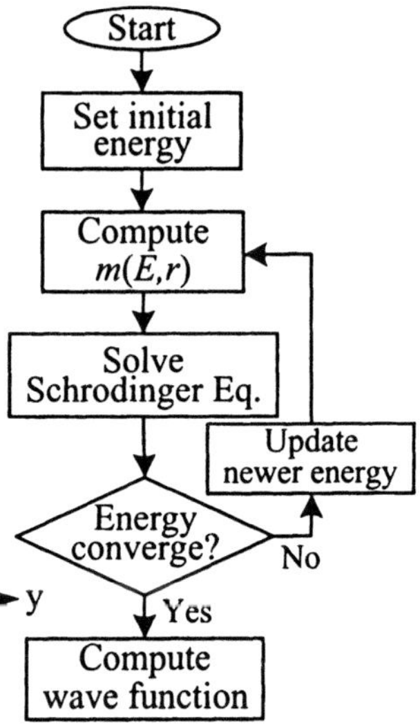

Fig. 2. A solution procedure for QD simulation.

3 Results and Discussions

The proposed nonlinear iterative method converges for various QD structures. It takes about 8-10 iteration loops to reach a stopping criterion in maximum energy error less than 10^{-7}. The global convergence mechanism is due to the effective mass is a monotone function in energy E. Fig. 3 shows the computed ground state ($l=0$) energies with parabolic and nonparabolic band effective mass approximations for R_0 = 10.0 nm (a) disk, (b) ellipsoid, and (c) conical shape QD, respectively. Similarly, Fig. 4 gives the excited state ($l=1$) energies for various QD shapes.

4 Conclusions

A novel computational technique for QD simulation has been proposed. With the developed QD simulator, we have found that the widely used the parabolic band approximation lead to a large discrepancy in calculation results for electron energy states in various small QDs. This parabolic approximation can be used only for large size QDs in the calculation of ground and excited state energies. The modelling, numerical method, and study presented here not only provide a novel way to calculate the energy levels of QD but also are useful to clarify principal dependencies of QD energy states on material band parameter and QDs size for various QD shapes.

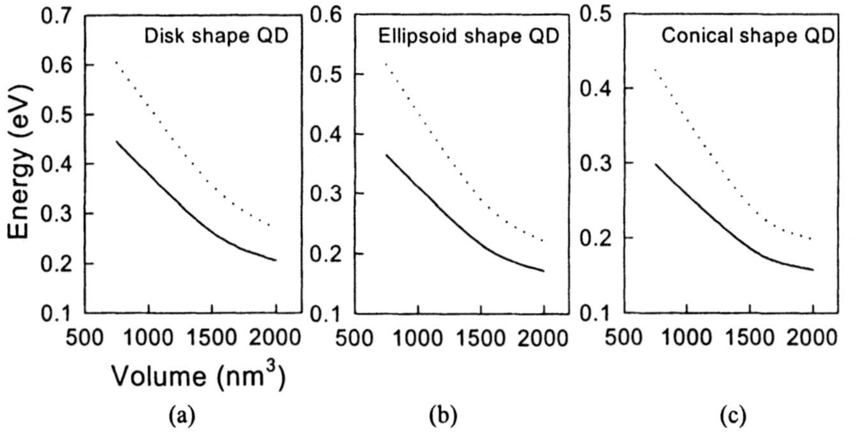

Fig. 3. Ground state ($l=0$) energy with nonparabolic (solid line) and parabolic (dot line) band effective mass approximations for (a) disk, (b) ellipsoid, and (c) conical shape InAs/GaAs QD.

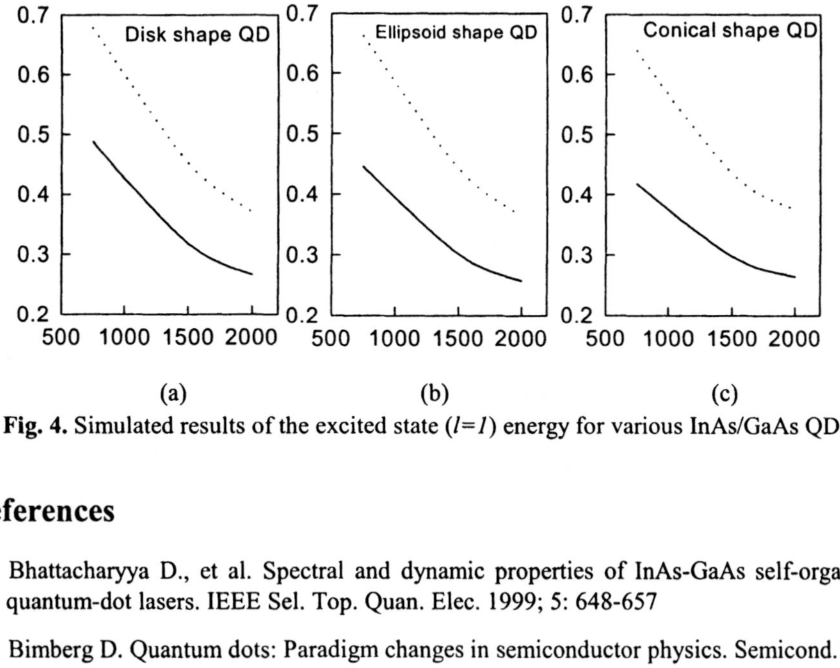

Fig. 4. Simulated results of the excited state ($l=1$) energy for various InAs/GaAs QDs.

References

[1] Bhattacharyya D., et al. Spectral and dynamic properties of InAs-GaAs self-organized quantum-dot lasers. IEEE Sel. Top. Quan. Elec. 1999; 5: 648-657

[2] Bimberg D. Quantum dots: Paradigm changes in semiconductor physics. Semicond.1999; 33: 951-955

[3] Pryor C. Geometry and material parameter dependence of InAs/GaAs quantum dot electronic structure. Phys. Rev. B 1999; 60: 2869-2874

[4] Bastard G. Wave Mechanics Applied to Semiconductor Heterostructures, Les Ulis, 1990

[5] Watkins D. QR-like algorithms for eigenvalue problems. JCAM. 2000; 123: 67-83

The modeling of a SOI microelectromechanical sensor

C. Ravariu[1], F. Ravariu[2], A. Rusu[1], D. Dobrescu[1], L. Dobrescu[1]

[1] "POLITEHNICA" University of Bucharest – Faculty of Electronics and Telecommunications 313, Splaiul Independentei, 77206, Bucharest, Romania. Phone: +40-1-4104740; Fax: +40-1-4104740; e-mail: cristir@mcma.pub.ro
[2]Institute for Research and Development in Microtechnologies, Romania

Abstract

The goal of this paper is to model a SOI pressure sensor based on piezoelectric effect. A PZT film deposited onto a SOI wafer acts like the transistor's gate. The transducer element is an SOI-MOSFET. The electrical simulation made with ATLAS presents the static characteristics of the device. An analytical model for sensitivities was presented to offer a designing rule. Mechanical simulations with ANSYS establish some mechanical characteristics.

1 Introduction

The classical pressure sensors, even SOI sensors, work on other principles (piezoresistive effect, capacitive effect) [1]. Piezoelectric layers represent an alternative to piezoresistors.

The combination between SOI technology and PZT layers brings some advantages: (1) electrical isolation of the device; (2) higher temperature of operation (from 120^0C in conventional technologies where reverse current become significant, up to 300^0C because SOI structure avoids reverse biased junctions for isolation); (3) the Curie temperature of PZT is suitable for this range ($T_C > 300^0$C); (4) possible miniaturization of the sensor because PZT layer represents the SOI-MOSFET's gate; (5) the buried insulator is an excellent etch stop layer. Selecting a piezoelectric layer with better parameters (e.g. PXE5 instead of PZT) yields a better sensitivity.

In this paper is proposed an analytical model for a pressure sensor with PZT as the gate of a SOI-MOSFET. The model allows a study of this kind of pressure sensor, establishing a bi-univocal correspondence between the mechanical pressure and an equivalent gate voltage (V_G). This formal parameter, V_G, has not any practical sense, but is helpful to design the sensor, because some device simulations with ATLAS can be done and some analytical models are available [2]. The mechanical simulations with ANSYS verify the mechanical resistance of the structure.

2 The analytical model for the sensor sensitivity

In figure 1 is presented the proposed sensor architecture.

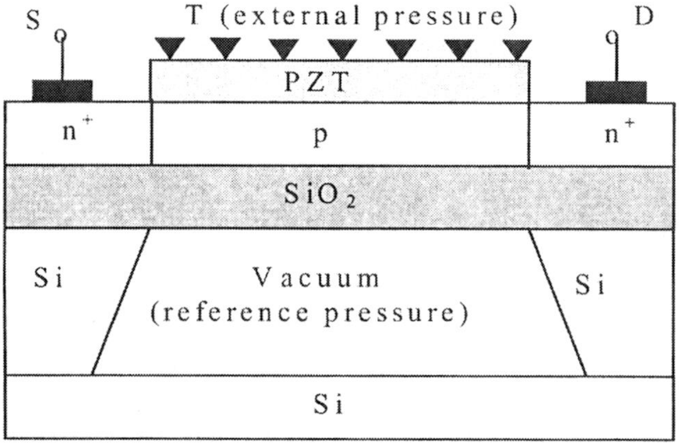

Fig.1. The sensor architecture.

Because the pressure exerts on vertical direction, the incoming pressure, T, induces a polarization state in PZT and depletes the film. An equivalent situation occurs, when the same amount of electric charge is obtained in the absence of any pressure, but with an equivalent positive gate voltage, V_G. Starting from the Poisson's equation integration, in non-depletion approximation, the following global charge enclosed in the transistor body, Q_S, versus the surface potential, ϕ_S results:

$$Q_S(\phi_S) = \sqrt{\frac{2\varepsilon_{Si} q N_A}{\beta}}\left[\exp\beta(\phi_S - 2\phi_F) - 1 + \beta\phi_S\right]^{1/2} \tag{1}$$

On vertical direction and Gauss's law, results:

$$D = d_{PZT}T = \varepsilon_{PZT}E = Q_S \tag{2}$$

where $C_{PZT} = \varepsilon_{PZT}/x_{PZT}$ – the specific capacitance of PZT; ϕ_F – the Fermi potential of Si-film; N_A - the doping film; ε_{Si} – the dielectric permittivity of silicon; x_{PZT} – the PZT thickness; ε_{PZT} – the dielectric permittivity of PZT on vertical axis; d_{PZT} – the component of piezoelectric coefficient on vertical axis, T – the pressure, E – the intensity of the electric field, β=q/kT (is a notation). In strong inversion the lasts terms in eq. (1) are negligible in respect with the exponential function. So, combining (1) and (2) results:

$$V_G = \frac{d_{PZT}T}{C_{PZT}} + \frac{2}{\beta}\ln\frac{d_{PZT}T}{\sqrt{2\varepsilon_{Si} q N_A/\beta}} + 2\phi_F \tag{3}$$

The function (3) establishes a bijection between V_G and T (because their first order derivatives preserve constant sign). Consequently, this expression can be used in the sensor sensitivity computing. The current through the inversion channel, in quasi-linear regime, depends on the gate voltage:

$$I_{D2} = \frac{f_g C_{PZT} \mu_{0n}}{1+\theta_0 (V_G - V_T)} \cdot \left[(V_G - V_T) \cdot V_D - \frac{V_D^2}{2} \right] \tag{4}$$

where f_g is a geometrical parameter, θ_0 is the mobility attenuation factor, μ_{n0} is the "pure" electron mobility. So, the sensitivity in current for the pressure sensor, S_i, is:

$$S_i = \left| \frac{dI_D}{I_D} \frac{T}{dT} \right| = \frac{\frac{d_{PZT} T}{C_{PZT}} + \frac{2}{\beta}}{[1+\theta_0 (V_G - V_T)](V_G - V_T)} \tag{5}$$

3 Electrical simulation with ATLAS

The electrical simulations with ATLAS use the equivalent gate voltage (eq 3).

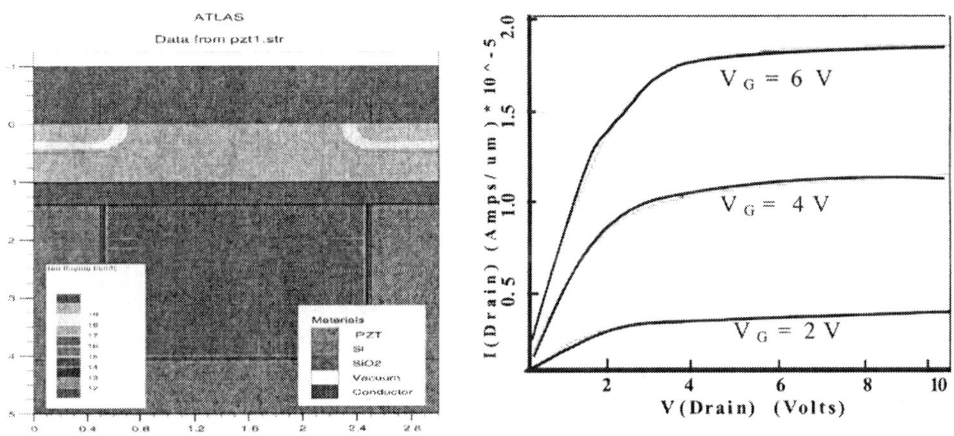

Fig. 2. The simulated structure and the current dependence on equivalent gate voltage, performed by ATLAS.

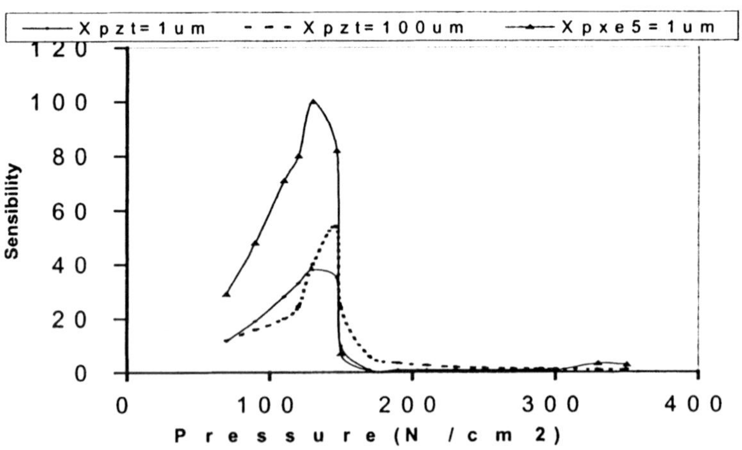

Fig. 3. The dependence of the sensitivities versus pressure in different cases.

The transistor had: $N_A=2\times10^{17}cm^{-3}$ (film), $N_D=10^{20}cm^{-3}$ (n+ source/drain), $N_A=10^{15}cm^{-3}$ (substrate), $Q_{SS1}=3\times10^{10}cm^{-2}$, $Q_{SS2}=10^{11}cm^{-2}$, $V_S=0V$, $V_D=V(3) \in (0,10V)$. From figure 2 a $V_D=1V$ is suitable for the sensor working in quasi-linear regime. The variation of V_G corresponds to a variation of pressure, T, from 0 to 350 N/cm^2. ATLAS has an especial command EXTRACT to extract the model parameters: $\theta_0=0.093V^{-1}$, $V_T= 1.08V$. With these parameters and PZT data from [3], accordingly with eq (5), an analysis was performed (see fig.3).

4 Mechanical simulations with ANSYS

From ANSYS simulations, the pressure that break the sensor was evaluated: $T_{lim} > 350N/cm^2$. In figure 4 is presented the simulations results regarding the distributions of the mechanical tensile into the structure when $T=100N/cm^2$, that correspond to the maximum sensitivity.

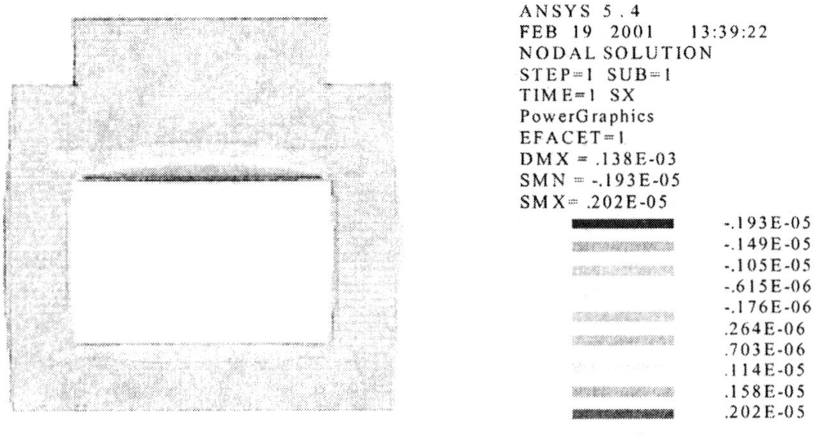

Fig. 4. The ANSYS structure at $T=100N/cm^2$.

5 Conclusions

The ANSYS analysis has a great impact since to consolidate the sensor shape by technological methods. The analytical model can be used for electrical simulations. The designing rule presented here represents a starting point in developing of this kind of devices.

References

[1] Sze S.M. Semiconductor sensors, John Wiley&Sons Inc., 1995, pp 104-111.

[2] Ravariu C et all, IEEE Int. Conf. on Modeling and Simulation of Microsystems, MSM'2000, San Diego, S.U.A., pp 404-407.

[3] *** Philips Data Handbook, 1970, pp G9.

A Computationally Efficient Model for Three-dimensional Monte Carlo Simulation of Ion Implantation into Complex Structures

Di Li, Geng Wang, Yang Chen, Gaurav Shrivastav, Stimit Oak,
Al Tasch and Sanjay Banerjee
Microelectronics Research Center, University of Texas at Austin
Austin, Texas 78758 USA
Email: dli@ece.utexas.edu

Abstract

A new three-dimensional Monte Carlo simulator has been developed based on UT-MARLOWE. Unbalanced Octree algorithm was used for spatial decomposition. A new trajectory replication scheme was developed and implemented to enhance computational efficiency. More than two orders of magnitude savings on CPU time have been achieved.

Introduction

Monte Carlo simulations of ion implantation into three-dimensional complex structures typically lead to severe challenges in term of computational efficiency. In order to obtain the same level of statistical significance for the overall structure, the number of incident ions at each specific grid cell must be kept approximately at the same level as for 1-D and 2-D simulations. *Approximately, for each additional dimension, the required number of ions for Monte-Carlo simulation should increase by about two orders of magnitude.*

New 3-D Trajectory Replication Scheme

Although not adequate to be applied to three-dimensional simulation directly, lateral trajectory replication scheme [1] provides a good insight into the problem. The key is to cleverly reuse the physically-computed trajectories for similar topography.

The lateral trajectory replication scheme is shown in Fig. 1. It starts by physically calculating all the required scattering events and the stopping mechanisms experienced by one particular ion until it stops. The algorithm then introduces an offset and replicates the model trajectory at the new incident position. At the same time, replication error is examined according to two factors: whether the material in the model trajectory matches that in the target trajectory and whether the damage level is comparable between them.

A schematic of how the new replication scheme proceeds is shown in Figure 2. The implant window is divided into small rectangles or squares. The first ion is set to implant into the left bottom corner segment and the trajectory is physically-computed and recorded. The initial relative position within the segment is determined randomly. Replication then proceeds up and to the right of the original trajectory. When an error occurs, replication stops, the trajectory is recomputed and the model trajectory is replaced with the newly calculated one. Thus, the simulator is able to *intelligently* reuse physically-calculated trajectories in two directions. Since the CPU time required to replicate is much less than the time required for the actual physical computation, large savings are expected with the application of the new replication scheme without sacrificing accuracy, as shown in Figures 3, 4.

Structure Decomposition

As the ions are propagating through the structure, the determination of the material type is crucial and could be very CPU time consuming. An unbalanced Octree decomposition scheme is adopted in this new simulator. Similar to the Quadtree decomposition scheme [1,2], Octree also recursively subdivides space until the desired resolution is reached. Octree structure offers two major advantages over other structured and unstructured meshes. First, it simplifies the structure since the resulting grid is coarse where the material is uniform, and fine near the material interfaces. Secondly, the tree data structure makes the tracking more efficient (proportional to logN, instead of N), and reduces the zoning time accordingly.

Model Verification

The physical models used by the new simulator are based on UT-MARLOWE and have been extensively verified.

In this simulator, two models have been provided to simulate ion implantation into poly-silicon. The first model uses amorphous silicon to represent poly-silicon, and the second one, or the worst case channeling scenario, uses a deterministic Monte Carlo model and simulates implant into a (110) wafer. Implantation into a (100) wafer with both tilt angle and rotation angle set to 45 degrees is equivalent to implantation into (110) wafer with both tilt angle and rotation angle set to 0 degrees. Therefore, we can compare the converted 1-D profile with SIMS data for the same implant condition into a (100) wafer with both tilt and rotation set to 45 degrees. As shown in Figure 5, for As 50keV implantation, the SIMS data and the converted simulation results match very well. In addition, As channeling through a three-layered structure, polysilicon, gate oxide and crystalline silicon was studied as shown in Figure 6.

References

1. B. Obradovic, etc, IEDM 1998 Technical Digest, 513.
2. G. F. Carey, etc. "Computer Simulation of Semiconductor Processes and Devices-- Mathematical and Numerical Aspects", John Wiley and Sons, 1996.

Figure 1. Side view of the replication algorithm proceeding from left to right. Light trajectories are physically computed, gray ones show where replication error occurs, and black ones are replicated trajectory.

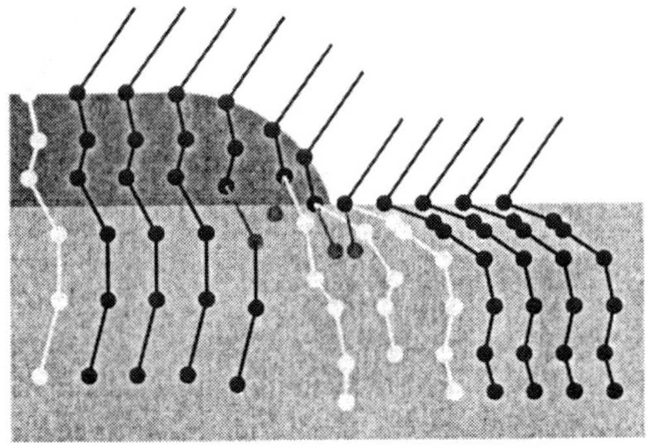

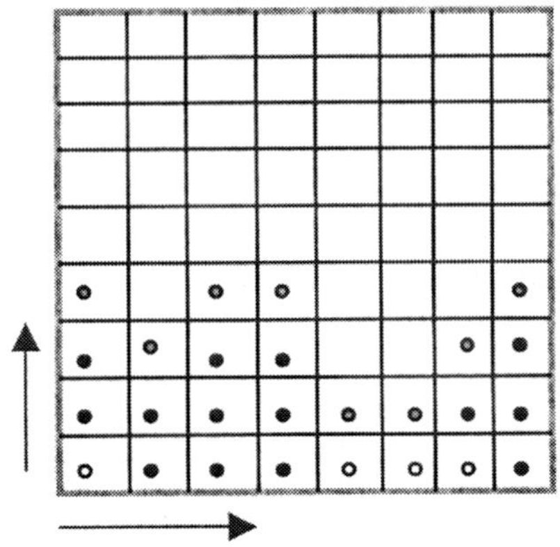

Figure 2. Top view of the simulation domain illustrating the replication algorithm. Hollow circles are physically computed, gray ones show that replication errors are encountered and trajectories are therefore physically computed, and black ones are replicated. Replication starts from the left bottom corner and proceeds up and to the right of the original trajectory.

Figure 3 (a). Simple trench topography for replication scheme demonstration. The length of the whole volume is 800 lattice constants. Depth is 700 lattice constants.

Figure 3 (b). Simulating boron 5keV implant into structure shown in Figure 3(a), tilt-7°, rotation-30°, dose-1×10^{14}cm^{-2}. CPU time is 72 minutes using 5,000,000 ions. The iso-concentration contours range from 1×10^{15}cm^{-3} to 1×10^{17}cm^{-3}. Estimated CPU time *without replication* is over 20 hours.

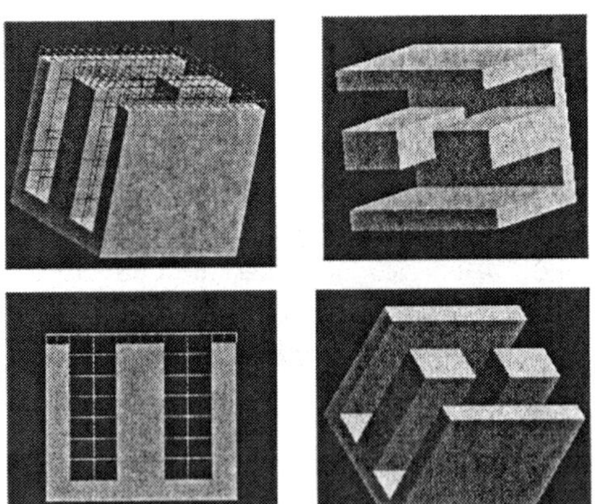

Fig. 2

Fig. 3

Figure 4 (a). Source/Drain structure with shallow trench isolation. Side length is 600 lattice constants.

Figure 4 (b). Simulating arsenic 15 keV implant into structure show in Figure 4(a), tilt-0°, rotation- 0°, dose $1\times10^{13}cm^{-2}$. CPU time is 88 minutes, using 5,000,000 simulated ions. The iso-concentration contours range from $1\times10^{16}cm^{-3}$ to $1\times10^{19}cm^{-3}$ Estimated CPU time *without replication* is over 20 hours.

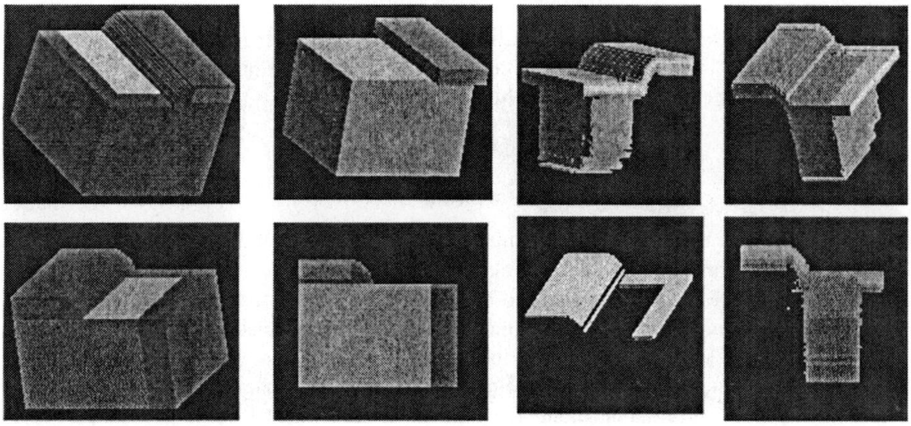

Figure 5. Profiles of a 50keV As on-axis implant into a (110) Si wafer with a dose of $2\times10^{12}cm^{-2}$ to simulate the worst case channeling scenario of an implant into poly-silicon.

Figure 6. Profiles of As on-axis implant into a three-layered structure (100nm polysilicon, 2.7nm gate oxide and 700nm crystalline silicon) at 1, 3, 8, 15, and 30 keV with a dose of $1\times10^{15}cm^{-2}$

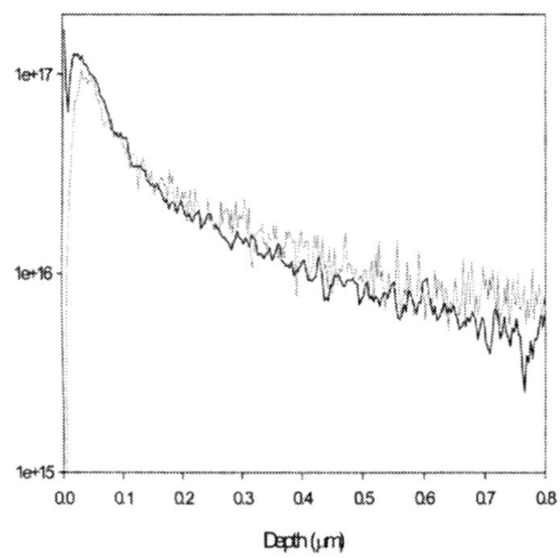

Fig. 5

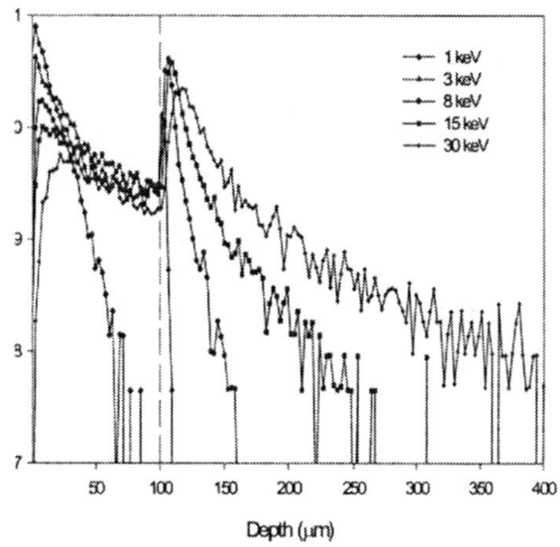

Fig. 6

A Full-Wave Analysis for Multi-Level Interconnects Using FDTD-PML method

Yountae Kim, Sukin Yoon, Ikjun Choi, Sechun Park, Ohseob Kwon
and Taeyoung Won
Computational Electronics Center,
School of Electrical and Computer Engineering, Inha University
253 Yonghyun-Dong, Nam-Gu, Inchon, Korea 402-751
E-mail: kyt@hsel.inha.ac.kr, twon@hsel.inha.ac.kr

Abstract

FDTD(finite difference time domain) method is a 3-D full-wave numerical algorithm for solving various electromagnetic phenomena and interaction problems. PML(perfectly matched layer) method is very efficient for absorbing the electromagnetic waves and then for solving unbounded problems. This paper deals with the application of FDTD-PML technique to the solution of wave-structure interaction problems. It is shown that the excited Gaussian pulses were propagated along the multi-level interconnect metal line as the increase of time and the interaction phenomena occurred. The absorbing effectiveness of PML as a function of the time step is also demonstrated.

1 Introduction

At high-speed VLSI, electromagnetic wave propagation and radiation effects become significant and consequently limit the performance of on-chip and off-chip VLSI interconnects. To accurately analyze parasitic electromagnetic wave phenomena, vectorial Maxwell's equations modeling tools are needed. The accuracy, efficiency, and overall utility of FDTD has been demonstrated in recent years for a variety of complex high-speed interconnect design problems[1]. In this paper, we used the FDTD method, which is a 3-D full-wave numerical algorithm for solving various electromagnetic phenomena and interaction problems.

2 FDTD - PML method

The FDTD method is a computationally efficient approach for modeling sinusoidal or impulsive electromagnetic wave interactions with arbitrary three-dimensional structures. It is an explicit grid-based technique for the direct solution of the Maxwell's curl equations with the six components of the electric and magnetic vectors. As illustrated in FIG. 1, the six components of the electric and magnetic vectors are staggered in space by dx/2, dy/2, dz/2 and in time by dt/2. The electric components (Ex, Ey, Ez) are located on the center of the edge with the tangential direction. And the magnetic components (Hx, Hy, Hz) are located on the center of the face with the normal direction. Using the Yee grid, all components of EM fields at cartesian coordinate system can be expressed [2]. The Maxwell's curl equations with the six components of the electric and magnetic vectors are discretized using central finite differences. Upon centered finite-difference approximations, we obtain the following expression:

$$E_x^{n+1}(i+\frac{1}{2},j,k) = E_x^n(i+\frac{1}{2},j,k) + \frac{\Delta t}{\varepsilon}\cdot\left[-\sigma E_x^n(i+\frac{1}{2},j,k) + \frac{H_z^{n+\frac{1}{2}}(i+\frac{1}{2},j+\frac{1}{2},k) - H_z^{n+\frac{1}{2}}(i+\frac{1}{2},j-\frac{1}{2},k)}{\Delta y} - \frac{H_y^{n+\frac{1}{2}}(i+\frac{1}{2},j,k+\frac{1}{2}) - H_y^{n+\frac{1}{2}}(i+\frac{1}{2},j,k-\frac{1}{2})}{\Delta z}\right] \quad (1)$$

To minimize the reflection of outgoing waves, we also used the PML technique, which is a new technique of free-space simulation to be used with the FDTD method. The PML is very efficient for absorbing the electromagnetic waves and then for solving unbounded problems. The PML technique consists in surrounding a computational domain with the absorbing layer described in FIG.2. The outgoing waves become weak through the PML, which requires the simulation of lossy regions near grid boundaries. To absorb the propagation waves at any angle of incidence, six electric and magnetic components are split into 12 subcomponents as follows: [3]

$$\varepsilon_o\frac{\partial E_{xy}}{\partial t} + \sigma_y E_{xy} = \frac{\partial(H_{zx}+H_{zy})}{\partial y}, \quad \varepsilon_o\frac{\partial E_{xz}}{\partial t} + \sigma_z E_{xz} = -\frac{\partial(H_{yx}+H_{yz})}{\partial z} \quad (2)$$

FIG. 3 shows the components of electric conductivity in each plane of the computational domain. In this paper, we used 10 layers of the PML and assumed respectively n=2 and $R(\theta)=10^{-7}$.

3 Simulation and Results

In this paper, the structure of four microstrip transmission lines is used to demonstrate the FDTD simulation, as illustrated in FIG. 4. The 150×40×40 of cells are used in the region of FDTD simulation and 472704 cells are applied in the PML region. The length, width and thickness of copper microstrip (ε_r=1.0, $\sigma=5.8\times10^7$ S/m) are respectively 300 μm, 3 μm and 1μm. Located between the microstrip and the ground plane is 1 μm-thick low loss dielectric substrate (ε_r=4.0, $\sigma=0.5\times10^{-3}$ S/m). The region above the microstrip is filled with air (ε_r=1.0006, σ=0.0). An electric field is modeled at one end of the microstrip; the other end is terminated with an open-circuit. The excited waveform is a baseband Gaussian pulse:

$$E_{z\_exicite} = E_0 e^{-(t-t_0)^2/\tau^2} \tag{3}$$

Where $t_0=3\tau$ and $\tau=0.5$ ps. FIG. 5 shows the source excitation plane. To minimize numerical dispersion errors and thereby ensure numerical accuracy, the time increments must be small enough relative to the space increments. The time increments must be determined using the Courant stability condition. The 2-D contour and contour surface of electric potentials at initial time step, which is calculated by central differential method, is illustrated in FIG. 6. Using the calculated electric potentials, the distribution of electric fields is extracted from the computational region, as illustrated in FIG. 7(a). We apply central difference to the Maxwell's curl equation for solving the distribution of magnetic fields. FIG. 7(b) shows the simulated magnetic fields from the electric fields. The FDTD simulation results are illustrated in FIG. 8.

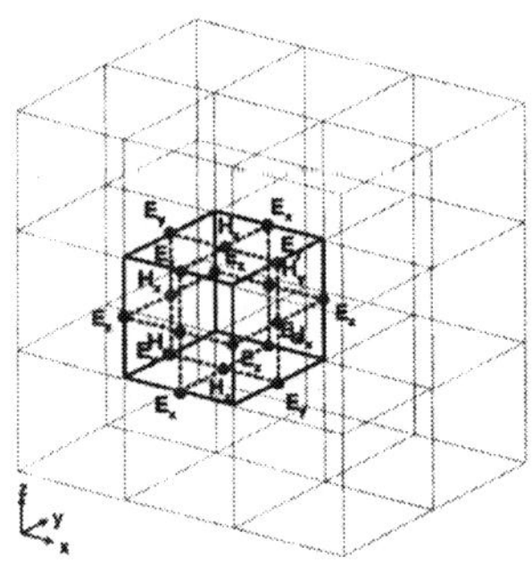

Fig. 1. Yee lattice with 6 components of electric and magnetic fields

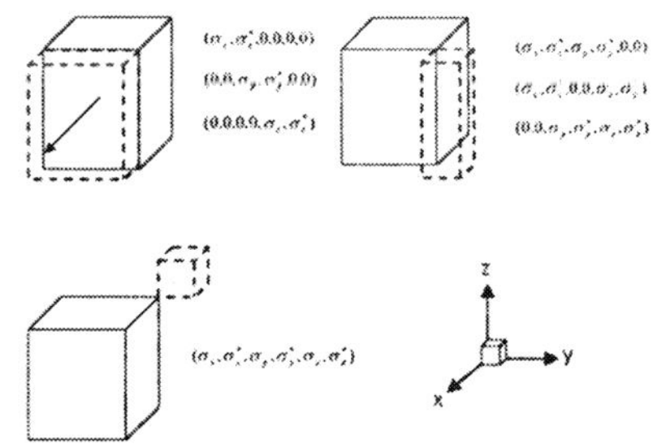

Fig. 3. The electric conductivity profiles in each plane of the computational domain

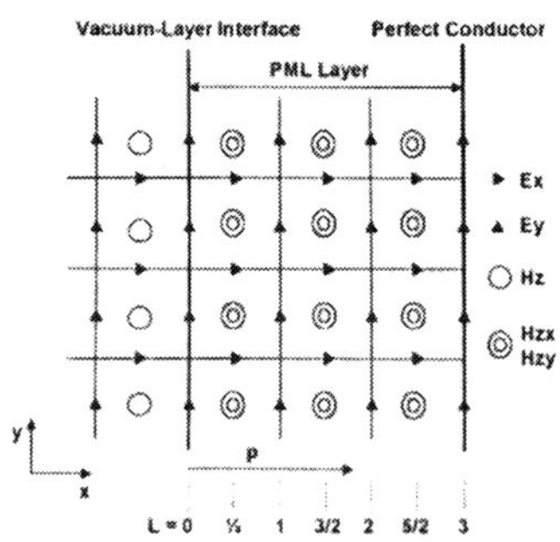

Fig. 2. Right side of a domain surrounded by a PML layer

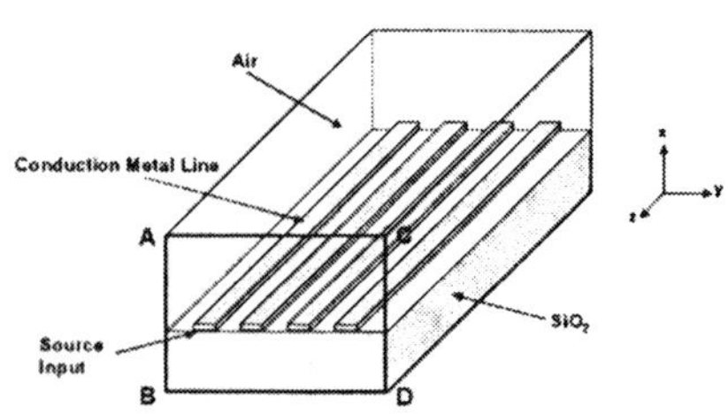

Fig. 4. The structure of microstrip lines

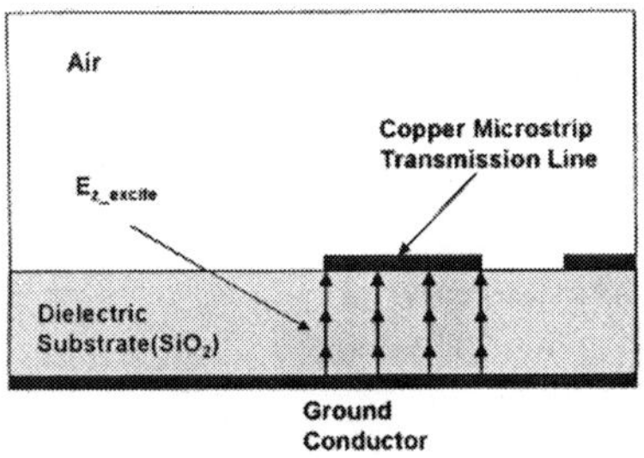

Fig. 5. Source excitation

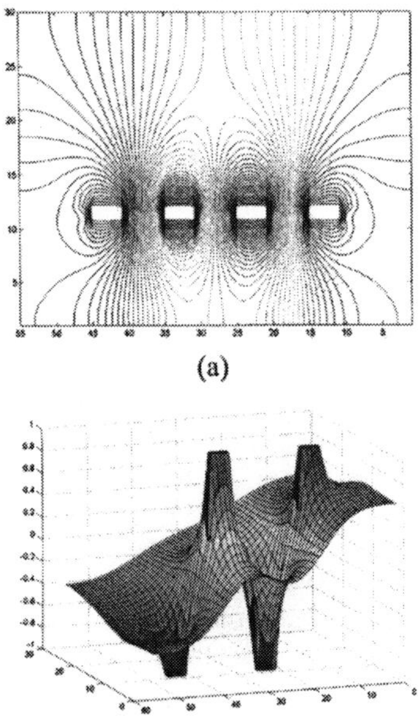

(a)

(b)

Fig. 6. Initial electric potential distribution: (a) 2-D contour, (b) contour surface

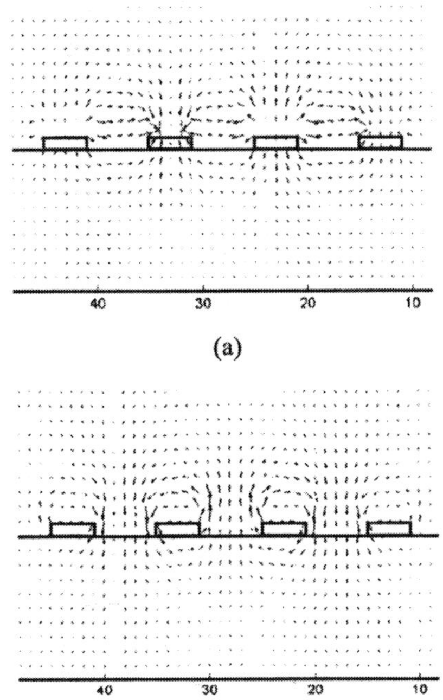

(a)

(b)

Fig. 7. Distribution of (a) electric field vectors and (b) magnetic field vectors

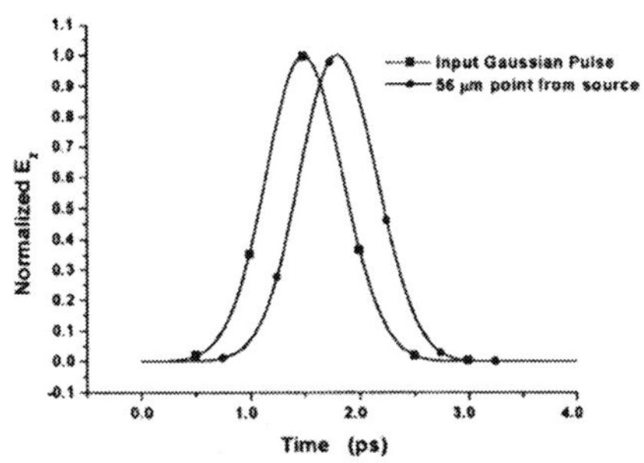

Fig. 8. Normalized Ez field observed 56 μm from the source.

References

[1] Charlie C.-P. Chen et al., ICCAD, 3D-3, 2000

[2] K. S. Yee, IEEE Trans. Antennas Propagation, vol. 14, no.3, pp.302-307, 1966.

[3] Jean-Pierre Berenger IEEE Trans. Antennas Propagation, vol. 44, no.1, pp.110-117, 1996.

Differences Between Quantum-Mechanical Capacitance-Voltage Simulators*

C.A. Richter†, E.M. Vogel, A.M. Hodge, and A.R. Hefner
Semiconductor Electronics Division
National Institute of Standards and Technology
Gaithersburg, Maryland, 20899-8121, USA

Abstract

We present an extensive benchmarking comparison of an ensemble of the most advanced quantum-mechanical (QM) capacitance-voltage (CV) simulators available. Quantitative differences in the accumulation capacitance of p-channel and n-channel devices as large as 20% are found in a systematic comparison. Some of the underlying physics and models that lead to the observed differences are described.

1 Introduction

Increasingly complex and sophisticated methods to simulate CV curves are needed to describe the additional physics (such as QM quantization and poly-Si depletion) associated with ultrathin gate dielectrics which make it difficult to predict MOS CV curves accurately[1-8,10]. Recently, we showed for n-channel (p-substrate) capacitors, that an ensemble of advanced QM simulators have similar performance accounting for polysilicon depletion and QM confinement; however, there are quantitative differences of up to 20% in the accumulation capacitance for ultrathin gate dielectric devices[1]. These differences indicate that parameters extracted from experimental CV curves (such as the effective oxide thickness, EOT[1]) are dependent upon the analysis software; therefore, it is critical that the inter-relationship of the simulators be well-characterized. This paper presents, for the first time, a systematic comparison of QM simulators for p-channel (n-substrate) devices, and it discusses and illustrates some of the underlying physical and modeling differences that lead to observed discrepancies between simulators. A method to extend the comparison to include 2D simulators is also discussed.

2 Simulator Ensemble

Six of the most advanced, one-dimensional, QM CV software packages were compared in this study: (1) the Quantum Mechanical CV Simulator[2] from the Device Group at UC-Berkeley (Berkeley), (2) NEMO, the Nanotechnology Engineering Modeling Program[3], (3) CVC[4], by Hauser at NCSU (NCSU), (4) UTQuant[5], from UT-Austin, (5) SCHRED[6], from an Arizona State/Purdue team, and

* Official Contribution of the National Institute of Standards and Technology; not subject to U.S. copyright. Work funded in part by the NIST Office of Microelectronics Programs.
† contact information: Curt A. Richter, NIST, Building 225, Room A305, Gaithersburg, MD 20899-8121, USA, Email: Curt.Richter@NIST.gov, Phone: (301) 975-2082.

(6) IBM's Tqm_v6[7]. In addition, we have developed in-house CV code[8] that is used to gain insights into how different physical approximations affect CV simulation results. Three of these simulators (Berkeley, UTQuant, and SCHRED) are based upon self-consistently solving the Schrödinger and Poisson equations. NEMO is a non-equilibrium Green's function solver, while NCSU and the in-house code are based upon models containing physics approximations enabling rapid calculation. IBM's Tqm_v6 is a QM CV analysis program based upon the results of IBM's extensive QM simulations. Simple n-channel and p-channel poly-Si gated MOS capacitor test structures were used to compare the various simulators.

3 Results and Discussion

A matrix of CV curves was created for comparison by varying the parameters (substrate doping, N_d, poly-Si doping, N_{poly}, and ideal SiO_2 thickness, d_{ox}). Figure 1 shows typical CV simulations for p-channel capacitors with $d_{ox} = 2.0$ nm[9]. Similar results are obtained for the n-channel devices. Figure 1 illustrates that the overall shape of the CV curves is the same; i.e., the simulators are qualitatively similar. However, there are important differences between the simulators. The largest difference is in the accumulation region capacitance, which leads to CV curves that appear to be from devices having different d_{ox}.

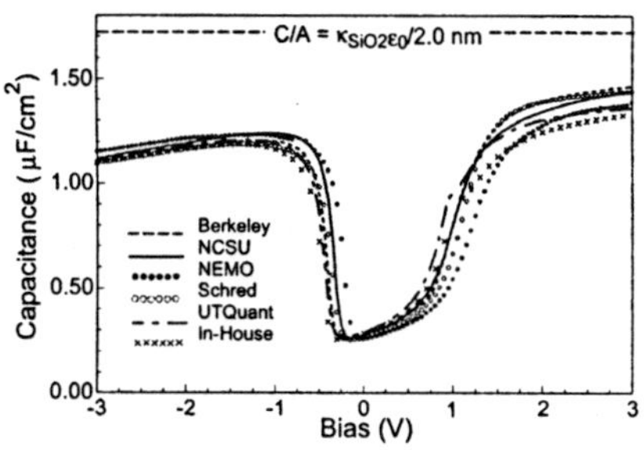

Figure 1: p-channel CV curves including QM confinement and poly-Si depletion. $d_{ox} = 2.0$ nm (1.901 nm for NEMO), $N_d = 1 \times 10^{18}$ cm^{-3}, and $N_{poly} = 1 \times 10^{20}$ cm^{-3}.

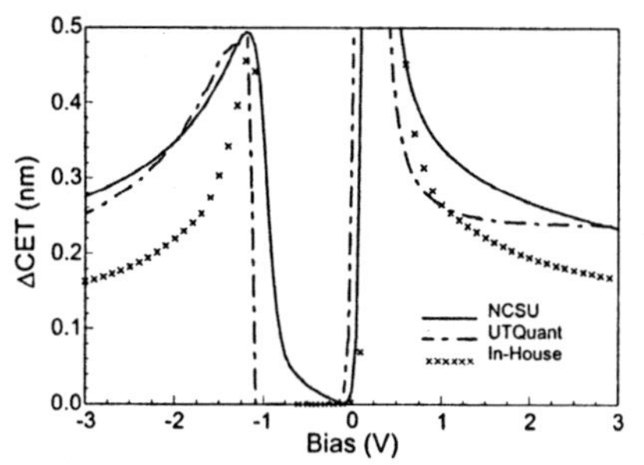

Figure 2: ΔCET for three different simulators. n-channel, $d_{ox} = 2.0$ nm, Nd = 1×10^{18} cm^{-3}.

We have used Tqm_v6, IBM's QM CV analysis program, to extract a reduced set of thickness parameters (or assessment criteria) for each simulated CV curve as a quantitative method to compare the results of the various simulators. Table 1 shows the extracted EOT and capacitive effective thickness, CET($|V| = 2.1$ V), for both n- and p-channel capacitors. $CET(V) = (\varepsilon_0 \kappa_{SiO2} A)/C(V)$; ε_0 is the permittivity of free space, κ_{SiO2} is the dielectric constant of SiO_2, and C(V) is the capacitance at bias voltage, V. There is a maximum difference (determined from comparing $\Delta d_{ox} = d_{ox}^{simulated}$ - EOT(Tqm_v6) for the various simulators) of 0.251 nm (0.220 nm) for p-channel (n-channel) devices simulated with $d_{ox} = 2.0$ nm. This discrepancy does not scale with thickness (the maximum difference is 0.238 nm (0.227 nm) when $d_{ox} = 1.0$ nm) and therefore is more problematic for thinner gate dielectrics.

One metric determining the exact outcome of a particular implementation of QM effects is the difference between CET simulated with and without QM effects

(ΔCET = CET^{QM} - CET^{classical}). Figure 2 shows ΔCET for three different simulators. This figure shows the result of modeling QM effects for these simulators, and quantitatively illustrates the (Å-scale) differences that arise between these implementations of QM.

Simulator	n-channel		p-channel	
	CET (nm)	EOT (nm)	CET (nm)	EOT (nm)
NCSU	2.465	1.836	2.461	1.828
NEMO (1.901 nm)	2.537	1.902	2.555	1.916
UTQuant	2.606	1.965	2.576	1.935
Berkeley	2.494	1.862	2.398	1.770
SCHRED	2.438	1.811	2.430	1.800
In-House	2.666	2.020	2.659	2.012

Table 1: Thickness parameters extracted by using Tqm_v6. d_{ox} = 2.0 nm (1.901 nm in NEMO), $N_d = 1 \times 10^{18}$ cm^{-3}, and $N_{poly} = 1 \times 10^{20}$ cm^{-3}.

There are many factors associated with the various modeling approaches and details of the physics that contribute to the observed differences. There are a number of ways to implement electron and hole quantization that can be divided into two primary categories: (1) simulations based on self-consistent solutions to the Schrödinger and Poisson equations, and (2) modification of the classical inversion and accumulation charge.

Some simulators in category (1) solve the Schrödinger equation throughout the entire structure while others limit the solution to the substrate and force the wave function to disappear at the interface. Wave-function penetration into the gate dielectric will cause higher predicted capacitances because the carriers are effectively closer to the interface. Recently, Mudanai, et al.[10] have reported this increased inversion-layer capacitance (of NMOSFETs) is insignificant except when $N_d \approx> 1 \times 10^{20}$ cm^{-3} and $d_{ox} <\approx 1.0$ nm.

Category (2) simulations can also be implemented in different ways. Some category (2) simulators modify the surface potential term in the classical calculations to include the extra band bending necessary to reach a given inversion or accumulation charge while others modify the bandgap near the interface by changing the intrinsic carrier concentration. The choice of carrier statistics when changing inversion and accumulation charge, i.e., Maxwell-Boltzmann (MB) or Fermi-Dirac (FD), leads to different results. This is most easily illustrated (Figure 3) for the classical, metal-gate, CV curves that are the starting baseline for category (2) simulators.

Currently, implementations of QM effects in 2D simulators are not as advanced as those found in many of the 1D simulators. In addition, QM effects are typically implemented as a locally 1D effect, and the role of true 2D quantization is yet to be determined. Benchmarking 2D simulators requires the development of a 2D structure to validate the 1D QM effects. Then it must be determined that these simulators effectively simulate 2D devices, properly accounting for effects such as fringe-fields and source/drain overlap. To achieve the first step, a 2D structure must be made large enough to emulate a 1D device. After a gate length calibration process, we found (for d_{ox} = 2.0 nm) that a 4 μm gate length MISFET is sufficiently long that the scaled capacitance per unit area has reached its 1D asymptotic limits (demonstrated in Figure 4). Although the classical 2D and 1D CV curves show good agreement in the accumulation region and inversion region capacitance for both metal and poly-Si gate devices, the 2D simulators studied thus far do not have sufficiently accurate QM corrections to be compared with the 1D results.

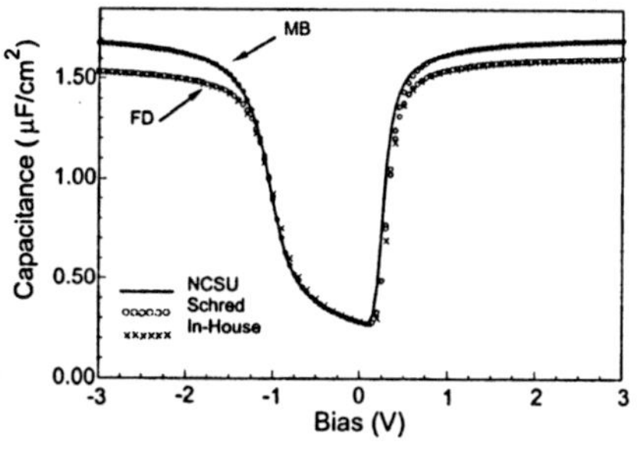

Figure 3: Classical CV curves with Fermi-Dirac (FD) statistics (Schred, in-house) and Maxwell-Boltzmann (MB) statistics (Schred, NCSU). n-channel, d_{ox} = 2.0 nm, Nd = 1 x 10^{18} cm^{-3}.

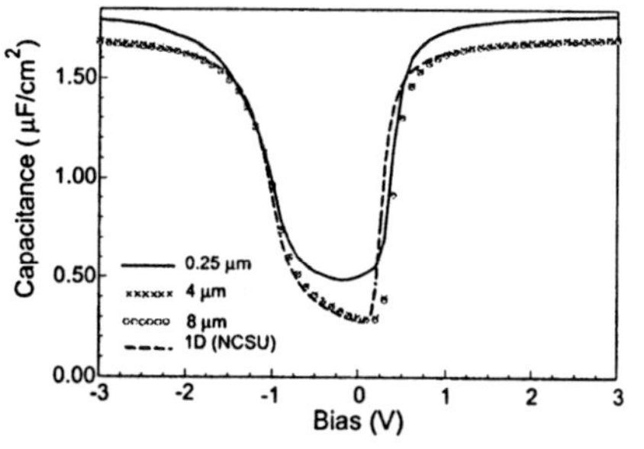

Figure 4: Scaled classical CV curves from 2D FETs as a function of gate length compared with 1D results. n-channel, d_{ox} = 2.0 nm, Nd = 1 x 10^{18} cm^{-3}.

4 Conclusions

We have performed an extensive comparison of QM CV simulators. As previously shown for n-channel devices, large differences (up to 20% for ultra-thin gate dielectrics) in the accumulation capacitance of p-channel capacitors are observed. We have illustrated some of the factors leading to this discrepancy. The observed differences are due to a complex interplay of many of these factors. Further improvements in QM corrections are needed in 2D simulators to emulate 1D MOS capacitors, and further work is required to develop an effective benchmark for 2D QM effects.

References

[1] C.A. Richter, A. Hefner, and E.M. Vogel, *IEEE Electron Device Letters*, **22**, pp. 35-37 (2001).

[2] www-devices.eecs.berkeley.edu/~kjyang/qmcv/index.html.

[3] R. Lake, et al., *J. Appl. Phys.*, **81**, 7845-7869, 1997. D.K. Blanks, et al., *Proc. IEEE 24th Int. Sym. on Compound Semiconductors*, (IEEE, New York, 1998) pp. 639-642.

[4] J.R. Hauser and K. Ahmed, in *Characterization and Metrology for ULSI Technology*, Seiler, et al., eds. (AIP, Woodbury, NY 1998) pp. 235-239, 1998. and W.K. Henson, et al., *IEEE Electron Device Lett.* **20**, 179-181, 1999.

[5] W-K Shih, et al., *UTQUANT 2.0 User's Guide*. Austin, TX: Univ. Texas, Oct. 1997.

[6] SCHRED was obtained from PUNCH, the Purdue University Network Computing Hubs; http://punch.ecn.purdue.edu/.

[7] S.-H. Lo, D.A. Buchanan, and Y. Taur, *IBM J. Res and Devp.* **43**, 327-337 (1999).

[8] E.M. Vogel and C.A. Richter, unpublished.

[9] A minimum mesh of the silicon lattice constant (0.271547 nm) is most physically realistic and gives consistent results for NEMO.

[10] S. Mudanai, et al., *IEEE Electron Device Letters*, **22**, pp. 145-147 (2001).

Investigation of a Novel Rapid Thermal Processing Concept Using an Electro-Optically Controlled Radiation Cavity

N.E.B. Cowern[1], F. Roozeboom and P. van der Sluis
Philips Research, Prof. Holstlaan 4, 5656 AA Eindhoven, The Netherlands
E-mail: nick.cowern@philips.com

Abstract

We present numerical investigations of a novel technique for reducing the thermal cycle during lamp-heated rapid-thermal processing (RTP). In our concept the reflective reactor-wall surface is replaced by an electro-optically switchable film, which allows efficient heating and subsequent radiative cooling of processed wafers. Our approach provides an efficient solution to ultrashallow junction requirements for forthcoming CMOS generations.

1 Introduction

The principal advantage of RTP is its low thermal budget and reduced process cycle time. In most RTP reactors, wafers are heated by lamp radiation in the red to near-infrared range [1]. The reactor walls are provided with highly reflective metallic coatings to minimize radiative losses. This enables rapid heating up to the set processing temperature, and minimal energy use once this temperature is reached. However, at the end of the thermal cycle the reflective wall presents a disadvantage as it slows the rate of radiative cooling of the wafer. At this stage of the thermal cycle one would actually prefer a heat-absorbing wall to accelerate wafer cooling.

Several attempts have been made to accelerate wafer cooling by changing the optical properties of the RTP chamber [2-4]. One approach is to use a chamber with an *absorptive cavity* [2,3]. This achieves rapid cooling, but an ultrahigh-power lamp (150-300 kW) is needed to counterbalance the increased radiative losses from the wafer during processing. Typical cooling rates for absorbing chambers are reported to be 180 °C/s while the best standard chambers cool at 90 °C/s [2,3].

A reduced thermal cycle, using conventional lamp power, can be achieved by switching the reactor wall from reflective, during heating, to absorptive during cooling. This can be done by mechano-optical switching [4], or electro-optical switching [5,6], the latter being of particular interest as it does not require moving parts. In this paper we present calculations showing the potential benefits of this concept. We consider a wafer radiating inside a symmetrical planar-bounded cavity with variable wall reflectivity. Between the wafer and the wall is a partially

---

[1] New address: Surrey University, Guildford, U.K. E-mail: n.cowern@surrey.ac.uk

transparent window (e.g. quartz), which absorbs significantly at wavelengths emitted by the wafer (mid- to far-infrared) thus modifying the impact of the switchable wall. In order to obtain analytical formulas for wafer cooling we make the following approximations, accurate to within a few per cent in the range of wavelengths and temperatures (600-1100 °C) of interest: (1) radiation is the dominant heat-loss process, (2) the wafer is opaque to its emitted infrared radiation, (3) the wafer emissivity is independent of wavelength and temperature, (4) the specific heat of the wafer is independent of temperature, (5) the intermediate window does not significantly reflect or re-emit the radiation from the wafer. Furthermore, since window absorption is not the main concern of this paper, we assume for convenience that this is independent of the infrared wavelength emitted by the cooling wafer.

2 Analysis of radiative cooling

Figure 1 shows schematically a wafer surface with reflectivity ρ, emitting radiation through a partly transmitting window of absorbance a, to a reactor wall with reflectivity r. To determine the net radiant power emitted by the wafer surface at absolute temperature T, consider the effect of the multiple absorption and reflection events that occur before each radiated photon is absorbed, either within the window or at the reactor wall. To do this we note the probabilities of absorption/reflection of the emitted photon at each encounter with a solid (wafer, window or wall), and sum over all possible reflections. The total radiated heat per unit area per second emitted from the surface at absolute temperature T, can then be calculated as

$$dQ/dt = [1-r\,(1-a)^2].[1 + \rho r(1-a)^2 + (\rho r)^2(1-a)^4+.....]\;\varepsilon\,\sigma\,T^4. \quad (1)$$

In Eq. (1), the factor $1-r\,(1-a)^2$ is the probability that the photon is absorbed during the first cycle of reflection between wafer and wall (solid arrowed lines in Fig. 1), and the series in square brackets takes into account all possible successive reflections (dashed lines in Fig. 1).

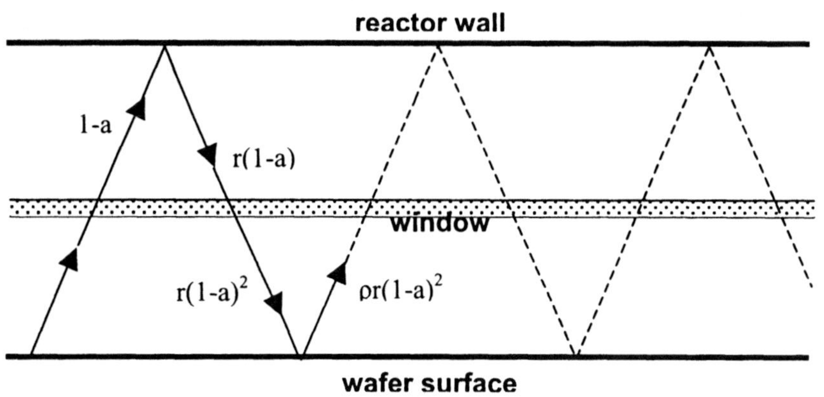

Figure 1: Schematic of radiation transport in the RTP reactor.

Bearing in mind Kirchhoff's law for opaque wafers, $\rho = (1-\varepsilon)$, the formula

$$1 + x + x^2 + x^3 + \ldots = 1/(1-x)\,, \tag{2}$$

and the relation between the heat capacity and the cooling rate of the wafer,

$$dQ/dt = (dQ/dT)\,.\,(dT/dt) = C_{Si}.\,\rho_{Si}.w\,.\;dT/dt\,, \tag{3}$$

where C_{Si} and ρ_{Si} are the specific heat and density of Si and w is half the wafer thickness (assuming symmetry), we can rewrite Eq. (3) in terms of the cooling rate,

$$dT/dt = -\varepsilon\,\{[1-r(1-a)^2]\,/\,[1-r\,(1-a)^2(1-\varepsilon)]\}\;\sigma\,T^4\,/\,(C_{Si}\,\rho_{Si}\,w)\,. \tag{4}$$

In a conventional RTP reactor the wall is highly reflecting and impedes wafer cooling. Radiation leaving the wafer has a high probability of being reflected back to the wafer and reabsorbed. If there are no absorbers within the chamber (i.e. a=0), the rate of cooling can be remarkably slow. For example, in the case of 90 % reflectivity, the cooling rate according to Eq. 4 is an order of magnitude slower than would occur in the absence of a reflecting wall. This is much slower than is actually found experimentally in a RTP reactor. If one now allows for a significant amount of absorption in the window, the prediction changes substantially. To illustrate this we consider a typical case where the window (a few mm of quartz) absorbs 20 % of the IR radiation from the wafer on a single pass, corresponding to the value a=0.2.

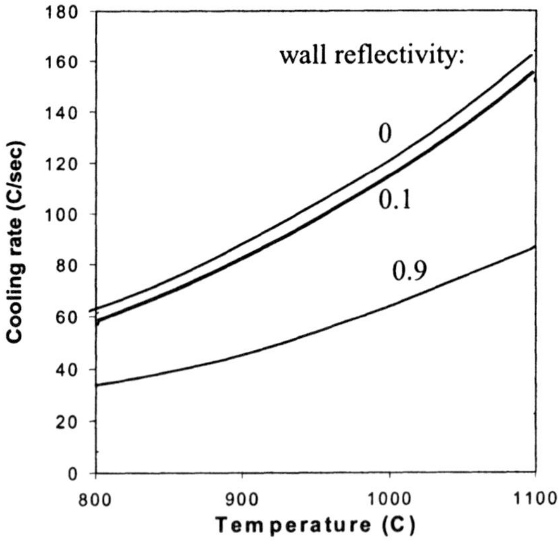

Figure 2: Cooling rate versus temperature for three values of r, with a=0.2, representing the effect of a quartz window between the wafer and the wall.

It is clear from Fig. 2 that, despite the damping effect of window absorption, the rate of cooling is enhanced by a factor of approximately two when the reactor wall is switched from reflective to absorbing. This leads to a significant decrease in thermal budget during 'spike' anneals used to fabricate ultrashallow junctions.

3 Conclusions

The implementation of a switchable-mirror coating for the RTP reactor wall, such that the reflectivity can be reduced to a low value, e.g. 10 %, during wafer cooling, offers approximately a factor of two gain in wafer-cooling rate compared with values reached in conventional lamp-heated RTP reactors. The performance is comparable to that in RTP designs that use high lamp power combined with a passive absorbing wall. The reduction in thermal cycle significantly extends the potential for RTP to form ultra-shallow junctions in forthcoming CMOS generations, while retaining the well-tested lamp-based method of wafer processing.

In order to model the thermal performance of the RTP reactor to a high level of accuracy, a more precise quantitative approach to modelling the cooling part of the RTP cycle may be required in the future. For example, it may become necessary to account for the wavelength dependence of infrared absorption, particularly in the quartz window material, and to consider the detailed reactor geometry. However the simple analysis presented here is a useful guide to the contributions of window and wall absorption to wafer cooling in RTP systems, and it clearly shows the value of the switchable-window approach.

References

[1] Roozeboom F, in *Advances in Rapid Thermal and Integrated Processing*, Roozeboom F (ed.), Kluwer Academic Publishers, Dordrecht, The Netherlands, 1996 pp. 1-34

[2] Fiory A.T., Camm D.M., Lefrançois M.E., McCoy S.P. and. Agarwal A, *Proc. 7th Int. Conf. Advanced Thermal Processing of Semiconductors (RTP'99)*, pp. 273-280

[3] Camm D.M. and Lefrançois M.E., Electrochem. Soc. Symp. Proc. **2000-9** (2000) 429

[4] Bussmann E., German Patent DE 4,142,466 (30 Nov., 1995)

[5] Van der Sluis P., Appl. Phys. Lett. **73** (1998) 1826.; Van der Sluis P. *et al.*, Appl. Phys. Lett. **70** (1997) 3356

[6] Based on U.S. patents 6,047,107 (4 April, 2000) and 6,173,116 (9 Jan., 2001) by Roozeboom F., Duine P.A. and Van der Sluis P; see also Roozeboom F., Van der Sluis P. and Cowern N.E.B., Electrochem. Soc. Symp. Proc. **2001-8** (2001) 41

A Shared Architecture for a Dynamic Technology Simulation Repository

Michael G. Khazhinsky, Alexander Hoefler, Michael Stockinger, David J. Collins, Iuval Clejan, Karl Wimmer, William Taylor, Mark Foisy, J.M.Higman
Motorola, Inc., 6501 William Cannon Drive, Austin, TX 78735, U.S.A.

Lars Bomholt, Christian Clémençon, Olga Zuyakova, Wolfgang Fichtner
ISE Integrated Systems Engineering AG, Zurich, Switzerland

Abstract

In this paper, the design of a shared technology simulation repository is described. This system allows the download, archival, and simultaneous translation of equipment recipe and run data into a shared, revision controlled repository, as well as the automated generation of Technology Computer Aided Design (TCAD) input. Based on this system Computer Integrated Manufacturing (CIM) data and integrated circuit layout data can be combined to provide rapid technology optimization, enabling new methods of technology development.

1. Introduction

The application of TCAD in an industrial product/process development environment presents unique challenges, aside from those traditionally attacked by the academic community. This paper focuses on the management of the broad range of complex and dynamic information one must pull from the semiconductor fabrication facility (fab).

Inside the fab, the CIM system manages the information, which, as a paradigm, has matured in the semiconductor industry on a local (fab) level. However, several challenges lie ahead. Time to market is a leading criterion for success in a marketplace that requires cost effective and rapid development and optimization of new technologies. Market conditions will likely force semiconductor companies into more scalable nonlocal paradigms, some of which are already utilized today, such as contract manufacturing, remote process equipment utilization (Losleben et al. 1996), rapid development-fab to production-fab transfer, and technology licensing.

TCAD has long been accepted as a cost effective development tool for new technologies, if problem-specific calibration methods (Mar 1996 and Park et al. 1997) are used. However, to collect relevant data, TCAD often relies on human intervention, although all of the necessary data exists on computers somewhere in the fabs. A key to the continued success of TCAD is the adoption of a methodology that provides a tight link to the CIM environment of the (possibly remote) process development site. Currently, a considerable amount of time is wasted in simply looking for process recipes, and even worse, in running simulations or calibrations using incomplete or incorrect recipes. In this manual approach, human errors are inevitable and the incorporation of changes requires a great deal of engineering time.

The availability of a database that contains process and TCAD related information makes it easy for a TCAD engineer to leverage the calibration work of other individuals or groups and to avoid duplication of efforts. It also provides an essentially self-documenting system to track changes. To this end, both the feasibility of technology simulation based equipment control (Fallon et al. 1999) and a technology simulation archive (Wimmer et al. 1995) have been demonstrated. Our architecture described in this paper implements the convergence of those two concepts.

2. System overview

The purpose of this system is to provide up-to-date and revision controlled simulation projects for TCAD (Figure 1).

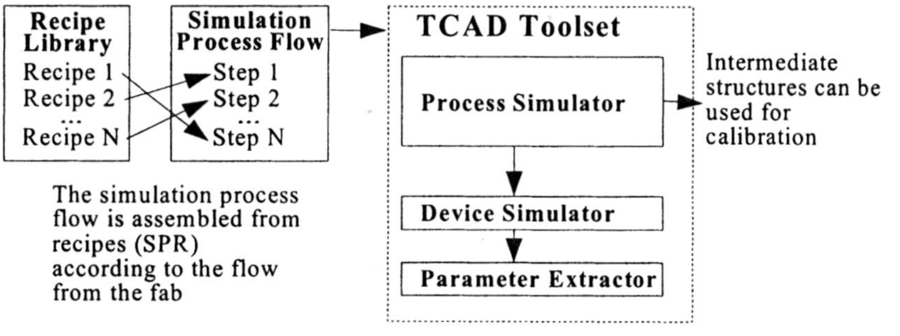

Fig. 1. The Use of Process Recipe Information in the TCAD Tool Environment.

Here, the challenge is an up-to-date library of recipes in the fab. The completeness of the recipes, the automatic downloading, and the direct translation greatly minimizes the chance of obtaining incorrect or out-of-date process flow information and vastly improves the prospect of meaningful process simulation.

3. Process Information Management

Up-to-date process flow information is readily available in a spreadsheet format from the fab's MES (Manufacturing Execution System). For the recipe library, data is extracted directly from the equipment if the MES under consideration does not have a comprehensive integrated recipe management system. Data access can be achieved via the network, through FTP to controller station, standard equipment interface, or sometimes as part of a regular back-up of recipes. In some cases *in situ* measured furnace temperature, gas flows, heater powers, *etc.* are available. For advanced process steps, these data are superior to recipe targets.

The recipes and additional data are placed into a recipe library, translated, and archived under revision control. For the process information, the SPR (Simple Process Representation) language of a proprietary input deck assembly tool (ISE 2001) is used as an intermediate representation, providing more separation between "physical" process conditions and simulator-specific options. SPR is an ASCII human-readable command language which provides macros in order to maintain the natural hierarchy of the information as it comes from the system. The recipe translation presently comprehends implantation and thermal steps but the approach described here can be extended to other computer-driven tool sets in the fab.

4. Simulation Library

The simulation library is a central part of the system, organized as a simple set of directories in a file system. The library contains a subdirectory with a collection of recipes (called the recipe library). In addition, the simulation library contains other files necessary for a successful operation of the system. The simulation library is structured as shown in Figure 2.

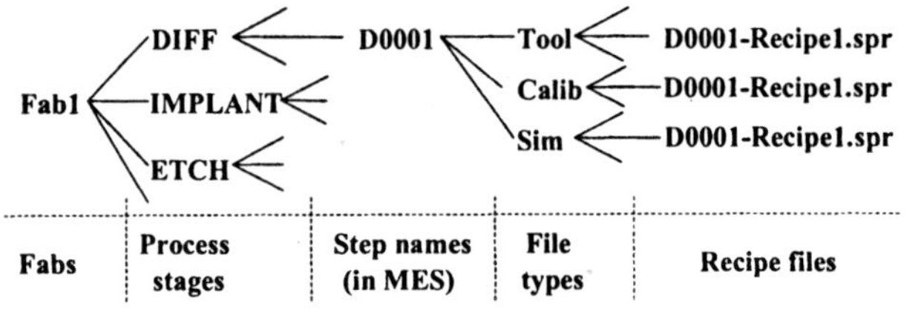

Fig. 2. The Structure of the Simulation Library.

The simulation library is revision controlled with CVS, the Concurrent Versions System, an open-source, network-transparent version control system (CVS 2001). This provides an opportunity for different engineers to collaborate while working on the library. Engineers can download recipes and update their local copies with the latest versions, or access previous versions of recipes by version number or date.

In most cases, translated recipes downloaded from the equipment are not quite ready for simulation. TCAD specific information, such as the type of physical models to be used, model parameters, etc. may have to be fine-tuned or added. Also, recipes can be optimized with regard to incompatible objectives, e.g., simulation speed or accuracy. The recipes, therefore, usually undergo calibration by a TCAD expert.

5. Assembly

The process flow, recipe information, and TCAD specific data are assembled into a complete simulation flow, again represented in SPR (Figure 3). The assembly tool, in turn, further translates process information together with layout and TCAD-specific information into process simulator input, or, possibly, other representations. All these pieces have been integrated into a proprietary TCAD simulation framework (ISE 2001). This framework also allows the use of process simulation together with other simulation elements, such as device simulation and extraction of electrical parameters.

6. Benefits

The wide accessibility of up-to-date and revision controlled TCAD data minimizes the turn-around time for the simulations and often enables simulations that would otherwise be unfeasible. In particular, the CVS repository allows to reconstruct the process flow and corresponding recipes for an earlier date although changes to the current process may have already been implemented.

Beyond TCAD, the comprehensive collection of data is also useful to other communities. Engineers in the development fab now have point-and-click access to a wide range of recipes, to time-temperature and gas flow data for all thermal steps. Engineers doing process transfers from one fab to another have access to all of the recipes on a specific tool. The overall benefit of the system becomes even more pronounced for the nonlocal paradigms of manufacturing.

The system is currently implemented at one of Motorola's R&D fabs. Recipe files are automatically downloaded from furnaces and implanters and translated. This approach is portable throughout development and production fabs with computer-driven tools and is beneficial to many engineering disciplines in the semiconductor manufacturing environment.

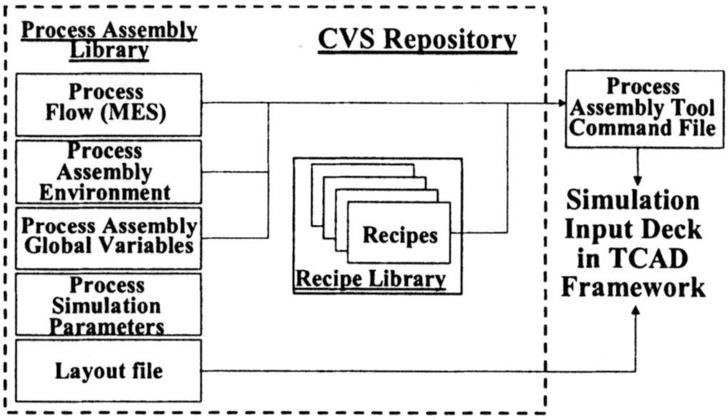

Fig. 3. Assembly of a Process Simulation Input Deck.

References

[1] CVS (2001), Concurrent Versions System, www.cvshome.org

[2] Fallon, M.; Findlater, K.; McGinty, J.; Rankin, N.; Yarr, A. (1999): Automatic furnace downloading to SUPREM format. In: Proceedings of the SPIE - The International Society for Optical Engineering, SPIE' 99 Vol.3742, pp.105-111

[3] ISE TCAD 7.0 (2001), ISE Integrated Systems Engineering AG, Zurich, Switzerland. www.ise.ch

[4] Losleben P., Boning, D.S . (1996): A new semiconductor research paradigm using Internet collaboration. In: Proceedings of the International Conference on Simulation of Semiconductor Processes and Devices, SISPAD' 96, pp.9-12

[5] Mar J. (1996): The application of TCAD in industry. In: Proceedings of the International Conference on Simulation of Semiconductor Processes and Devices, SISPAD' 96, pp.139-145

[6] Park H., Bafleur, M.; Borucki, L.; Sudhama, C.; Zirkle, T.; Wild, A. (1997): Systematic calibration of process simulators for predictive TCAD. In: Proceedings of the International Conference on Simulation of Semiconductor Processes and Devices, SISPAD' 97, pp. 273-275

[7] Wimmer, K.; Noell, M.; Taylor, W.J.; Orlowski, M.; Ryssel, H.; Pichler, P. (1995): A data-model for a technology and simulation archive. In: Proceedings of the International Conference on Simulation of Semiconductor Processes and Devices, SISPAD' 95, pp. 54-57

Level Set Modeling of Profile Evolution during Deposition Process

Ohseob Kwon* and Taeyoung Won**
Computational Electronics Center,
School of Electrical and Computer Engineering, Inha University
253 Yonghyun-Dong, Nam-Gu, Inchon, Korea 402-751
E-mail: kos@hsel.inha.ac.kr*, twon@hsel.inha.ac.kr**

Abstract

In this paper, we report three-dimensional modeling of the sputter deposition process for ULSI interconnects. The numerical method in this work is based upon the level-set scheme for accurately tracking the moving boundaries of the deposited profiles. A new approach is proposed in an effort to reduce the calculating CPU time during the calculating step of the deposition rate. Our simulation incorporates three-dimensional direct and/or indirect flux distributions and shadow-effects as well as the dependence of sticking coefficient that affects not only the thickness of film at different position but also the initiation of the creation of a void. In this work, we present several numerical examples for copper deposition process, which include L-shaped trenches and contact holes with different aspect ratios.

1 Introduction

Sputter deposition is one of the most essential processes for forming ULSI interconnects. Due to the increasing complexity of devices and interconnects, the control of the uniformity of the deposited film thickness is required in the sub-half-micron era. Modeling and simulation of sputter deposition processes is quite important for understanding the characteristics of the topographical evolution in the reliable formation of densely packed interconnect metal lines.

In recent years, a great deal of research effort has been made on the development of three-dimensional simulator for sputter deposition process with taking the equipment modeling into account [1]. In this work, we employ the level set method [2] with the novel methods to investigate the fast and accurate modeling of the topography evolution for a deposition process.

2 Feature Profile Evolution Model

To calculate a deposition rate of each surface patches, one point 'B' in patch was selected. Fig. 1 shows that all intersection points between edge of cell and zero level set are selected. In case the zero level set is located on the edge of cell, both vertexes of edge are selected as the intersectional points. In Fig. 1, the central point

'A' was calculated using four intersectional points and then the nearest point 'B' from central point 'A' is found in zero level set. The nearest point B is used for the calculation of the deposition rate of each surface patches.

The sputter yield from the target has been calculated by Monte Carlo method with the binary encounter model [3]. The sputter yield, $Y(\Omega)=Y(\theta,\varphi)$ as a function of incident energy and incident angle, as well as the distribution of atomic emission, can be calculated for wide range of 10 eV ~ 100 keV. In this work, Ar ions, which normally incident on the copper target with energy of 1 keV, was chosen to calculate the sputter yield of copper atoms. The maximum emission of sputtered copper atoms is at 45° from the axis without azimuth dependence.

In the simulation of deposition process with low sticking coefficient, as we assumed the reflection with cosine distribution, the ratio of the flux at i-th and $(i+1)$-th reflection would converge to a constant value. Therefore, the flux at arbitrary position can be calculated when the convergence is generated at n-th reflection. The calculation of the reflected flux at arbitrary position is expressed by

$$F^{sum}_{reflected} = \sum_{i=1}^{n} F^{i}_{reflected} + \frac{F^{i+1}_{reflected}}{1 - F^{i+1}_{reflected} / F^{i}_{reflected}}$$

A novel approach to reduce the CPU time was made by reducing the number of the level set points for calculating the deposition rate. We put the deposition rate for calculating the level set equation at each grid equivalent to the deposition rate of cell including a nearest surface point from (i, j, k) grid, as shown in Fig. 2. And then, the deposition rate at all grid of narrow band is determined. Fig. 3 shows the number of points used for calculation the deposition rate to the various grid sizes. Our method has very good computational efficiency to the old method. In case width of narrow band is twice grid width, the number of points is about five times more than presented method. Therefore, for N points, it requires N^2 evaluation for calculating the deposition rate.

3 Results and Discussion

In Fig. 4 is shown the profiles for copper sputter deposition performed on L-shaped trench and contact-hole structures with the aspect ratio of '1' and '2', respectively. In this simulation, the width of trench was chosen as 0.13 μm and the size of contact-hole was chosen as 0.13 μm×0.13 μm. The flux due to the surface reflection was taken into account with an assumption that the sticking coefficient Γ is 0.5. In Fig. 4, as the aspect ratio increases, the deposited film thickness on the sidewall decreases due to the decrease of the visible solid angle. In Fig. 4(c), (d), (e) and (f), the deposited film of left side is thicker than right side. It is due to the difference of amount of incident flux, i.e., the size of the visible solid angle. In case of the results with the aspect ratio of '2', the thickness of film deposited at the sidewall and bottom is thinner than the aspect ratio of '1'. We find that as increasing the aspect ratio, the amount of incident flux decrease.

Fig. 5(a) shows the eight different positions indicated on the cross-sectional view, AA' of L-shaped trench and on the cress-sectional view, BB' of contact hole in order to analysis the variation of the thickness ratio (i.e., ratio of the film thickness at a particular position to the flat-surface film thickness w) in Fig. 4. Fig. 5(b) and (c) illustrates the variation of the thickness ratio at eight different positions indicated in Fig. 5 as a function of normalized time and different aspect ratio of Fig. 4. The film thickness at the sidewall does not exhibit a monotonously increasing dependence on the depth of trench or the contact hole. The amount of the incoming flux is the largest at top surface, while the flux being the smallest at the corner of bottom. Therefore, the film topography on the sidewall tends to be bowed and the thickness ratio becomes worse gradually as the simulation time continues.

Fig. 6 illustrates the void creation of L-shaped trench and contact hole. The aspect ratio is '1' and '3', respectively, and, other parameters are equal to Fig. 4. In Fig. 6, the width of void at L-shape trench is smaller than at contact hole. It is because the ratio of incident flux (i.e., the bottom to the flat surface without the geometrical shadowing) at L-shape trench is bigger than at contact hole. Furthermore, the void is created earlier in a contact hole with a large aspect ratio than that with a small aspect ratio due to the visible solid angle.

4 Conclusions

In this paper, we report the topography simulator for the sputter deposition process of copper. We discussed more detail about the geometrical effects in three-dimensional simulation of the profile of a L-shape trench and a contact hole with taking into account the physical parameters such as the visible solid angle and the surface reflection. As the aspect ratio increases, the deposited film thickness on the sidewall decreases due to the decrease of visible solid angle, and it affects the time of the void creation. Additionally, the thickness ratio of overall surface becomes worse gradually as the simulation time continues.

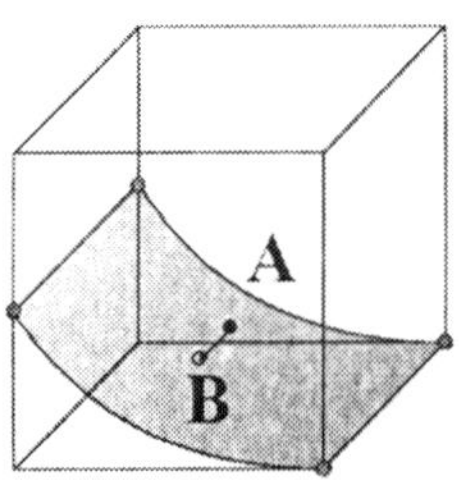

Fig. 1. Schematic diagram of determining the point 'B' for calculating the deposition rate of each surface patches.

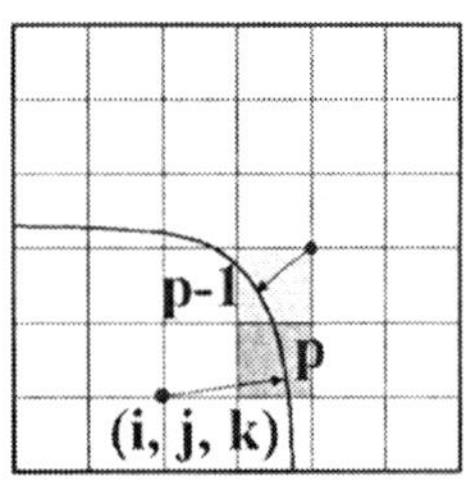

Fig. 2. Schematic diagram of determining the deposition rate of level set function.

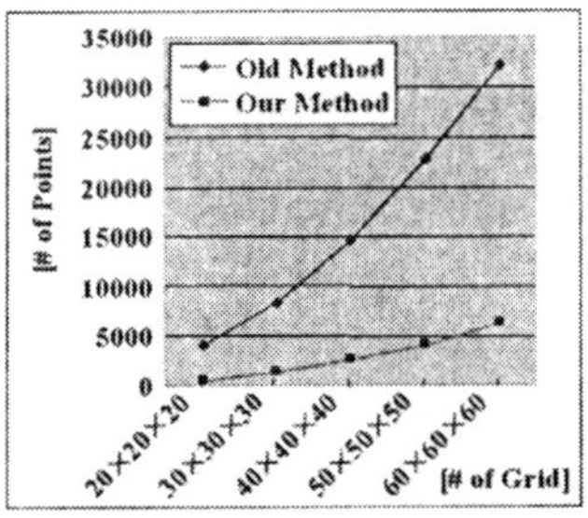

Fig. 3. Plot illustrating the comparison of our new method with old method.

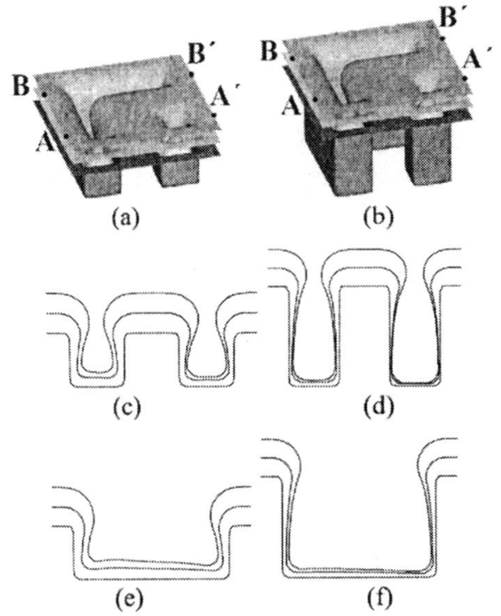

Fig. 4. Plots showing (a) and (b) 3D views, (c) and (d) the cross-sectional view along AA', and (e) and (f) the cross-sectional view along BB' with AR=1 and AR=2, respectively.

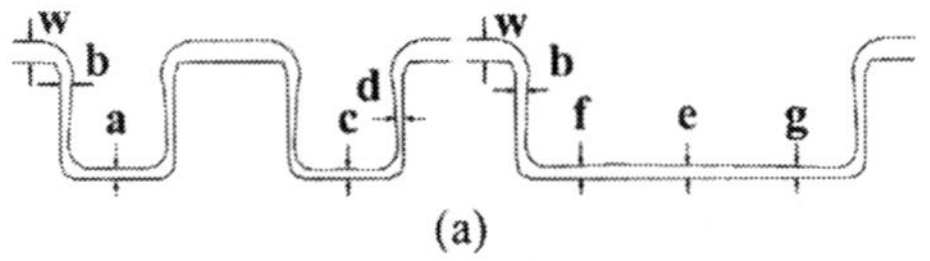

(a)

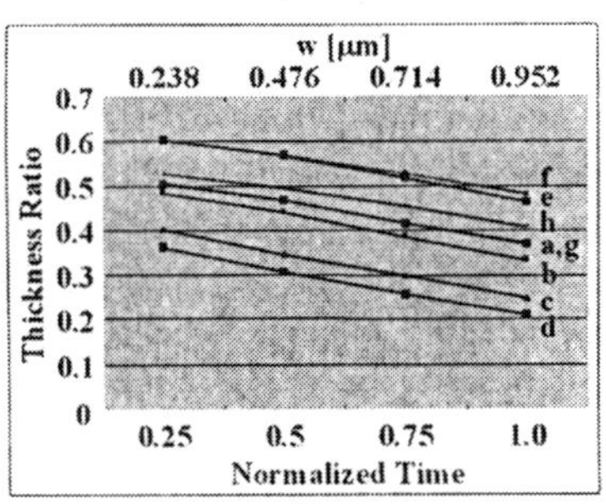

(b)

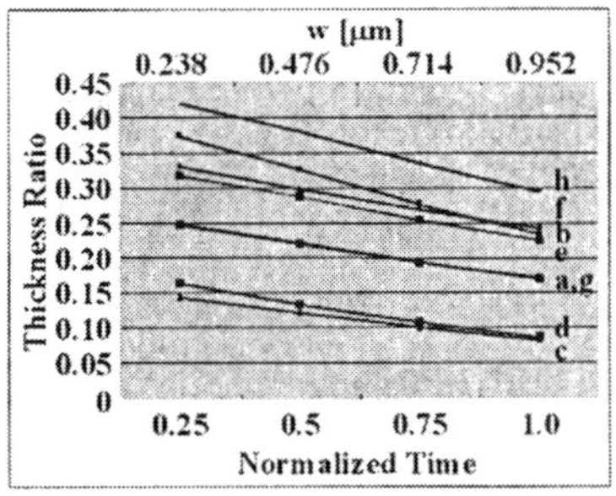

(c)

Fig. 5. Schematic diagrams illustrating (a) the locations of our interest for comparison of the thickness ratios, (b) the thickness with the aspect ratio of 1, and (c) the thickness ratio with the aspect ratio of 2.

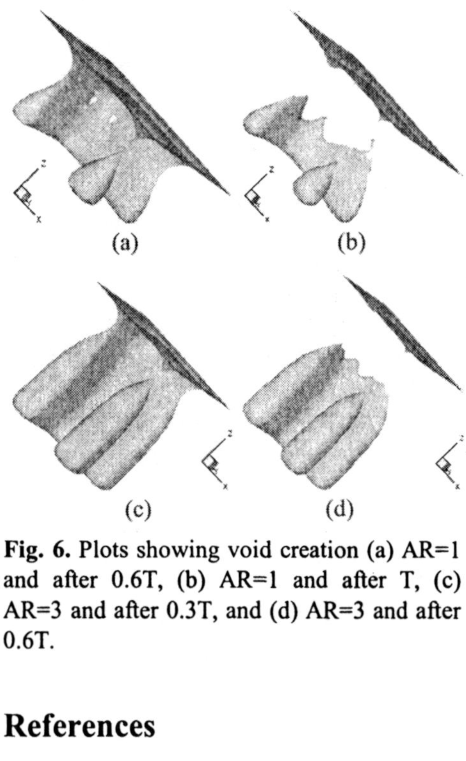

Fig. 6. Plots showing void creation (a) AR=1 and after 0.6T, (b) AR=1 and after T, (c) AR=3 and after 0.3T, and (d) AR=3 and after 0.6T.

References

[1] E. Bär and J. Lorenz, Proc. SISPAD '99, 75 (1999).

[2] J. A. Sethian, Level Set Methods (Cambridge University Press, USA, 1996).

[3] Youngchan Ban, *et al.*, J. Korean Phys. Soc., 35, S829 (1999).

Parameter and Coupling Ratio Extraction for SPICE-Compatible MACRO Modeling of Source Side Injection (SSI) Flash Cell

Sang-Pil Sim[1), 3)], Ping Guo[2)], Al Kordesch[2)], Ben Lee[2)], Chun-Mai Liu[2)], Kwyro Lee[3)], and Cary Y. Yang[1)],

[1)] Microelectronics Lab., Santa Clara University, 500 El Camino Real, Santa Clara, CA 95050, Tel. 408-554-6817, Fax 408-554-5474, e-mail ssim@scu.edu

[2)] Winbond Electronics Corp. America, 2727 North First Street, San Jose, CA 95134

[3)] Dept. of EECS, Korea Advanced Institute of Science and Technology, Korea

Abstract

We present a new methodology to generate a two-transistor MACRO model of a SSI Flash cell based on a practical cell partitioning and a systematic and rigorous parameter extraction scheme. Through extensive TCAD simulation and precise measurement techniques, BSIM3 parameters of the individual transistors and coupling ratios of the cell were extracted, yielding a SPICE-compatible MACRO model for a three-poly split-gate Flash cell.

1. Introduction

Source Side Injection (SSI) Flash cell has drawn much attention during the past few years for its high injection efficiency and low programming current, as a leading candidate for low-voltage operation or multi-bit storage technology [1, 2]. To optimize the cell operation and to facilitate the array control circuit design, better understanding of the cell behavior and accurate prediction of the I-V characteristics of the cell are necessary. In this paper, we present a general methodology to create a MACRO model for a split-gate Flash cell using a two-transistor representation. Through judicious use of TCAD simulation and precise measurement techniques, BSIM3 parameters of the individual transistors and coupling ratios (CR's) of the cell were extracted, yielding a SPICE-compatible MACRO model for a three-poly split-gate Flash cell [3].

2. MACRO Cell Representation of SSI Cell

Fig. 1 (a) shows a cross-sectional view of our three-poly split-gate Flash cell and its operating parameters are shown in Table 1. The cell is approximated by a representation of two transistors and five coupling capacitors[1)] as shown in Fig. 1 (b). Device simulation shows that this 2-transistor representation is reliable under most operating conditions except when both transistors are in deep saturation. Here current flows penetrate deep into the bulk between SG (select gate) and FG (floating gate), and the node ***C*** between the two transistors does not have a well-defined position and hence is a virtual node. Further, interaction between M_{FG} and M_{SG} transistors can

[1)] We note the "gap" transistor is not included here and the capacitors are assumed to be bias-independent to yield fixed coupling ratios occur. For instance, the drain voltage of M_{FG}, which is applied to a deep junction, can substantially affect V_{th} of M_{SG}. Thus, the effect of the drain voltage of M_{FG} on the current of M_{SG} should be considered. We have studied this effect and found it to be small for this specific cell and hence it was not modeled here.

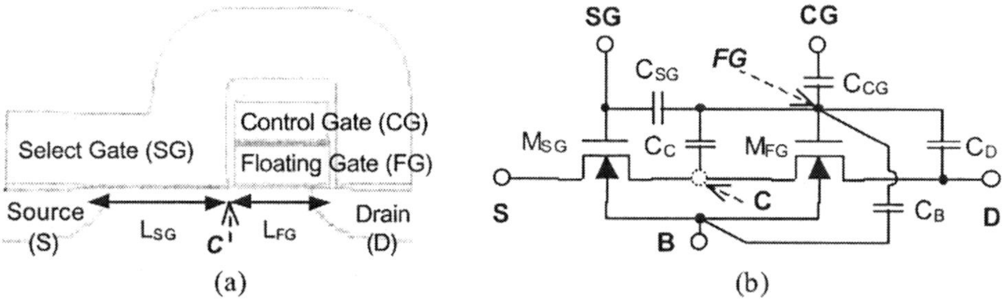

Fig. 1: Cross-sectional schematic of SSI Flash cell (a) and Equivalent circuit of MACRO cell (b). ***C*** represents a virtual node between the two transistors. The ***FG*** node is connected to 5 capacitors. L_{SG}=0.5μm, L_{FG}=0.4μm.

	V_{SG}	V_{CG}	V_S	V_D	V_B
Write	1.0	10	0	5	0
Erase	0	-10	Open	5	0
Read	3.3	1.2	1.2	0	0

Table. 1: Operating conditions of 3-poly split-gate Flash cell. Values shown are in units of [V].

3. Parameter and Coupling Ratio Extraction

For developing a reliable MACRO model, accurate and systematic parameter extraction for all these elements is critical. But the series-connected transistor configuration makes it difficult to extract parameters of one transistor independently of the other. Moreover, for each transistor, one of the junctions is physically absent. As shown in Fig. 2 (a), under the stated biasing condition, the drain of M_{SG} can be considered as an induced junction extended from the inversion layer of M_{FG}. Consequently the DIBL effect of M_{SG} is expected to be smaller than that of its stand-alone counterparts while the body effect is larger[2)]. Measurements results shown in Fig. 2 (b) confirm these projections. The parameter extraction procedure is given schematically in Fig. 4. Here a divide-and-conquer strategy is deployed. First we extracted a full set of BSIM3 parameters for M_{SG} and M_{FG} via their respective groups of stand-alone transistors. Then using the accessible gate cell[3)], some key parameters of each component-transistor M_{SG} and M_{FG} are extracted without any significant effect from the other transistor. Using a device simulator, in which we can arbitrarily exclude the effect of the series-connected transistors, we found that the following

[2)] This entails the modeling parameters of each component-transistor will be dependent on the direction of the current flow.

[3)] In accessible gate cell, FG and CG are tied.

device parameters--effective channel length, body effect, DIBL, and channel length modulation--of the component-transistor are mostly different from those of a stand-alone transistor. Then the 2 sets of parameters are combined to yield an optimized parameter set to fit the split-gate transistors I-V characteristics using a commercial parameter optimizer, UTMOST [4].

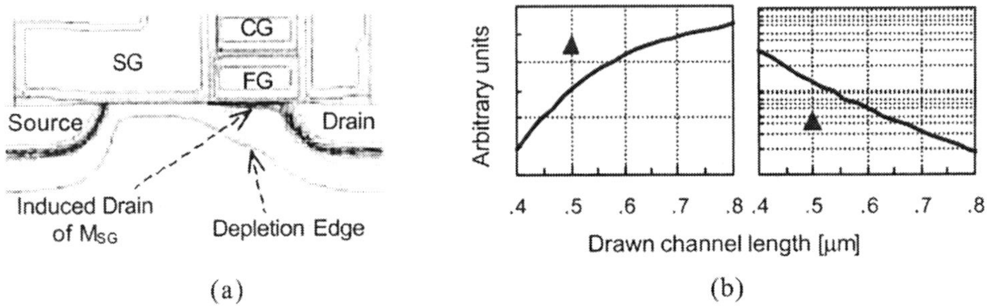

Fig. 2: (a) Simulated electron density contours with high V_{FG}. This bias condition is used to investigate M_{SG} without series resistance effect from M_{FG}. (b) Measured body effect (left) and DIBL (right) of M_{SG}. Component transistor of the split-gate cell (▲) shows much different characteristics from stand-alone transistor (line).

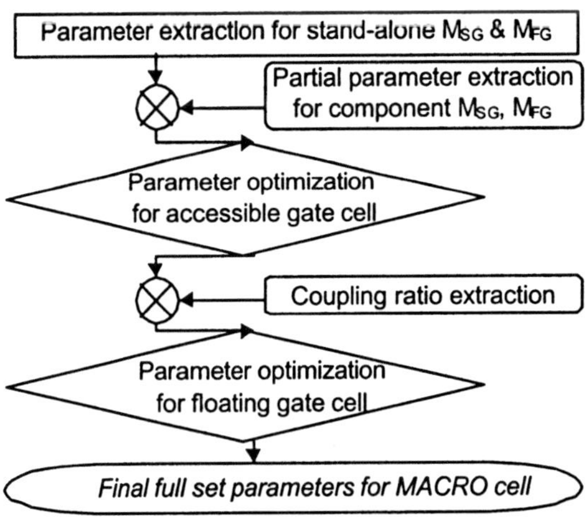

Fig. 3 : Flow diagram for parameter extraction.

To extract CR's, a method based on source follower voltage (V_{sf}) measurement as illustrated in Fig. 4 was used [5]. V_{sf} has a unique relationship with the floating gate potential (V_{FG}), which can be accurately measured using an accessible gate reference cell. The change in V_{FG}, which results in change in V_{sf}, is given by,

$$\Delta V_{FG} = \sum \alpha_i \Delta V_i + (\Delta Q_{FG} / C_{total})$$

where coupling ratio α_i is defined as C_i/C_{total} for each coupling node with voltage V_i and capacitance C_i.

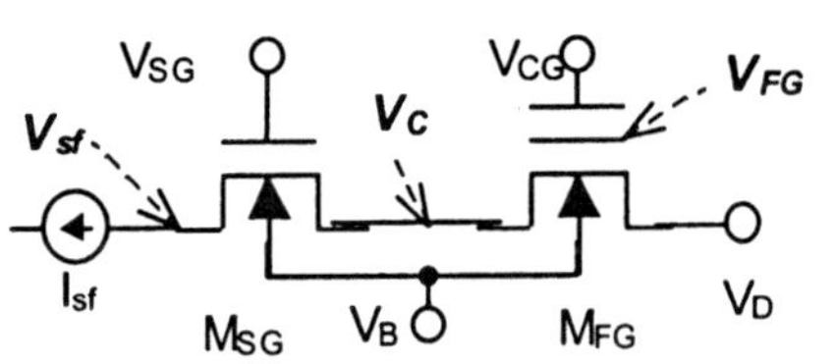

	α_{SG}	α_{CG}	α_C	α_D	α_B
CR	0.12	0.56	0.14	0.15	0.03

Fig. 4 : Measurement set-up where $V_{sf} \cong V_C = (V_{FG}$ - V_{th} of $M_{FG})$ and extracted coupling ratio (CR) values under "Read" operation.

Measuring V_{sf} by ramping each of V_{SG}, V_{CG}, V_D, and V_B sequentially, we obtain 4 equations containing 5 unknown coupling ratios α_i's. With $\sum \alpha_i = 1$ being the fifth equation, all 5 α_i's can be computed.

4. Results and Conclusion

Using the optimized BSIM3 parameters and the extracted coupling ratios, our MACRO cell model was implemented under a variety of conditions. Figs. 5 and 6 show excellent agreement between SPICE simulation and measurement results. In summary, we presented a SPICE-compatible MACRO model of a SSI Flash cell based on a methodology of practical cell partitioning and a systematic and rigorous parameter extraction scheme, augmented and guided by extensive TCAD simulation.

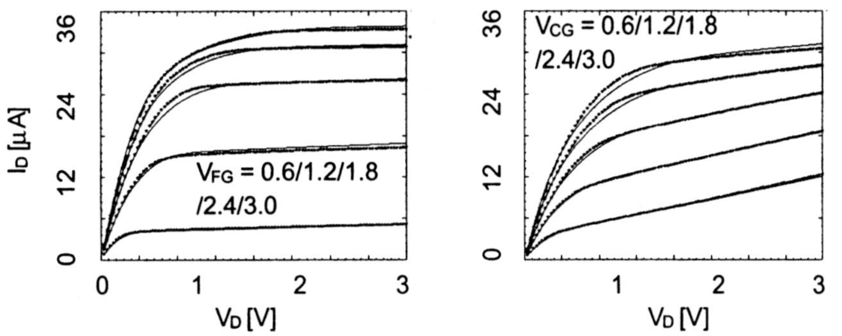

Fig. 5: Comparison of I-V characteristics between measurement (solid line) and SPICE simulation (dotted line) for accessible gate cell (left) and for floating gate cell (right) at $V_S = V_B = 0$, $V_{SG} = 1.2$[V].

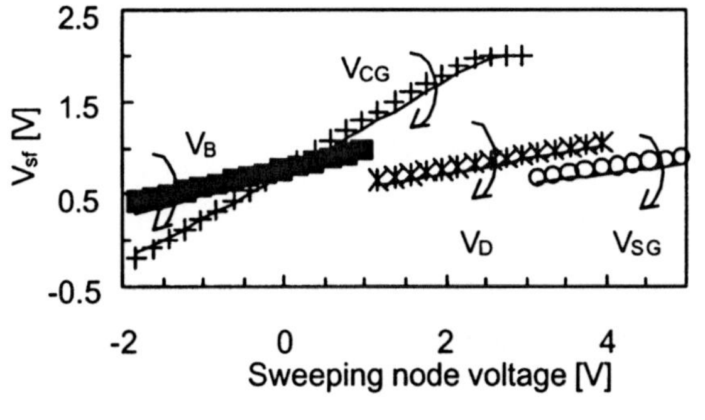

Fig. 6 : V_{sf} as functions of sweeping voltages is well matched between measurements (symbols) and simulation (line).

References

[1] S. Kianian, et al., VLSI Tech Symp., 1994, pp.71~72
[2] Chun-Mai Liu, et al., VLSI Tech. Symp., 1999, pp.187~188
[3] Y. Ma, et al., VLSI Tech. Symp., 1994, pp.49~50
[4] From SILVACO International.
[5] Chun-Mai Liu, et al., Ext. Abs. of SSDM, 2000, pp.288~291

Theoretical calculation of a charged particle detector's response fabricated by Semi Insulating (SI) GaAs

Vasilios G. Theonas and G. Papaioannou
University of Athens, Solid State Physics Section
Panepistimiopolis Zografos, 15784 Athens, Greece
mailto:vtheonas@cc.uoa.gr

Abstract

In the present work we have conducted theoretical calculations of the transient response of SI GaAs radiation detectors. The detector under consideration is a typical reverse biased Schottky diode of high purity undopped GaAs or compensated material. The electrical characteristics of the "dark" equilibrium state, derived by the Poisson and continuity equations, indicate that in the compensated detector is established a wider active region [1]. We considered that the detector operates in pulse or current mode. In pulse mode, a single charged particle (i.e. proton, a–particle) incidents the detector on either the Schottky or ohmic contact, whereas in current mode the detector is irradiated by a short-lived particle beam.

1 Ionization of the detector – C.C.E and Energy Resolution

The effect of the charged particles interaction with the detector's material is the ionization of the detector. This ionization (direct and due to recoils) is calculated by the use of Monte Carlo simulation (IBM SRIM v.2000.39)[2] and consists of e–h pairs. The electric field applied to the detector by the reverse bias, forces the generated carriers to the collecting contacts.

1.1 Energy deposition – Bragg curves

The plot of specific energy loss along the track of a charged particle is known as a Bragg curve. For most of the track the charge of an a–particle is two electronic charges and the specific energy loss increases roughly as 1/E (E: a–particle energy). Near the end of the track, the charge is reduced through electron pickup and the curve falls off. The Bragg curves for a single a–particle track and for the average behavior of a parallel a–particle beam, are somehow different because of the energy straggling phenomena. Figure 1 represents the Bragg curves of a 100 protons and a–particle parallel beams incident on a bulk GaAs detector.

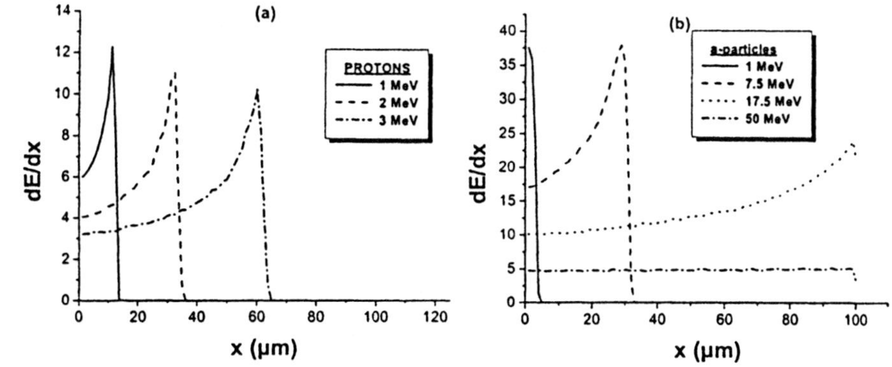

Fig. 1. Bragg curves indicating the energy deposition of (a) protons in a 125μm thick bulk GaAs detector, (b) a–particles in a 100μm thick bulk GaAs detector. The ionization of the detector is proportional to the particle's energy loss dE/dx.

1.2 Detector's charge collection efficiency and energy resolution

The c.c.e achieved by the detector is calculated by the Ramo's theorem and the evaluation of energy resolution is based on the calculation of the mean value of the "induced charge" to the detector's contacts and it's typical deviation. The results of these calculations are illustrated in Figure 2. We concluded that there is a difference in the results for c.c.e and energy resolution when the detector is irradiated from the front (Schottky) and the back (ohmic) contact, especially when the incident a-particle energy is in the range of 1 to 60 MeV.

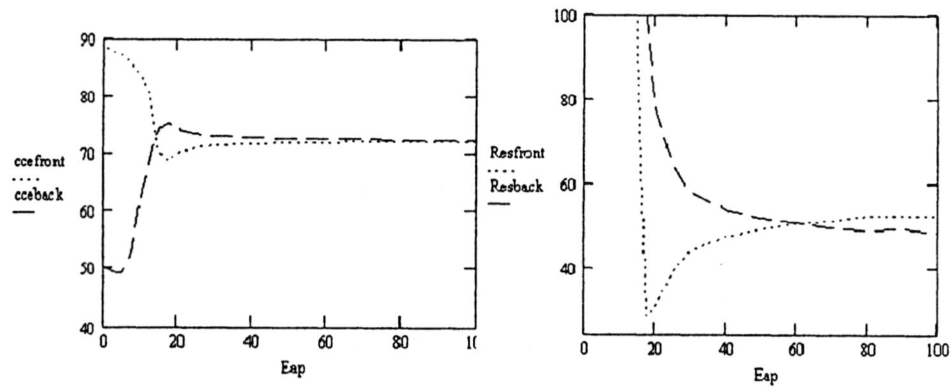

Fig. 2. Energy dependence of the c.c.e (left) and energy resolution (right) calculated for a bulk SI GaAs 100μm thick detector irradiated by a–particles with energy range 1–100MeV.

The difference is significant in this energy range, due to the dissimilar distribution of the photo-carriers within the detector (Fig.1) and the width of the specific active region established. The best results are accomplished when the detector is irradiated by 17.5MeV a–particles (detector's width 100μm, V_{bias}=100V, active region 66μm), hence the choice of the irradiated contact is meaningless for particle energies grater than 60 MeV. In this specific energy the c.c.e achieved during the back irradiation is better than the one achieved during the front irradiation, hence the energy resolution behavior is adverse.

2 Dynamic behavior

The dynamic behavior of the detector was obtained by the use of finite element analysis, taking under consideration the generation – recombination processes, the drift of the carriers to the collecting contacts and their collection. We divided the detector's volume into a considerable large number of thin strips and calculated the amount of carriers contained in each strip for a specific time step. The time step was taken considerably low, in order to ensure that the maximum drift of the carriers contained in a strip is limited to the next strip and that the recombination process will be practically negligible in this time. The detector's current pulse response is obtained by integrating the charge of all the strips, for a time interval greater than the irradiating beam duration. Figure 3 illustrates the calculated current pulse produced by a 100μm thick bulk GaAs detector when operating in pulse and current mode, irradiated by 17.5MeV a-particles.

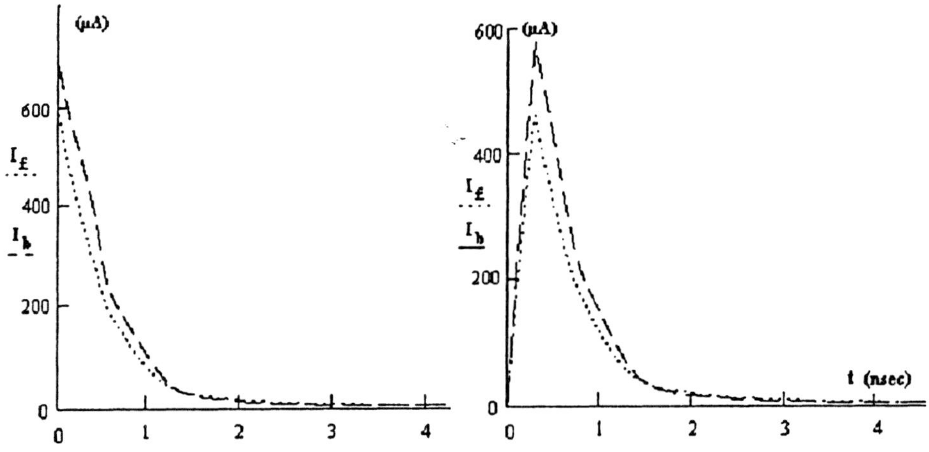

Fig. 3. Calculated current pulse produced by a GaAs detector when irradiated from the Schottky contact (doted) or the ohmic contact (dashed) when operating in pulse mode (left) and current mode (right).

In this calculation we considered the contribution of all the generated carriers to the "induced charge", because the formation of the current pulse is not only due to the collected charge but is also due to those carriers that never reach the contacts because of the trapping and recombination phenomena.

The pulse formation illustrated above explains the adverse behavior of the detector's c.c.e and energy resolution during irradiation by 17.5 MeV a-particles. During front irradiation c.c.e is worst but the photo-carriers collection rate is better resulting a lower amplitude current pulse but sufficiently narrower than the one achieved during back irradiation.
The current pulse shaping dependence is explained in Figure 4. The pulse rise is due to (faster) electrons and the abrupt fall shape indicates the collection of the electrons generated near the ohmic contact. The "knee" that follows is due to the lateness collection of the electrons generated away from the collecting contact and those that have been trapped and re-emitted by the native GaAs traps. Finally the slow current component is due to the hole drift and collection to the Schottky contact.

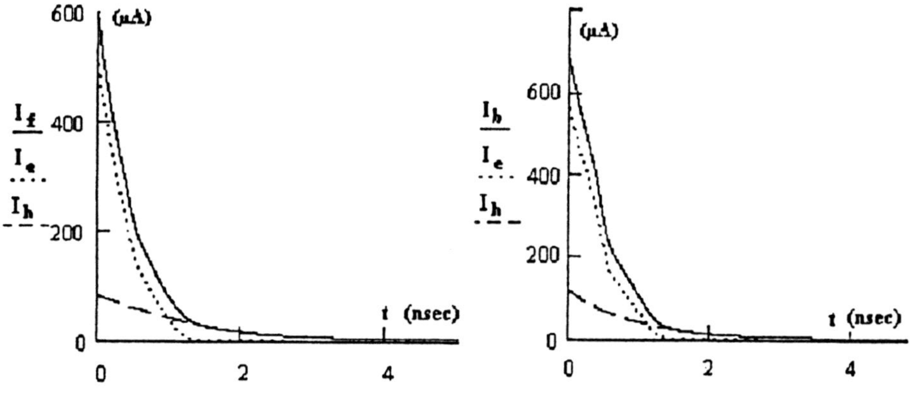

Fig. 4. Simulated current pulse during the 17.5 MeV front (left) and back (right) irradiation of a 100μm thick bulk GaAs detector operating in pulse mode. The electron (dot) and hole (dash) current components resultant is the total current (solid).

The current pulse shape and amplitude are significant information for both pulse and current mode operation. In pulse mode, this information is useful to derive the pulse height spectra i.e. the energy of the incident particle. In current mode operation this information is useful to derive the transient characteristics of the incident particles.

References

[1] V. Theonas et al, XVI Greek Conference on Solid State Physics Proceedings Volume 1 (2000) 510-513

[2] J. F. Ziegler, J. P. Biersack and U. Littmark, The Stopping and Range of Ions in Solids, Pergamon Press, New York, 1985 (new edition in late 1996).

[3] J.W.Chen et al, Nuclear Instruments and Methods in Physics Research A 365, (1995) 273–284

[4] T.E.Schlesinger–R.B.James Semiconductors for Room Temperature Nuclear Detector Applications, Semiconductors and Semimetals V.43, Academic Press (1995)

Stress Modelling Of Multi Level Interconnect Schemes For Future Deep Submicron Device Generations

Carlos Gonzales Montes De Oca
Baltimore Technologies, Dublin 8, Ireland

Sean Foley
Cypress Semiconductors, Cork, Ireland

Alan Mathewson, James F. Rohan
NMRC, Cork Ireland

Abstract

Copper and low dielctric constantant (k) materials are poised to become the dominant interconnect scheme for integrated circuits for the future because of the low resistance and capacitance that they offer which can improve circuit performance by more than 30 % over conventional interconnect schemes. This paper addresses the thermomechanical stresses in the Cu/Low k interconnect scheme through numerical simulation and identifies the locations of maximum stress in the structure with view to providing information on the impact that different dielectric materials have on the stress distribution in the interfaces between metals and dielectric layers.

1 Introduction

Copper is poised to take over as the main on-chip conductor in Ultra Large Scale Integrated circuit (ULSI) fabrication. Compared to aluminium, which was used exclusively up to this point as interconnect material, copper has a lower resistance. This is critically important in high-performance microprocessors and fast static RAMs, since it enables signals to move faster by reducing the RC time delay. Furthermore, the capacitance of the interconnect dielectric also can be reduced using low dielectric constant materials (low-k) for the Inter Metal Dielectric, and this also results in a reduction in the crosstalk between tracks as well as introducing a beneficial reduction of dielectric thickness requirement as a side effect.

Copper and low-k dielectrics represent a completely new material set for IC manufacturers. There is an absence of knowledge and experience of this material set and for this reason there is an understandable level of concern over the impact that it's introduction in an advanced technology could have. Furthermore, with the increase of device packing density and the decrease of device dimensions, mechanical stress has become an critical reliability issue which needs to be addressed as early as possible in the development of the copper low-k schemes.

1.1 Interconnect structure

In this work, copper and low-k multi level metal interconnect schemes considering different low-k dielectrics (SiLK and Aerogel), encapsulations for copper (tantalum, titanium, titanium nitride and silicon nitride) as well as different process geometries (barrier thickness and Aspect Ratio variations) have been studied. Finite Element Modelling simulations and associated materials characterisation has been used in order to evaluate the potential impact on device reliability due to thermomechanical stress. In these structures, stress levels which could easily lead to potential delamination or stress migration problems, have been identified and localised using two and three dimensional simulation techniques.

	Young's modulus (GPa)	TCE (10^{-6}/°C)	Poissons Ratio
Silicon	**130.2**	**3.0**	**0.28**
Copper	**129.8**	**17.0**	**0.343**
Ti	**116**	**8.6**	**0.32**
TiN	**600**	**9.4**	**0.25**
Ta	**186**	**6.3**	**0.34**
Si_3N_4	**150**	**1.0**	**0.25**
Aerogel	**0.005**	**3**	**0.2**
SiLK	**2**	**66**	**0.34**

Table .1. Mechanical properties of chosen materials

1.2 Experimental Details

A set of material properties were established from literature[1,2]and numerical 2D simulations were performed using the general purpose finite element package ANSYS[3-4]. A set of typical single damascene[5,6]. structures were implemented (see Fig. 1 and 2) combining copper as interconnect layer with encapsulation of titanium, titanium nitride and tantalum as metal barriers with thicknesses from 0.04 to 0.1μm, and silicon nitride as the dielectric barrier and thickness of 0.1μm. This encapsulation is surrounded by low-k (SiLK) or ultra-low-k (Aerogel) dielectrics, laying on top of a silicon substrate which is extended below the area of interest.

Figure 3 presents an example of stress results in the Y direction from simulations of a typical single damascene copper interconnect layer with 1:4 Aspect Ratio. In all the simulations the stress in the copper layer at room temperature tends to be tensile ($\sigma>0$) while all barriers, specially at corners, tend to be in compression ($\sigma<0$). This means that the higher levels of stress will be at the interface where the copper film and encapsulating barriers are in contact as well as in corners or where two different barriers are present at the same interface with copper. Delaminations at these interfaces have been observed to reduce the overall lifetime of the Cu interconnect to significantly less (~x2) than the x10 improvement expected over Al based interconnect schemes,[6,7]. In general the stress levels achieved in the whole copper layer can be summarised in Table 2, where typical stress ranges are given.

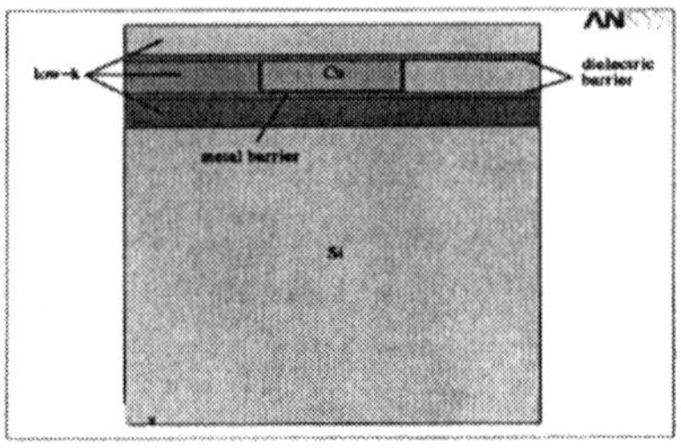

Figure 1. Representation of a typical 'Single Damascene' 2-D simulation

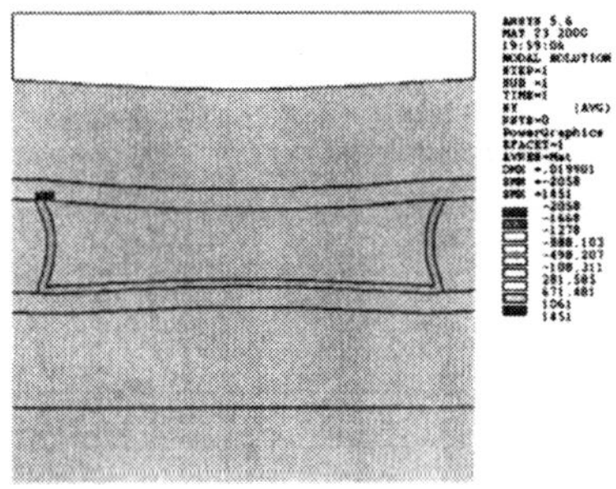

Figure 2 Stress in the Y direction for a copper layer of..1:4Aspect VIA , encapsulated by 40 nm of Ti and100 nm Si_3N_4 barriers using Aerogel as the intermetal dielectric

Encapsulation			σ_x (MPa)	σ_y (MPa)
Ti	Si_3N_4	SiLK	(140,480)	(-88,260)
TiN			(172,614)	(176,495)
Ta			(125,533)	(72,317)
Ti		Aerogel	(-234,200)	(-108,281)
TiN			(-334,414)	(-159,796)
Ta			(-46,151)	(-171,400)

Table 2. Stress range distribution at the copper layer for different encapsulations.

2 Conclusions

From this work, it is clear that the most reliable scheme would be the one made of copper, titanium or tantalum with SiLK. The use of titanium nitride leads to higher levels of tensile stress in the copper layer, because it is by far the most rigid material. Higher levels of stress in any structure multiply this effect where the structure presents special features (at interfaces, corners, etc.). In all schemes, the layer at the interface with barrier is in tension (σ>0) all along the centre of the layer and it is in compression (σ<0) at its corners (where the effect of barriers become dominant). So, in the Cu/TiN interface, the copper layer is experiencing differences in stress in the order of Giga Pascals over a few microns, while for the Cu/Ta interface this is reduced to a few hundreds of Mega Pascals, almost one order of magnitude smaller. This can be seen to be a more relaxed scheme and has been identified as a more reliable structure. Stress levels in copper are always close to its ultimate tensile stress (221 MPa),. Therefore this situation may lead to a delamination occurring at the interfaces between the Cu and the barrier Also it can be shown that in general, and in both x and y directions, the use of Aerogel instead of SiLK as the intermetal dielectric has an effect on the stress distribution of copper, where part of the film, well beyond the interface, is also in compression. This can be explained by the intrinsic softness of Aerogel as a material, and because it is very fragile this leads to a general reduction in the stability of the scheme.

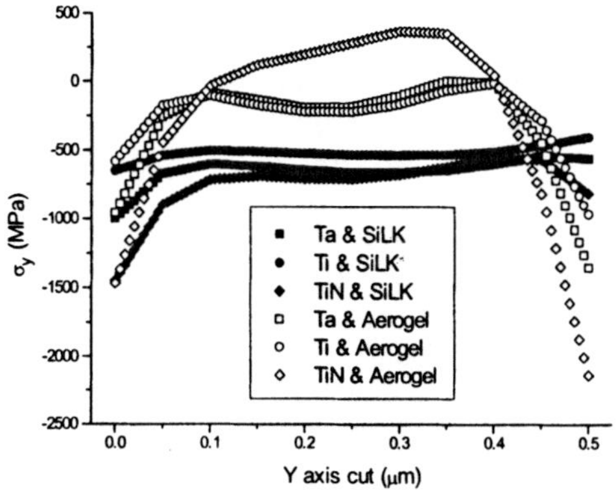

Figure 3 Stress distribution at the Y direction in 1:4 Aspect Ratio copper film at the interface with the metal barriers (Ta, Ti and TiN) for both intermetal dielectric (SiLK and Aerogel).

3 References

[1] C.G.M.DeOca M.Eng.Sc Thesis, National University of Ireand–Cork (NUIC) (2000)

[2] W.E.BeadleJ.C.C.Tsai, R.D.Plumber. Quick Reference Manual For Silicon Integrated Circuit Technology, John Wiley&SOns New York 1985

[3] Swanson Analysis Systems, Inc. P.O.Box 65, Johnston Road Houston,PA 15324 – 0065 Ansys Users Manual.

[4] ANSYS Inc 2001 Inc. P.O.Box 65, Johnston Road Houston,PA 15342 – 1300 Ansys Workbook July 1996.

[5]E.M.Zielinsky,S.W.Russell,,R.S.List,A.M.Wilson,C.Jin,K.J.Newton,J.P.Lu,W.Y.Hsu,V.C ordasco. Proc. IEEE Hong Kong Electron Devices Meeting 1997 pp 936-938

[6] A. I. Sauter, , PhD Thesis, Leand Stanford Junior University, USA. Feb. 1991

[7] S.Foley PhD Thesis NUIC . Irl 2000

Robust Method for Fast and Accurate Simulation of Random Dopant Fluctuation-Induced Vth Variation in MOSFETs with Arbitrary Complex Doping Distribution

Indranil De, Andrei Shibkov, Sharad Saxena
PDF Solutions, Inc.
San Jose, USA
indranilde@yahoo.com

Abstract

We present a robust, fast, and accurate method for prediction of threshold voltage fluctuations induced by random placement of dopants in the MOSFETs. It is used to accurately predict the threshold voltage fluctuation for an analog transistor design. The method allows easy coupling to a process and device simulation framework, and, hence, evaluation of dopant induced threshold voltage fluctuation at the transistor design stage. The methodology is also used to compare the threshold voltage fluctuation in ultra-short channel devices having different profile architectures.

1 Introduction

Random dopant fluctuation-induced variation of threshold voltage in MOSFETs is a dominant source of mismatch in closely laid out analog transistors for current technologies. It is also one of major sources of parametric fluctuation in high-performance short-channel transistors and DRAM cell transistors, which have aggressively scaled gate areas and high doping concentration in channel. It is necessary for device designers to be able to predict the impact of changes in transistor design on sensitivity of the device characteristics to random dopant fluctuation. Accurate estimate of variance of device parameters induced by statistical doping fluctuations must also be obtained early in development cycle so that design manufacturability can be assessed.

2 New Method

Previous work in this area did not allow for fast and accurate prediction of dopant-induced variability in device structures with arbitrary complex doping distributions and meshes. Process development in most technologies takes place with the aid of TCAD where calibrated process and device simulators are coupled to evaluate impact of process changes on transistor performance. Process simulation of short channel devices in any technology result in devices where channel length (Leff) is much shorter than gate length (Lgate), and where complex doping profiles exist due

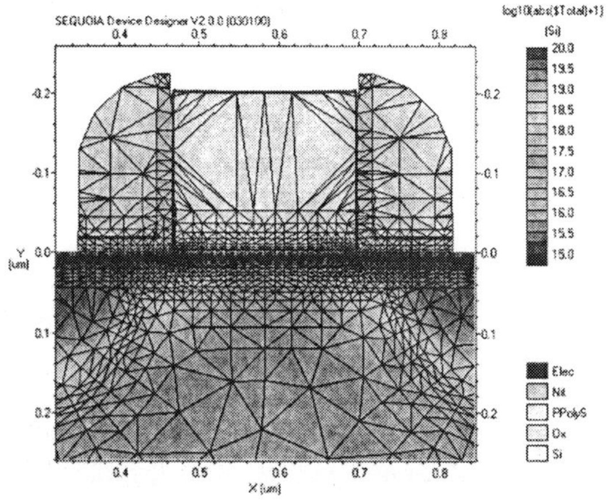

Fig.1. Device structure of a 0.25um gate length logic device showing rapidly varying doping profiles and a non-orthogonal simulation mesh.

to pocket implants and dopant diffusion mediated by extension defects. Furthermore, simulation grid in a realistic device structure obtained from process simulation must typically be replaced by non-uniform general triangular grid optimized for device simulation so that the device simulation is fast, robust and free from mesh-induced noise [2]. A typical device simulation mesh is shown in Fig. 1. Analytical models [3] cannot be used to derive parameter variation in such structures. Other more accurate methods involving use of 3D device simulation [1] are too slow for practical applications. Our method addresses both accuracy and speed issues and can be used with arbitrary doping profiles. It derives dopant-induced fluctuation from arbitrarily complex doping distributions and can be used with any mesh suitable for device simulation. Our algorithm calculates statistical doping fluctuation for each grid node based on the doping concentration at the node using binomial statistics (Fig. 2). Node area is calculated in a manner consistent with that used for descretization of PDE's in device simulator, and device width is

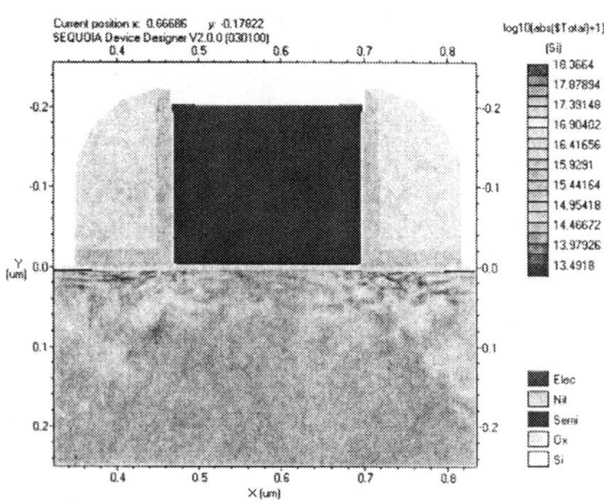

Fig.2. Random component of doping distribution obtained for structure in Fig.1 using method described here. This is superimposed on the nominal structure.

provided as user input. Nominal device structure is repeatedly subject to statistical perturbations followed by device simulation to extract important device parameters until sufficient statistical sample is obtained. Solution obtained from nominal device simulation is used as initial guess for perturbed device structures, allowing considerable speedup in extraction of device parameter variability.

3 Results

To demonstrate accuracy and efficiency of our method it was used to extract σ(Vt) for a 0.25um technology, using device profiles extracted using carefully calibrated process and device simulation. Excellent match between experimental and simulated data is observed (Fig.3). Quantum effects must be included to get proper match between experimental and simulated data. Note that the standard deviation of the Vt does not really scale at 1/Sqrt(Lgate), because Leff is typically much smaller than Lgate for shortest channel devices (see Fig. 4). The simulated variation with respect to width predicts a 1/sqrt(width) dependence.

As a practical application of this technique, impact of random dopant fluctuation on transistor characteristics for a 0.1um technology are used to compare robustness of several different device design options as shown in Fig. 5. Vertically graded profiles (retrograde) and laterally graded profiles (halo) are used to control short channel effects in ultra-short channel transistors. When compared at the same off-state leakage and threshold voltages, retrograde channel devices have lower dopant induced fluctuation, which is consistent with earlier predictions from analytical models [2]. Similar comparisons performed with realistic pocket profiles show that the dopant-induced fluctuation is a strong function of the pocket profile itself. If the pocket profile implanted at 30 or 45 degrees, and maintains it position as is common with heavier species like As or In, the induced fluctuations are smaller than with uniform profiles. However, if there is large dopant pileup at the surface, as is common with lighter pocket species like B, the induced fluctuations are much larger than in uniformly doped profiles. The results clearly indicate that there is much room for device designers to suitably tailor their device profiles to minimize such Vt fluctuations.

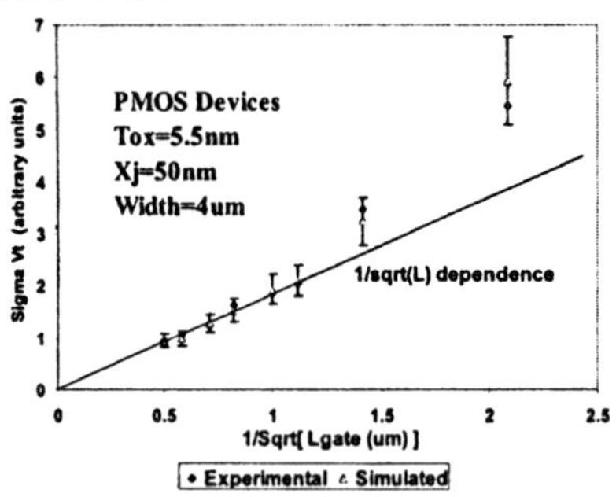

Fig.3. Experimental and simulated σ(Vt) versus length for device profiles descibed in Fig.1. Deviation from 1/sqrt(Lgate) trend at small channel

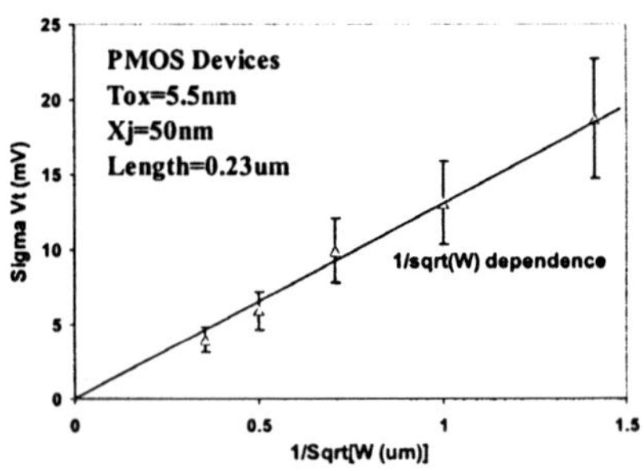

Fig.4. Binomially distributed fluctuation of dopants at each device node predicts dependence of σ(Vt) on width to be 1/sqrt(W)

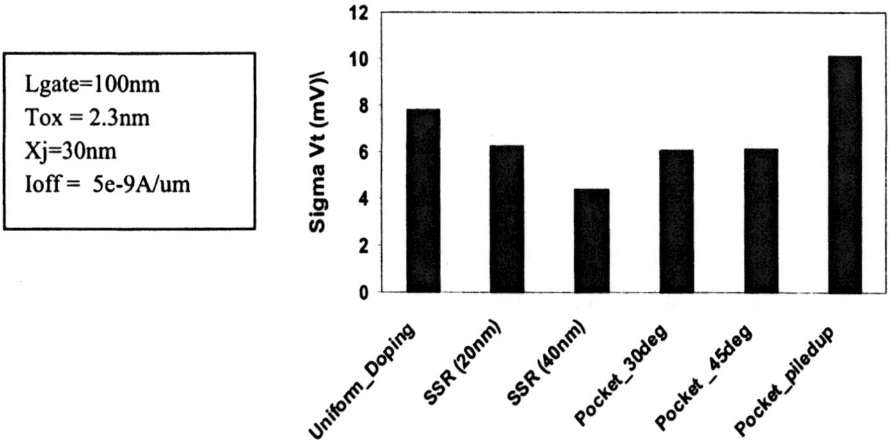

Fig.5. σ(vt) for realistic device profiles for a 100nm device, where doping levels have been adjusted in all profiles to have the same off-state leakage. Compared to uniform doping profile, retrograde channel with peak buried at 20nm has lower σ(vt). Increasing the peak depth to 40nm further reduces the Vt fluctuation. Profiles typical of angled pockets have lower σ(Vt) than uniformly doped devices. However, when the pocket dopant is allowed to couple with extension implant defects and diffuse, it piles up at the surface, and the resulting σ(Vt) is much higher.

4 Summary

A new method for fast and accurate simulation of the effect of statistical dopant fluctuations on MOSFET device characteristics has been developed. The method was used to simulate device performance variability of several different device architectures and proved to be both accurate and flexible.

References

[1] Asenov, A., Subhash, S., (1999) : Suppression of random dopant-induced threshold voltage fluctuations in sub-0.1-μm MOSFET's with epitaxial and delta-doped channels : IEEE Trans. Elec. Dev. 46 : 1718-1724

[2] Axelrad, V., Duane, M., IWSM Honolulu, HI, 1998

[3] Takeuchi, K., Tatsumi, T., Furukawa, A., (1997) : Channel Engineering for reduction of random-dopant-placement-induced threshold voltage fluctuation : International Electron Devices Meeting, IEDM' 97, Washington, DC : 841-844

An agent-based common software platform applied to multi-scale device and process simulations

Shirun Ho, Yasuyuki Ohkura*, Masato Ikegawa*, Nobuyuki Mise*,
Real World Computing Partnership, *Hitachi, Ltd., Tokyo, Japan
ho@crl.hitachi.co.jp, okura-yasuyuki@sic.hitachi.co.jp,
{mtikega, mise}@gm.merl.hitachi.co.jp

Joshi Prasad**, Yuko Kawashima*** and Shoichi Kubo***
**TATA ELXSI, Ltd, Bangalore, India,
***Hitachi ULSI Systems Co., Ltd., Tokyo, Japan
Prasad@tataelxsi.co.in, {Kawashima-Yuko, skubo}@hitachi-ul.co.jp

Abstract

This paper describes a common software platform that allows users to combine multi-scale physical simulations in the form of agents. To demonstrate the benefit of combing multiple simulations, hydrogen desorption at the interface of gate oxide and Si-substrate due to hot electrons in a deep sub-micron device is analysed by coupling device and lattice Monte Carlo simulation. Deposition profiles on a large-scale wafer are studied by using particle method hybridised with continuum simulation. The desorption probability of hydrogen and the coverage of deposition layer are shown to be strongly dependent on the position under the gate and on the wafer respectively.

1 Introduction

It becomes increasingly important to study multi-scale and multi-physics phenomena[1],[2] in ultra-small devices for improving performance and reliability such as transconductance degradation due to defect generation and microscopic manufacturing process on a large-scale Si-substrate. However, the technique of a seamless computing to couple multiple simulations needs to be much improved to reduce time and effort, which essentially involves implementing spatial and time-scale transformation. This paper describes composite analyses using an agent-based common software platform that allows interdisciplinary simulations to be efficiently combined into a single program. The first demonstrates stochastic desorption of H-terminated dangling bonds due to hot electrons, in which a device window Monte Carlo (DMC) simulation is combined with a lattice Monte Carlo (LMC) simulation to analyse desorption probability of H bonds on the Hitachi SR2201 parallel machine and WS clusters. Furthermore, deposition profiles in sub-micron trenches distributed on a Si-substrate using plasma chemical vapour deposition (CVD), in which hydrodynamic simulation to analyze diffusive gas flow in a reactor is hybridised with dynamic simulation Monte Carlo (DSMC) method near the trenches for rarefied gas regime.

2 Agent-based common software platform

To mask the complexities involved in composite parallel simulations that employ FDM, FEM and particle method, the system provides agent library calls to define correlative specifications such as nearest points, first nearest neighbours(1NN) and in-sphere correlation between mesh points and particles. Using standard message passing library such as MPI(Message Passing Interface) and PVM(Parallel Virtual Machine), as schematically shown in Fig.1, the agent associated with each simulation process can automatically determine to which communication is required and transform physical values associated with correlated discrete points while conserving density, momentum and energy. The performance of agents on the Hitachi SR2201 to combine two parallel simulations which employ different kinds of N discrete points and are decomposed into the same M domains is shown in Fig. 2. The agents can detect the correlated discrete points automatically, of which transaction time is designed to be $O(N\log_2 N)$ using a bucket method[3].

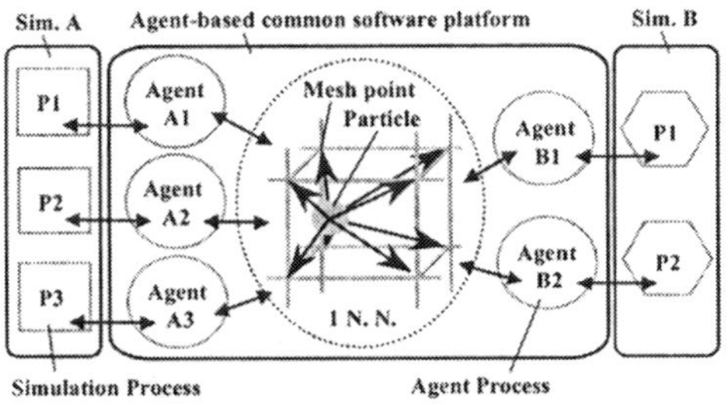

Fig. 1 Combination between different parallel simulations, in which physical values associated with discrete points are transferred and transformed by agents.

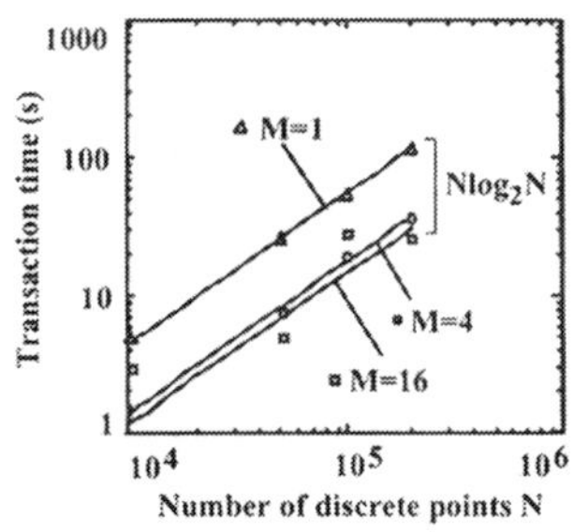

Fig. 2 Transaction time to detect correlated points of 1NN when simulations are decomposed in M domains.

3 Multi-scale composite simulations

The combination between device and lattice Monte Carlo simulations can be effectively realized using 34 agent library calls, together with some additional statements of 240 steps to convert the formats of the discrete points. The deposit simulation can be combined by using 24 agent library calls together with some additional statements of 100 steps, both of which are less than 1/10 of conventional development. Each simulation is easy to reuse and the optimum combination can be rearranged to improve physical accuracy and execution time on user demands.

3.1 Hydrogen desorption due to hot-electrons and multi-phonon interaction

The desorption probability P_{des} of H bonds is studied at the interface of Si/SiO_2 in deep sub-micron device of L_{eff} = 0.18um and oxide thickness of 5nm. In a window DMC, non-equilibrium electron distribution using 15000 electrons is calculated at V_d = 1.5V, V_g = 1.0V for 67ns. Desorption and diffusive migration of H atoms of 10^{17} m^{-2} due to hot electrons near the interface region, which is discretized by 250x15x30 FDM meshes of 1nm width, are analysed in the LMC. When combining two simulations, the position of H bonds and defects is transferred to the DMC by the agents and used to identify the scattered electrons attributed to phonon absorption and emission at H bonds. As shown in Fig.3, the position of the scattered electrons is transferred back to the LMC on the two-way iterative communication. At low V_d less than the threshold energy E_{th} of 3.7eV for bond breaking, a multi-phonon process excited by hot electrons is considered to be one of the major mechanisms for H desorption leading to defects[4]. P_{des} due to hot electrons and multi-phonon interaction can be derived as a following equation.

$$P_{des} = \sum_m f(m,n) \{P_u(N_u, \varepsilon_u)P_d(N_d, \varepsilon_d)\}^{\frac{m}{2}} \left\{\frac{P_u(N_u, \varepsilon_u)}{P_d(N_d, \varepsilon_d)}\right\}^{\frac{n}{2}}$$

Where f(m,n) represents number of pathways from ground phonon state to n state at m times of electron scattering. P_u and P_d means effective scattering probabilities associated with emission and absorption of electrons, of which number is N_u, N_d and average energy is ε_u, ε_d respectively. H-Si bending phonon, of which energy is 0.29eV and n becomes 13, is assumed to be the typical phonon. In Fig. 4, P_{des} is dependent on the position under the gate and shows the maximum value at the drain side. Transconductance degradation can be evaluated using the position-dependent P_{des}, desorption time of H bonds and recombination rate with defects.

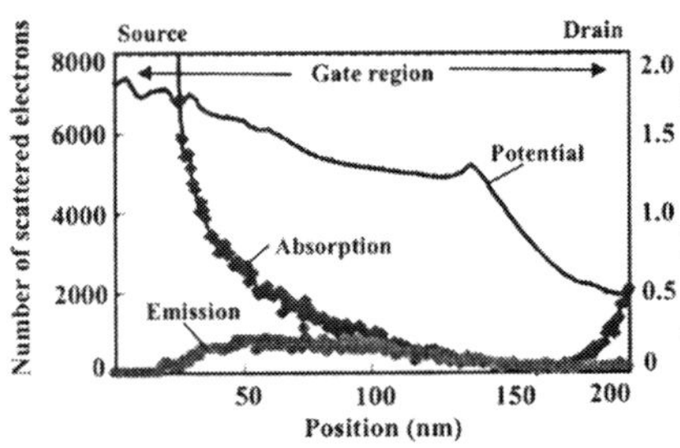

Fig. 3 Number of scattered electrons by a H bond at the interface of Si/SiO$_2$ under gate region

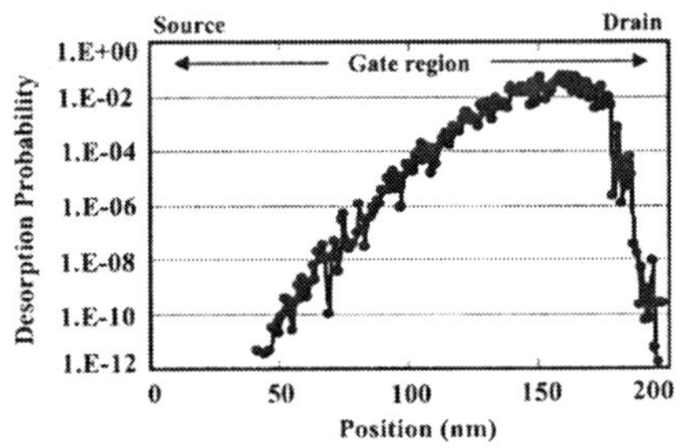

Fig. 4 Dependence of H desorption probability on the position under the gate electrode

3.2 Deposition profiles in sub-micron trenches on a large-scale Si-substrate

In the second composite simulation, the hydrodynamic simulation based on the Navier-Stokes equation is discretized by 5076 FEM meshes. As shown in Fig.5, a source region supplies raw reactive gases and a Si-substrate of 300mm exists at the bottom of a reactor. SiH_4 and reaction products of SiH_2 become major materials to form thin film in trenches. The partial pressure of SiH_2 at 973K is 6~7 orders smaller than that of SiH_4. On the other hand, the sticking coefficient of SiH_2 is 0.7 which is 6 orders larger than SiH_4 of 9.4×10^{-7}. The analytical region near the 0.1um trenches on the substrate should be analysed by the DSMC method[5], because the mean free path of molecules λ becomes larger than the size of the trenches D where $K_n = \lambda/D > 0.01$. Therefore, the DSMC is parallelized to different domains associated with each trench, which enables to analyse deposition profiles distributed on the Si-substrate simultaneously. Using the agent-based system, the partial pressure of SiH_4 and SiH_2 obtained from the hydrodynamic simulation can be automatically transferred and transformed into 5000 gas molecules at each domain in the DSMC according to the Maxwell distribution. Because the normal component of gas flow at the bottom of the reactor is assumed to be zero in the hydrodynamic simulation. Fig.6 shows that coverage of the deposition profiles increases at the edge regime. Partial pressure of reactive production SiH_2 is too high at the central region to form a thicker layer on the surface due to high sticking coefficient. The composite simulation can be used to analyse microscopic manufacturing process by designing such as temperature and flow rate of reactive gases from hydrodynamic simulation.

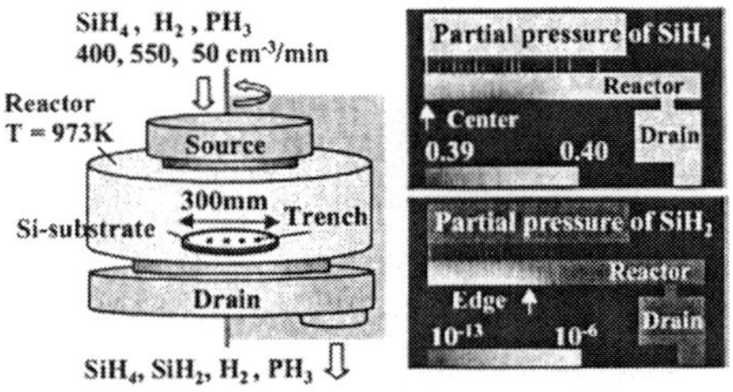

Fig. 5 Partial pressure of SiH_4 and SiH_2 obtained from hydrodynamic simulation using CVD

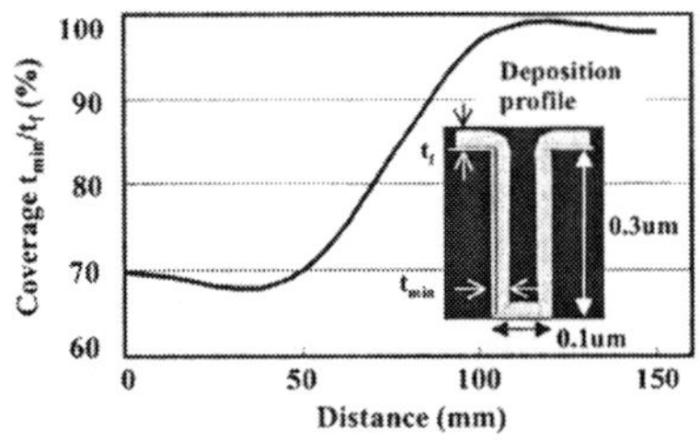

Fig. 6 Dependence of coverage in 0.1um trenches on the distance from the center of a Si-substrate

4 Conclusions

Taking account of the multi-phonon mechanism, the desorption probability of H bonds due to hot electrons is shown to be strongly dependent on the position under the gate electrode. The coverage of deposition profiles changes drastically according to the position on the large-scale substrate. It is shown that macroscopic characteristics and microscopic mechanisms are so intertwined that the agent-based software platform can be an efficient tool to analyse multi-scale simulations.

References

[1] Abraham, F. F., Broughton, J. Q., Bernstein N., and Kaxiras, E., "Spanning the length scales in dynamic simulation," in Computer in Physics, Vol. 12, pp.538-546, 1998.

[2] CISPAR: Coupling of Industrial Simulation codes on PARallel systems, ESPRIT No. 20161, http:/www.research.GermanLloyd. de/Projects/CISPAR.

[3] Ho, S., Itoh, S., Ihara, S., and Schlichting, R. D., "Agent middleware for heterogeneous scientific simulations", Proceedings of the 1998 ACM/IEEE SC98 conference, http://www.supercomp.org/sc98/TechPapers/sc98_FullAbstracts/Ho693

[4] Hess K., Kizilyalli. I. C., Lyding J. W., IEEE Transactions on Electron Devices Vol.45 PP.406-416, 1998

[5] Oran, E.S., Oh, C.K., and Cybyk, B. Z., "Direct simulation Monte Carlo: recent advances and applications", Annu. Rev. Fluid Mech. 30, pp.403-441, 1998

Simulation of Dark Count in Geiger Mode Avalanche Photodiodes

J. C. Jackson, B. Lane, A. Mathewson
National Microelectronics Research Centre,
Lee Maltings, Prospect Row, Cork, Ireland
cjackson@nmrc.ie

A. P. Morrison
Department of Electrical and Electronic Engineering
University College Cork
Cork, Ireland

Abstract

Silicon avalanche photodiodes operated above breakdown, in Geiger mode, can be sensitive enough to allow single photon detection. An inherent limit to GM-APD sensitivity is the noise caused by thermally generated carriers. This noise manifests itself as extraneous counts in the absence of light and is termed the dark count. The presence of the dark count in GM-APD detectors reduces the signal to noise ratio and increases the integration times that are necessary for photon detection. A 1-Dimensional model of the dark count in GM-APD detectors was developed and dark count was found to depend predominantly on carrier generation through Shockley-Read-Hall (SRH) generation centres and secondly on thermal diffusion of minority carriers in the bulk. Simulations performed show that minimisation of the dark count is limited by bulk diffusion of minority carriers. The reduction of process induced damage minimises the dark count and allows theoretically minimum dark counts to be achieved.

Structure

GM-APD detectors are specially designed and fabricated p-n diodes that are operated at biases above the breakdown voltage. Avalanche breakdown of a diode is a statistical process and requires a charge carrier in the depletion region for initiation. A GM-APD can be biased above the breakdown voltage for a finite period of time until an electron or hole enters the depletion region. Depending on the applied bias and the separate ionisation rates of electrons and holes, a probability exists that the incident charge carrier will initiate a self sustaining breakdown. Once a self sustaining avalanche has occurred, quench and reset circuits must be used to reduce the diode bias below the breakdown voltage and allow for the removal of charge within the depletion region. After a suitable hold off time, necessary for the removal of trapped charge, the GM-APD can be biased above breakdown again. In this manner, of breakdown-quench-reset a GM-APD can be used to detect single photons of light. The major limiting factor of device operation is the dark count from thermally generated carriers. To reduce the dark count, commercial detectors rely on large breakdown voltages and complex doping profiles to ensure the junction electric field is kept below the breakdown field. These devices are not compatible with CMOS technology. CMOS compatible shallow junction GM-APD detectors were fabricated as shown in Figure 1.

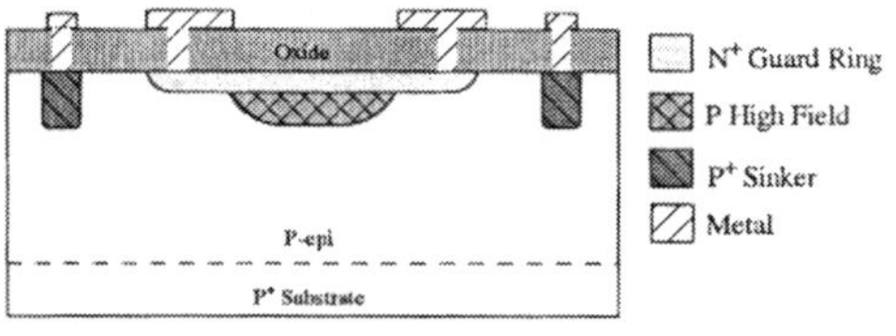

Figure 1: Schematic of shallow junction GM-APD used in this study.

1.1 Dark Count

The room temperature dark-count performance for 10μm and 20μm diameter GM-APD detectors is shown in Figure 2. For these devices, dark counts of less than 100 counts s^{-1} were measured and the dark count scaled with area. The dark count for 50μm diameter devices does not scale with area. The increased dark count can be attributed to defects present in the active area [1]. To determine the theoretical dark count limit for GM-APD operation a device simulator capable of predicting dark counts was necessary.

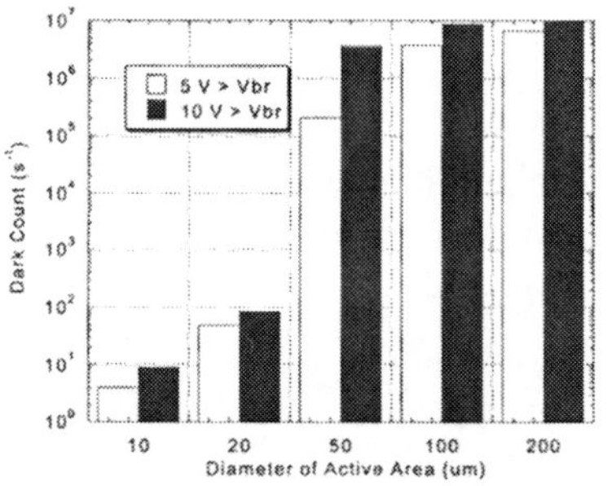

Figure 2: Dark Count for various GM-APD detectors at two different voltages above breakdown voltage.

Commercially available process and device simulators can be used to accurately predict diode breakdown voltage but cannot directly predict the dark count in GM-APD devices. Analytical solutions exist to calculate darkcount and were used to form the basis for the calculation of the minimum theoretical dark count [2]. In this paper a simple 1-D model based upon the local electric field model and commonly accepted impact ionisation coefficients is proposed. This model uses information provided by commercial process and device simulators and extends it to simulating darkcount in GM-APD devices.

1.2 Simulation & Results

A calibrated process simulation was used to obtain doping profiles within the GM-APD detector. The model requires the local electric field, minority carrier concentration and SRH generation rate data. Defects are introduced by modifying the Shockley-Read-Hall generation lifetime or through the inclusion of multiple defect levels in the device simulator. Poisson's equation was modified to account for added charge and the local electric field was then solved. The local electron and hole impact ionisation rates were calculated using data from van Overstraeten and De Man [3]. Electron and hole breakdown initiation probabilities were calculated by solving the differential equations presented by Oldham *et al.* [4]. These are shown in Figure 3 for various voltages above the breakdown voltage.

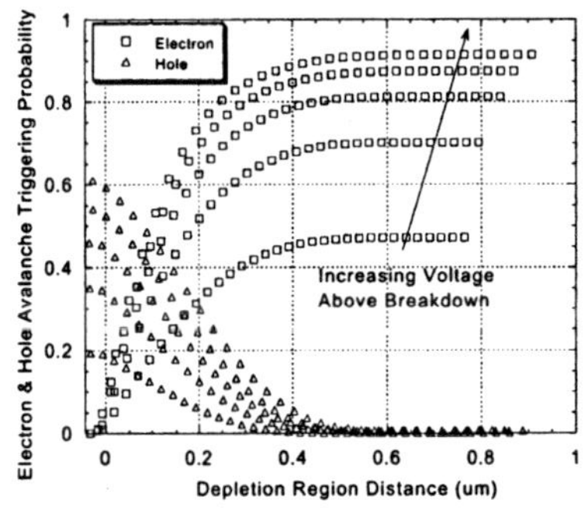

Figure 3: The position dependent electron and hole avalanche triggering probability.

The avalanche initiation probabilities were then combined with the minority carrier densities from the device simulator to yield dark count contributions for each region of the device as shown in Figure 4.

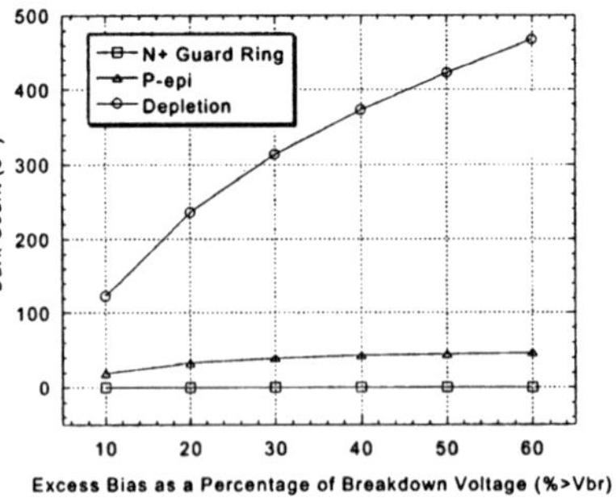

Figure 4: Major contributions to dark count density in a GM-APD detector.

The local generation rate data was used with the local avalanche initiation data obtained from Figure 3 to calculate the contributions of carriers generated in the depletion region. The addition of defects has been incorporated in the device simulator and compared with measured GM-APD dark count performance in Figures 5. A simulation incorporating zero defects has been shown as well, in Figure 5, which predicts the theoretical dark count limit based on a defect free depletion region.

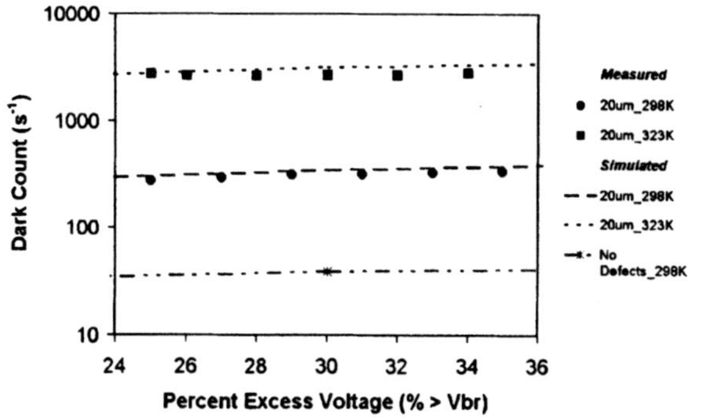

Figure 5: Measured and simulated dark count for a 20µm diameter GM-APD versus percent excess voltage at 298K and 323K A defect free detector with SRH generation removed is shown to predict the theoretical minimum dark count.

Conclusions

In summary a 1-D model based on the local electric field model has been developed to accurately predict dark count performance of GM-APD detectors. The inclusion of defects and trapping centres was accounted for within the model. It was shown that Shockley-Read-Hall generation and minority carrier bulk diffusion contribute to the dark count present in GM-APD detectors. This technique provides a simple and accurate method for predicting real and ideal GM-APD performance.

References

[1] A. Zanchi, F. Zappa, M. Ghioni, A. Giudice, A. P. Morrison and V. S. Sinnis. "Probe detectors for mapping manufacturing defects", *Proceedings of 3rd International Caracas Conference on Devices, Circuits and Systems* (ICCDCS2000).

[2] W. J. Kindt and H. W. van Jeijl, "Modeling and fabrication of Geiger mode avalanche photodiodes", *IEEE Transactions on Nuclear Science*, vol. 45, no. 3, 715-719, 1998

[3] R. Van Overstraeten and H. De Man, "Measurement of the ionization rates in diffused silicon p-n junctions", *Solid-State Electronics*, vol. 13, 583-608, 1970

[4] W. G. Oldham, R. S. Samuelson and P. Antognetti, "Triggering phenomena in avalanche diodes", *IEEE Transactions on Electron Devices*, vol. 19, no. 9, September 1972

Device Simulation and Measurement of Hybrid SBTT

K. Matsuzawa, K. Uchida, A. Nishiyama, T. Numata
Advanced LSI Technology Laboratory, Toshiba Corporation
8, Shinsugita-cho, Isogo-ku, Yokohama 235-8522, Japan
kazuya.matsuzawa@toshiba.co.jp

M. Noguchi
Semiconductor Company, Toshiba Corporation

Abstract

A hybrid Schottky barrier tunneling transistor is assessed by device simulations and measurements. n^+ regions are formed in Schottky contact regions of source and drain. The unified simulation technique is used to simulate ohmic and Schottky contact natures. The devices are fabricated by utilizing the conventional extension process. It is shown that potential modulation by n^+ regions reduces drain leakage current and enhance the tunneling probability at the source contract.

1 Introduction

The Schottky Barrier Tunneling Transistor (SBTT) has been studied as one of candidates for sub-0.1μm MOSFET structure (Tucker et al. 1994) because of its immunity to the short channel effect (SCE). However, there are problems, namely the high source contact resistance and the high leak current of majority carriers at the drain electrode. In this work, a Hybrid SBTT (HSBTT) designed to suppress the problems was assessed by device simulations and measurements.

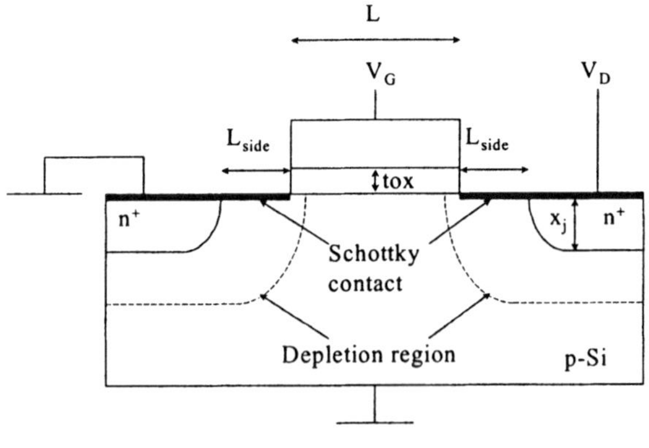

Fig. 1. Schematic of HSBTT.

2 Device simulation of HSBTT

Fig. 1 shows a schematic of the HSBTT. The potential distribution around the edge of the Schottky electrode is modulated by n^+ region. In order to carry out device simulations for the structure, the unified simulation technique for ohmic and Schottky contacts (Matsuzawa et al. 2000) was necessary. Fig. 2 shows the forward bias characteristics of a Schottky barrier diode (SBD) simulated by using the contact model. It was confirmed that the results for SBD get nearer to those of ideal ohmic contact as the impurity concentration increases. Fig. 3 shows the gate bias dependence of the drain current for three kinds of MOSFETs. Although the SCE immunity of HSBTT is inferior to that of SBTT, it is superior to that of the conventional MOSFETs. $ErSi_2$ was used in the simulations as the Schottky contact to reduce the source contact resistance for electrons. It was found that the resistance could be reduced also by adding n^+ region to the source contact region, which is shown experimentally in the following section. Fig. 4 shows the gate bias dependence of the substrate current. The substrate current of HSBTT is one-order lower than that of SBTT. This is because the depletion region spreading from the drain n^+ region relaxes the electric field around the drain electrode and suppresses the tunneling of holes.

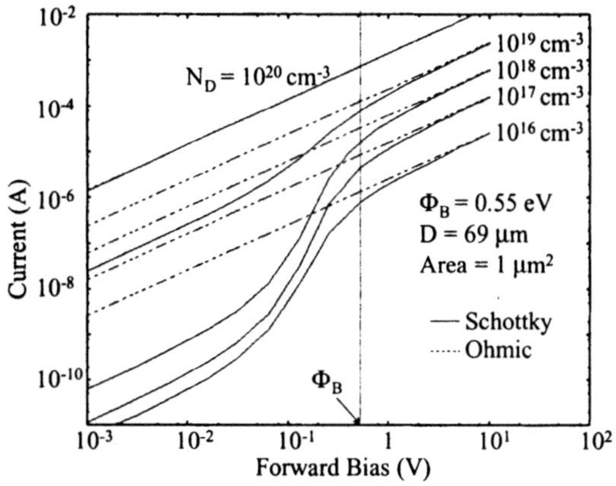

Fig. 2. Forward bias characteristics of SBD.

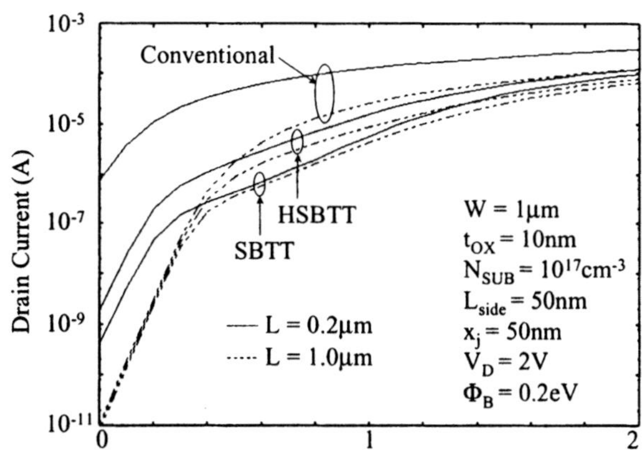

Fig. 3. Gate bias dependence of drain current for three kinds of MOSFETs.

3 Measurement of HSBTT

Fig. 5 shows the fabrication process of HSBTT using a typical material, $TiSi_2$, in our clean room. The distance between the edge of n^+ regions and the edge of source/drain electrodes L_{side} was controlled by the deposition thickness of the polysilicon, t_{poly}. It was confirmed that the fabricated Schottky contact exhibited a reasonable rectifying characteristic, as shown in Fig. 6. Fig. 7 shows measurements of the gate bias dependence of the source current. The SCE of HSBTT is sufficiently suppressed comparing with that of the extension structure. It was found that the source current of HSBTT could be controlled by t_{poly}. This is because the depletion region spreading from the n^+ region modulates the shape of the tunneling barrier at the source electrode and the potential profile depends on L_{side} as shown in Fig. 8.

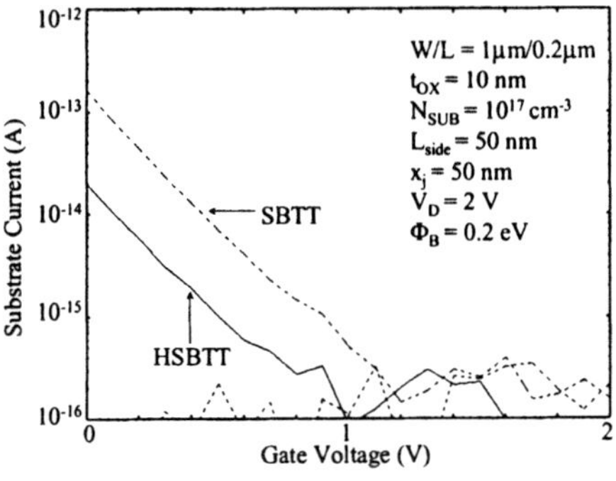

Fig. 4. Gate bias dependence of substrate current for SBTT and HSBTT.

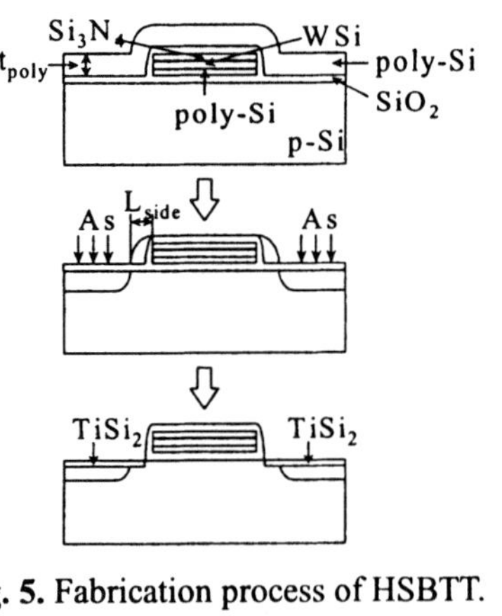

Fig. 5. Fabrication process of HSBTT.

4 Conclusion

The electrical characteristics of HSBTT were assessed by device simulations and measurements. The immunity of HSBTT to the short channel effect was superior to that of the conventional MOSFET. The source contact resistance was controlled by the position of the high impurity region in the Schottky electrode region. The leak current at the drain electrode of HSBTT was one-order lower than that of SBTT.

References

[1] Tucker, J.R., Chinlee Wang, Carney, P.S. Silicon field-effect transistor based on quantum tunneling. Appl. Phys. Lett. 1994; 65: 618-620

[2] Matsuzawa, K. Uchida, K. Nishiyama, A. A unified simulation of Schottky and ohmic contacts. IEEE Trans. Electron Devices 2000; 47: 103-108

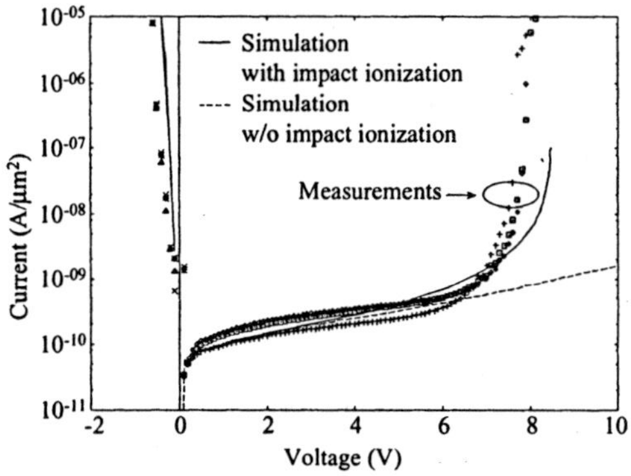

Fig. 6. Rectifying characteristic of fabricated Schottky contact.

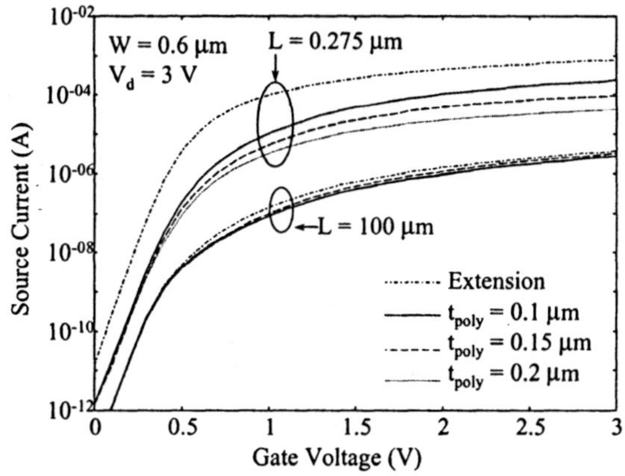

Fig. 7. Gate bias dependence of source current for fabricated HSBTT.

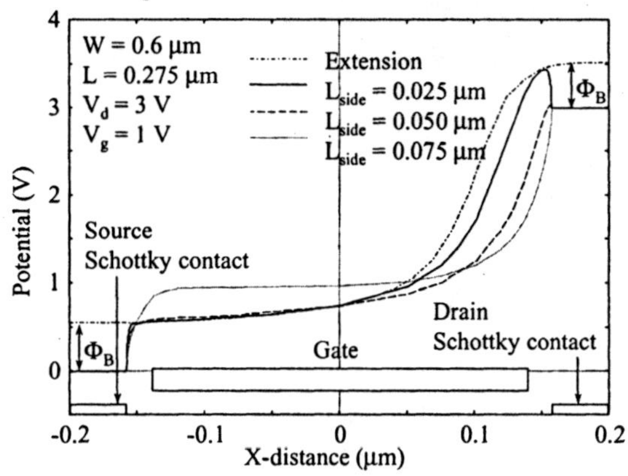

Fig. 8. Potential distribution at substrate surface.

Two-Dimensional Diffusion Characterization of Boron in Silicon using Reverse Modeling

Eitan N. Shauly
Faculty of Materials Engineering, TECHNION-Israel Institute of Technology, Haifa 36000, Israel and Tower Semiconductor Ltd, Migdal Ha'Emek 10556, P.O.B. 619, Israel, (email: cezsmh@TowerSemi.com)

Richard Ghez and Yigal Komem
Faculty of Materials Engineering, TECHNION-Israel Institute of Technology, Haifa 36000, Israel

Abstract

This work deals with the simulation of two-dimensional impurity diffusion in CMOS silicon devices. The Reverse Modeling method was used to determine the diffusion coefficient (D_I), surface recombination rate of defects (K_I) and the characteristics of the injecting source. Analysis showed similarity between D_I in 2-D system compared with the value obtained from non-patterned samples. The results for D_I and K_I are very well described by the Arrhenius expressions. D_I was found to be related to the substrate type e.g. EPI or CZ. The values of K_I related to the interface type, oxidizing or non-oxidizing (SiO_2 or Si_3N_4).

1 Introduction

Simulation of semiconductor device structures requires modeling the underlying physics of semiconductor fabrication processes. The need for accurate simulation becomes more critical with the growing complexity of design and the continuing reduction in device dimensions. For this reason much effort has been invested in determining the values of model parameters, which give good predictive capability. The primary goal of this work is to develop a Reverse Modeling concept, which can be used to determine model parameters that cannot be measured directly. This method was used to study 2-D impurity diffusion in CMOS silicon devices, and the interaction between point defects and diffusing species. The parameters determined were the diffusion coefficient (D_I) and surface recombination rate of interstitials (K_I) and the characteristics of the injecting source, θ, that represents the amount of Si atoms consumed during oxidation and injected as intertistals down into the bulk.

2 The Reverse Modeling concept

The Reverse Modeling concept enables studying process parameters and is based on statistical comparison between 2-D process and device simulations and various electrical and physical measurements of sub-micron devices. A process simulator was used to compute the physical characteristics of a CMOS device. Using this data, a device simulator was used to predict electrical characteristics, which were compared with measurements of real devices. The comparison was used as feedback to update the process simulator, until a reasonable agreement was reached. Unlike function fitting, this concept does not require a direct measurement of the studied parameters.

3 Parameters for point defect diffusion and recombination

The physical mechanism of atomic diffusion is similar in all semiconductor systems. However, the complexities of devices, introduce new boundary conditions for the diffusing elements. For example, in cases where part of an implanted layer was covered by nitride capping before oxidation, three types of diffusion conditions need to be studied: under the cover, in the oxidized area and in the transient zone around the blocking edge.

It is generally accepted that injection of silicon atoms from the growing oxide lead to oxidation enhanced diffusion (OED). The injected interstitials will diffuse, like any other element, having temperature depended diffusivity (D_I). Some interstitials that reach an interface, will recombine at the "recombined surfaces" having a surface recombination rate K_I. Assuming no generation or presence of vacancies, one can specify a linear combination of flux and field at the boundary:

$$D_I \nabla C_I - K_I\left(C_I - C_I^*\right) = 5\times 10^{22}\theta\left(\frac{dx}{dt}\right)^S \tag{1}$$

where C_I and C_I^* are the local and the equilibrium concentrations of interstitials, respectively. The left hand-side of equation 1 is the "thermodynamic force" (∇C_I) or the "driving force for diffusion" ($D_I \nabla C_I$) into the silicon bulk [1]. The right hand-side, represents the surface injection flux, where θ is the percentage of consumed silicon lattice atoms that are injected back into the silicon bulk as interstitials, and dx/dt is the oxide growth velocity. The power S depends on the oxidation conditions with typical values of 0.3-0.6 [2].

The values of D_I, K_I and θ are hard to obtain, since they can found only through modeling of another phenomena, like OED [3, 4] or Oxidation Induced Stacking Faults (OISF) [5]. K_I, values can only be extracted using special test structure [12, 6]. A literature review for the values of D_I, K_I and θ showed considerable scatter in results up to several orders of magnitude.

4 Experimental details and calculation methods

The study was initiated by tuning the process and device simulators by 1-D process modeling adjustment, using results of 1-D SIMS, thickness, and sheet resistance measurements. 2D electrical modeling for the mobility and threshold voltage was adjusted based on long channel device measurements.

The bulk of the work included experiments performed under different process conditions namely heat treatment under N_2 or dry O_2 atmospheres, at different temperatures. Submicron CMOS devices were fabricated using 0.8μ technology, with gate lengths in the range of 1.6μ to 0.56μ, full details are given by Shauly [7]. 2-D process and device simulations were executed using SUPREM-IV [8] and MEDICI [9], respectively. A large matrix of simulations was run using the principles of statistical design of experiments, to find the values of D_I, K_I and θ. This matrix enabled fitting a quadratic model for each one of the electrical parameters to quantify it's sensitivity to the post implantation heat treatment conditions. A large number of electrical testing (ET) parameters were measured including transistor drive current, threshold voltage, transistor peak substrate current and more. For each of the matrix simulation runs, the response value (Δ) was calculated, to evaluate the difference between simulation and experimental results.

$$\Delta = \frac{\left(\mathrm{O}_{\text{simulation}} - \mathrm{O}_{\text{experimental}}\right) - \left(\mathrm{N}_{\text{simulation}} - \mathrm{N}_{\text{experimental}}\right)}{\text{SCALE}} \tag{2}$$

where "O" refers to the values obtained under oxidation conditions, and "N" to the values obtained at N_2 ambient conditions. To facilitate a joint solution for more than one ET parameter, the observed Δ was also scaled by dividing by a predetermined scaling factor for each ET parameter.

5 Results and discussion

The accuracy of the simulation system was verified using a series of transistors implanted with different *p*LDD doses and annealed under N_2 ambient. The measured and simulated values of electrical parameters of these transistors were equal within experimental error. This gave confidence that simulation system describes well the devices fabricated with N_2 anneal. Based on this, it was concluded that the results obtained to describe OED using the Reverse Modeling concept do reflect the effects of O_2 in the anneal ambient.

Using surface response method and setting Δ equal to 0, values for D_I, K_I and θ were found. Over the temperature range of 800°C - 950°C, the results for D_I and K_I are very well described by the Arrhenius expressions:

$$D_I = 1.06\times10^{-4}\exp\left(\frac{-1.32\text{eV}}{kT}\right) \qquad \text{cm}^2/\text{sec} \qquad (3)$$

$$K_I = 1.26\times10^{-11}\exp\left(\frac{+1.65\text{eV}}{kT}\right) \qquad \text{cm}/\text{sec} \qquad (4)$$

The typical value of θ in this temperature range was found to be 0.01. The parameter showed weak dependence on temperature. To validate these results, B profiles after oxidation, as measured using SIMS, were compared with 1-D simulated profiles and D_I was extracted. Analysis showed similarity between D_I in 2-D system compared with the value obtained from non-patterned samples (Fig 1). Since the intestinal diffusivity in the bulk does not depend on the nature of the interface surrounding the bulk, the values obtained for D_I were similar (see lines c& d). Comparison with results from the literature showed similarity to results obtained by some researchers (lines e & f), and pronounced differences compared with others (lines a & g). Careful examination of the data showed that D_I depends on the substrate material. We verified experimentally this assumption and showed that D_I in samples having epitaxial (EPI) top layer, is higher by a factor of 3 compared to Czochralski (CZ) samples. A model for bulk trapping which affects D_I was proposed. Interactions of interstitials with traps located in the bulk have reduced D_I. The fact that the bulk traps density is lower in EPI samples compared to CZ, explained the higher value of D_I found.

The activation energy for the surface recombination rate was found (Fig 2). The variations between the values of K_I were attributed to the dependence of K_I on interface type (Table 1). The difference was explained for the first time, by stress induced from the upper layers, into the silicon substrate. Experiments showed that K_I of samples having a strained Si_3N_4/Si layer was half an order of magnitude lower compared to samples with SiO_2/Si interface An explanation for the reduction in K_I with a rise in temperature was given by decomposition of SiO_2 molecules to Si+O_2. Some of the Si atoms have diffused and were trapped to the Si/SiO_2 interface, reducing K_I.

Finally, the equilibrium concentration of interstitials was calculated based on the $D_I C_I^*$ relation developed by [10]. This gave the ability to present a full set of parameters needed for simulation of 2-D diffusion of B in Si.

Table 1: K_I values at 900°C and activation energy values for various trapping interfaces

Reference	Interface Structure	K_I (cm/sec)	*Ea* (eV)
[2]	Si/SiO2	6.3×10^{-4}	-1.75
This work	Gate=150A, Poly=3250A	1.0×10^{-4}	-1.65
[4]	Pad~800A, Nitride=1200A	3.5×10^{-6}	-1.58
[6]	Pad=400A, Nitride=800A	1.5×10^{-7}	-0.89
[5]	Oxidation Induced Staking Faults	2.3×10^{-8}	+2.4

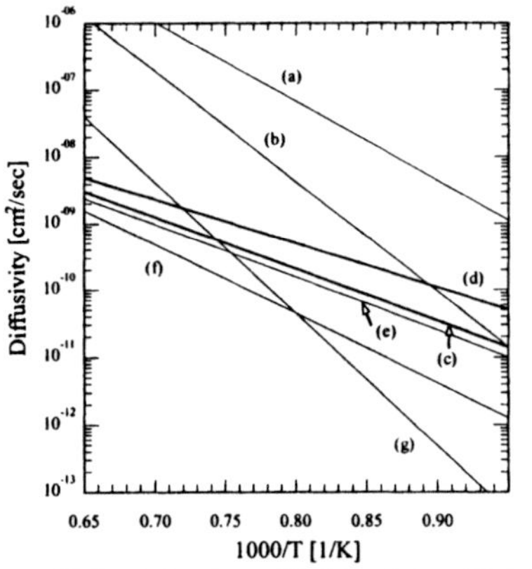

Fig. 1: Values for the self diffusivity (D_I) from the literature and this work. (a) [11]; (b) [12]; (c) 1-D, This work; (d) 2-D, This work; (e) [13, 8]; (f) [6]; (g) [5].

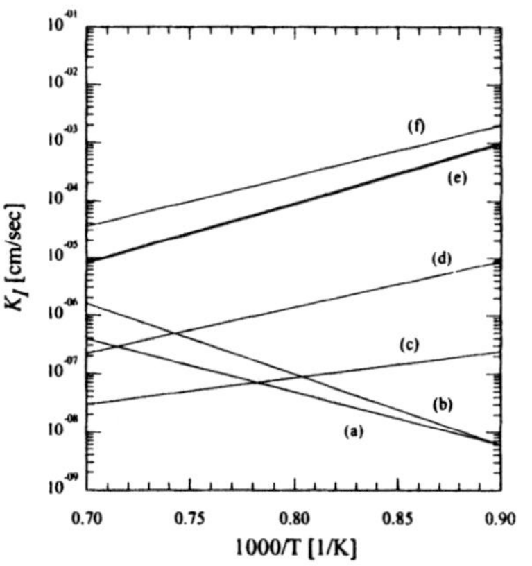

Fig 2: Interstitial recombination velocities, K_I, as determined experimentally at this work, and by other researchers. (a) [14]; (b) [5]; (c) [6]; (d) [13]; (e) 2-D, This work; (f) [2, 8].

6 Acknowledgements

The generous support of Tower Semiconductor Ltd. is gratefully acknowledged. The authors would also like to thank the Fund for the Promotion of Research at the Technion for partial support.

7 References

[1] Ghez, (1990): A primer of diffusion problems. John Wiley & Sons, New York.

[2] Dunham S. T. (1992): Interactions of silicon point defects with SiO2 films. J. Appl. Phys., 71:685-696.

[3] Griffin P. (1990): Physics and modeling of Two-Dimensional Diffusion in SUPREM-IV. PhD thesis, Stanford University.

[4] Kump, M. R., Dutton, R. W. (1988): The efficient simulation of coupled point defect and impurity diffusion. IEEE Trans. on Computer Aided Design, 7:191-204.

[5] Taniguchi, K., Antoniadis, D. A., Matsushita Y. (1983): Kinetica of self-Interstitials generated at the Si/SiO_2 interface. Appl. Phys. Lett., 42:961-963. See also: Taniguchi, K., Antoniadis, D. A., Matsushita Y. (1983): Kinetics of self-interstitials generated at the Si/SiO_2 interface. Appl. Phys. Lett., 42:961-963.

[6] Griffin, P. B., Plummer, J. D. (1986): Process physics determining 2D impurity profiles in VLSI devices. In Technical Digest, Intl. Elec. Dev. Mtng, Los Angeles, pp. 522-523.

[7] Shauly, E. N., (2001): A study of two-dimensional diffusion of dopants in CMOS devices. PhD thesis, Technion -Israel Institute of Technology, Haifa.Israel.

[8] SUPREM-IV, (1997): TSUPREM-IV Two-Dimensional Process Simulation Program ver. 6.5 User's Manual. Technology Modeling Associates Inc., Sunnyvale, Ca.

[9] MEDICI, (1997): MEDICI IV Two-Dimensional Device Simulation Program User's Manual Ver. 4.0. Technology Modeling Associates Inc., Sunnyvale, Ca.

[10] Morehead, F., (1988): Defects in electronic materials. Mat. Res. Soc. Symp. Proc., Pittsburgh, Pa., 104:9

[11] Bronner, G. B. & Plummer, J. D., "Gettering of gold in silicon: A tool for understanding the properties of silicon interstitials," J. Appl. Phys, 61(12), (1987), pp. 5286-5298.

[12] Griffin, P. B., Fahey, P. M., Plummer, J. D., Dutton, R. W. (1985): Measurement of silicon interstitial diffusivity. Appl. Phys. Lett., 47:319-321.

[13] Kump, M. R. (1988) Two-dimensional computer simulation of diffusion in silicon. Ph.D. Thesis, Stanford University, Palo Alto, Ca.

[14] Collard, D., Taniguchi, K. (1986): IMPACT - A point defect based two dimensional process simulator: Modeling the lateral oxidation enhanced diffusion of dopants in silicon. IEEE Tans. Elec. Dev., ED-33:1454-1462.

Simulation of Advanced n-MOSFET Emphasizing Quantum Mechanical Effects on 2-D Characteristics

Yutao Ma[(1)*], Lifeng Chen[(1)], Bo Jiang[(1)], Min Zhang[(1)], Lilin Tian[(1)], Zhiping Yu[(2)], Litian Liu[(1)], Zhijian Li[(1)]

[(1)] Institute of Microelectronics, Tsinghua Univ. Beijing, P.R.China 100084

[(2)] Integrated IC Lab. Stanford Univ. Stanford CA. U.S.A.

[*] Email: mayt@263.net

Abstract

Advanced n-MOSFET structure with featured size of 90nm channel length is simulated using a newly developed Quantum Mechanical (QM) correction model based on Modified Airy Function (MAF) method. The influences of Quantum Mechanical Effects (QMEs) on the carrier distribution in the whole channel is included and the output as well as the transfer characteristics are compared with and without QM correction. It is demonstrated that QMEs result in more severe short channel effects such as threshold voltage roll off and DIBL effects .

1 Introduction

The Quantum Mechanical Effects (QMEs) on the characteristics of modern MOSFET's with thin gate oxide and high substrate doping level become more and more significant. Simulation of 2-D characteristics of MOSFET including QMEs is necessary. In the existing 2-D QM correction models, some are too simple to give accurate information of carrier distribution and subband structure and carrier occupation information[1].Some other models are too time consuming to be applicable to practical 2-D device simulation due to the high computation burden of fully numerically solving Schrodinger equations as an eigenvalue problem[2].

In this work, Schrodinger equation is solved with Modified Airy Function (MAF) method[3] and a new QM correction model in the whole channel region in MOSFET is developed. Due to the semi-analytical nature of MAF method, the numerical computation task is substantially reduced. The output characteristics of 90nm channel length Well-Tempered MOSFET[4] is simulated with and without QMEs to manifest the influences of QMEs on device characteristics. A series of MOSFET structures[5] with different gate lengths are simulated and it is shown that QMEs renders more severe short channel effects.

2 Quantum Mechanical Correction Model based on MAF

Accurate determination of carrier distribution and potential profile in quantized inversion and accumulation layer in MOS structure requires solving coupled Schrodinger and Poisson equations. Usually, Schrodinger equation is solved numerically as an eigenvalue problem. But it is very time-consuming. MAF method provides the semi-analytical solution of Schrodinger equation and was widely used in wave guide analysis. Accurate energy levels and wave functions of arbitrary number of subbands can be obtained while keeping satisfied efficiency due to the analytical nature of the method.

The wave function from solution of Schrodinger equation by MAF method is[3]:

$$\psi_i(x) = A\cdot\left(d\xi_i/dx\right)^{1/2} Ai\left(-\xi_i(x)\right) \tag{1}$$

where Ai is the Airy function, A is constant and:

$$\xi_i(x) = \left\{1.5\cdot\int_x^{x_0}\left[\left(2m_i^*/\hbar^2\right)\left(E_i - V(t)\right)\right]^{1/2} dt\right\}^{2/3} \tag{2}$$

in which $V(\cdot)$ is potential, E_i is energy level. x_0 is the position where $E_i = V(x_0)$. The energy level E_i is obtained from the boundary condition of $\psi_i(0)=0$. That is, one needs to solve E_i from the equation:

$$Ai_0^i = \left\{1.5\cdot\int_0^{x_0}\left[\left(2m_i^*/\hbar^2\right)\left(E_i - V(t)\right)\right]^{1/2} dt\right\}^{2/3} \tag{3}$$

in which Ai_0^i is the i-th zero point of Ai.

In order to testify the accuracy of MAF, two approaches are used to solve Schrodinger equation. One is using fully numerical method to calculate the eigenvalues and eigenfunctions of discretized Schrodinger equation, the other is suing MAF method. Then MOS devices are simulated. Fig.1 shows the energy levels calculated by MAF method and by fully numerical method as a function of inversion layer carrier sheet density. E_{ij} in the legends denotes the j-th level in the i-th valley. Carrier density profiles in the channel region calculated by MAF method and by fully numerical method are shown in fig.2. It is clearly shown that MAF has satisfied accuracy.

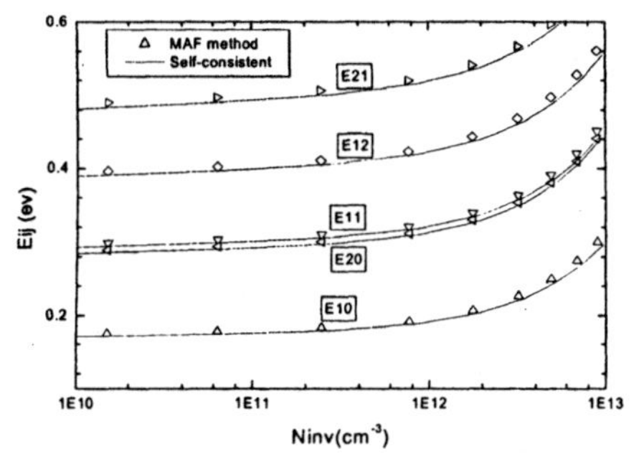

Fig.1: Energy levels calculated by MAF method and by full numerical solution of Schrodinger equation

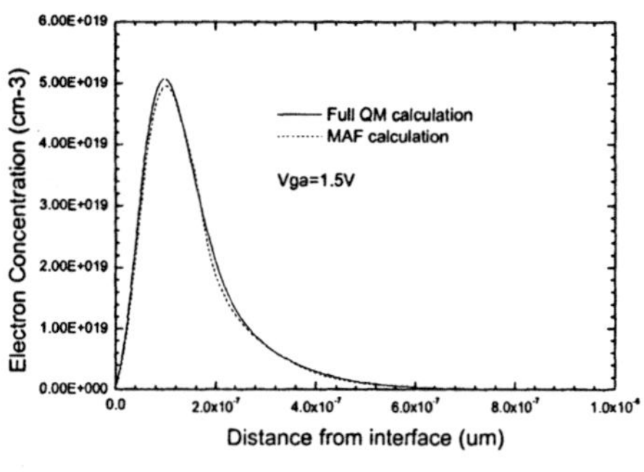

Fig.2: Carrier profile calculated by MAF method and by full numerical solution of Schrodinger equation.

In our 2D model, Schrodinger equation is solved with MAF method to calculate the subband energies and carrier wavefunctions and then obtain the quantum mechanical electron profile in the channel. Then Poisson equation is solved to get new voltage potential. After convergence of the iteration between Schrodinger and Poisson equations, the current continuity equations for electrons and holes are solved to adjust the current density.

3 Simulation results of short channel MOSFET's

The well-tempered MOSFET [4] with featured channel length of 90 nm is used as the test structure. The subband information in the whole channel region is presented in Fig.3. The results show clearly the asymmetric of the subband structure along the channel especially with large source drain bias. The output and transfer characteristics (fig.4 and fig.5) of the device are presented as illustration of the application of the simulator. It is shown that the reduction of output current due to QMEs can be as large as 10 percent of the total value. The results in this work clearly demonstrate the necessity of including QMEs in simulation of advanced MOSFET.

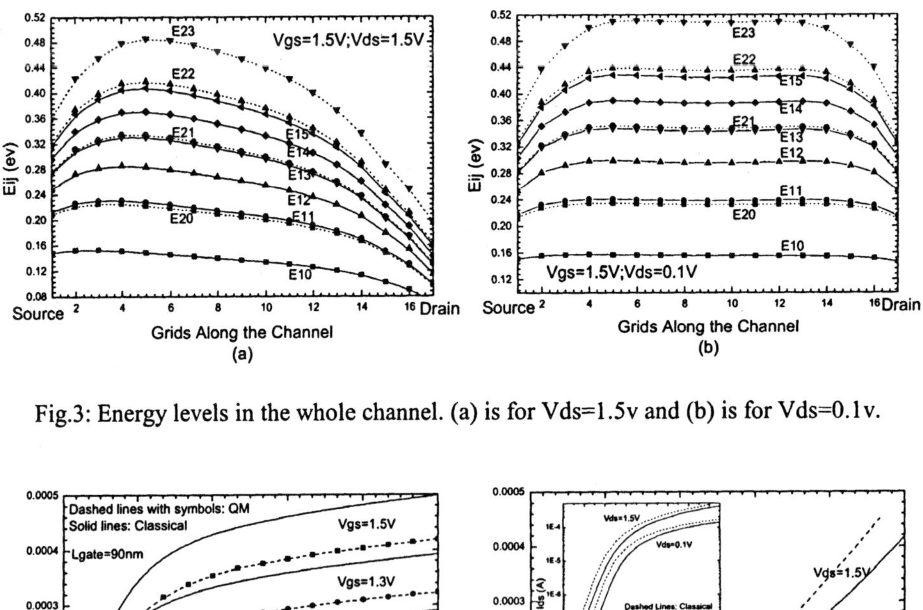

Fig.3: Energy levels in the whole channel. (a) is for Vds=1.5v and (b) is for Vds=0.1v.

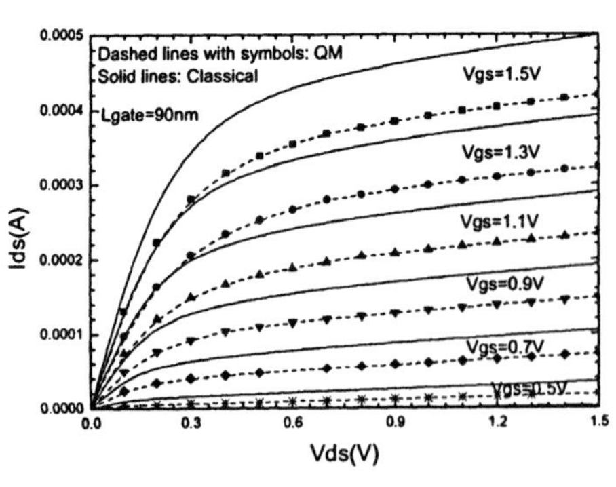

Fig.4: Output characteristics of MOSFET with and without QMEs.

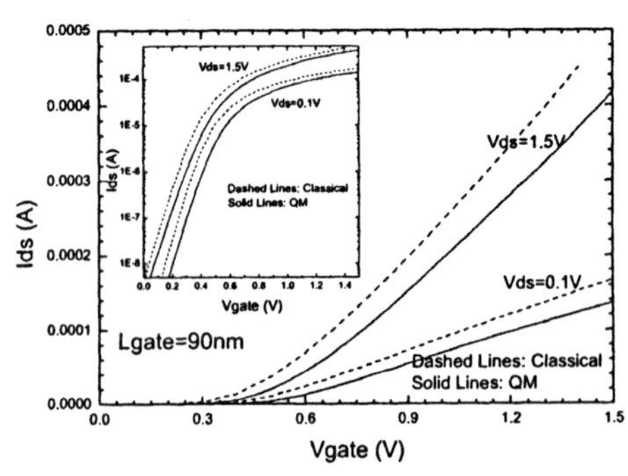

Fig.5: Transfer characteristics of MOSFET with and without QMEs.

A group of devices [5] with different gate length and all the same other parameters are simulated to investigate the short channel effects including QMEs (Fig.6~8).

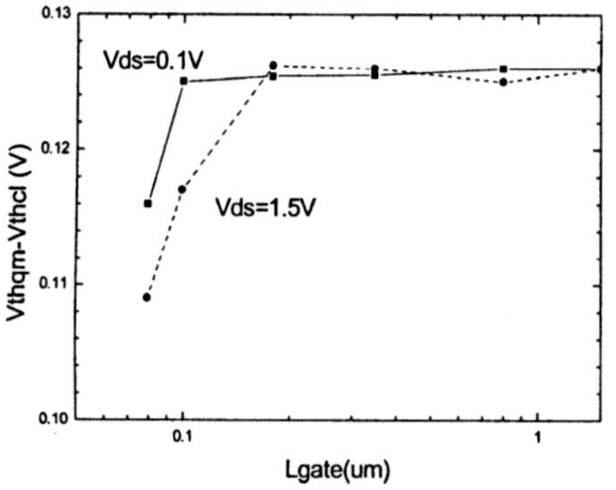

Fig.6: Threshold voltage shift due to QMEs for devices with different gate lengths.

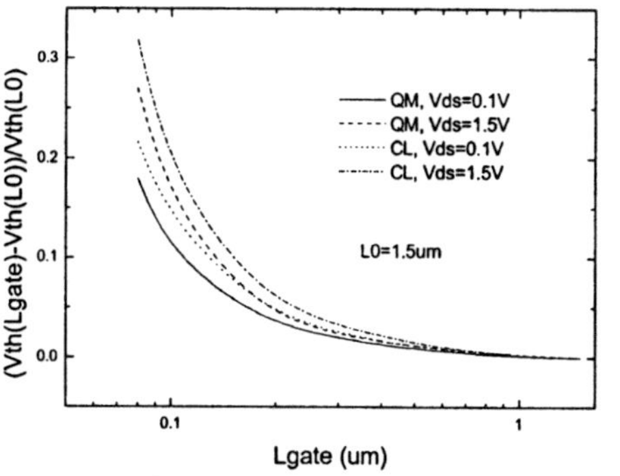

Fig.7: Vth shift due to SCE with and without QMEs.

Fig.6 gives the threshold voltage shift due to QMEs in devices with different gate lengths. Fig.7 shows the results of Vth roll off due to short channel effects in QM cases and semi-classical cases respectively. Fig.8 gives the results for DIBL effects. It can be seen from the figures that short channel effects are quite different with and without consideration of QMEs. So, in order to accurately characterize short channel MOSFET's, it is necessary to take into consideration of the influences of QMEs.

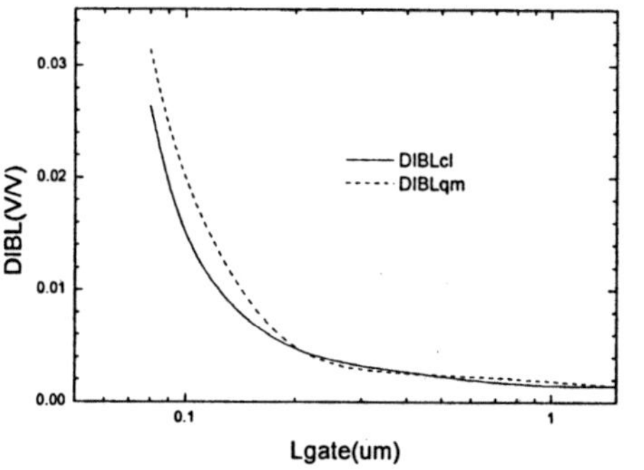

Fig.8: DIBL effects with and without QMEs.

References

[1] van Dort, M.J.; Woerlee, P.H.; Walker, A.J., Solid-State Electron., vol.37, no.3, p.435, 1994.

[2] Spinelli, Alessandro S.; Benvenuti, Augusto; Pacelli, Andrea. IEEE Transactions on Electron Devices. v 45, n 6, pp 1342-1349, 1998.

[3] A.K.Ghatak, et.al., Modified Airy Function and WKB Solutions to the Wave Equation. NIST Monograph.

[4] http://www-mtl.mit.edu/Well-tempered-mosfet\Well-Tempered Bulk-Si NMOSFET Device Home Page.htm

[5] G.Timp, et.al. ,"The ballistic Nano-transistor". IEDM99, p48.

V_{th} Model of Pocket-Implant MOSFETs for Circuit Simulation

D. Kitamaru, H. Ueno, K. Morikawa[1], M. Tanaka, M. Miura-Mattausch
H. J. Mattausch[1], S. Kumashiro[2], T. Yamaguchi[2], K. Yamashita[2], N. Nakayama[2]
Graduate School of Advanced Sciences of Matter,
[1]Research Center for Nanodevices and Systems,
Hiroshima University 1-4-1, Kagamiyama, Higashi-Hiroshima, 739-8527, Japan
Phone: +81-824-24-7637 Fax: +81-824-22-7195 E-mail: hq6@hiroshima-u.ac.jp
[2]Semiconductor Technology Academic Reserch Center,
3-17-2, Shin Yokohama, Kanagawa 222-0033

Abstract

A new threshold voltage (V_{th}) model has been developed for the pocket-implant technology. The model extracts the threshold condition from the entire mobile charge concentration in the channel with only two parameters; the maximum doping concentration (N_{subp}) of the pocket profile and the penetration length (L_p) into the channel. The model reproduces the measured V_{th} *vs.* gate-length (L_{gate}) characteristics with an average error of a few mV under any bias conditions.

1 Introduction

The pocket-implant technology enables L_{gate}-reduction of MOSFETs down to the sub-100nm region, because a strong reverse-short-channel effect (RSCE) suppresses the short-channel effect (SCE) on V_{th} [1]. However, the complicated V_{th}-characteristic is not well understood and strongly dependent on the pocket-implant profile. This leads to 2 major concerns for practical applicability:

- a precise V_{th}-model for circuit simulation, to enable correct design of analog and digital circuits
- an efficient method for pocket-implant optimization

Here we present a new V_{th}-model, which is mainly targeted at resolving the circuit-simulation issue. In addition, the model can be used also for a first step optimization of the pocket profile.

2 Model description

Fig. 1 shows measured $V_{th} - L_{gate}$ characteristics. Specific feature is the strong RSCE suppressing SCE down to $L_{gate} = 100nm$. To model this feature we determine V_{th} with the average carrier concentration (n_{av}) along the channel, which is required to be equal to that at threshold condition (n_{th}). Under the drift-diffusion approximation, the current density (j_n) can be described by the gradient of the quasi-Fermi potential (ϕ_f) along the channel direction y [2]

$$j_n = -q\mu n \frac{d\phi_f}{dy}.$$

This leads to the drain current (I_{ds}) equation as a function the drain voltage (V_{ds}) as

$$I_{ds} = qW_{ch}\mu n_{av}\frac{V_{ds}}{L_{ch}} = qW_{ch}\mu\frac{V_{ds}/L_{ch}}{\left(\int_0^{L_{ch}} n_x^{-1}dy\right)/L_{ch}} \quad (1)$$

where n_x is the integrated concentration along the depth direction. The inverse of the denominator of the right-hand side equation is defined as n_{av}. Parameters q, L_{ch}, W_{ch} and μ are the electron charge, the channel length, the channel width, and the electron mobility, respectively.

The pocket profile is modeled by a linearly graded function along the channel with the parameters L_p and N_{subp} as depicted in Fig. 2. The bulk concentration is homogeneous and fixed to N_{subc}.

Fig. 3 shows simulated carrier concentrations vs. V_{gs} with MEDICI for different channel concentrations, where n_c and n_p are homogeneous cases with $N_{subc}(< N_{subp})$ and $N_{subc}(= N_{subp})$, respectively. As can be seen, different N_{sub} concentrations lead to different V_{gs} at which the threshold condition is fullfilled. In the technology studied here, N_{subc} and n_{th} are $5 \times 10^{17} cm^{-3}$ and $1.5 \times 10^{10} cm^{-2}$, respectively. For analytical description of n_{av} we propose the equation

$$n_{av} = An_c \tag{2}$$

with

$$A = L_{ch}/(L_{ch} + 2L_p(n_c/n_p - 1)) \leq 1$$

Eq. (2) means that $n_{c,th}$ must become larger than n_{th} at the threshold condition to preserve $n_{av,th} = n_{th}$ for the pocket-implant case, as schematically depicted in Fig. 3. Consequently $n_{c,th}$, the carrier concentration of the non-pocket channel region at V_{th}, varies with the pocket profile parameters and channel length. If L_p and N_{subp} are given, n_c and the ratio n_c/n_p in eq. (2) can be calculated by solving the Poisson equation. The final analytical V_{th}-description is

$$V_{th(pocket)} = \phi_{th} + V_{fb} + \frac{\sqrt{2qN_{subc}\epsilon_{si}(\phi_{th} - V_{bs})}}{C_{ox}} \tag{3}$$

where the threshold surface potential ϕ_{th}, giving the $n_{c,th}$ value in eq. (2), is obtained by solving the Poisson equation under the Gauss law.

Though all descriptions are given for the n-channel case, they are valid for the p-channel case as well.

3 Results and discussion

The derivation of eqs. (1)-(3) assumes a small V_{ds} ($\leq 50mV$). When V_{ds} increases, a surface potential increase occurs only in the drain-side pocket for long-channel cases as demonstrated in Fig. 4.

Fig. 5a shows the subthreshold characteristics simulated with MEDICI for different channel conditions and V_{ds} values. With only a drain-side pocket an enhanced SCE is observed, which can be modeled by a reduced gate-length to pocket-penetration length L_p and a linear increasing channel concentration added in the conventional V_{th}-model [3]. With only a source-side pocket the V_{th} dependence on V_{ds} due to the SCE is totally suppressed as can be seen in Fig. 5b. With pockets on both sides an averaged V_{th} dependence on V_{ds} is observed. Therefore the SCE is present in pocket-implant MOSFETs up to the long-channel range, preventing the desirable voltage saturation for large V_{ds}.

Calculated $V_{th} - L_{gate}$ characteristics are depicted by solid lines in Fig. 1. Fig. 2 also shows the pocket-profile obtained from TSUPREM by inverse modeling. The model profile is about the average of the real pocket profile along the vertical direction. Though the simple linearly-graded model profile shows deviations from reality for extremely short L_{gate}, agreement is in general good as demonstrated in Fig. 6. For $L_{\text{gate}} \leq 0.3\mu m$, the pockets from the source side and the drain side overlap, and the channel concentration eventually becomes even higher than N_{subp}. This is also included in our model by adding two pocket concentrations according to the strength of their overlap.

The model describes the lateral distribution of the pocket profile as shown in Fig. 2. However, the vertical profile plays also a role for the V_{th} characteristics. The vertical profile can be determined in addition by adjusting measured V_{th} *vs.* $\sqrt{\phi_{\text{th}} - V_{\text{bs}}}$ characteristics as demonstrated in Fig. 7 [4]. For reduced L_{gate}, deviation of calculated V_{th} *vs.* $\sqrt{\phi_{\text{th}} - V_{\text{bs}}}$ becomes obvious. The reason is the impurity-concentration increase of the pocket profile in the direction vertical to the channel. We use this deviation to extract the vertical profile for obtaining an accurate impurity concentration below the gate oxide, which is important for the on-current characteristics.

4 Conclusion

We have developed a new V_{th} model for both circuit simulation and pocket-profile optimization. The basic idea is to introduce an average carrier concentration to determine V_{th}. This concentration is well defined by the maximum concentration of the pocket and its penetration length into the channel. The model reproduces measured V_{th} within an accuracy of a few mV.

References

[1] Y. Taur and E. J. Nowak, IEDM Tech. Dig. p215, 1997.
[2] S. Sze, "Physics of Semiconductor Device", Wiley, 1981.
[3] M. Miura-Mattausch et al., IEEE Trans. CAD/ICAS, vol. 15, 1, 1996.
[4] M. Suetake et al., Proc. SISPAD, p207, 1999.

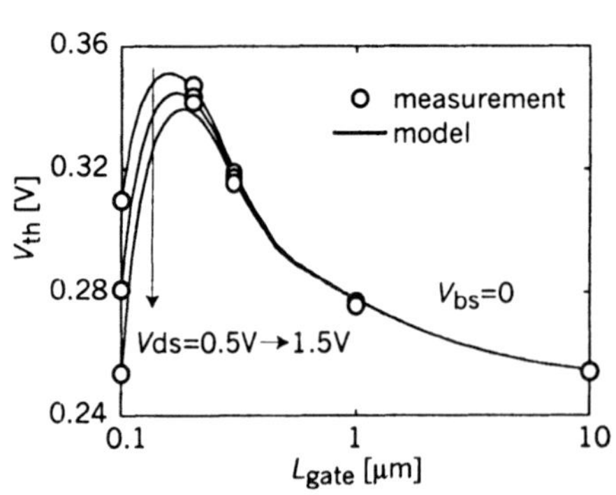

Fig. 1. V_{th} vs. L_{gate} for a pocket implant technology. Symbols are measurements and solid lines are model results.

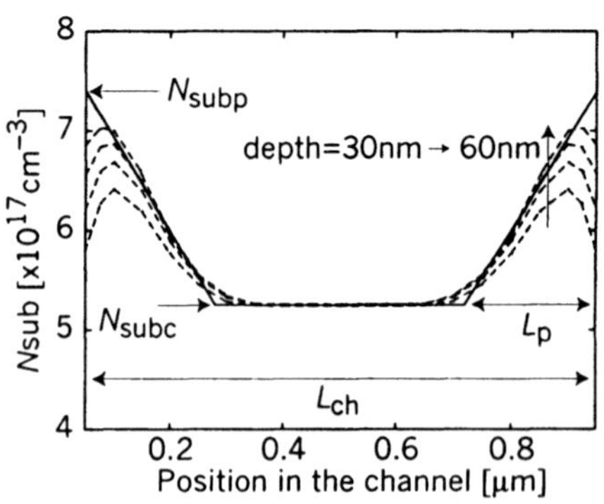

Fig. 2. Extracted pocket-profile model depicted by a thick solid line. The dashed curves are TSUPREM results for various depths.

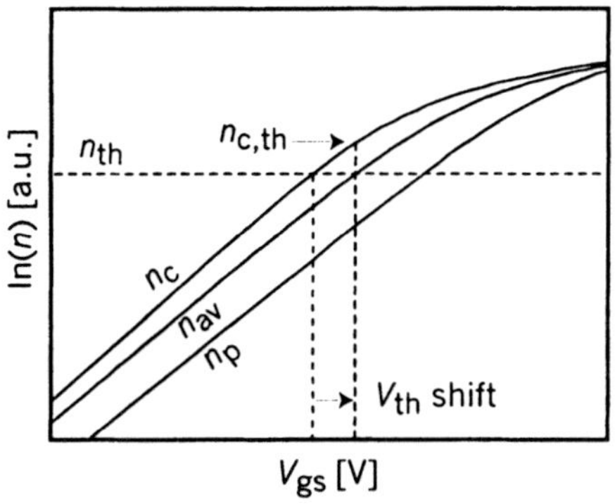

Fig. 3. Simulated carrier concentrations as a function of the gate voltage V_{gs} for two homogeneous substrate concentrations without pocket implantation (n_c and n_p) and with the pocket (n_{av}).

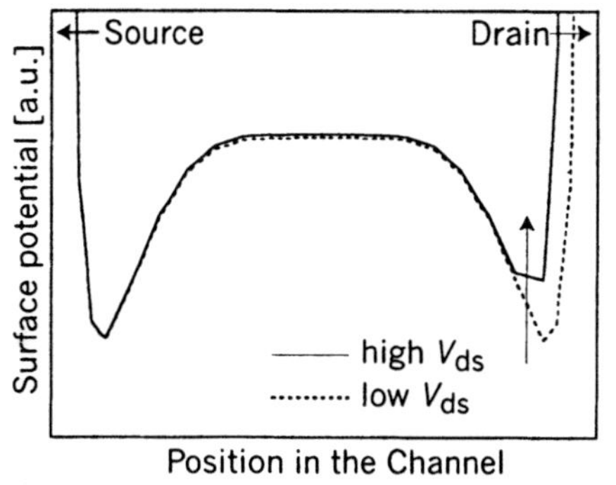

Fig. 4. Simulated surface potential distribution along the channel for two V_{ds} values.

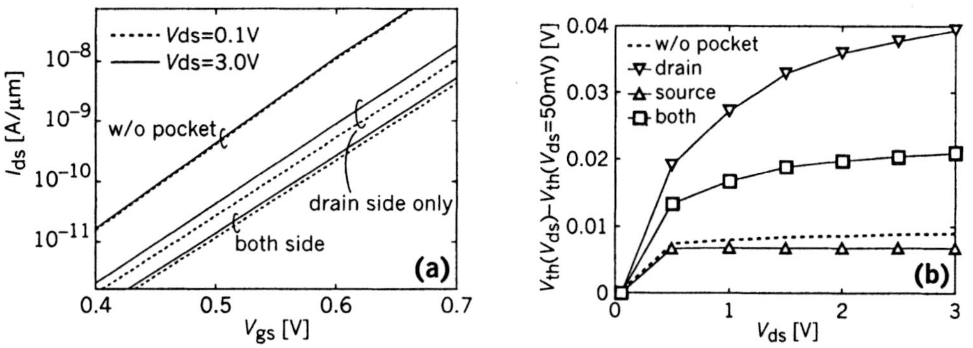

Fig. 5. (a) Simulated I-V characteristics of $L_{gate} = 1.0\mu m$ with MEDICI for different substrate conditions, without the pocket implantation, the pocket only at the drain side, and at the both sides. (b) Simulated V_{th} shift from V_{th} at $V_{ds} = 50mV$ as a function of the drain voltage V_{ds}.

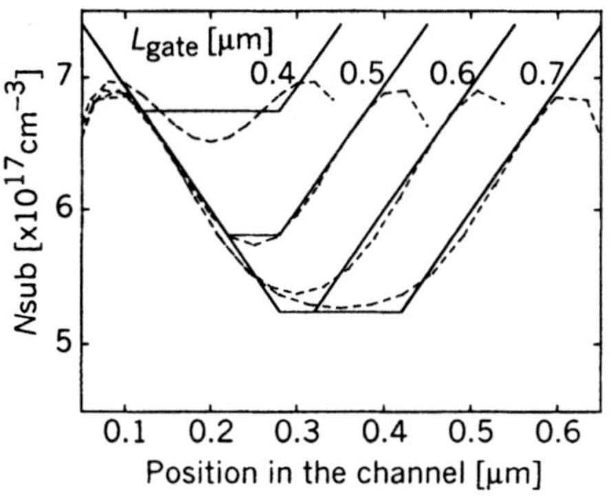

Fig. 6. Extracted pocket profile along the channel for different L_{gate} lengths (solid line). Dashed curves are results with TSUPREM at the depth of 50nm.

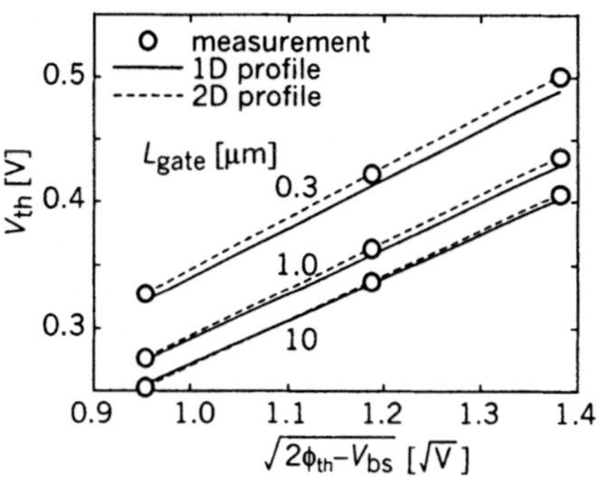

Fig. 7. Comparison of calculated V_{th} body coefficients. Symbols are measurements and thick solid lines are results with the pocket-profile model shown in Fig. 6. Dashed lines are results with the 2D pocket profile including the gradient to the depth direction.

Electrothermal Device Simulation of an ESD Protection Structure Based on Bipolar DC Characteristics

A. Icaza Deckelmann, G. Wachutka,
Institute for Physics of Electrotechnology,
Munich University of Technology, D-80290 Munich, Germany;
Phone: ++49-89-289-23103 Fax: ++49-89-23134 mailto:icaza@tep.ei.tum.de

G. Groos, W. Kanert,
Infineon Technologies
Balanstr. 73, D-81541 Munich, Germany;
Phone: ++49-89-234-27615, Fax: ++49-89-234-712016
mailto:Gerhard.Groos@infineon.com

Abstract

The high current characteristics of an ESD protection structure fabricated in a smart power technology is calculated by the use of electrothermal device simulation. We find reasonable agreement with experimental results after careful calibration of the technology-dependent transport parameters. The still existing differences between measurement and simulation results are discussed, together with some conclusions for the application of our simulation strategy in future.

1 Introduction

Nowadays ESD protection structures are conceived as a must in view of the high reliability requirements of integrated circuits. Several protection principles are known and widely used [1]. However, the development of devices matching the protection needs for a given technology and application field still relies mostly on expensive and time-consuming experiments. One of the reasons is that numerical simulations are based on physical model parameters which require a very careful calibration for the consistent and reliable modeling and prediction of high current characteristics of protection devices. In this paper we demonstrate the simulation of an ESD protection structure fabricated by means of an industrial

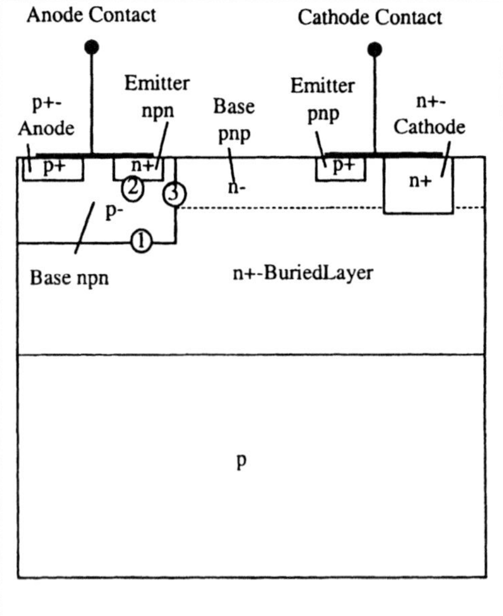

Fig. 1 Schematic view of the ESD protection structure investigated

smart power technology. The physical parameters are extracted by comparing DC measurements with isothermal simulations of DC characteristics. Yet, also the transient high current behavior is reasonably reproduced using these parameters.
A schematic view of the device under investigation is drawn in Fig. 1. The anode contact is assumed to be at ground potential during ESD stress, while the cathode contact is connected to the discharge path. The discharge through the device forces the cathode voltage to rise rapidly, thus driving junction 1 into avalanche breakdown. As a consequence of the strong increase of the hole density in the p^--base, junction 2 gets forward-biased, allowing the vertical npn-transistor (with the n^+-buried layer as collector) to drive an increasing amount of the device current. The cathode voltage is reduced drastically with increasing transistor current. This snap-back behavior is observed until the terminal voltage attains its minimum value (holding voltage V_H). The lateral pnp-transistor also conducts part of the current, and, thus aids to further reducing the terminal voltage. Once V_H is reached, the terminal voltage rises again with increasing current due to high injection.

2 Transport Parameters of the Simulation Model

The simulations were performed using the commercial device simulator DESSIS. Two devices were selected for calibration: a quasi-1D reverse-biased pn-diode and a cylindrical npn-bipolar transistor operated in active mode. Both device types play a fundamental role in the function of the full protection structure, as explained in section 1.
The breakdown voltage V_{BD} of the pn-diode was measured for 233K, 298K and 423K, in order to adjust the coefficients controlling the temperature dependence of the impact ionization coefficients according to Chynoweth's law (see e.g. [3]). After calibration, the simulations conform well to the experiment, as it is shown in Table 1.
The transistor characteristics $I_c(V_e)$ and $I_b(V_e)$ as well as the current gain $\beta(I_c)$ were measured at room temperature with base and collector at fixed potentials and varying emitter voltage V_e. Low current conditions (small V_e) in the base and collector current characteristics were used to calibrate the lifetimes in the Schockley-Read-Hall-model and the carrier mobilities (see Fig. 2). For high current conditions (large V_e) the static current gain $\beta=I_c/I_b$, shown in Fig. 3 as a function of collector current I_c, is correctly reproduced after calibration of the model parameters for trap-assisted- and band-band-Auger recombination, and for bandgap narrowing. The latter quantities have a very sensitive influence on the device characteristics because of the exponential dependence of the carrier concentrations on the bandgap energy. Proper adjustment of these parameters yield excellent agreement of the simulation results with the measured data (Fig. 3.).

Table 1 Comparison between measured and simulated breakdown voltages

T	*233K*	*298K*	*423K*
V_{BD}, *measured*	41,1	43,6	47,1
V_{BD}, *simulated*	43	44	47

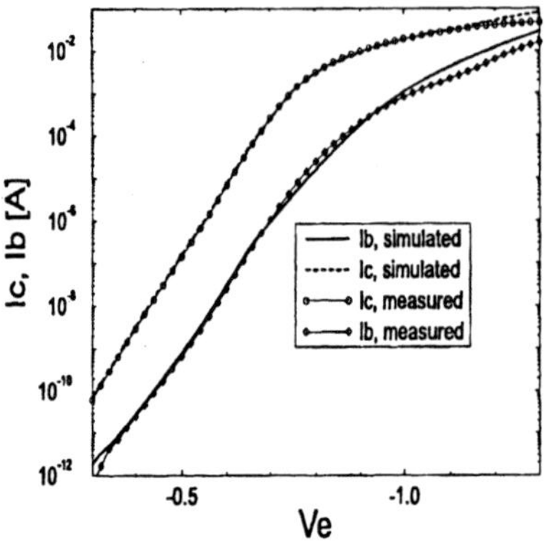

Fig. 2 Collector and base current vs. emitter voltage for the bipolar transistor

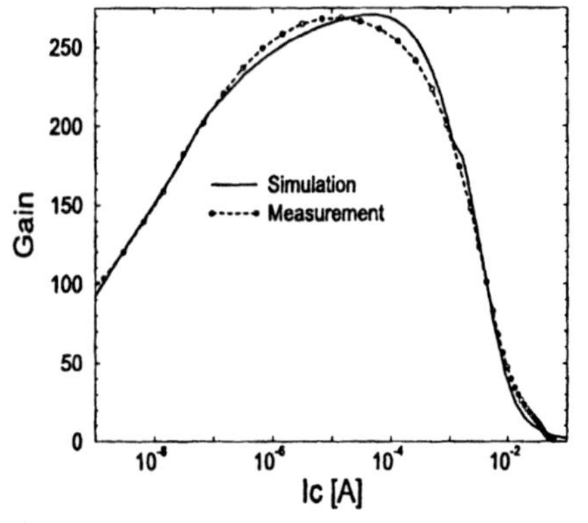

Fig. 3 Current gain vs. collector current for the bipolar transistor

3 Measurement and Simulation of High Current Characteristics

Experimentally, the I-V-curve is determined by forcing rectangular current pulses through the device, with varying current height but fixed duration of 100ns (transmission line pulse, TLP). For a given current, the voltage measured between the terminals is averaged, thus defining one point of the high current characteristics.

For the simulation of these measurements, the current pulse through the device is simulated using the drift-diffusion approach extended by the heat transport equation, in order to account for the self-heating of the device. Fig. 4 shows the calculated high current characteristics. The simulated holding voltage of 30V is significantly larger than the experimental value of 22 V.

Two 'hot spots' are observed in the simulation, which are located near the base-collector junction of both the npn- and pnp-transistors (junctions 1 and 3 in Fig.1). The maximum temperature is 585 K at a current level of 1 A.

4 Discussion

Compared to the calibration strategy pursued in [4] and [5], we followed a more comprehensive approach by also using the IV characteristics and current gain of the npn bipolar transistor for calibration. But, compared to those results, the agreement of TLP simulations and measurements is not as good. The breakdown voltage comes out correctly in the simulation, while the holding voltage is overestimated, as can be concluded from Fig. 4. Since the ESD protection device is primarily operated as a bipolar transistor in the high current range, and the simulation had been successfully calibrated with reference to such a device, one should have expected a better agreement.

The simulation of the current pulses also reveals that the lattice temperature rises up to 585K inside the device at 1 A. As the calibration was performed for room temperature, the temperature dependence of the physical parameters is possibly not correctly incorporated in the models used. When the npn-transistor reaches the high-injection regime, the free carriers in the space-charge region (SCR) at the collector-base junction cannot be neglected anymore. Considering the npn-transistor in only one dimension, the collector current density can then be calculated by means of

$$j_C = \mu_e \cdot n_C \cdot E_C, \qquad (1)$$

while for the emitter current density,

$$j_E = \frac{2 \cdot q \cdot D_a \cdot n_i}{W_B} \cdot \exp\left(\frac{V_{BE}}{2 \cdot k \cdot T/q}\right) \qquad (2)$$

holds if $W_B \ll L_a$ (see e.g. [2]). Here D_a and L_a denote the ambipolar diffusion coefficient and diffusion length, E_C and n_C are electric field and electron density in the collector-base SCR, respectively, and W_B is the effective base width).

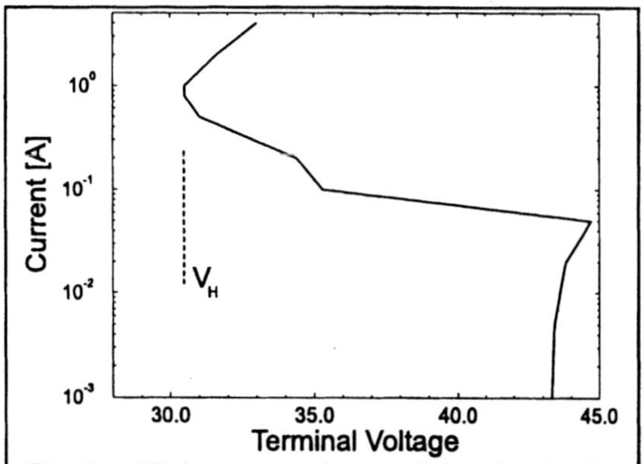

Fig. 4 High current characteristics for the investigated device. The simulated holding voltage V_H is indicated.

These equations reveal that a number of transport parameters account for the overall device behavior, and that the local device temperature, as well as manufacturing processes like implantation and annealing (influencing e.g. the recombination rates) may cause strong variations of those parameters in space and time. This is why careful calibration of all model parameters is so important to get the possibility of predictive device simulation.

Our results show that the behavior of ESD-structures in the high current regime cannot be represented by two separated devices, namely a reverse biased diode and an operated BJT. The observed discrepancies have to be investigated in more detail to identify their origin and to develop test structures for the extraction of the relevant model parameters. These investigations have to include the temperature dependence of the relevant model parameters up to the temperatures that arise in these devices under high current stress conditions.

References

[1] A. Amerasekera and Ch. Duvvury: ESD in Silicon Integrated Circuits; J. Wiley & Sons, Chichester (1995)

[2] D.J. Roulston: Bipolar Semiconductor Devices; McGraw-Hill, New York (1990)

[3] S. Selberherr: Analysis and Simulation of Semiconductor Devices; Springer-Verlag, Wien (1984)

[4] K. Esmark et al.: Simulation and experimental study of temperature distribution during ESD stress in smart-power technology ESD protection structures; Proc. IRPS 2000, pp. 304-9

[5] K. Esmark et al.: Advanced 2D/3D ESD Device Simulation – A Powerful Tool Already Used in a Pre-Si Phase; Proc. EOS/ESD 2000, pp. 420-9

Characterization of Low-Frequency Noise of MOSFETs Using 2-D Device Simulator

Hyunchul Nah, Young June Park, and Hong-Shick Min
School of EECS, Seoul National University, Seoul, Korea
woods@amity.snu.ac.kr

Chanho Lee
School of EE, Soongsil University, Seoul, Korea

Hyungsoon Shin
School of IEE, Ewha Womans University, Seoul, Korea

Abstract

Low-frequency noise of the MOSFET was simulated using the transfer impedance method together with a 2-D device simulator. In the simulation, various generation-recombination (GR) components such as SRH recombination centers and surface traps of various lifetimes are taken into account. It is shown that lifetimes of traps determine the characteristics of low-frequency noise. Also is shown that the GR process not only at the surface but also in the bulk may be responsible for the low-frequency noise.

1 Introduction

As the MOSFET becomes an important active device in the RF applications, the prediction of the noise characteristics is a challenging issue. Recently, a noise simulation algorithm using the PDE based 2-D device simulator has been developed [1]. It is based on the transfer impedance method [2], in which the Langevin type noise sources for both diffusion and GR noise are considered as the fundamental forcing terms of the fluctuation. In the previous papers, the simulation results of the high-frequency noise of the MOSFET [1] and the GR noise of simple resistors [3] have been reported

The low-frequency noise, however, has been rarely mentioned although it has been known to be the major cause of the various phase noises in the oscillators or mixers. Very often the low-frequency noise spectra of devices can be explained in terms of various GR noise components. It is known that the low-frequency noise characteristics can be properly predicted if all the fluctuation mechanisms of the carrier population in a device are clarified.

In this paper, GR noise sources with various lifetimes and locations in the device are considered in the framework of transfer impedance method implemented in the SNU-2D [4] to predict the comprehensive characteristics of the low-frequency noise of MOSFET devices.

2 Theory

The physical model considered as a basis for the noise analysis is the bipolar drift-diffusion model and the trap continuity equation together with the appropriate boundary conditions [1];

$$\nabla^2\varphi + \frac{q}{\varepsilon}(p-n+C) + \frac{\rho_t}{\varepsilon} = 0$$

$$\frac{\partial n}{\partial t} - \frac{1}{q}\nabla \bullet \mathbf{J}_n + R_n - G_n = \gamma_n$$

$$\frac{\partial p}{\partial t} + \frac{1}{q}\nabla \bullet \mathbf{J}_p + R_p - G_p = \gamma_p$$

$$\frac{\partial n_t}{\partial t} = R_n - G_n - R_p + G_p + \gamma_t,$$

where the symbols are given their usual meanings and n_t and γs indicate the electron density in trap states and the corresponding Langevin forcing terms, respectively.

The last equation can be back-substituted into the first three equations using the relations between n_t and ρ_t, R_n, G_n, R_p, and G_p [5]. With these equations (the Poisson, and electron and hole continuity equation), the noise formulae can be established by the transfer impedance method [1][3] to obtain the Green function describing the coupling between the distributed noise sources and the device terminals. The fluctuation sources in the resulting equations are transformed to

$$\tilde{s}_\varphi = \Gamma_{\varphi t}\tilde{\gamma}_t, \quad \tilde{s}_n = \tilde{\gamma}_n + \Gamma_{nt}\tilde{\gamma}_t, \quad \text{and} \quad \tilde{s}_p = \tilde{\gamma}_p + \Gamma_{pt}\tilde{\gamma}_t,$$

where subscript φ, n, and p represent the Poisson and electron and hole continuity equation. The expressions for the Γ's can be found in [3].

In the consideration of the multi-trap, we have assumed that the traps are independent to other traps, *i.e.* there are no direct carrier transitions between traps. To consider the silicon-oxide interface where the very fast (dominant in DC) and the very slow (believed to be the cause of $1/f$ noise) traps coexist, the interface states with different surface recombination velocities has been included.

The cross power spectral density between the open circuit voltage δv_i and δv_j at terminal i and j is given by [1][3]

$$S_{\delta v_i,\delta v_j} = \sum_{\alpha,\beta=\varphi,n,p} \int_\Omega \tilde{G}_\alpha(\mathbf{r}_i,\mathbf{r},\omega) K_{\tilde{s}_\alpha,\tilde{s}_\beta}(\mathbf{r},\omega) \tilde{G}^*_\beta(\mathbf{r}_j,\mathbf{r},\omega) d\mathbf{r},$$

where G and K are the scalar transfer impedance and the microscopic local noise source, respectively.

The microscopic local noise source (K) can be obtained from the equations for s_φ, s_n, and s_p [3] whose expressions are not repeated here.

The correlation matrix of the noise current power spectrum can be obtained from the calculated correlation matrix of the noise voltage power spectrum and the AC admittance matrix $\mathbf{Y}$ as:

$$\mathbf{S}_I = \mathbf{Y}\mathbf{S}_V\mathbf{Y}^\dagger,$$

where $\mathbf{Y}^\dagger$ is the complex conjugate of **Y**.

3 Simulation results and discussions

Using the fore-mentioned scheme, an n-channel MOSFET was simulated. Its structure has 0.8 μm gate length with 200 Å gate oxide thickness. The threshold voltage of the device is about 0.3 V.

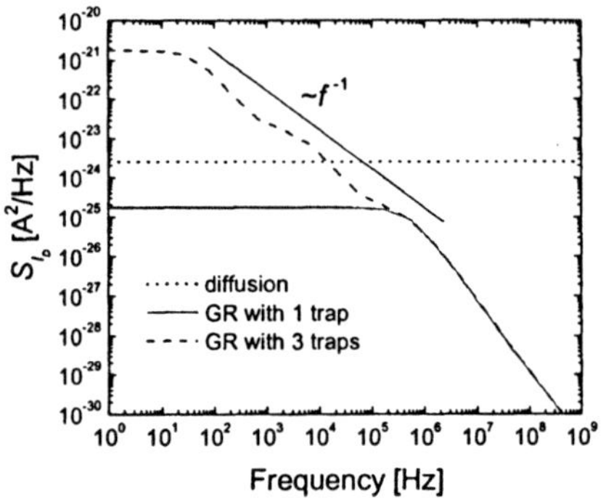

Fig. 1. The simulated noise spectrum of n-channel MOSFET at $V_D = V_G = 3$ V. The solid line is from one-trap with $\tau_n = \tau_p = 10^{-7}$ s. The dashed line is three-traps with 10^{-7} s, 10^{-5} s and 10^{-3} s

Fig. 1 shows the simulated power spectrum of the drain noise current (S_{id}) at $V_G = V_D = 3$ V. The solid line represents the GR noise due to the single trap whose lifetimes for electron and hole and the trap density are 10^{-7} s and 10^{11} cm^{-3}, respectively. Generally, there exist traps with various lifetimes in the device. Usually, the only fastest trap is taken into account when the DC characteristics of the device is to be obtained, because it dominates the GR rate. For the AC and noise behaviour, however, the slower traps can be no longer ignored. The dashed line in the figure shows the data including three kind of traps whose lifetimes are 10^{-3} s, 10^{-5} s and 10^{-7} s. The spectrum can be considered as the sum of three Lorenzians due to each traps and the envelope roughly looks like $1/f$ spectrum as expected. It can be seen that the corner frequency which is an important parameter of $1/f$ noise can properly predicted by the careful decision of the trap parameters.

In the above figure, no surface recombination has been considered; only bulk recombination gives the $1/f$ spectrum. Apparently, there are rare probability for very slow traps (*ex.* trap of lifetime of 1 s) to exist in bulk [6]. So the measured $1/f$ spectrum having $1/f$ characteristics at very low frequency may be the sum of bulk GR noise (intermediate frequency) and the surface GR noise (very low frequency).

The bias dependence of $1/f$ noise is shown in Fig. 2. Though the accurate level of the noise spectrum cannot be predicted, the bias dependency of the simulated results are reasonable. Many experiments report that in the strong inversion and linear region, the $1/f$ spectrum is almost constant along with the gate bias (left) [7]. The right figure of Fig. 2 is the V_D dependence of the $1/f$ noise. S_{ID}/I_D^2 is expected to be nearly flat in the linear region, increase in between the linear and saturation region and become flat again in the saturation region [8]. The simulated result basically follows the trend of the measurement but the point at which the spectrum flattens rather deviates from the saturation bias point, whose reason is the topic of the further work.

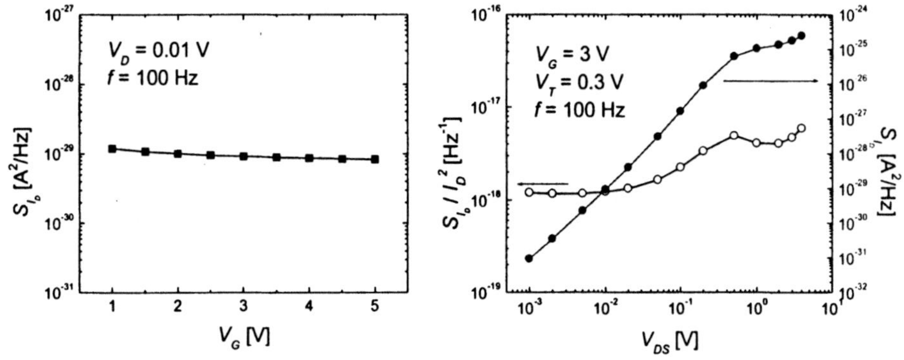

Fig. 2. V_G (left) and V_D (right) dependence of $1/f$ noise.

Acknowledgement

This work was supported by the National Research Laboratory Project of the Ministry of Science and Technology, a Collaborative Project for Excellence in Basic System IC Technology, and Brain Korea 21 Project.

References

[1] F. Bonani, G. Ghione, M. R. Pinto and R. K. Smith, *IEEE Trans. Electron Devices*, vol. 45, no. 1, p. 261, 1998

[2] C. M. van Vliet, *IEEE Trans. Electron Devices*, vol. 41, no. 11, p. 1902, 1994

[3] F. Bonani and G. Ghione, *Solid-State Electronics*, vol. 43, p. 285, 1999

[4] W. S. Choi, J. K. Ahn, Y. J. Park, H. S. Min and C. G. Hwang, *IEEE Trans. CAD*, vol. 13, p. 899, 1994

[5] A. Gnudi, P. Ciampolini, R. Guerrieri, M. Rudan and G Baccarani, *Proceeding of NASCODE V*, p. 207, 1987

[6] A. van der Ziel, *Noise in Solid State Devices and Circuits*, John Wiley and Sons, 1986

[7] C. H. Park and Y. J. Park, *Solid-State Electronics*, vol. 44, p. 2053, 2000

[8] C. H. Park, H. S. Min and Y. J. Park, *Proceedings of ICNF 2001,* to be published

Electron Beam Lithography Simulation for Subquartermicron and High Density Patterns

Ioannis Raptis, Nikos Glezos
Institute of Microelectronics, NCSR "DEMOKRITOS"
15310 Ag. Paraskevi Attikis, GREECE
raptis@imel.demokritos.gr

Abstract

A fast simulation tool for electron beam lithography is applied for the prediction of energy deposition and the resist profile of high resolution patterns. For the exposure part, an analytical solution based on the Boltzmann transport equation is used. The energy deposition has been combined with analytical functions describing the resist development and complete simulation of dense layouts in the sub-quarter-micron range has been carried out. The simulation results have been compared with experimental ones and found in very good agreement.

1 Introduction

Electron beam lithography (EBL) has been used intensively for the development of prototype devices and for the exploration of future devices with very small critical dimensions (nanotechnology). Additionally, EBL plays an important role in mask making processes for ICs with critical dimension in the sub 0.25μm range. In the future, its use is expected to increase through the application of EBL tools in mass production in mix and match lithography schemes and in nanotechnology. Therefore the understanding of electron beam interactions with matter and the physicochemical changes associated with these interactions are of vital importance for the optimization of the whole lithographic process.

Lithography modeling has been proved to be an adaptable tool in the efforts for the improvement of optical and e-beam lithography over the last years. However since the resolution required in mask making was not so challenging in the past, the main target of EBL simulation tools was the correction of the proximity effect. In this work, simulation results for thin resist layers for a wide range of e-beam energies using a novel electron beam lithography simulator [1] are presented and compared with experimental ones. As it will be proved, the proposed simulator is capable for simulation of dense patterns with critical dimension in the subquartermicron range.

2 Structure of the simulation tool

The simulation tool is structured in 4 modules (Figure 1). In the first module the energy deposition EDF(r) in the resist film due to a "point" beam electron source is calculated [2]. EDF(r) is the fundamental quantity for EBL simulation since it is independent of the layout to be exposed, the beam shape, dimensions and exposure strategy. Thus, the EDF(r) can be applied in all simulation cases where the beam energy and the composition of the resist and substrate films are the same.

The simulation tool is structured in 4 modules (Figure 1). In the first module the energy deposition EDF(r) in the resist film due to a "point" beam electron source is calculated [2]. EDF(r) is the fundamental quantity for EBL simulation since it is independent of the layout to be exposed, the beam shape, dimensions and exposure strategy. Thus, the EDF(r) can be applied in all simulation cases where the beam energy and the composition of the resist and substrate films are the same.

In the second module the beam shape and dimensions are taken into account (Single Pixel Exposure, SEDF(r)) through appropriate convolution of the EDF(r). In the third module the SEDF(r) is convolved with the actual layout LEDF(r). This module is the most CPU time consuming of the whole simulation procedure, since the energy deposition to a specific point depends strongly on the layout mainly through the contribution of backscattered electrons. The influence range of these electrons depends strongly on the composition of the substrate and the beam energy therefore the CPU time required depends on these parameters. The LEDF(r) is used as input for the simulation of resist development (fourth module) where an appropriate development model and development rate is applied and the final resist profile in two and three dimensions is calculated for the assumed development conditions.

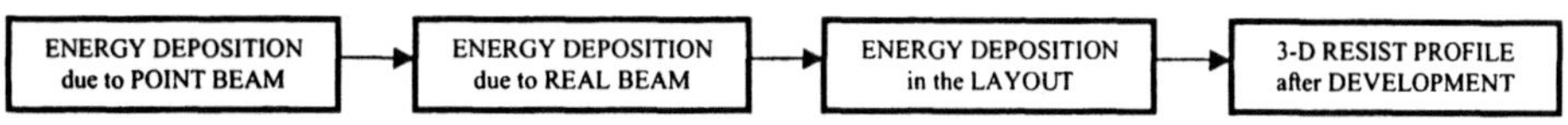

Fig. 1. Process flowchart of EBL simulation tool in the case of conventional resists

3 Energy deposition calculations

The energy deposition is the most important quantity for EBL simulation in accordance with the importance of aerial image in the simulation of optical lithography. The method applied in first module of this simulation tool is based on the Boltzmann transport equation and it is well documented in the literature [e.g 1-2]. This method can be used for the calculation of EDF(r) in thin resist films on homogeneous substrates (i.e. bulk Si) and on substrates consisting of horizontal layers of different materials (multi layer). Since EDF(r) is calculated using analytical functions the required CPU time is very small allowing the use of a very fine grid. The CPU time for the calculation of EDF(r) in the case of 0.4 μm PMMA on bulk Si using 10X10 nm grid and 12 μm lateral distance at 50KeV was 26 sec using a Pentium III – 866MHz computer. The CPU time does not depend on the substrate composition (homogeneous or multi layer); this is not the case for Monte Carlo method where the existence of an interface implies the application of time consuming if–statements.

In figure 2 the EDF(r) for 20, 50, 75 keV e-beam energy in the case of 0.4 μm PMMA on bulk Si is presented. From this figure it is obvious that the higher energy causes a more uniform backscattering contribution (larger area) while the contribution from the forward scattered electrons is localized closer to the Z - axis. Due to this behavior the resolution is superior in the case of higher energies while

the required dose in this cases increases. The EDF(r) and SEDF(r) are also valuable for proximity effect correction algorithms. The energy deposition is fitted with a sum of Gaussians. The coefficients and standard deviations of these Gaussians are used as input in the proximity effect correction software.
In figure 3 the effect of resist thickness on the energy deposition is presented. At the bottom of the resist layer the standard deviation of I_f (energy deposition due to forward scattered electrons) is significantly larger compared to the according value at the top. Thus for the optimum resolution the combination of thinner resist film – higher e-beam energy – best focus beam must be used.

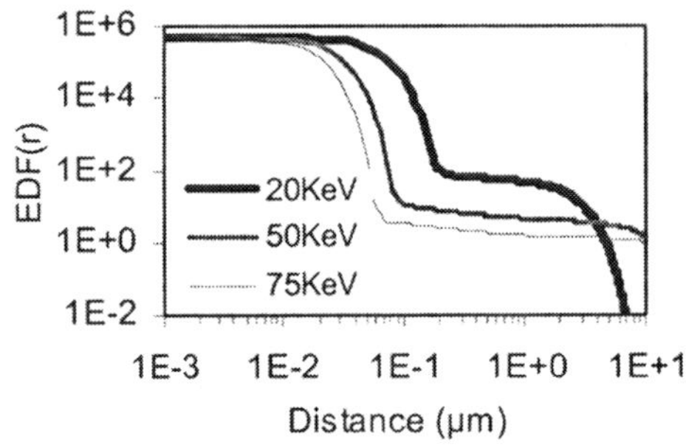

Figure 2: EDF(r) for various e-beam energies.

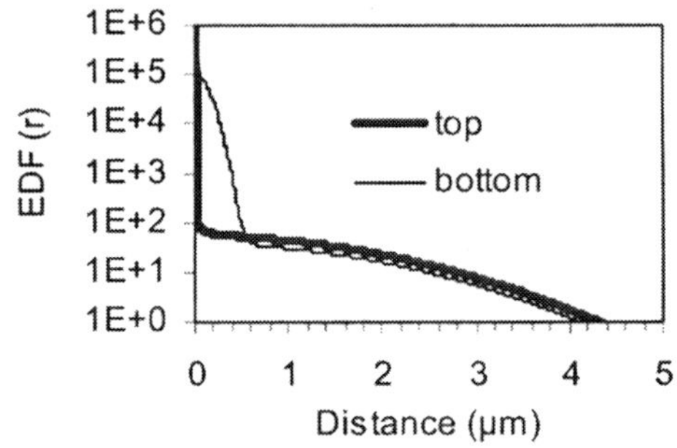

Figure 3: EDF(r) in the case of 0.4 μm PMMA on bulk Si for 50 KeV electrons.

4 Resist Development

The LEDF(r) is used for the prediction of the resist profile under specific development conditions. The EDF(r) and consequently LEDF(r), do not depend strongly on the resist material however the development mechanism of each resist is different and generally it is difficult to develop a simple function that delineates the development mechanism and can be used for the prediction of the resist profile. For instance in some specific resists the developer penetrates the resist film and causes swelling that deteriorates the resolution significantly while in other cases the development time does not affect the resist profile. Therefore for effective simulation a great effort must be given for the development of such a function.
For PMMA a simple function has been developed [3] and used successfully for a long time. In this function the development rate depends on the local energy deposition and on some other parameters that can be measured by experimental techniques. In the present work, these parameters have been calculated using a Dissolution Rate Monitor [4] and have been found different from those mentioned in the literature. More specifically the exponential factor (α) was calculated 4.5 instead of 2.0. In figure 4 the PMMA profile (0.8 μm thick) of an isolated 0.2 μm line is presented for three development times. It is clear that development time of 120 sec is not enough for proper development. On the contrary this development time is enough in the case of 0.4 μm PMMA exposed with the same dose.
By using the development rate function mentioned above, extensive simulations have been carried out for the PMMA resist at different resist thicknesses. The layout

consists of numerous very long lines with nominal linewidth (NLW) of 250 nm and 150 nm with spaces 500nm and 300nm respectively. In figure 5 the simulation and experimental results are compared. The accuracy of simulation in all cases is

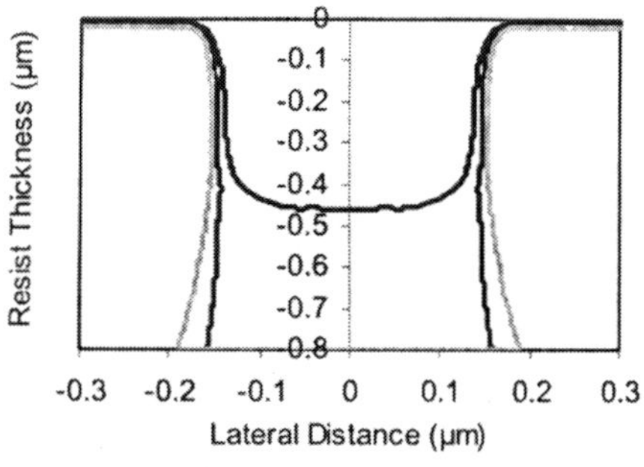

Figure 4: Resist profile for 0.2 μm isolated line in 0.8 μm thick PMMA. The development times are 120, 180, 240 sec

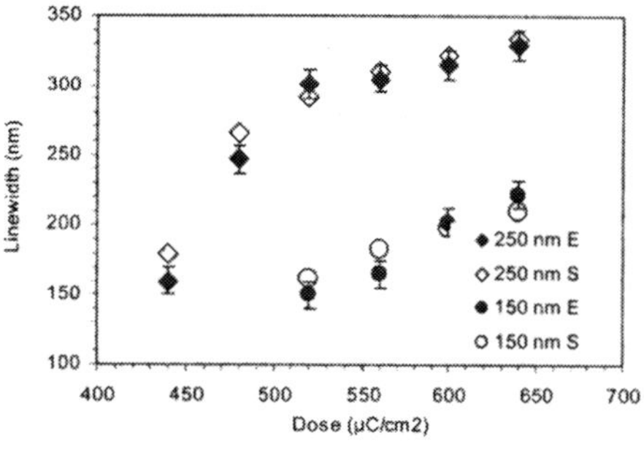

Figure 5: Comparison of experimental width simulation results for 0.4 μm PMMA on bulk Si (50 KeV)

satisfactory. The necessary exposures for the calculation of the development rate function and the high resolution patterning were performed by a Leica EBPG-3 e-beam exposure tool operational at 20 and 50 KeV. For the metrology a high resolution SEM (LEO 440) was used.

5 Conclusions

In the current work a fast electron beam lithography simulator has been applied for the simulation of dense patterns in the 100 – 250 nm NLW range at high energies using the conventional positive resist PMMA. Using that simulation tool the effects of beam energy, resist thickness, development time, exposure dose have been studied in depth in order to extract the necessary values for high resolution patterning. In order to improve the simulation's accuracy the development rate parameters have been measured experimentally and found different from those published in the literature. The validity of these values has been verified from the simulation, where the use of literature values produces results far away from the ones observed experimentally.

References

[1] Glezos N. Raptis I. *A fast electron beam lithography simulator based on Boltzmann transport equation* IEEE Trans. CAD 1996 15::92-100

[2] Raptis I. Ph.D thesis University of Athens 1996

[3] Neureuther A. Kyser D. Ting C. *E-beam resist edge profile simulation* IEEE Trans. Electron Devices 1979 26:686-691

[4] Raptis I., Velesiotis D., Vasilopoulou M., Argitis P. *Development mechanism study by dissolution monitoring of posistive methacrylate resists* Microelec. Eng. 2000 53:489-492

Green's Function Approach for Three-Dimensional Diffusion Simulation of Industrial High Voltage Applications

J. Cervenka, M. Knaipp*, A. Hössinger, and S. Selberherr

Institute for Microelectronics, TU Vienna, Gusshausstrasse 27–29, A-1040 Vienna, Austria
Phone: +43-1-58801/36029, Fax: +43-1-58801/36099, Email: Cervenka@iue.tuwien.ac.at
*Austria Mikro Systeme International AG, A-8141 Unterpremstätten, Austria

Abstract

In todays high voltage processes the optimization of process and layout design is a key point to get competitive products. Effects like punch-through between two junctions and breakdown near the surface of the wells make it necessary to analyze complex three-dimensional process steps by simulators which give accurate answers to the process engineers.

1 Introduction

In multiple well technology the relatively low dopant concentration and the high applied voltage cause large space charge regions, which results in a big waste of space and an increase of the on-resistances.

The investigated structure is part of a high voltage PMOS transistor. To optimize the two-dimensional section of the PMOS transistor, it is necessary to ensure that no other three-dimensional effects dominate the device characteristic. In the actual device the critical three-dimensional part is the area at the tip of a drain finger where the drain PTUB is placed inside a Shallow/Deep NTUB combination (SDNTUB) which is located above a p-doped substrate. In normal operation the PTUB is biased up to 90V so that the PTUB/SDNTUB junction is biased in reverse direction and a space charge fills the area between the PTUB/NTUB junction and the SDNTUB/substrate junction. The complete device area is located in a shallow NTUB. To achieve the required PTUB/NTUB reverse voltage, a DNTUB mask should enlarge the distance of the PTUB/NTUB and the SDNTUB/substrate junctions. The layout of the well masks is shown in Fig. 1. A small DNTUB finger mask is located inside the PTUB mask. The long diffusion of the NTUBs finally leads to an NTUB/substrate junction which is located outside of the PTUB mask. The three-dimensional simulation is necessary because the spherical diffusion of the DNTUB implant dilutes the DNTUB concentration in the area of the finger's tip. This leads to punch-through at low bias between the PTUB and the substrate.

2 Simulation Method

The conventional simulation procedure is to simulate the whole ion implantation process first [1] and then the three-dimensional transient diffusion [2]. Thereby both steps require a particularly refined grid to achieve the needed accuracy [3] and, therefore, the vast amount of memory and huge calculation times constitute prohibitive demands in practice.

Alternatively we have chosen a method, specially adapted to this problem. Because of the long diffusion ranges, the exact simulation of the ion implantation process could be neglected and the implanted ions were assumed as Dirac impulses only located at the

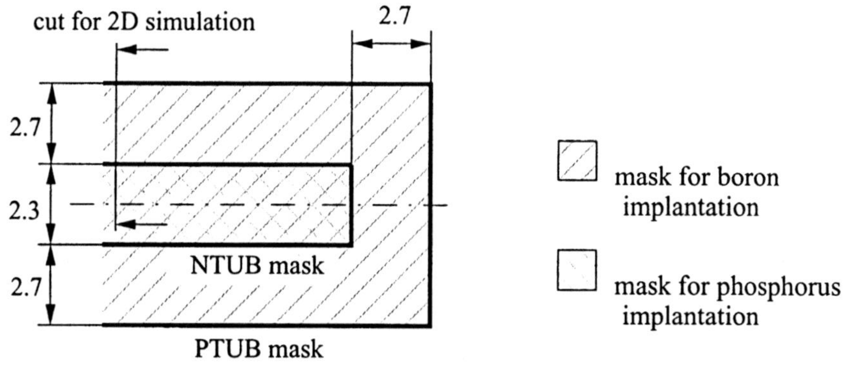

Figure 1: Layout of the well masks, units in μm

wafer top. With this simplification the final diffusion profile inside the wafer can be calculated as the sum of some partial diffusion processes, represented by the Green's function of the diffusion equation [4].

Attention must be placed to preserve the dose of the implanted ions and therefore a dose integration after implantation and after simulation must be carried out. In addition these models have to be calibrated by the two-dimensional simulation results which are available far away from the tip of the finger. For the full three-dimensional simulation a sufficiently fine grid in the areas of high diffusion gradients must be granted and therefore the simulation time was enormous, whereas for the simplified algorithm a grid is only necessary at the surface of the wafer and the resulting doping can be calculated at any point of interest. It is to note that a limitation of this method occurs, if the ranges of the implantation depth and diffusion width get in the same size.

The assessment criterion of the new layout parameters is given using the fact that the dopant concentration of the PTUB/SDNTUB junction at the surface of the wells is the same for the two-dimensional case and the three-dimensional finger case. This ensures that the breakdown at the surface in the three-dimensional structure takes place in the same voltage range as compared with the two-dimensional structure.

3 Results

The simulation results show that the spherical out-diffusion of the DNTUB is larger than expected because of the large NTUB depth. This depth is about 7.5 μm in the two-dimensional simulation. The spherical diffusion length is also in the same size from the top of the DNTUB finger to the direction of the two-dimensional case. In fact the two-dimensional situation is given when the DNTUB mask is enlarged by about 7 μm as compared to Fig. 1. This means that the DNTUB mask even can exceed the PTUB mask. However an enlargement of 7 μm would cause impact ionization near the surface's PTUB top. So the limiting case of the DNTUB enlargement is the dopant concentration of the two-dimensional simulation at the surface of the junction. This critical concentration is given when the DNTUB mask is shifted by 2 μm towards the PTUB mask. This is shown in Fig. 2, where the upper one represents the original mask layout and the lower one shows the enlarged finger behavior.

Another interesting effect is that the punch-through in the three-dimensional case does not occur directly under the symmetry line of the finger. The explanation is that the DNTUB dopant diffuses in spherical coordinates while the PTUB dopant diffuses in cylindrical coordinates. The punch current therefore has its maximum density near

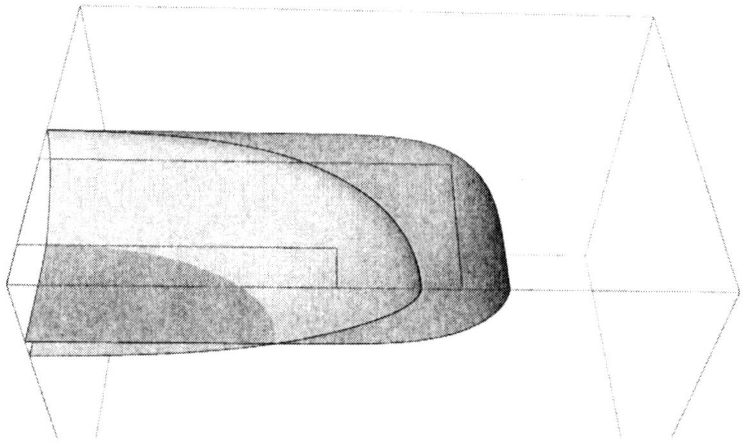

a)

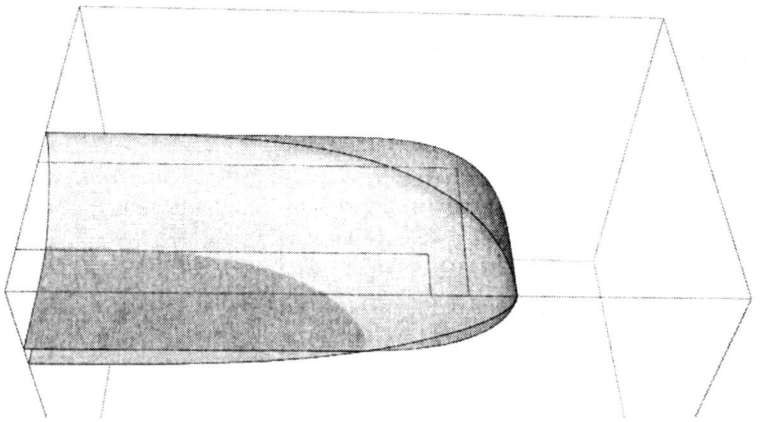

b)

Figure 2: Relevant iso-surfaces of the phosphorus and boron dopings, with a simulation domain of 15.7μm x 10μm x 10μm

the edge of the PTUB mask, confer Fig. 3, a). With the enlargement of the finger the punch-through is completely avoided as shown in Fig. 3, b).

4 Conclusion

With these careful considerations the device has been optimized to fulfill electrical strength, particularly with regard to punch-through between the junctions and breakdown by impact ionization. Without the outlined simulation methodology it would not have been possible to fully optimize the structure.

Acknowledgment

The authors acknowledge support from the "Christian Doppler Forschungsgesellschaft", Vienna, Austria.

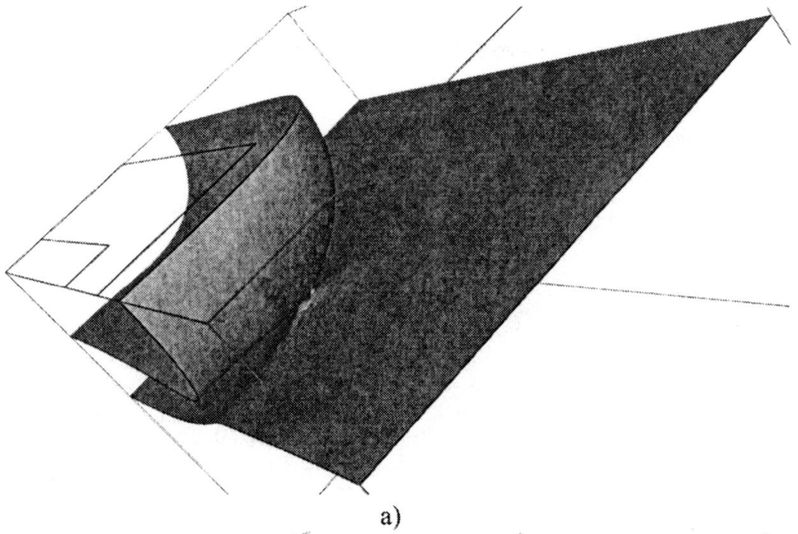

a)

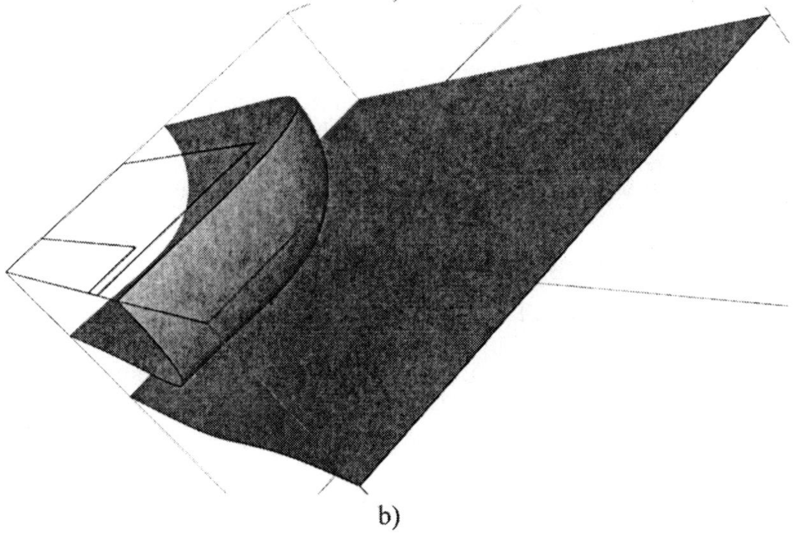

b)

Figure 3: Surfaces surrounding the space charge regions

[1] A. Hössinger, S. Selberherr, M. Kimura, I. Nomachi, and S. Kusanagi, "Three-Dimensional Monte-Carlo Ion Implantation Simulation for Molecular Ions," in *Electrochemical Society Proceedings*, vol. 99-2, pp. 18–25, 1999.

[2] M. Radi, E. Leitner, E. Hollensteiner, and S. Selberherr, "Analytical Partial Differential Equation Modeling Using AMIGOS," in *Proc. IASTED International Conference Artificial Intelligence and Soft Computing*, (Banff, Canada), pp. 423–426, July 1997.

[3] P. Fleischmann, R. Sabelka, A. Stach, R. Strasser, and S. Selberherr, "Grid Generation for Three-Dimensional Process and Device Simulation," in *Simulation of Semiconductor Processes and Devices*, (Tokyo, Japan), pp. 161–166, Business Center for Academic Societies Japan, 1996.

[4] H. Dirschmid, *Einführung in die mathematischen Methoden der theoretischen Physik*. Vieweg, 1976.

Analysis of Ultra Short MOSFETs with High-k Gate Dielectrics

K. Dragosits, Youri Ponomarev*, Charles Dachs*, and S. Selberherr

Institute for Microelectronics, TU Vienna, Gusshausstrasse 27–29, A-1040 Vienna, Austria
Phone: +43-1-58801/36018, Fax: +43-1-58801/36099, Email: Dragosits@iue.tuwien.ac.at
*Phillips Research Leuven, IMEC, Kapeldreef 75, Leuven, Belgium

Abstract

The introduction of alternative gate dielectrics into advanced CMOS devices has quite an impact on the device performance. In this work transient analyses of this type of devices are made for the first time. The analyses are made by mixed-mode simulation of a ring oscillator. It is shown that this method allows a deeper insight into the device properties than mere static analysis. Utilizing the results, first design rules for advanced CMOS devices are extracted.

1 Introduction

According to the International Technology Roadmap for Semiconductors (ITRS) [1], the gate oxide becomes one of the critical issues for short channel devices. If SiO_2 is used the proposed oxide thicknesses are getting smaller than 1nm thus leading to massive direct tunneling from the gate into the substrate. To reduce this parasitic effect alternative materials with higher k values and consequently higher oxide thickness have to be investigated.

2 Analysis of Ultra Short MOSFETs

The first basic analyses of the effects of the increased oxide thickness were carried out in recent years by investigation of the transfer characteristics of CMOS devices [2][3][4]. Basically a degradation of the device performance stemming from an increase of the parasitic capacitances was observed. In this paper these analyses of the device properties are extended to the transient behavior of ultra short devices.

2.1 Transient Simulation Approach

At first a useful criterion that allows the qualification of the transient device properties has to be found. The most interesting quantity concerning the transient behavior of any CMOS circuit is its delay time t_d. The usual definition of the delay time is the time difference between the 50% points of the input and output waveforms [5][6]. This definition should be used with care because the result strongly depends on input waveform and output load [7]. Therefore, the design of the experiment should guarantee a similar waveform to the one expected at actual circuit operation. This can be achieved with a ring oscillator, i.e., an inverter chain with an odd number n of stages, where the output of the last stage is fed back to the input of the first stage. This structure is schematically outlined in Fig. 1.

This circuit will oscillate at a frequency $f = 1/(2nt_\mathrm{d})$. Consequently the delay time reads as $f = 1/(2nf) = t_\mathrm{osc}/(2n)$, where t_osc is the period of the observed oscillation.

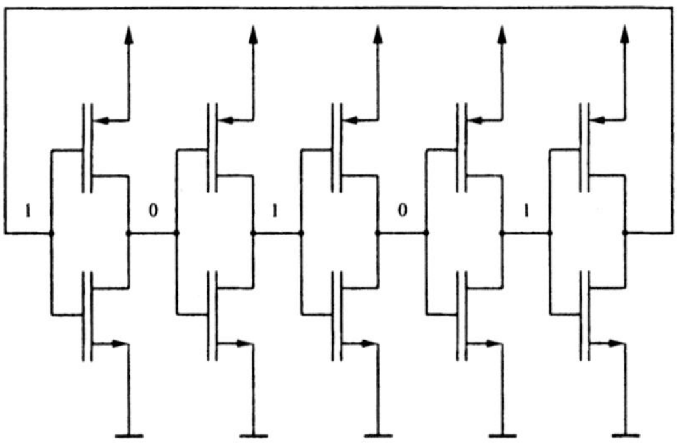

Fig. 1: Ring Oscillator with Five Stages

This experimental setup can be straightforwardly simulated utilizing the mixed-mode capabilities of MINIMOS-NT [8][9]. The ring oscillator used for the calculations is identical to the one outlined in Fig. 1 and consists of an inverter chain with five stages. The PMOS transistor of the inverter was assumed to be symmetric to the NMOS transistor.

2.2 Results

For our research advanced CMOS structures with a gate length of 50nm was chosen. Geometry and doping was chosen according to the ITRS specifications. For our analyses three different high-k device structures, all of them with an equivalent oxide thickness of 0.7nm were used. A device with SiO_2 as gate dielectric was considered as a reference. The high-k structures were chosen with respect to their simplicity in order to make the results as meaningful as possible. Structure I is the simplest of these structures, in which the high-k material completely replaces the SiO_2. Structure II is a device which has a planar high-k layer on top of the Si-substrate outlined in Fig. 2. In Structure III the high-k layer is only under the gate, as shown in Fig. 3. Typically, this structure is processed by removing the gate dielectric and subgate oxide and regrowth of the high-k material.

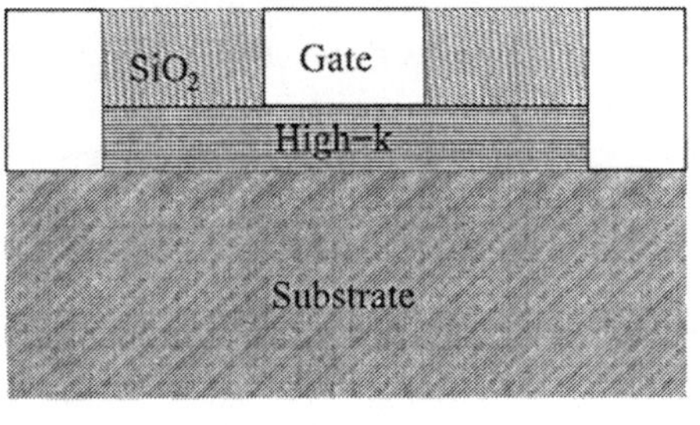

Fig. 2: Structure II

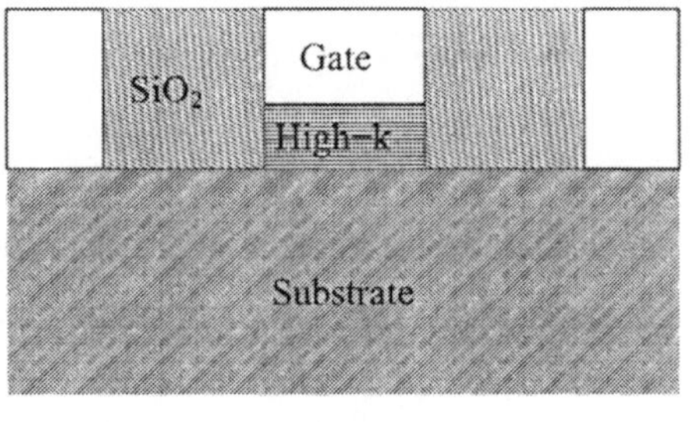

Fig. 3: Structure III

2.3 Static Analysis

At first, static analyses were carried out with Structure I, as this design shows the biggest deviation from the low-k device [4]. The different geometry of the high-k structure has a significant impact on the transfer characteristics of the device, which can be seen in Fig. 4. The increase of parasitic capacities leads to a reduction of the channel control. Depending on the k value of the high-k dielectric two effects can be observed, which will be analyzed in the following: A reduction of the slope of the transfer characteristics and a decrease of the threshold voltage.

An increase of the k-value increases the drain to channel capacitance, which adds an additional charge to the channel independently from the applied gate voltage. The additional charge causes a shift of the transfer characteristics to the left. This effect can be compensated by other measures like workfunction engineering in the gate or modification of the channel profile. So materials with an 'only-shifting' k value can be considered as alternative gate oxides.

The decrease of the slope can be interpreted as a general loss of channel control by the gate as the fringing fields and consequently the gate to drain/source capacitance increase. This effect can only be compensated by a variation of the stack geometry.

Fig. 5 shows the slope as function of k values for the IRTS device. The slope was extracted at an operating point with $V_{Gate} = 1.0V$. The degradation limit lies in the range of k = 50.

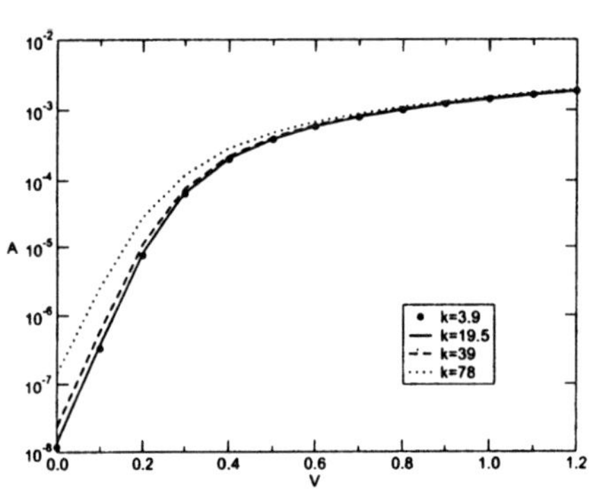

Fig. 4: Transfer Characteristics

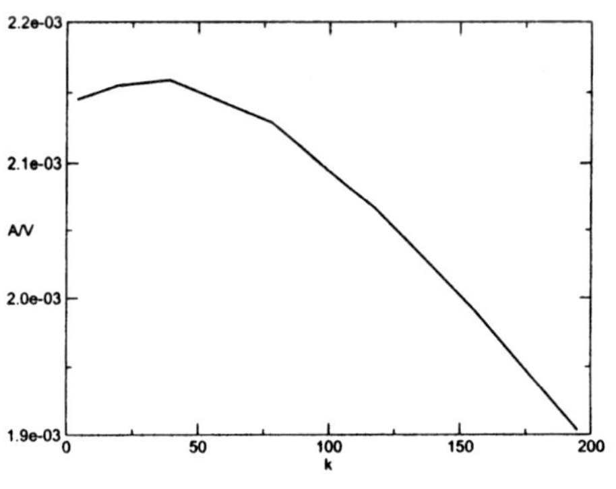

Fig. 5: Slope as Function of k-Value

2.4 Transient Analysis

The question of interest is, if the analysis of the transfer characteristics is significant for the transient device performance or not. The k-value used for the following simulation is 19.5. The deviance of the transfer characteristics is negligible in this case.

The delay time is closely linked to the contact capacities of the device, consequently an influence of the geometry on the obtained result is expected. The results show that the gate geometry has a significant impact on the delay time (Fig. 6), which increases by about 30% for Structure I. This result is remarkable as it cannot be expected from the static transfer characteristics, which showed almost no deviation between the results for

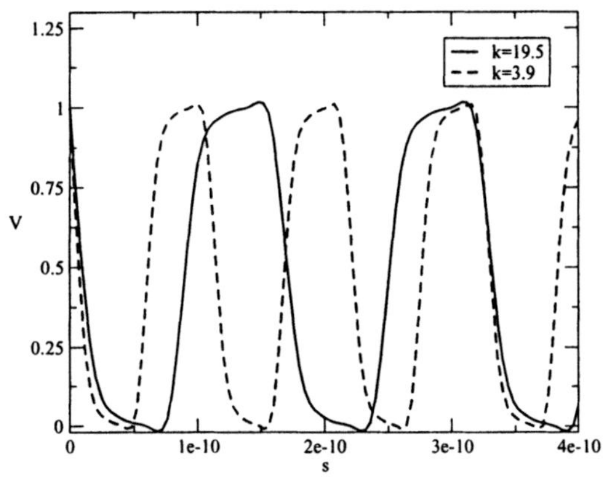

Fig. 6: Node Voltage for Different k-Values

the two devices (Fig. 4). It can be explained by an increase of the capacities between the gate contact and the drain/source doping due to the higher k-value of the gate oxide.

Due to the high computational time, the analyses for the other two structures were stopped at the first time step, but still the slope of the node voltage curve gives a clear indication of the device performance. The results of the replaced gate structure are identical to the simulations of the SiO_2 device ($\Delta V/\Delta t = 70.78\text{V}/\mu\text{s}$), but the planar structure already shows a clear decrease of the performance ($\Delta V/\Delta t = 67.82\text{V}/\mu\text{s}$), which is still comparatively high to the slope obtained for Structure I ($\Delta V/\Delta t = 51.28\text{V}/\mu\text{s}$), thus making Structure II and Structure III interesting candidates for the actual design of advanced CMOS devices.

3 Conclusion

It was shown that transient analyses offer a deeper insight into advanced CMOS devices compared to the static extraction of transfer characteristics. First analyses have been performed and it is expected that this method can serve as a powerful support for the development of future device generations.

[1] Semiconductor Industry Association, *The National Technology Roadmap for Semiconductors*. 1999.

[2] S. Krishnan, G.-F. Yeap, B. Yu, Q. Xiang, and M. Lin, "High-k Scaling of Gate Insulators: An Insightful Study," in *Proc. Conf. on Microelectronic Device Technology*, vol. 3506, pp. 65–72, SPIE, September 1998.

[3] B. Cheng, M. Cao, R. Rao, A. Inani, P. V. Voorde, W. Greene, J. Stork, Z. Yu, P. Zeitzoff, and J. Woo, "The Impact of High-k Gate Dielectrics and Metal Gate Electrodes on Sub-100nm MOSFETs," *IEEE Trans.Electron Devices*, vol. 46, no. 7, pp. 1537–1544, July 1999.

[4] A. Inani, R. V. Rao, B. Cheng, and J. Woo, "Gate Stack Architecture Analysis and Channel Engineering in Deep Sub-Micron MOSFETs," *J.Appl.Phys.*, vol. 38, no. 4B, pp. 2266–2271, April 1999.

[5] H. Bakoglu, *Circuits, Interconnections and Packaging for VLSI*. Addison-Wesley, 1990.

[6] N. Weste and K. Eshraghian, *Principles of CMOS VLSI Design*. Addison Wesley, second ed., 1993.

[7] G. Schrom, *Ultra-Low-Power CMOS Technology*. Dissertation, Technische Universität Wien, 1998. http://www.iue.tuwien.ac.at/diss/schrom/diss-new/diss.html.

[8] T. Binder, K. Dragosits, T. Grasser, R. Klima, M. Knaipp, H. Kosina, R. Mlekus, V. Palankovski, M. Rottinger, G. Schrom, S. Selberherr, and M. Stockinger, *MINIMOS-NT User's Guide*. Institut für Mikroelektronik, 1998.

[9] T. Grasser, *Mixed-Mode Device Simulation*. Dissertation, Technische Universität Wien, 1999. http://www.iue.tuwien.ac.at/diss/grasser/diss-new/diss.html.

TCAD Analysis of Gain Cell Retention Time for SRAM Applications

A. Gehring, C. Heitzinger, T. Grasser, and S. Selberherr

Institute for Microelectronics, TU Vienna, Gusshausstrasse 27–29, A-1040 Vienna, Austria
Phone: +43-1-58801/36016, Fax: +43-1-58801/36099, Email: Gehring@iue.tuwien.ac.at

Abstract

We present simulations of a recently published SRAM memory gain cell consisting of two transistors and one MOS capacitor, representing an alternative to conventional six transistor SRAMs. Inverse modeling is used to fit a given device characteristic to measurement data. To account for de-charging due to tunneling, we use a simple, non-local tunneling model and calibrate it with data from literature. By optimization, we find values for the contact voltages in the off-region at which the retention time is a maximum.

1 Introduction

To meet the demand for fast, nonvolatile memory, a further increase in the density of SRAM cells is inevitable. However, common SRAM technology still relies on the conventional six transistor cell, where downsizing is difficult. Recent papers present new approaches to increase the packing density of SRAM cells using advanced cell layouts. An example is found in [1] where a MOS capacitance is used as storage node and two access transistors serve for independent read and write operations. The schematics of such a cell is shown in Fig. 1. The contacts are denoted by WWL (write word line), WBL (write bit line), RBL (read bit line) and RWL (read word line), respectively. While it is not a real SRAM circuit because of the volatility of the charge on the storage node, it offers the possibility of non-destructive read out due to capacitive coupling of the read transistor: when the storage node is charged, a positive voltage at the RWL contact suffices to open the read transistor, leading to a high current at the sense contact RBL. The sensing current is thus delivered by the RWL contact and does not reduce the charge on the storage node. The cell can be fabricated with standard process steps and consumes much less die area as compared to a six transistor SRAM.

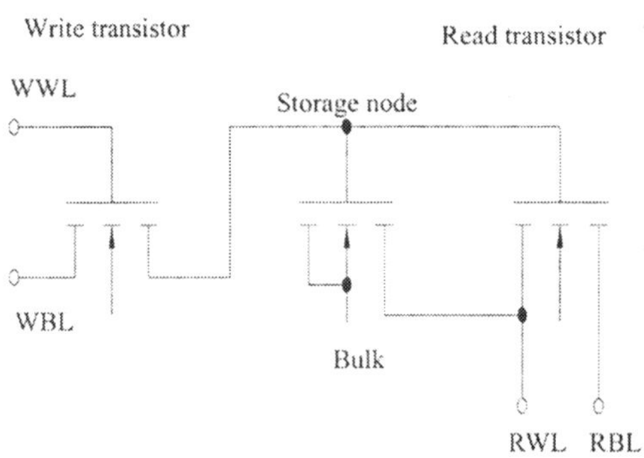

Fig. 1: Schematic of the proposed gain cell.

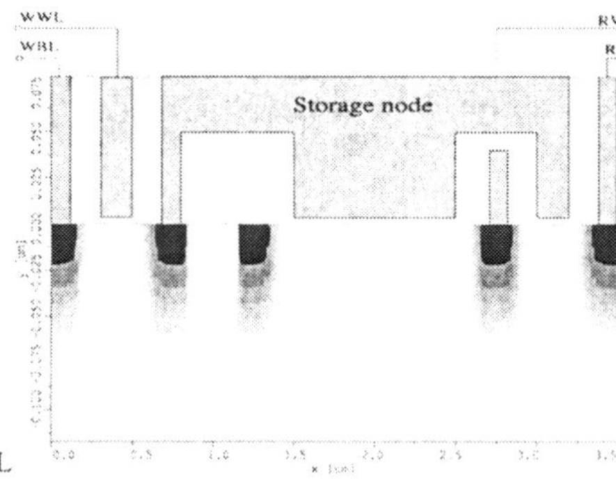

Fig. 2: Doping profile of the memory cell.

For device simulation a complete three-dimensional doping profile of the cell is not necessary. Using standard process simulation tools, single MOSFET device profiles can be used for the creation of the doping profile of the whole cell. The two-dimensional doping profile of the simulated cell is shown in Fig. 2. The doping profile was produced using the doping profile of a standard 0.2μm gate length n-type MOSFET and the process simulation tool prost2d . For such a memory cell two effects can be identified to affect the retention time: the leakage current through the access transistors and the gate tunneling current through the large gate area of the storage node.

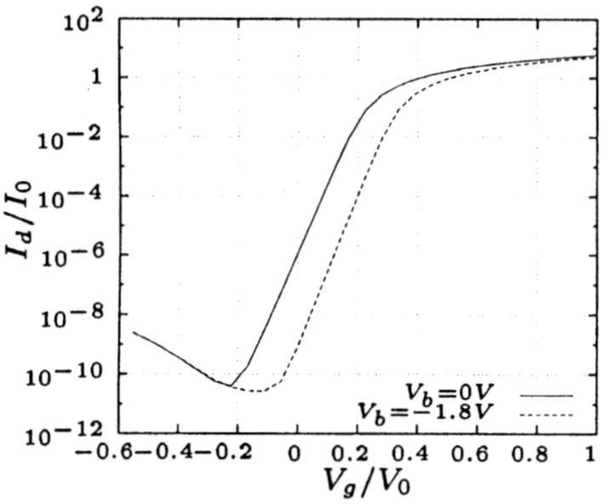

Fig. 3: Transfer chracteristics of the NMOS device.

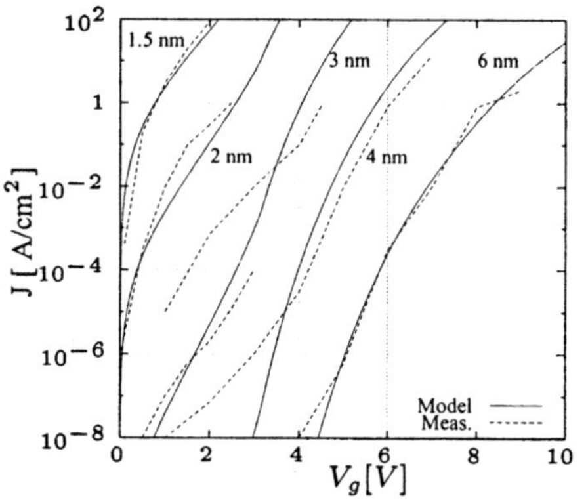

Fig. 4: Gate Current density for different gate oxide thicknesses.

2 Inverse Modeling

We used the optimization tool SIESTA [2] and the device simulator MINIMOS-NT [3] to fit the simulation models for measured transfer characteristics at different bulk biases of the NMOS device. We chose the following parameters for inverse modeling: the bulk doping for accurate modeling of the bulk voltage induced shift of the threshold voltage, the band-to-band tunneling and Shockley-Read-Hall parameters to model the drain current increase for negative gate voltages, and the work function difference. The transfer characteristics is shown in Fig. 3 for different bulk voltages, the obtained agreement to measurement data is perfect. We tried several optimization schemes (genetic optimization, gradient based optimization and simulated annealing) and achieved the best fit using simulated annealing.

3 Modeling Tunneling Currents

Due to the large size of the MOS capacitor, the storage node is de-charged by gate tunneling currents. To give some order-of-magnitude estimations, we made use of a simple non-local, electric-field based tunneling model. For low gate voltages, the SiO_2 barrier is of trapezoidal shape. Using the standard WKB approximation, the tunneling current density evaluates as

$$J = \frac{q^3 m_{ox}}{8\pi h m_0 \Phi_b} \left(\frac{V_{ox}}{t_{ox}}\right)^2 \cdot \exp\left[-\frac{4 t_{ox}\sqrt{2m_{ox}}}{3qV_{ox}\hbar}\left(\Phi_b^{3/2} - (\Phi_b - qV_{ox})^{3/2}\right)\right] \quad (1)$$

where V_{ox} is the oxide voltage, t_{ox} the gate oxide thickness and Φ_b the barrier height. The other symbols have their usual meanings. The derivation of this equation can be found in [4], a similar equation is used in [5].

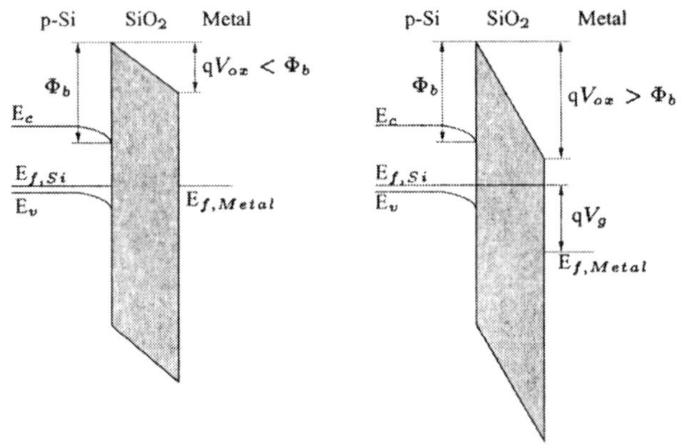

Fig. 5: Tunneling barrier for the case of direct tunneling (left) and Fowler-Nordheim tunneling (right).

As shown in Fig. 3, (1) applies for $qV_{ox} < \Phi_b >$. Otherwise, the barrier is triangular, giving rise to Fowler-Nordheim tunneling causing a gate current density of

$$J = \frac{q^3 m_{ox}}{8\pi h m_0 \Phi_b} \left(\frac{V_{ox}}{t_{ox}}\right)^2 \cdot \exp\left(-\frac{4\sqrt{2m_{ox}}}{3qF_{ox}\hbar}\Phi_b^{3/2}\right) \tag{2}$$

which is the expression usually found in literature. We used literature data to calibrate the model, with the electron mass in the oxide m_{ox} as fitting parameter and Φ_b set to 3.1 eV. In the simulation we extracted the potential values at the oxide interfaces at equidistant lateral positions beneath the gate contacts. In Fig. 4 the tunneling current densities are plotted for an oxide mass of $m_{ox} = 0.47m_0$. It shows that the model, despite of its simplicity, is accurate enough to estimate the order of magnitude of the tunneling current density over a wide range of gate voltages. The measurement values were taken from [5], [6], and [7].

4 Contact Voltage Optimization

In addition to the gate tunneling current also the voltages at the contacts in the off-region have some influence on the retention time which is caused by the leakage current of the write transistor. The leakage current of the read transistor plays a minor role since only its gate is connected to the storage node. The write transistor leakage current shows no clear dependence on the contact voltages, as shown in Fig. 6. Also the transfer characteristics in Fig. 3 justifies the assumption that there exists a minimum of the leakage current. However, since there are five independent contact voltages (WWL, WBL, RWL, RBL and the bulk contact), optimization 'by hand' becomes difficult. Thus, we again used SIESTA for optimization, employing different optimization strategies. One constraint was that all voltages in the off-region had to stay below 0.5V. For the read transistor, it turned out that positive voltages at the turned-off contacts can increase the retention time. Also, a positive bias on the bulk contact leads to a higher retention time.

5 Transient Simulation

Finally, we show results of transient simulations using the gate current model and the optimized contact voltages. In Fig. 7 the cell de-charging curves are shown for the case of optimized and not optimized contact voltages and different gate oxide thicknesses. Only with gate oxides thinner than 2nm, a significant reduction in retention time can be seen. For optimized contact voltages, the retention time can be increased by nearly three orders of magnitude. This emphasizes the need for proper chosen contact voltages when using such devices.

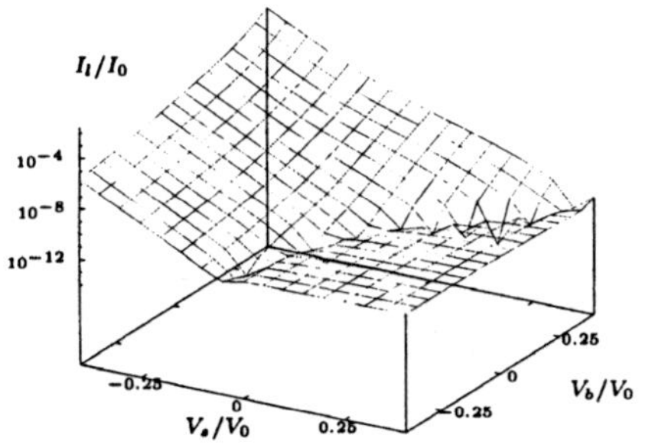

Fig. 6: Effect of source and bulk voltage on the leakage current of the write transistor.

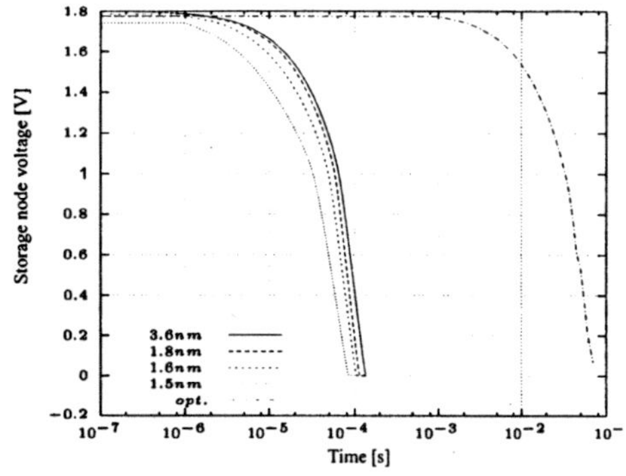

Fig. 7: De-charging curves for different oxide thicknesses (left) and optimized curve (right).

6 Conclusion

We presented simulations of a new SRAM cell consisting of two transistors and one capacitor. We used inverse modeling and optimization techniques together with rigorous device simulation to analyze the retention time of the device. We showed that, even with a very simple gate current model, measured data can be fitted to some accuracy, valuable for first estimations. However, gate current induced charge loss is crucial only for gate oxides thinner than 2nm. Of higher importance is the right choice of the contact voltages, which can increase the retention time by orders of magnitude. Without TCAD based optimization it would have been cumbersome, if possible at all, to find the right optima.

Acknowledgment

The authors acknowledge support from the "Christian Doppler Forschungsgesellschaft", Vienna, Austria.

[1] N. Ikeda, T.Terano, H. Moriya, T. Emori, and T. Kobayashi, "A Novel Logic Compatible Gain Cell with two Transistors and one Capacitor," in *Symposium on VLSI Technology Digest of Technical Papers*, pp. 168–169, 2000.

[2] C. Heitzinger and S. Selberherr, "An Extensible TCAD Optimization Framework Combining Gradient Based and Genetic Optimizers," in *Proc. SPIE International Symposium on Microelectronics and Assembly: Design, Modeling, and Simulation in Microelectronics*, (Singapore), pp. 279–289, 2000.

[3] T. Simlinger, H. Kosina, M. Rottinger, and S. Selberherr, "MINIMOS-NT: A Generic Simulator for Complex Semiconductor Devices," in *25th European Solid State Device Research Conference* (H. de Graaff and H. van Kranenburg, eds.), (Gif-sur-Yvette Cedex, France), pp. 83–86, Editions Frontieres, 1995.

[4] J. P. Shiely, *Simulation of Tunneling in MOS devices.* Dissertation, Duke University, 1999.

[5] K. F. Schuegraf, C. C. King, and C. Hu, "Ultra-thin Silicon Dioxide Leakage Current and Scaling Limit," in *Symposium on VLSI Technology Digest of Technical Papers*, pp. 18–19, 1992.

[6] N. Yang, W. K. Henson, J. R. Hauser, and J. J. Wortman, "Modeling Study of Ultrathin Gate Oxides Using Direct Tunneling Current and Capacitance-Voltage Measurements in MOS Devices," *IEEE Trans.Electron Devices*, vol. 46, July 1999.

[7] S. H. Lo, D. A. Buchanan, Y. Taur, and W. Wang, "Quantum-Mechanical Modeling of Electron Tunneling Current from the Inversion Layer of Ultra-Thin-Oxide nMOSFETs," *IEEE Trans.Electron Devices*, vol. 18, May 1997.

Optimization for TCAD Purposes Using Bernstein Polynomials

C. Heitzinger and S. Selberherr

Institute for Microelectronics, TU Vienna, Gusshausstrasse 27–29, A-1040 Vienna, Austria
Phone: +43-1-58801/36035, Fax: +43-1-58801/36099, Email: Heitzinger@iue.tuwien.ac.at

Abstract

The optimization of computationally expensive objective functions requires approximations that preserve the global properties of the function under investigation. The RSM approach of using multivariate polynomials of degree two can only preserve the local properties of a given function and is therefore not well-suited for global optimization tasks. In this paper we discuss generalized Bernstein polynomials that provide faithful approximations by converging uniformly to the given function. Apart from being useful for optimization tasks, they can also be used for solving design for manufacturability problems.

1 Introduction

Automated TCAD optimization is difficult since the evaluation of the objective function is usually very computationally expensive. There are two main approaches: the first is to optimize the given objective function, and the second is to optimize an approximation of the objective function. Both approaches are implemented in the SIESTA (Simulation Environment for Semiconductor Technology Analysis) framework [1, 2]. The second approach relies on how good an approximation was chosen, and that it can be evaluated much faster than the original objective function so that conventional optimization algorithms requiring many more evaluations can be applied.

In the RSM (response surface methodology) [3] almost exclusively polynomials of degree two (or less) are used. This method, however, suffers from the fact that there is no reason why such an approximation should preserve the global properties of the given function: the set of of all polynomials of degree two or less is not dense in $C(X)$, $X \subset \mathbb{R}^p$ compact. Moreover, evaluating the objective function at more and more points does generally not improve the RSM approximation – these evaluations are wasted. A simple example for this fact are the functions $e_\lambda : x \mapsto \mathrm{e}^{\lambda x}$ which are ubiquitious in TCAD applications. Other examples are functions containing transitions from exponential to linear behavior.

Although the RSM approach can be improved by transforming the variables before fitting the polynomials, it has to be known a priori which transformations are useful and should be considered. If this knowledge is available, it can of course be applied to other optimization approaches as well.

To overcome the shortcoming of the RSM approach, we propose using generalized Bernstein polynomials for approximating objective functions.

We also note that a good approximation resembling the global properties of the objective function can be used for solving design for manufacturability problems. Furthermore, this method of computing approximations evidently gives rise to a recursive optimization algorithm. After a first approximation either further approximations of interesting areas are computed, or – if needed – the first approximation is refined using additional points.

2 Properties of Bernstein Polynomials

In this section we discuss some important properties of (generalized) Bernstein polynomials. In order to keep the formulas simple we will concern ourselves with functions defined on the (multidimensional) intervals $[0,1] \times \cdots \times [0,1]$. Using affine transformations it is straightforward to apply the results to arbitrary intervals.

The following theorem is due to Sergei N. Bernstein.

2.1 Theorem *Let $f : [0,1] \to \mathbb{R}$ be a continuous function. Then the Bernstein polynomials*

$$B_{f,n}(x) := \sum_{k=0}^{n} f\left(\frac{k}{n}\right) \binom{n}{k} x^k (1-x)^{n-k}$$

converge uniformly to f for $n \to \infty$.

A proof can be found in [4, p. 339]. If f even satisfies a Lipschitz condition, a stronger result can be shown giving an error bound.

2.2 Theorem *If f additionally satisfies a Lipschitz condition $|f(x) - f(y)| < L|x-y|$, then the inequality*

$$|B_{f,n}(x) - f(x)| < \frac{L}{2\sqrt{n}}$$

holds.

Additional to uniform convergence, also the derivatives of the approximation converge to those of the given function.

2.3 Theorem *If f has a continuous i-th order derivative $f^{(i)}(x)$ on $(0,1)$, then $B^{(i)}_{f,n}(x)$ converges uniformly to $f^{(i)}(x)$ on $(0,1)$.*

The proof for this theorem is still elementary but requires more careful analysis.

The generalization for a function of two variables is obtained by first approximating one variable and then the second. But using this straightforward method we can only prove pointwise convergence.

2.4 Theorem *Let $f : [0,1] \times [0,1] \to \mathbb{R}$ be a continuous function. Then the two-dimensional Bernstein polynomials*

$$B_{f,n}(x,y) := \sum_{k=0}^{n} \sum_{\ell=0}^{n} f\left(\frac{k}{n}, \frac{\ell}{n}\right) \binom{n}{k} \binom{n}{\ell} x^k (1-x)^{n-k} y^\ell (1-y)^{n-\ell}$$

converge pointwise to f for $n \to \infty$.

This method can of course be applied recursively.

2.5 Theorem *Let $f : [0,1] \times \cdots \times [0,1] \to \mathbb{R}$ be a continuous function of m variables $x_1,\ldots,x_m$. Then the multi-dimensional Bernstein polynomials*

$$B_{f,n}(x_1,\ldots,x_n) := \sum_{k_1,\ldots,k_m=0}^{n} f\left(\frac{k_1}{n},\ldots,\frac{k_m}{n}\right) \prod_{j=1}^{m} \left(\binom{n}{k_j} x_j^{k_j} (1-x_j)^{n-k_j} \right)$$

converge pointwise to f for $n \to \infty$.

3 Examples

In this section we discuss two examples illustrating the properties of Bernstein polynomials, namely an analytical function and a two-dimensional inverse modeling example.

The example of the function $f : [0,1] \times [0,1] \to \mathbb{R}$,

$$f(x,y) := (1/2)\mathrm{e}^{-10((x-1/2)^2+(y-1/2)^2)} + \mathrm{e}^{-50((x-1)^2+(y-1)^2)}$$

shows that approximation using generalized Bernstein polynomials resembles the global properties of a given function better than using multivariate polynomials of degree 2 or less, even when using a small number of lattice points. The two approaches are compared in Fig. 1. In the left hand figure, f is plotted at the $11 \cdot 11$ lattice points that were used for calculating the two-dimensional Bernstein polynomial $B_{f,10}(x,y)$ and the least squares fit $rsm(x,y)$ of degree 2. f and $B_{f,10}$ have two local maxima on $[0,1] \times [0,1]$, whereas rsm has only one. Their respective values are (up to six digits): $f(0.5,0.5) = 0.5$, $f(0.999661, 0.999661) = 1.00338$; $B_{f,10}(0.500674, 0.500674) = 0.331634$, $B_{f,10}(1,1) = 1.00337$; $rsm(0.696706, 0.696706) = 0.283076$.

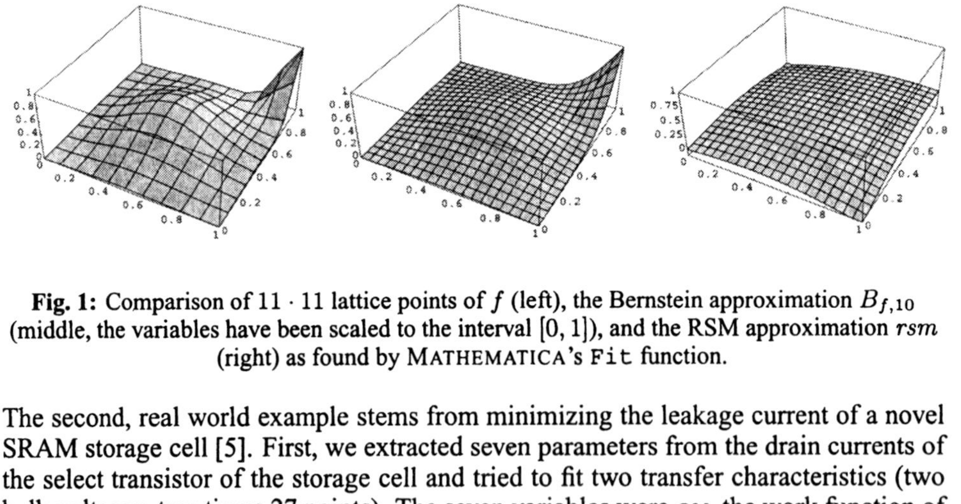

Fig. 1: Comparison of $11 \cdot 11$ lattice points of f (left), the Bernstein approximation $B_{f,10}$ (middle, the variables have been scaled to the interval $[0,1]$), and the RSM approximation rsm (right) as found by MATHEMATICA's `Fit` function.

The second, real world example stems from minimizing the leakage current of a novel SRAM storage cell [5]. First, we extracted seven parameters from the drain currents of the select transistor of the storage cell and tried to fit two transfer characteristics (two bulk voltages, two times 27 points). The seven variables were `ew`, the work function of the gate material, `sr`, the source resistance, `f`, a parameter controlling the doping, and four variables pertaining to the Shockley–Read–Hall model [6, page 71]. In the second step the extracted values were used when minimizing the leakage current.

In the course of the inverse modeling task it was found that two variables, namely the parameter of the gate material (`ew`) and the parameter controlling the doping (`f`), have a major influence on the result. For further investigations, these remaining variables were

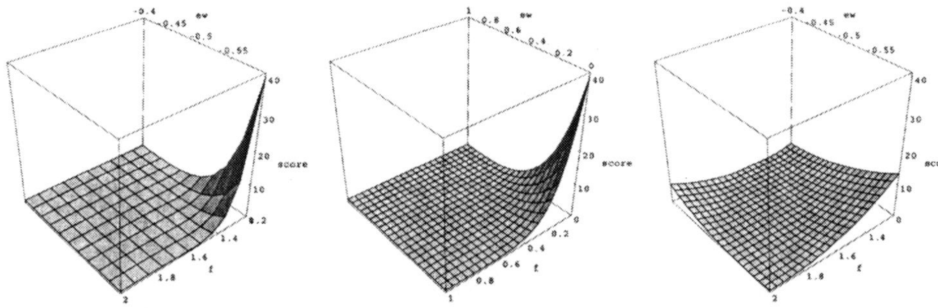

Fig. 2: Comparison of the computed lattice points (left), the Bernstein approximation (middle), and the RSM approximation (right) as found by MATHEMATICA's `Fit` function.

then fixed at the values of the minimum found, and the objective function was evaluated at $11 \cdot 11$ lattice points with these two most sensitive parameters (cf. Fig. 2, left). Using these points, two approximations were calculated: the two-dimensional Bernstein polynomial (where the variables were scaled to the interval $[0, 1]$), and the least squares approximation from the set of all polynomials of degree two or less (cf. Fig. 2). Again the RSM approximation is misleading.

4 Conclusion

For optimization tasks involving computationally expensive functions, we propose using multivariate Bernstein polynomials for approximating objective functions instead of the conventional RSM approach of using polynomials of degree two or less. We show that this approach is mathematically sound and present two examples illustrating its advantages.

Acknowledgment

The authors acknowledge support from the "Christian Doppler Forschungsgesellschaft", Vienna, Austria.

[1] C. Heitzinger and S. Selberherr, "An Extensible TCAD Optimization Framework Combining Gradient Based and Genetic Optimizers," in *Proc. SPIE International Symposium on Microelectronics and Assembly: Design, Modeling, and Simulation in Microelectronics*, (Singapore), pp. 279–289, 2000.

[2] R. Strasser, R. Plasun, and S. Selberherr, "Practical Inverse Modeling with SIESTA," in *Simulation of Semiconductor Processes and Devices*, (Kyoto, Japan), pp. 91–94, Sept. 1999.

[3] G. Box and N. Draper, *Empirical Model-Building and Response Surfaces*. New York: Wiley, 1987.

[4] I. Berezin and N. Zhidkov, *Computing Methods*, vol. 1. Pergamon Press, 1965.

[5] N. Ikeda, T.Terano, H. Moriya, T. Emori, and T. Kobayashi, "A Novel Logic Compatible Gain Cell with two Transistors and one Capacitor," in *Symposium on VLSI Technology Digest of Technical Papers*, pp. 168–169, 2000.

[6] T. Binder, K. Dragosits, T. Grasser, R. Klima, M. Knaipp, H. Kosina, R. Mlekus, V. Palankovski, M. Rottinger, G. Schrom, S. Selberherr, and M. Stockinger, *MINIMOS-NT User's Guide*. Institut für Mikroelektronik, 1998.

Advanced Hybrid Cellular Based Approach for Three-Dimensional Etching and Deposition Simulation

A. Hössinger, T. Binder, W. Pyka, and S. Selberherr

Institute for Microelectronics, TU Vienna, Gusshausstrasse 27–29, A-1040 Vienna, Austria
Phone: +43-1-58801/36027, Fax: +43-1-58801/36099 Email: Hoessinger@iue.tuwien.ac.at

Abstract

For the simulation of etching and deposition processes the cellular method is very popular, because of its high robustness compared to alternative methods like the level set or the moving boundary approach. We present a method for the simulation of topography processes based on the cellular method that overcomes the problem of loss of information when using a cellular based simulator in combination with simulators using polygonal data representations. Additionally the new method allows to increase the resolution of the topography simulator without a significantly higher memory requirement.

1 Introduction

For three-dimensional simulation of etching and deposition processes cellular based algorithms [1] [2] have advantages to polygonal based algorithms and level set algorithms due to their high robustness. For instance the formation of voids which is a very serious problem for polygonal based algorithms is implicitly handled. Nevertheless the major drawback of the cellular based simulators is the cellular data format. On the one hand side the cellular resolution is not very high because the memory requirement increases dramatically by choosing a higher accuracy. But when a topography simulator is applied to front end process simulation also thin layers like the gate oxide in a MOS transistor have to be resolved accurately. This means that the cell size should be smaller than one tenth of the gate oxide thickness. For instance, if the volume of the simulation domain is 1 μm^3 and the gate oxide thickness is 5 nm 8 billon cells were necessary to discretize the simulation domain, while the error is still not less than 10%.

A way to reduce the memory requirement is to use more advanced discretization methods than a simple regular grid. A bucket octree based representation could be the method of choice, but it still suffers from the fact that the discrete representation is normally not compatible to the polygonal based data format of other simulators used for semiconductor process simulation. As a consequence, several conversions between the cellular and the polygonal data format are necessary, if the cellular based simulator is integrated into a process flow. Errors of the order of the resolution of the cellular representation are introduced by each conversion. As already mentioned, small structures (especially thin layers) present in the simulation domain may be seriously modified or even get lost in the worst case, although for instance, not even small modifications of the gate oxide thickness are tolerable.

We present an advanced hybrid cellular approach which attacks the discretization error problem from two sides. On the one hand side we avoid a discretization of the complete simulation domain by restricting the descritization to a layer around the surface attacked by the topography process and on the other hand side two levels of cell resolution are used. Both measures give a higher cellular resolution resulting in a smaller discretization error.

2 Internal Data Representations

During the simulation two internal data representations are used. The core of the topography simulator is based on a regular grid layed over the whole simulation domain. Each cell defined by this grid either can contain material information or another regular grid defining refined cells. Note that no physical memory is associated with this grid at the beginning. Just during the simulation the material information is generated and stored on grid cells.

The topology of the complete simulation domain is stored seperately in the so called Wafer State Server [3]. This Wafer State Server holds the structure of the simulation geometry in a thetrahedral volume mesh discretized format and provides functions to comfortably access information about the simulation domain. Thereby material cells used by the topography simulator can be easily generated during the simulation. To achieve an accaptable access time the mesh is stored on a finite octree inside the Wafer State Server.

3 Topography Process

First the upper surface (surface attacked by the topography process) of the simulation domain is extracted by the Wafer State Server and discretized pointwise. Around each discretized point, refined topography simulator cells (called refined cell in the following) are generated. The material type of the cells is determined by the Wafer State Server and as a consequence the size of the structuring element [2] related to each cell placed at a material interface.

After setting up the discretized surface the topography process is performed. Thereby all previously determined structuring elements are applied to the simulation domain. All cells covered by the smallest rectangular prisms containing the structuring elements are defined at that time. Defining a cell means that memory is allocated for the cell if necessary and all the informantion stored on the cell is prepared by the Wafer State Server. This procedure is indicated by the gray area in Fig. 1. All cells containing the surface, the front of the topography process and interfaces between different materials are additionally refined (as indicated for the surface cells in Fig. 1) to further increase the resolution in such sensitive areas.

The simulation proceeds similar as in the classical cellular approach, meaning that the material type of all cells attacked by the topography process is changed during the simulation. The only differences to the classical approach are that

- memory is successively allocated whenever a cell is attacked outside of an allocated memory area.
- cells close to the process front (e.g. etch front) become refined cells.
- cells moving away from the process front are converted back to simple cells if they contain refined cells just of one material type.

The use of a two level cell resolution does not only save memory but also increases the performance of the simulation, because generally each cell has to be checked whether it is inside or outside of a structuring element in order to determine if the material type of a cell has to be changed. If a cell which is not refined is completely contained within or outside of the structuring element, its subcells need not to be checked. This significantly reduces the number of test operations, if the size of the unrefined cells is smaller than the size of the structuring element.

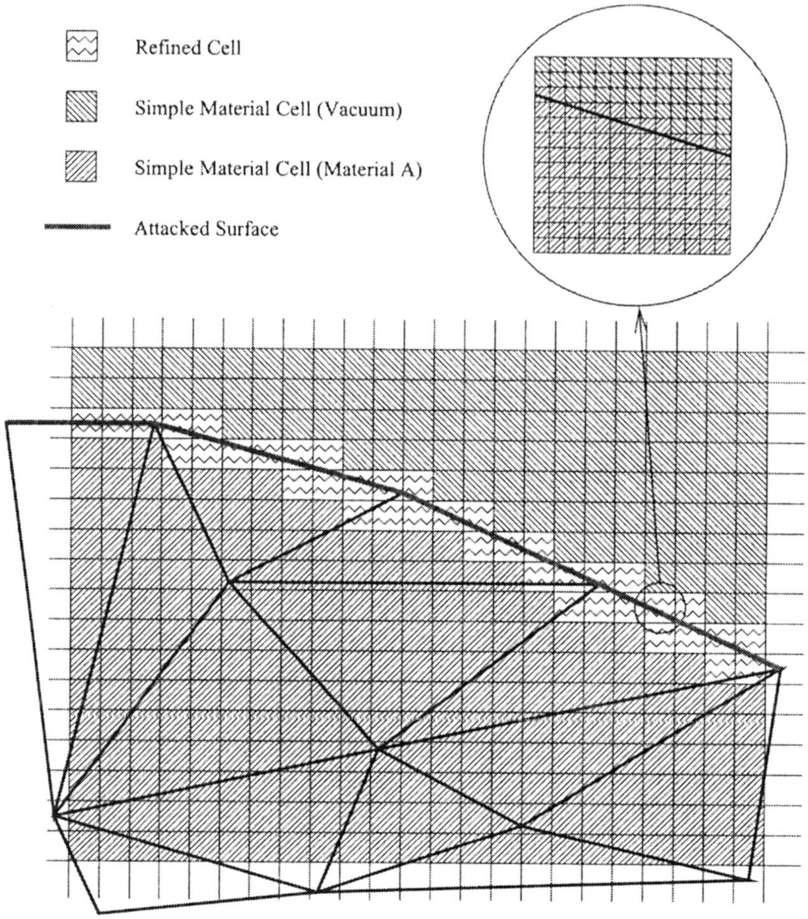

Fig. 1: Illustration of the descretization procedure.

4 Post Processing

After completing the topography simulation the topology of the modified simulation domain has to be generated. Therefore a triangulated representation of the etchfront is determined from the cellular data structure, simply by generating two triangles from all interfaces between a material cell and vacuum cells. Since all cells around the front of the topography process are refined cells just triangles with half the area of one side of a refined cell are generated. In order to reduce the huge number of triangles and to improve the quality of the triangulated front two postprocessing routines are applied.

On the one hand side the manhatten like structure is smoothed by projecting each point into a plane with the minimum distance from points surrounding the point which has to be projected. By restricting the offset due to projection to the half length of the diagonal of a refined cell the generation of illegal structures (intersecting triangles) is avoided. Thereby the angle between triangles which should be coplanar is made quite small.

On the other hand side the number of triangles is reduced by removing edges [4] which are surrounded just by approximately coplanar triangles (the angle between the triangles is below a minimum tolerable angle). The number of triangles is usually significantly reduced and, additionally, more vestiges of the manhatten like structure are removed.

Finally a new wafer has to be set up which contains all the information available before the topography process, modified just by the effects of the topography process. At that point the major advantage of coupling the topgraphy simulator with the Wafer State Server shows up. During the simulation the Wafer State Server keeps track of mesh element attacked by the topography process. Thereby the Wafer State Server finally knows which segments have been attacked by the topography simulation and just these segments are modified. The modification is performed by constructing new segments from the old ones by cutting the segment surfaces with the etch front and remeshing the resulting topography. This procedure is schematically depicted in Fig. 2. In case of a deposition simulation no existing segments have to be modified. Just new segments are created by meshing the space between the initial wafer surface and the final deposition fronts. Worth mentioning is that more than one front may have been generated during the simulation due to the formation of voids.

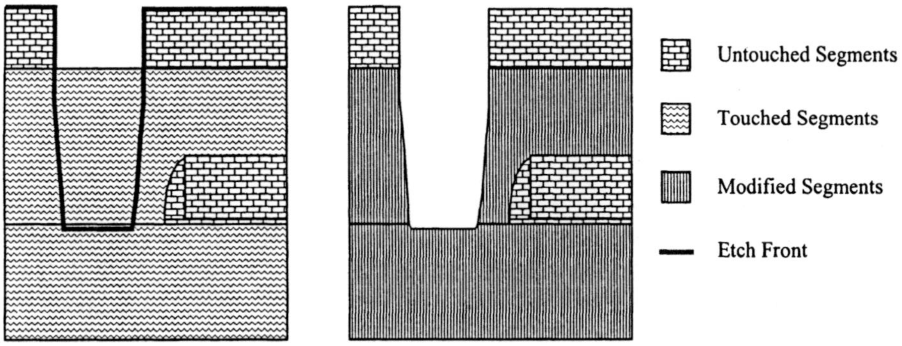

Fig. 2: Illustration of the modification of the wafer state data due to the etch process.

5 Conclusion

We have presented a method for the simulation of etching and deposition processes which makes use of the robust cellular algorithm. The new method overcomes the problem of conversion errors when a cellular based simulator interacts with simulators based on polygonal data structures. Additionally the memory requirement is drastically reduced so that also large structures can be simulated with sufficiently high resolution.

Acknowledgment

This work is supported by LSI Logic and by APART (Austrian Program for Advanced Research and Technology) from the Österreichische Akademie der Wissenschaften.

[1] W. Pyka, R. Martins, and S. Selberherr, "Optimized Algorithms for Three-Dimensional Cellular Topography Simulation," *IEEE J.Technology Computer Aided Design*, 2000. `http://www.ieee.org/products/online/journal/tcad/accepted/Pyka-March00/`.

[2] E. Strasser and S. Selberherr, "Algorithms and Models for Cellular Based Topography Simulation," *IEEE Trans.Computer-Aided Design*, vol. 14, no. 9, pp. 1104–1114, 1995.

[3] T. Binder and S. Selberherr, "Object-Oriented Wafer-State Services," in *14th European Simulation Multiconference* (R. V. Landeghem, ed.), (Ghent, Belgium), pp. 360–364, Society for Computer Simulation International, May 2000.

[4] P. Lindstrom and G. Turk", "Fast and Memory Efficient Polygonal Simplification," in *Proceedings of IEEE Visualization*, vol. 98, pp. 279–286, 1998.

A Methodology for Deep Sub-Quartermicron CMOS Technology Characterization

V. Palankovski, N. Belova*, T. Grasser, H. Puchner*, S. Aronowitz*, and S. Selberherr

Institute for Microelectronics, TU Vienna, Gusshausstrasse 27–29, A-1040 Vienna, Austria
Phone: +43-1-58801/36017, Fax: +43-1-58801/36099, Email: Palankovski@iue.tuwien.ac.at

*LSI Logic Corporation, 3115 Alfred Street, Santa Clara CA-95054, USA

Abstract

We present a novel methodology for characterization of sub-quartermicron CMOS technologies. It involves process calibration, device calibration employing two-dimensional device simulation and automated Technology Computer Aided Design (TCAD) optimization, and, finally, transient mixed-mode device/circuit simulation. The proposed methodology was tested on 0.25 μm technology and applied to 0.13 μm technology in order to estimate ring oscillator speed. The simulation results show an excellent agreement with available experimental data.

1 Introduction

The manufacturing process with shrinking technology is becoming so complicated that using simulation in a predictive manner has been recognized as an integral part of any advanced technology development. In order to satisfy predictive capabilities the simulation tools must capture the process as well as device physics. Before going to production one can optimize the process steps and estimate device performance characteristics such as threshold voltage, saturation current, leakage current and circuit speed.

Several tools for simulation of semiconductor technology (e.g. [1, 2]) as well as semiconductor devices (e.g. [3]–[6]) are well established for device engineering applications. The two-dimensional device simulator MINIMOS-NT is equipped with an extensive mixed-mode circuit capability including modeling of distributed devices [7]. It works in an automated device optimization framework [8]. This allowed creation of a novel methodology for very deep sub-micron technology characterization.

2 Methodology

The technology characterization system implemented at LSI Logic has been successfully used for 0.25 μm and 0.18 μm technologies. It included technology process and device calibration using TSUPREM4 [1] and MEDICI [3], respectively, sensitivity analysis and Monte Carlo statistical analysis, and, finally, gate delay estimation using a compact physical model in HSPICE [9]. For 0.13 μm technology the HSPICE physical model failed to fit the I-V characteristics extracted by MEDICI and it was impossible to estimate the gate delay characteristic of that technology using the old methodology.

In our new approach the process calibration part is kept the same while device simulation is performed by MINIMOS-NT in the SIESTA optimization TCAD framework. The same distributed devices are then employed in a transient mixed-mode device-circuit simulation [7]. The gate length L_g, gate width W_g, optical gate oxide thickness T_{ox}, and applied voltage V_{dd} for the investigated technologies are summarized in Table 1.

Technology	L_g	W_g	T_{ox}	V_{dd}
0.25 μm	0.2-1.0 μm	20 μm	4.7 nm	2.5 V
0.13 μm	0.115-0.7 μm	10 μm	2.3 nm	1.5 V

Table 1: Key parameters for the technologies considered in this work

2.1 Device Calibration

We use the SIESTA TCAD framework to perform an automated device calibration without user interaction during the optimization process. The setup allows simultaneous calibration of the saturation current $I_{D,sat}$ and the threshold voltage V_t to the measured values for several devices of different gate lengths with a minimum global error [10]. For the 0.25 μm technology calibration, 11 NMOS and 11 PMOS devices and for 0.13 μm technology 8 devices of each type, respectively, were used. Only a few technology dependent physical model parameters which can vary with the process were used for calibration. Such parameters are the workfunction difference E_w which is calibrated to account for the unknown density of surface states, or the surface mobility parameters μ_ν^{ref}, S_ν^{ref}, and γ_ν. The high-field mobility models are treated carefully, because their parameter values, e.g. the carrier saturation velocities v_ν^{sat} at 300 K generally must not be used as fitting parameters.

NMOS and PMOS device calibrations are performed separately because the model parameters are different. The proposed methodology of device calibration was tested on 0.25 μm technology. The new approach comprising of process calibration (TSUPREM4 - MEDICI) and device calibration (SIESTA - MINIMOS-NT) was applied to 0.13 μm technology. Device calibration was completed in about 8 CPU-hours. Good agreement with measured data was achieved both for the NMOS (Fig. 1) and for the PMOS (Fig. 2) devices. The final physical parameters from the optimization procedure are summarized in Table 2.

Parameter	0.25 μm NMOS	0.25 μm PMOS	0.13 μm NMOS	0.13 μm PMOS
E_w [eV]	-0.372	0.551	-0.433	0.407
μ_ν^{ref} [cm^2/Vs]	582	78	573	82
S_ν^{ref} [V/cm]	5.4e5	6.6e5	6.3e5	6.2e5
γ_ν	7.1	8.0	6.0	8.7
β_ν	1.18	1.06	1.1	1.1
v_ν^{sat} [cm/s]	9.8e6	9.8e6	1.2e7	1.18e7

Table 2: Mobility model parameters for the 0.25 μm and 0.13 μm technologies

The agreement achieved is within 2% for V_t and $I_{D,sat}$ versus gate length. All six model parameters remain close to their default values, while the saturation velocity resulting from the 0.13 μm technology calibration is $\sim$ 20% higher than the default value of $v_\nu^{sat} = 10^7$ cm/s. Therefore, we performed a comparative hydrodynamic (HD) simulation. A significant velocity overshoot governs the larger part of the channel area in the nominal 0.13 μm NMOS device compared to the overshoot observed in the 0.7 μm device. In drift-diffusion (DD) simulation this overshoot effect can be globally accounted for by increasing the v_ν^{sat} [11]. Such a change will only slightly influence the long channel devices, but will have a large effect on the short channel ones. Thus,

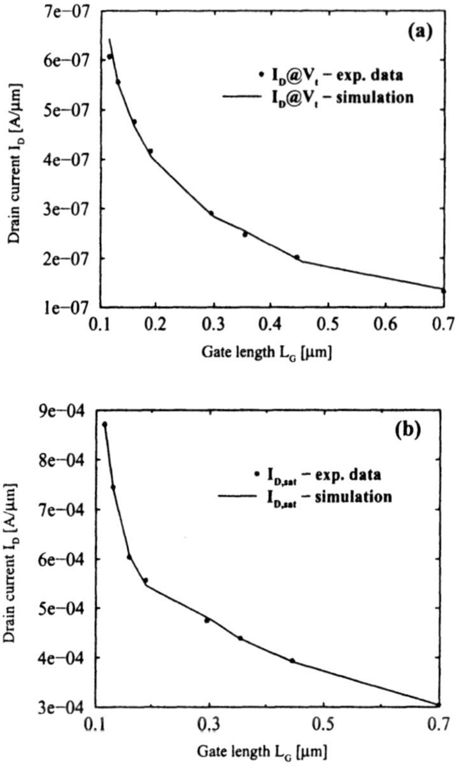

Fig. 1: Calibration of the threshold voltage V_t (a) and the saturation current $I_{D,sat}$ (b) for NMOS devices created with 0.13 μm technology

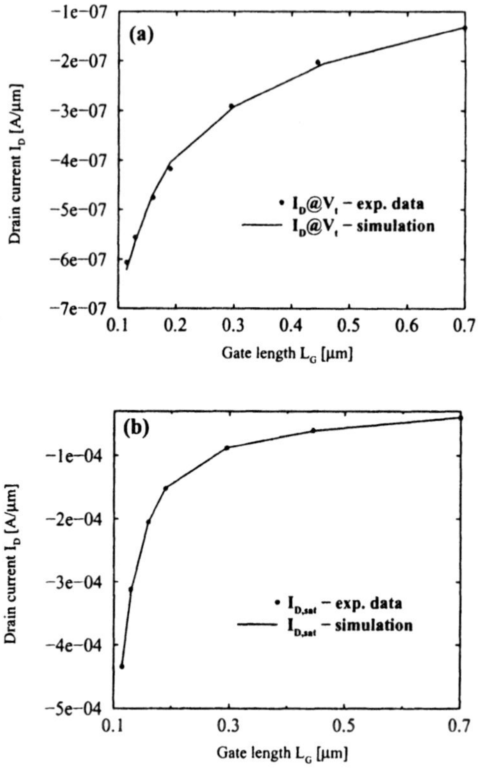

Fig. 2: Calibration of the threshold voltage V_t (a) and the saturation current $I_{D,sat}$ (b) for PMOS devices created with 0.13 μm technology

the high value of v_ν^{sat} in our calibration is justified considering that a DD simulation was performed.

2.2 Circuit Simulation

The calibrated model parameters were used for the circuit simulation. The ring oscillator simulation was carried out in transient mixed-mode using basic DD equations. The extracted gate delay for the 0.25 μm technology is 29.5 ps/stage which is in very good agreement with the experimental 32 ps/stage and with the 28.6 ps/stage extracted by the HSPICE physical model. To explore how many stages are sufficient to guarantee accurate results at minimum computational costs the ring oscillator circuits with three, five, and seven inverter stages were simulated. A simulation with five stages turned out to be sufficient to achieve the same results for the circuit speed as in simulations with more stages. The node voltages of the five stage ring oscillator circuit from the 0.13 μm technology are depicted in Fig. 3. The simulated inverter delay time is 15.2 ps/stage. In Fig. 4 the gate delay times extracted from several ring oscillators, each with 119 inverter stages (solid line represents the mean value) are compared with simulations (filled circles) performed for two calibrated wafers. The predicted gate delays are within the scattering range of the measured data.

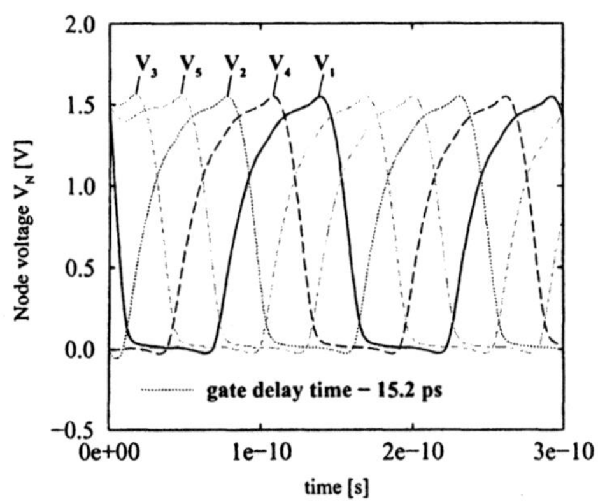

Fig. 3: The simulated node voltages of a five stage ring oscillator created with 0.13 μm technology

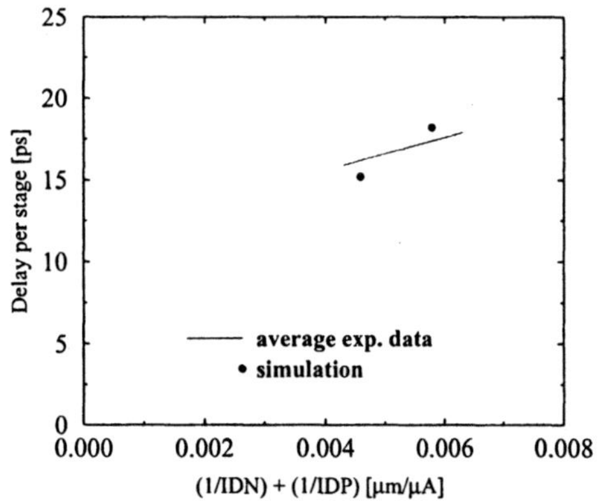

Fig. 4: The average experimental gate delay compared to simulation results obtained for two calibrated wafers

3 Conclusion

A new methodology was established for deep sub-quatermicron technology characterization. The new approach includes process calibration performed by TSUPREM4 - MEDICI, device calibration carried out by MINIMOS-NT in the SIESTA optimization framework and, finally, mixed-mode circuit simulation with distributed devices made by MINIMOS-NT. The methodology was tested and the tools were calibrated with a 0.25 μm technology. The approach was applied to a 0.13 μm technology characterization. Predicted ring oscillator speed is in excellent agreement with experimental data. The new methodology can be extremely beneficial in the early stages of process development for estimation of device performance.

[1] Avant! Corp., Fremont, CA, *TSUPREM-4, Two-Dimensional Process Simulation Program, Version 6.6*, June 1998.

[2] ISE Integrated Systems Engineering AG, Zürich, Switzerland, *DIOS-ISE, ISE TCAD Release 6.0*, July 1999.

[3] Avant! Corp., Fremont, CA, *MEDICI, Two-Dimensional Device Simulation Program, Version 4.1*, July 1998.

[4] ISE Integrated Systems Engineering AG, Zürich, Switzerland, *DESSIS-ISE, ISE TCAD Release 6.0*, July 1999.

[5] T. Binder, K. Dragosits, T. Grasser, R. Klima, M. Knaipp, H. Kosina, R. Mlekus, V. Palankovski, M. Rottinger, G. Schrom, S. Selberherr, and M. Stockinger, *MINIMOS-NT User's Guide*. Institut für Mikroelektronik, TU Wien, 1998.

[6] S. Beebe, F. Rotella, Z. Sahul, D. Yergeau, G. McKenna, L. So, Z. Yu, K. Wu, E. Kan, J. McVittie, and R. Dutton, "Next Generation Stanford TCAD—PISCES 2ET and SUPREM 007," in *Intl.Electron Devices Meeting*, San Francisco, pp. 213–216, 1994.

[7] T. Grasser, V. Palankovski, G. Schrom, and S. Selberherr, "Hydrodynamic Mixed-Mode Simulation", in *Simulation of Semiconductor Processes and Devices* (K. De Meyer and S. Biesemans, eds.), (Leuven, Belgium), pp. 247–250, 1998.

[8] Ch. Pichler, R. Plasun, R. Strasser, and S. Selberherr, "High-Level TCAD Task Representation and Automation", *IEEE J.Technology Computer Aided Design*, May 1997. `http://www.ieee.org/journal/tcad/accepted/ pichler-may97/`

[9] Avant! Corp., Fremont, CA, *HSPICE, Elements and Models User's Manual.*

[10] T. Grasser, R. Strasser, M. Knaipp, K. Tsuneno, H. Masuda, and S. Selberherr, "Device Simulator Calibration for Quartermicron CMOS Devices," in *Simulation of Semiconductor Processes and Devices* (K. De Meyer and S. Biesemans, eds.), (Leuven, Belgium), pp. 93–96, 1998.

[11] J.D. Bude, "MOSFET Modeling into the Ballistic Regime", in *Simulation of Semiconductor Processes and Devices*, (Seattle, USA), pp. 23-26, Sept. 2000.

A Review of Modeling Issues for RF Heterostructure Device Simulation

R. Quay, R. Schultheis*, W. Kellner *, V. Palankovski**, and S. Selberherr**

Fraunhofer-Institute of Applied Solid-State Physics, Tullastrasse 72, Freiburg, Germany
*Infineon Technologies AG, Wireless Products, Otto-Hahn-Ring 6, Munich, Germany
**Institute for Microelectronics, TU Vienna, Gusshausstrasse 27–29, A-1040 Vienna, Austria
Phone: +43-1-58801/36017, Fax: +43-1-58801/36099, Email: Palankovski@iue.tuwien.ac.at

Abstract

We give a summary of the state-of-the-art of heterostructure RF-device simulators for industrial application based on III-V compound semiconductors and compare critical modeling issues. Results from two-dimensional hydrodynamic simulations of High Electron Mobility Transistors (HEMTs) and Heterojunction Bipolar Transistors (HBTs) with MINIMOS-NT are presented. Simulation examples are chosen to demonstrate technologically important issues which can be addressed and solved by III-V device simulation. A summary of remaining modeling issues is provided.

1 Introduction

Heterojunction Bipolar Transistors (HBTs) and High Electron Mobility Transistors (HEMTs) are among the most advanced semiconductor devices. They match well today's requirements for high-speed operation, low power consumption, medium- integration, low cost in large quantities, and operation capabilities in the frequency range from 0.9 to 215 GHz. HBT ICs are used for microwave power and low power wireless communications applications, hand held communication, and high-speed digital data transmission. HEMT ICs are used for low power wireless communication applications, local multi-point distribution services for broadband Internet access (LMDS), for automotive cruise control (ACC) radar, and high-speed digital transmission (40 Gbit/s and beyond). The paper summarizes the state-of-the-art of device simulators, including the two-dimensional device simulator MINIMOS-NT, and discusses particular simulation results of such devices obtained with the same simulation tool, model set, and set of model parameters. It concludes with critical modeling issues remaining to develop for the simulation of advanced III-V semiconductor devices.

2 RF-Simulators

Commercial tools, e.g. [1, 2, 3, 4], and University-developed simulators, e.g. [5, 6, 7], are available for device engineering applications. However, for the fast development of the RF industry a stronger feedback from technological state-of-the-art process development to simulator development is needed. Table 1 summarizes features of III-V device simulators to the best of the authors knowledge.

Gateway Modeling offers a one-dimensional Schrödinger-Poisson solver POSES [3] for charge analysis in HEMT devices for process control. Several quasi-two-dimensional approaches are demonstrated, e.g. [7]. This approach, based on a one-dimensional current equation, stresses the interface to the microwave simulator as MDS/ADS. The emphasis is put on the extraction of compact large-signal models. This tool combines the advantages of a hydrodynamic (HD) transport model coupled with Schrödinger's equation, but has the drawback of the simplified one-dimensional current equation. The

Simulator	Dimension	Model	Features	Remarks
POSES	1D		Schrödinger-Poisson solver	
Leeds	quasi 2D	HD	Schrödinger equation Electrothermal model	1D current equations interfaces
Fast Blaze	quasi 2D	HD		1D current equations interfaces
ATLAS	2D	DD,ET	TE heterojunction model	no tunelling modeling
PISCES	2D	DD,ET	III-V models	
G-PISCES	2D	DD	full set III-V models	no HD
MEDICI	2D	DD,HD	anisotropic properties	mixed-mode interfaces
MINIMOS-NT	2D	DD,HD	full set III-V models TE/TFE model	
DESSIS	2D,3D	DD,HD	trap modeling, TFE model	III-V modeling

Table 1: Comparison of different device simulators. (TE): thermionic emission, (TFE): thermionic field emission.

two-dimensional device simulator PISCES has modeling capabilities for GaAs and InP based devices. The variant called G-PISCES from Gateway Modeling [3] has been extended by a more complete set of III-V models. Examples of MESFETs, HEMTs, and HBTs for InAlAs/InGaAs, AlGaAs/InGaAs, AlGaAs/GaAs, and InGaP/GaAs HBTs are demonstrated. A disadvantage is the lack of an energy transport model (ET) or HD transport model. The device simulator MEDICI from Avant!, also based on PISCES, offers simulation capabilities for SiGe/Si HBTs and AlGaAs/InGaAs/GaAs HEMTs. Advantages of this simulator are HD simulation capabilities and the rigorous approach to generation/recombination processes. An option treating anisotropic properties was announced. Weaknesses are found for the III-V materials modeling, for the heterostructure interface modeling and for mixed-mode device-circuit simulation. A quasi-two-dimensional tool is Fast Blaze from Silvaco, based on code from the University of Leeds, which together with the two-dimensional ATLAS [4] has addressed the simulation of AlGaAs/GaAs and pseudomorphic AlGaAs/ InGaAs/GaAs HEMTs. DESSIS from ISE [2] has demonstrated a rigorous approach to semiconductor physics modeling. Critical issues, as the above stated extensive trap modeling, are solved. A heterojunction framework has been included. In the following section simulation results with MINIMOS-NT are presented.

3 Selected Results for Industrially Relevant Devices

Self-heating simulations, especially of high-power multi-finger devices, require realistic boundary conditions taken from a three-dimensional circuit environment. For high-power HBTs self-heating is a major source of limitation. The simulated output device characteristics of an InGaP/GaAs HBT compared to measurements are shown in Fig. 1 with constant boundary conditions for the base current (I_B=0.1-0.5 mA). Note that agreement of simulation and measurement can be achieved introducing self-heating (SH) into the two-dimensional simulation with appropriate boundary conditions, while the isothermal simulation completely disagrees with the measurement. Fig. 2 shows a comparison between the simulated and measured Gummel plots for 400×400 μm^2 $Si_{1-x}Ge_x$ HBTs fabricated with the same geometry and layer specification and with x = 16%, 22%, and 28% Ge content in the base. All simulations were carried out with the same set of models, and no adjustment of model parameters was performed.

For the next generation of data transmission at 80 Gbit/s or high gain analogue applications InAlAs/InGaAs HEMTs high-speed transistors are used. Fig. 3 shows simulation and measurements for a gate length l_g = 150 nm $In_{0.52}Al_{0.48}As/In_{0.53}Ga_{0.47}As$ HEMT on a metamorphic buffer. High field effects such as impact ionization are considered [8]. The precise simulation for various low V_{DS} is useful to account for the low power RF capabilities. Fig. 4 shows an evaluation of the possible use of low-k dielectrics to reduce critical capacitances (e.g. C_{gd}) for highest frequency operation.

4 Critical Issues of Modeling Heterostructure Devices

A number of critical modeling issues for III-V devices can be solved with a precise physics-based small-signal parameter extraction [9]. An interface to a microwave simulator such as MDS/ADS from Agilent Technologies, e.g. in [7], allows the analysis of statistical variations of the large-signal models used for circuit design. A statistical sensitivity analysis is especially important for applications beyond 50 GHz, where speed, gain, and output power are very sensitive to the parasitic environment and the large signal properties are very sensitive to the statistical changes of mass production. The simulators have successfully demonstrated good agreement with measurements. Further, improved understanding of the processed semiconductors with respect to the large signal properties is needed. This includes the understanding of changes of transport and interface parameters due to the process. In particular: More attention is required for the process dependence of the insulator semiconductor surface quality and the development of models for non-ideal dielectrics. The impact of the ledge thickness and the surface charge density was studied for a one-finger InGaP/GaAs HBT with respect to reliability [10]. The contact modeling needs further attention since contact degeneration is a major source of device degradation. Non-local approaches for Ohmic and Schottky contacts have been demonstrated, such as [11], however due to the meshing required for application, they are not suitable for general purpose simulators. For the analysis of dynamic breakdown during large-signal operation, e.g., in pseudomorphic GaAs HEMTs, a study requires a combination of impact ionization, traps, and thermionic field emission effects. The treatment of anisotropic properties is needed for SiGe HBTs for the inclusion into a conventional simulator concepts. In contrast to Monte Carlo approaches, this offers the possibility to extract RF information. Quaternary material models are required for a number of III-V devices, i.e. for InGaAsP, AlGaAsSb, and InAlGaAs [12].

5 Conclusion

A review of critical modeling issues for simulation tools of heterostructure RF-devices is presented, specifically addressing remaining modeling issues. New simulation results for industrially relevant devices in good agreement with measured data are demonstrated. With an increasing number of stable, reliable heterostructure technologies available, a meaningful comparison between simulation results and statistically analyzed small and large signal data is possible and delivers model verification and valuable process information.

[1] Avant! Corporation, Fremont, CA, *Medici, Two-Dimensional Device Simulation Program*, Version 1999.2, 1999.

[2] ISE Integrated Systems Engineering AG, Zürich, Switzerland, *DESSIS-ISE, ISE TCAD Release 6.0*, 1999.

[3] POSES, in: R. Anholt, *Electrical and Thermal Characterization of MESFETs, HEMTs, and HBTs*. Boston: Artech House, 1995.

[4] Silvaco International, Santa Clara, USA, *"ATLAS Users Manual, Device Simulation Software, Ed. 6"*, 1998.

[5] S. Beebe, F. Rotella, Z. Sahul, D. Yergeau, G. McKenna, L. So, Z. Yu, K. Wu, E. Kan, J. McVittie, and R. Dutton, "Next Generation Stanford TCAD—PISCES 2ET and SUPREM 007," in *Intl.Electron Devices Meeting*, San Francisco, pp. 213–216, 1994.

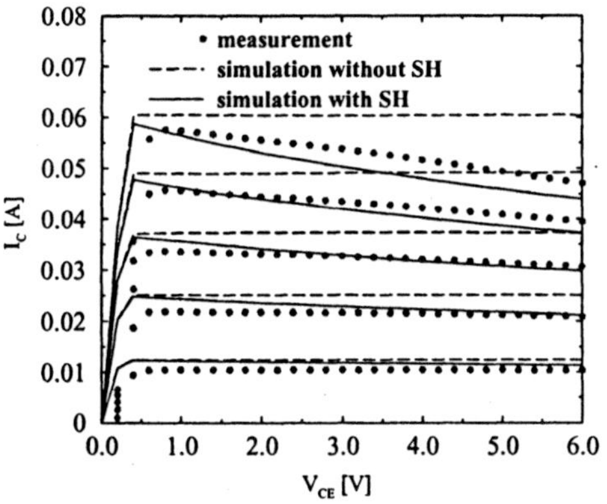

Fig. 1: GaAs HBT output characteristics. Simulation with and without self-heating (SH) compared to measurement data at constant I_B stepped from 0.1 mA to 0.5 mA.

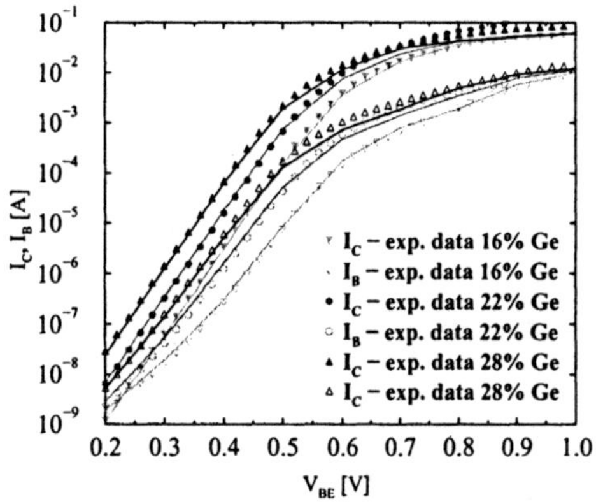

Fig. 2: Forward Gummel plots at $V_{BC} = 0$ V for SiGe HBTs. Comparison between simulation and measurement for different Ge contents.

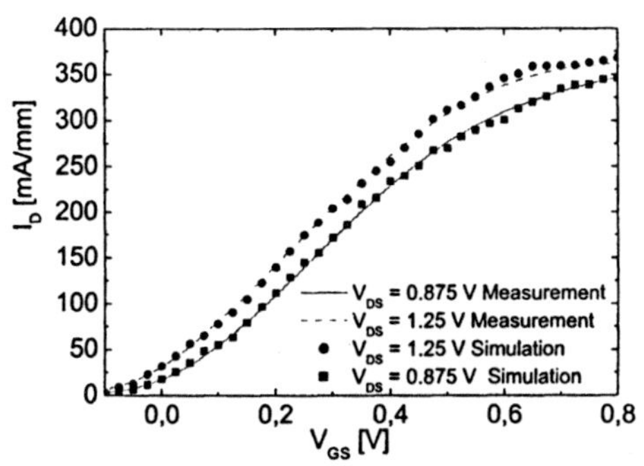

Fig. 3: Transfer characteristics of a metamorphic InAlAs/ InGaAs HEMT with $l_g = 150$ nm for two different biases.

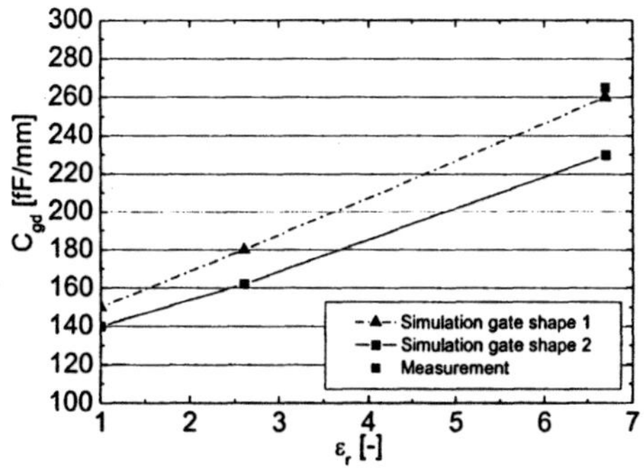

Fig. 4: Reduction of the gate-drain capacitance C_{gd} of an $l_g = 100$ nm InAlAs/InGaAs HEMT using different dielectrics.

[6] H. Kosina and S. Selberherr, "A Hybrid Device Simulator that Combines Monte Carlo and Drift-Diffusion Analysis," *IEEE Trans.Computer-Aided Design*, vol. 13, no. 2, pp. 201–210, 1994.

[7] C.G. Morton, J.S. Atherton, C.M. Snowden, R.D. Polard, and M.J. Howes, "A Large-Signal Physical HEMT Model," in *Dig. MTT-S Intl. Microwave Symp.*, San Francisco, vol. 3, pp. 1759–1762, 1996.

[8] R. Quay, V. Palankovski, M. Chertouk, A. Leuther, and S. Selberherr, "Simulation of InAlAs/InGaAs High Electron Mobility Transistors with a Single Set of Physical Parameters", in *Intl. Electron Devices Meeting*, San Francisco, pp. 186–189, 2000.

[9] R. Quay, H. Massler, W. Kellner, T. Grasser, V. Palankovski, and S. Selberherr, "Simulation of Gallium-Arsenide Based High Electron Mobility Transistors," in *Simulation of Semiconductor Processes and Devices*, Seattle, pp. 74–77, 2000.

[10] V. Palankovski, S. Selberherr, R. Quay, and R. Schultheis, "Analysis of HBT Degradation After Electrothermal Stress," in *Simulation of Semiconductor Processes and Devices*, Seattle, pp. 245–248, 2000.

[11] M.K. Ieong, P.M. Solomon, S.E. Laux, H.S. Wong, D. Chidambarrao, "Comparison of Raised and Schottky Source/Drain MOSFETs using a Novel Tunneling Contact Model", in *Intl. Electron Device Meeting*, San Francisco, pp. 733–736, 1998.

[12] C.S. Whelan, P. F. Marsh, W.E. Hoke, R.A. McTaggart, P.S. Lyman, P.J. Lemonias, S.M. Lardizabal,R.E. Leoni III, S.J. Lichwala, T.E. Kazior, " Millimeter-Wave Low-Noise and High-Power Metamorphic HEMT Amplifiers and Devices on GaAs Substrates", *IEEE J. Solid-State Circuits*, vol. 35, no. 9, pp. 1307–1311, 2000.

Modeling and Simulation of Charge Generation Events Caused by Ion Irradiation in High Voltage Power Devices

Winfried Kaindl and Gerhard Wachutka
Institute for Physics of Electrotechnology, Munich University of Technology
D–80290 Munich, Germany
kaindl@tep.ei.tum.de, wachutka@tep.ei.tum.de

Gerald Sölkner
Infineon Technologies AG, Power Semiconductors
D–81730 Munich, Germany
Gerald.Soelkner@infineon.com

Abstract

In this paper, we present a model that describes the initial generation of charge due to the loss of kinetic energy of an ion penetrating into a semiconductor device. 2D simulations of a reverse biased power diode visualize the evolution of carrier densities and electric fields in the interior of the device as initiated by the ion. For the case of irradiation with 17MeV ^{12}C ions, the simulations agree well with recent experimental findings.

1 Introduction

Silicon semiconductor devices can fail when they are exposed to highly ionizing radiation like heavy ions or cosmic rays [1]. These so–called single event effects (SEE) occur in digital circuits [2], as well as in power devices [3]. As it is very difficult to perform measurements of the extremly fast processes occuring in the interior of the device, much effort has been spent in simulating the effects of a penetrating ion on the device behaviour, in particular regarding logic or low power devices like SRAM cells or low power MOSFETs. But only little work has been done on simulating the impacts of irradiation effects on high power devices.
In this paper, we present the numerical analysis of an ion irradiation experiment that was recently performed by *Maier et al.* [4]. A high power diode was irradiated by several types of ions. The charge generated by one single ion was measured in dependence on the applied bias. For a small reverse bias, the charge corresponds to the total absorption of the ion's kinetic energy. Applying a sufficiently high reverse bias, charge multiplication of up to four orders of magnitude sets on.

2 Modelling

For a detailed numerical analysis of the experimental processes and the visualization of the dynamical behaviour in the interior of the device, we developed a model that describes the initial generation of charge due to the loss of kinetic energy of an ion penetrating into the device. Additional terms

$$G_{ion}(\vec{x},t) = \frac{A}{2\pi a s} \cdot \frac{dE}{dz} \cdot \exp\left(-\frac{r^2}{2a^2}\right) \cdot \exp\left(-\frac{1}{2}\left(\frac{t-t_0}{s}\right)^2\right)$$

have to be added to the right hand sides of the carrier balance equations for electrons and holes, respectively. Here, r denotes the distance from the track of the ion and t the time. The energy loss function dE/dz was computed using the simulator TRIM (Transport of Ions in Matter) [5]. We obtain estimates of the radial (a=0.1μm) and temporal (s=1ps) Gaussian widths of the initial generation profile by solving the equation of motion of the ion within the device [6]. The initial generation rate is proportional to the kinetic energy E of the ion. Implementing the described model in the device simulator DESSIS [7] allowed us to perform numerical simulations of the ion irradiation experiment.

3 Simulation

We performed simulations of a p$^+$n–diode of 450μm length and a rated reverse blocking voltage of about 3.5kV. The junction is formed by a 1.5μm deep Gaussian p–diffusion in a lowly doped n–base. The width of the structure is 100μm about the plane of incidence of the ion. The diode was simulated without any external circuit, assuming an ideal voltage source. Therefore, the voltage was kept constant during the simulation. Fig. 1 shows the temporal and spatial evolution of the electric field in the plane of incidence, when a ^{12}C ion with a kinetic energy of 17MeV impinges perpendicularly from the anode side at z=0μm.

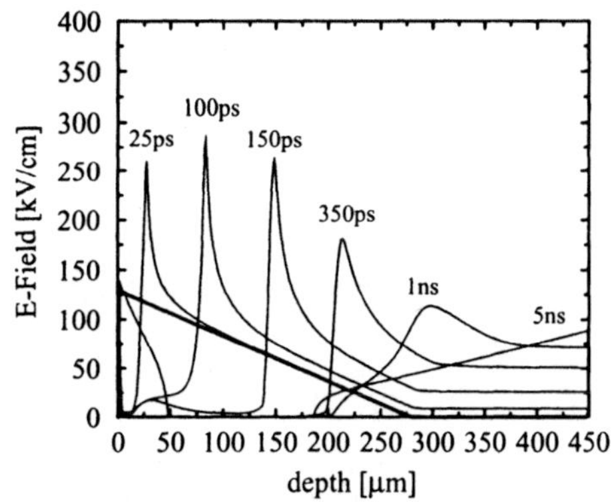

Fig. 1. Temporal and spatial evolution of the electric field in the plane of incidence for a reverse bias of 1800V.

After the ion impact, the generated electron/hole plasma distorts the initial triangular field and a steep field peak forms which starts propagating through the device. In the region left–sided of the peak plasma conditions occur. Here, the electric field is therefore nearly completely suppressed. For a reverse bias of 1800V, the field peak smoothes out while propagating towards the cathode contact (Fig. 1). The periode of a high electric field peak and a corresponding strong impact ionization is too short in order to enable the generation of a large amount of additional charge carriers. Therefore, the charge flowing out of the device corresponds to the total absorption of the kinetic energy of the ion.

A completely different situation occurs for a reverse bias of 2200V. Here, the initial electric field is already higher than that of the 1800V case in consequence of the higher reverse bias that has to be sustained. The steep field peak caused by the initial charge plasma injected by the ion is, accordingly, also higher. Finally, this leads to a field peak that propagates through the whole device with an undiminishing value of above 300kV/cm (see Fig. 2). It reaches the cathode contact after about 150ps.

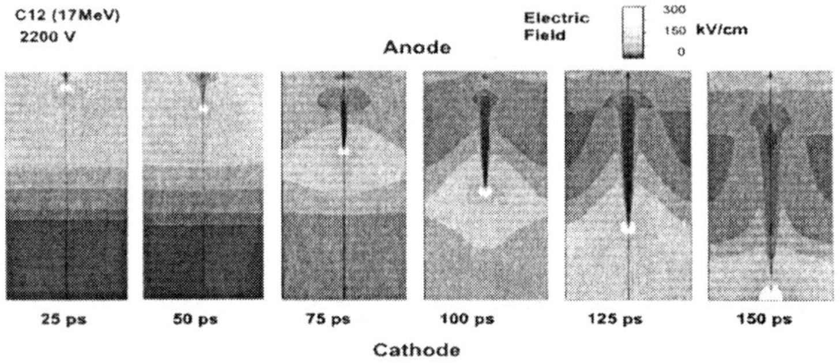

Fig. 2. 2D–plot of the propagating electric field peak for a reverse bias of 2200V.

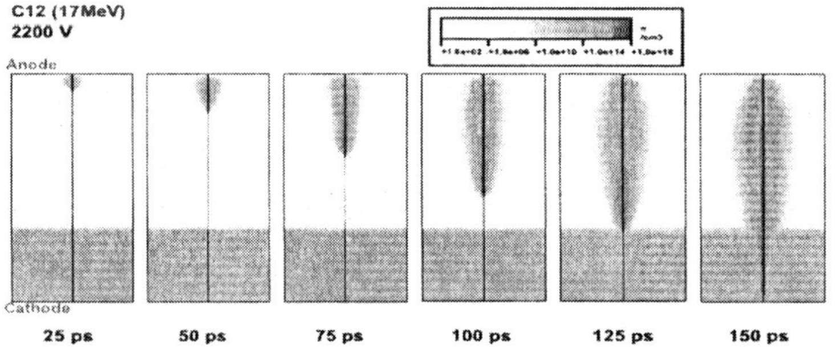

Fig. 3. 2D–plot of temporal and spatial evolution of the charge carrier density for a reverse bias of 2200V. A highly conductive filament has formed after 150ps.

Due to the corresponding strong avalanche multiplication caused by the high electric field the device is flooded with charge carriers along the whole plane of incidence. A highly conductive filament can form (see Fig. 3) which causes a short cut between anode and cathode after 150ps and, therefore, a high flow of current. As a consequence, charge can flow out of the device which is by serveral orders of magnitude higher than that in the absorption case.

4 Comparison between Simulation and Experiment

In Fig. 4 the simulated charge generated within the diode by one single ion is plotted and compared with the experimental findings of *Maier et al.* [5]. For the

17MeV ^{12}C ions simulation and experiment are in good agreement. The abrupt onset of charge multiplication of serveral orders of magnitude can be reproduced. But for the 270MeV ^{86}Kr ions the onset voltage is shifted towards higher voltages compared to the experiment. Here, the limitations of the 2D simulation become obvious. In 2D the field peak is not a real peak, but a field front. The third dimension in which the field can also be confined is missing. All in all, this leads to a lower peak value and, finally, to the shift of the onset of charge multiplication. The nature of the problem would require a full 3D simulation, but due to the limitions in computational resources available for us, this is not (yet) tractable.

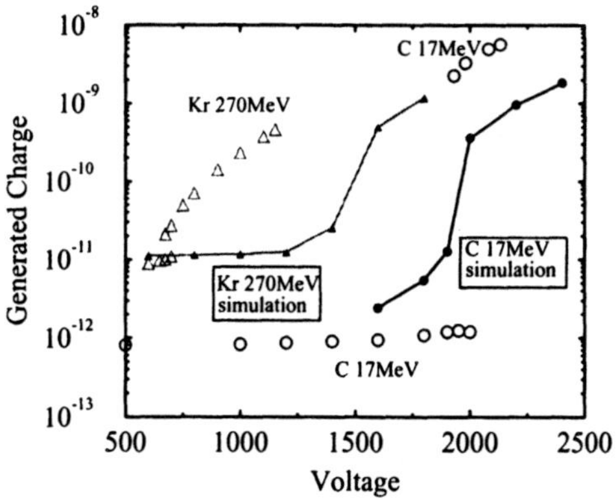

Fig. 4. Generated charge within the diode by one single ion. Comparison of experiment (symbols) and simulation (solid lines).

Even on the basis of 2D simulations, we were able to get a qualitative visualization of the intrinsic mechanisms of charge multiplication events occuring as consequence to the irradiation of high voltage devices by heavy ions. This knowledge about the physical processes in the interior of the device is indispensable for the design of new devices which are more resistant against cosmic rays.

References

[1] Terrestrial cosmic rays and soft errors, IBM J. of Res. and Devel., vol. 40, no. 1, 1996, pp. 1–136

[2] T.J. O'Gorman, The Effect of Cosmic Rays on the Soft Error Rate of a DRAM at Ground Level, Trans. Nucl. Sci., vol. 41, no. 4, 1994, pp. 553–557

[3] H. Kabza, et al., Cosmic radiation as a cause for power device failure and possible countermeasures, Proc. ISPSD, Davos, Switzerland, 1994, pp. 9–12

[4] K.H. Maier, A. Denker, P. Voss, H.-W. Becker, Nucl. Instr. and Meth. in Phys. Res., B146, 1998, pp. 596–600

[5] J.F. Ziegler, J.P. Biersack, U. Littmark, The Stopping and the Range of Ions in Solids, Pergamon Press, New York, 1985.

[6] W. Kaindl, G. Sölkner, P. Voss, G. Wachutka, Proc. of 3rd Int. FORTWIHR Conf. Erlangen, Germany, 2001

[7] ISE Integrated Systems Engineering AG, Dessis user manual 5.0, 1996

Determination of the Radiation Efficiency, Contrast and Sensitivity in Electron and Ion lithography

K.Vutova , G.Mladenov ,
Institute of Electronics, Bulgarian Academy of Sciences, Tzarigradsko shose 72
Sofia 1784, Bulgaria,
e-mail: vutova@ie.bas.bg mladenov@ie.bas.bg

I.Raptis
Inst.of Microelectronics, NCSR "Demokritos",15310 Aghia Paraskevi, Greece
e-mail: raptis@imel.demokritos.gr

Abstract

Knowledge of the solubility rate dependence on exposure dose *S(D)* gives possibility to estimate the values of the radiation efficiency, sensitivity and contrast characteristics in electron and ion lithography. The idea for interconnection between the sensitivity and the contrast characteristics for an arbitrary combination of resist and developer (i.e. at a given molecular weight, resist density and radiation efficiency of the charged particles) is a base for the sensitivity optimisation in order to achieve the required contrast at chosen development conditions. The contrast parameter value γ_s (related to the traditionally used contrast parameter γ_d) is determined by the slope of the dependence *S(D)* of the resist solubility rate *S* on the average exposure dose *D* .

1 Introduction

Microstructures for nano-electronics and photonics with critical dimensions in the order of 30 nm – 150 nm will be mastered during the next years. Electron and ion lithographies are the techniques, which meet the requirements of the challenge for further electron devices miniaturization. The quantitative estimation of the resist modification parameters [1] is an important step at optimisation of the electron and ion lithographies exposures and of the connected development processes.

The solubility rate changes upon irradiation due to the cross-linking or the scission of the polymer molecules of the resist layer. After a proper development process by suitable solvent (developer) relief microstructure in the resist layer can be observed. During the development process, the irradiated areas for the positive and opposite- the non-irradiated areas – for the negative resists are removed. The solubility rate variation of the irradiated spots depends on the used exposure dose. The sensitivity of the polymer resist to the radiation (with the electron and ion beams) is measured as the minimal dose D_0 for the development process of the exposed image. The resist sensitivity is determined by the radiation efficiency of irradiated particles that can be characterized by the mean number of chemical events (chain scission or destruction) per one unit (namely electron-volt) of the absorbed energy. In the field of high-resolution electron beam and ion beam lithography it is very important to characterize the developed profiles in the resist. and that can be done by the contrast parameter. The contrast parameter γ_d is defined for the remaining thickness of the utilised resist layer for given time and used dose D_0 by the dose interval between the initial exposure dose D_I (at which a start of development can be observed) and the mentioned dose D_0 (for full development of studied resist thickness). The value of the contrast parameter γ_d at the removed normalised thickness can be calculated from the relation $\gamma_d = [lg\,(D_I/D_0)]^{-1}$

Sensitivity and contrast values have been calculated experimentally for some average molecular weight of the used resist (measured for example in a.u.) and for a chosen developer. Data for the sensitivity D_0 and for the contrast γ_d of PMMA in the case of electron sub-micron lithography are given in [2,3]. The effect of the developer and the development conditions on these parameter values are discussed. A similar effect of the used developer on the contrast value γ_d for electron lithography of Amil Acetate layers is shown in [4] and in the case of PMPS & Novolak in [5]. In addition in the ref. [6, 7] data for the sensitivity and for the contrast γ_d in the case of some ion beam resists are given. Generally, the resist sensitivity of the polymer resists at the ion exposure is higher than for electron exposure [1,8]. The comparison is not precise due to the different conditions and the peculiarities of the figures of merit used. The sensitivity and the contrast are described usually as independent properties. Recently a definition of another contrast parameter $\gamma_s = (\Delta S/\Delta D)$ as the slope of the solubility rate dependence on the exposure dose $S(D)$ at a given exposure dose (i.e. sensitivity D_0) was accepted. Then these parameters γ_s and D_0 for a resist-developer combination are mutually connected [9,10].

Chemically amplified resists (CARs) are the most suitable candidates for future direct writing since they provide high resolution and high sensitivity. In CARs, acid is generated by the sensitiser during exposure which catalyse the necessary reaction for solubility change during post exposure bake (PEB) and thus alters the latent resist image after exposure. Modified sensitivity and contrast parameters for these resists can be evaluated. An acid diffusion coefficient as additional resist parameter can be evaluated too, using the time evolution curve $L(t_{PEB})$ of the image dimension L at different PEB conditions (t_{PEB} (s), T_{PEB} (^{0}C)) [11].

In this work an approach for calculating the radiation efficiency and the contrast characteristics for an arbitrary couple of a positive resist – developer in the field of electron and ion lithography is proposed. The approach is based on the theoretical approximation of the experimental dependence of the resist solubility rate S on the adsorbed energy of the irradiating charged particles beam in the resist. The methodology for evaluation of the radiation efficiency, contrast parameters at a radiation dose D (i.e. sensitivity D_0), uses Monte-Carlo calculation technique. The presented examples of evaluated contrast values and radiation efficiency values are for electrons or for various ions and at different energies.

2 Theoretical background and evaluation results

The evaluation of the resist parameters is based on the knowledge of the solubility rate S as a function of the exposure dose D. The resist solubility rate can be evaluated by the following empirical formula:

$$S = Ro + B/\overline{M_f}^{A} \quad , \tag{1}$$

where $\overline{M_f}$ is the average polymer molecular weight after radiation, and R_0, B and A are constants for a given resist-developer couple. The parameter R_0 has a physical meaning (the starting solubility rate), and A is the power coefficient. The molecular weight M_f depends on the radiation efficiency g and on the exposure dose D as well as from original molecular weight M_n. Then [13] the solubility rate S will be:

$$S = R0 + B\,[(\,1/\,Mn\,) + (gED/\rho NA)]A\,, \tag{2}$$

where E is the adsorbed energy in the resist , ρ is the resist density, N_A is Avogadro's number, g is the radiation efficiency.The curve $S/S_0(D)$ can be used instead of $S(D)$ (where S is the resist solubility rate after the exposure and S_0 is this rate without exposure). In the ion irradiation case, E presents the electronic and the nuclear components of the absorbed energy. According to [16] electronic stopping power is responsible for polymer modification. Some examples of evaluated coefficients values R_0 ,B and A for PMMA resist from eq. (1) and (2) are given in Table 1.

Table 1. The solubility rate parameters (eq. 1) for PMMA and for various developers at electron irradiation:

Solvent	R_0 (A/min)	B (A/min)	A	References
MIBK:IPA 1:1	0	$3{,}34.10^{6}$	1.5	[15] , [13]
IPA	0.008	6.10^{21}	5.31	[9] , [13]
MIBK	84	$3{,}14.10^{8}$	1.3	[14]
MIBK:IPA 1:1	0	$3{,}14.10^{8}$	1.19	[14]
MIBK:IPA 1:3	0	$3{,}14.10^{8}$	3.86	[14]

It can be found that the contrast parameter γ_s is related to the traditionally estimated contrast parameter γ_d. An example of the contrast parameter values γ_s calculated for the electron exposure of PMMA and for one of the developers is shown in Table 2. One can see a high sensitivity D_0 at a low contrast value γ_d and a high contrast value γ_d at lower sensitivity D_0.

Table 2. Contrast γ_s and sensitivity D_0 for PMMA at one of developers (MIBK:IPA mixed in ratio 1:1-see Table 1) for electron irradiation.

D_0 [C/cm^2]	2.10^{-5}	1.10^{-4}	$5.5\ .10^{-4}$
γ_s [A.cm^2/min.C]	0.001	0,3	2,8

The value of the radiation efficiency g, is specific for every chosen material and scanning particle. In Table 3 the radiation efficiency values are presented for various ions and for a wide range of energies, as they was evaluated on the base of the relation (2) by experimental data in ref.[13].

Table 3. The radiation efficiency values for PMMA (ion beam lithography).

Ion	Ion energy (keV)	PMMA thickness (μm)	g (eV^{-1})
Li	50	0.4	$1.7x10^{-2}$
Ga	50	0.15	$2.3x10^{-2}$
Li	60	0.4	$1.7x10^{-2}$
H	60	0.4	$0.6x10^{-2}$
Be	200	1.5	$0.993x10^{-2}$
Si	200	0.7	$0.6x10^{-2}$

3 Conclusions

The correct selection of the exposure and the development conditions in order to ensure the necessary resolution and the resist contrast is the general task of the nano-image's technology designer. Using the presented approach – on the base of the knowledge on the dependence S(D) of the solubility rate on the exposure dose - the contrast parameter and the sensitivity can be optimized at chosen development process parameters like: developer, development temperature, development time at a given resist thickness and polymer material characteristics.

Acknowledgments

The research was funded through Science for Peace NATO program (SfP 973718 project)

References

[1] Brewer G. Electron Beam Technology in Microelectronic Fabrication. Acad Press, 1980

[2] Greeneich J.S. J.Electrochem.Soc. (1975); 122:970

[3] Moreau W.M. Optical Engineering. (1983); 22:181.

[4] IBM, U.S. Pat. 3,931,435.

[5] Bowden M. J.Elecrochem.Soc. (1981); 128:1304.

[6] Brault R.G., Miller L.J. Polymer Engineering and Science (1980); 20:1064

[7] Karapiperis L., Adesida I., Lee C.A., Wolf E.D. J.Vac.Sci. & Technol. (1981); 19:1259.

[8] Ryssel H., Haberger K., Kranz H. J.Vac. Sci. & Technol. (1991); 19(4):1358.

[9] Mladenov G., Seyfarth H. Vacuum (1986): 36:649

[10] Vutova K., Mladenov G., Microelec. Eng. (accepted for publication) 2001

[11] Raptis I. Grella L. Argitis P. Gentili M. Glezos N. Microel. Eng.(1996); 30:295.

[12] Mladenov G. Dissertation for Dr.Sc.of Phys., Inst. of Electronics, Sofia, 1986, p.127.

[13] Vutova K, G.Mladenov. Thin Solid Films, (1991); 200:353-362.

[14] Greeneich J.S , J.Appl.Phys. (1974); 45:5264.

[15] Murata K., E.Nomura and K.Nagami, J.Vac.Sci, & Technol., (1979); 16:1734.

[16] Mladenov G., Emmoth B. Appl.Phys.Lett. 1981; 38:1000.

Author Index

Kristin De Meyer,
Serge Biesemans (eds.)

Simulation of Semiconductor Processes and Devices 1998

SISPAD 98

1998. XIV, 406 pages.
Hardcover DM 177,–, öS 1242,–, EUR 90,–*)
*) Recommended retail prices
All prices are net-prices subject to local VAT,
Euro-price valid as of Jan 2002.
ISBN 3-211-83208-4

This volume contains the proceedings of the 1998 International Conference on Simulation of Semiconductor Processes and Devices and provides an open forum for the presentation of the latest results and trends in modeling and simulation of semiconductor equipment, processes and devices.
Topics include:

- semiconductor equipment simulation
- process modeling and simulation
- device modeling and simulation of complex structures
- interconnect modeling
- integrated systems for process, device, circuit simulation and optimisation
- numerical methods and algorithms
- compact modeling and parameter extraction
- modeling for RF applications
- simulation and modeling of new devices (heterojunction based, SET's, quantum effect devices, laser based ...)

SpringerWienNewYork

A-1201 Wien, Sachsenplatz 4–6, P.O. Box 89, Fax +43.1.330 24 26, e-mail: books@springer.at, Internet: www.springer.at
D-69126 Heidelberg, Haberstraße 7, Fax +49.6221.345-229, e-mail: orders@springer.de
USA, Secaucus, NJ 07096-2485, P.O. Box 2485, Fax +1.201.348-4505, e-mail: orders@springer-ny.com
Eastern Book Service, Japan, Tokyo 113, 3–13, Hongo 3-chome, Bunkyo-ku, Fax +81.3.38 18 08 64, e-mail: orders@svt-ebs.co.jp

Springer-Verlag and the Environment

We at Springer-Verlag firmly believe that an international science publisher has a special obligation to the environment, and our corporate policies consistently reflect this conviction.

We also expect our business partners – printers, paper mills, packaging manufacturers, etc. – to commit themselves to using environmentally friendly materials and production processes.

The paper in this book is made from no-chlorine pulp and is acid free, in conformance with international standards for paper permanency.

Printed in the United States
151726LV00005B/4/P